ELEMENTARY NUMBER THEORY
AND ITS APPLICATIONS

AT&T

ELEMENTARY NUMBER THEORY
AND ITS APPLICATIONS

Third Edition

Kenneth H. Rosen
AT&T Bell Laboratories

ADDISON-WESLEY PUBLISHING COMPANY
Reading, Massachusetts • Menlo Park, California • New York
Don Mills, Ontario • Wokingham, England • Amsterdam • Bonn
Sydney • Singapore • Tokyo • Madrid • San Juan • Milan • Paris

Cover art: Jasper Johns, *Numbers in Color,* 1958–59, encaustic and newspaper on canvas, 66-1/2" x 49-1/2", Albright-Knox Art Gallery, Buffalo, New York, Gift of Seymour H. Knox, 1959.

Library of Congress Cataloging-in-Publication Data

Rosen, Kenneth H.
    Elementary number theory and its applications/Kenneth H. Rosen.
— 3rd ed.
        p. cm.
    Includes bibliographical references and index.
    ISBN 0-201-57889-1
    1. Number theory. I. Title.
QA241.R67 1992
512'.72—dc20   92-17471

This book was prepared using the UNIX® TROFF text processing system and the Linotronic 300 Image Setter at AT&T Bell Laboratories in **Murray Hill, New Jersey.**

ISBN 0-201-57889-1

9 10-MA-009998

Preface

Number theory has long been a favorite subject for students and teachers of mathematics. It is a classical subject with a reputation for being the "purest" part of mathematics, yet recent developments in cryptology and computer science are based on elementary number theory. The first edition of this book was the first text to integrate these important applications of elementary number theory with the traditional topics covered in an introductory course. This text remains unique in this regard, since no other number theory text presents elementary number theory and its applications is a truly integrated fashion.

This book is suitable as a text for an undergraduate number theory course at any level. There are no formal prerequisites for most of the material covered, although a moderate level of mathematical maturity is needed. In fact, even a bright high-school student with strong mathematical aptitude and interest could use this book. The text is also designed to be useful as a supplement for computer science courses, and as a number theory primer for computer scientists interested in learning about new developments in cryptography.

There have been many satisfied users of the first and second editions of this book, along with many favorable published reviews. This third edition preserves all the strengths of previous editions while adding many new features requested by users. My overall goal in the preparation of the third edition has been to enhance the teachability, flexibility, and richness of the text.

FEATURES OF THE THIRD EDITION

- ### A Development of Classical Number Theory

  A major goal of the text is to present classical number theory in a lucid, comprehensive, and captivating manner. At the core of this book is a solid, carefully conceived development of elementary number theory. I have started with the basic material on each important topic, developed it carefully, and then moved on to more difficult results on this topic.

- ### Applications

  This book is packed with applications of number theory. These are integrated with the development of the subject to illustrate the usefulness of the different parts of the theory. I have concentrated on two major application areas, cryptography and computer science. Extensive coverage is devoted to such topics as character and block ciphers, public key cryptography, pseudorandom number generation, fast multiplication of integers, and data storage. However, I have also included applications to many other areas, including scheduling, telephony, entomology, and zoology.

- **Unifying Themes**

Primality testing and factorization are important areas in both classical number theory and in applications to cryptography. I have used these topics as unifying themes, returning to them repeatedly. Almost every chapter has material devoted to these topics.

- **Accessibility**

This book has been designed with a minimum of prerequisites. The goal has been to present material so that it is as self-contained as possible. Students only need a solid background in college algebra to understand most of the book, although a degree of mathematical maturity is helpful. Also, all the computer science applications can be explained to students with a minimal background in this discipline.

- **Accuracy**

Thanks to the reviewers of the various editions of this book, as well as to the many users of the text, the examples, theorems, tables, and answers to exercises have a high level of accuracy. Extra care has been devoted to accuracy in this third edition, with several reviewers carefully checking the text. Being realistic, I realize that errors remain. I would appreciate that these errors, whether they be mathematical or typographical, be brought to my attention.

- **Building Mathematical Maturity**

The book begins with axioms for the integers and examples showing how they are used in proofs. This material gives students an appreciation for the importance of rigor and careful thinking in constructing proofs. Furthermore, the first chapter contains a section devoted entirely to mathematical induction, with many examples and a tremendous variety of exercises whose solutions require inductive proofs.

The text contains many different types of proofs, providing students the opportunity to learn about indirect proofs, proofs by contradiction, proofs by cases, existence proofs, and counterexamples.

- **Extensive Use of Examples**

Numerous examples are given to illustrate the basic skills and concepts developed in the text. The proofs of many theorems are preceded by numerical examples that illustrate the central ideas in these proofs. Finally, examples are used to show how theorems can be applied.

- **Carefully Motivated Proofs**

  Proofs in the text are presented in a careful, rigorous, and fully explained manner. Students can refer to the numerical examples illustrating the working of the proofs to help understand the more general statements being proved.

- **Algorithmic Thinking**

  I have stressed the algorithmic aspects of number theory in this book. Not only are many algorithms described, but also their complexities are analyzed. Examples of algorithms discussed in this book include algorithms for computing greatest common divisors, for primality testing, and for computer arithmetic.

- **Rich Exercise Sets**

  Student learn mathematics by doing exercises. Consequently this text contains extremely extensive and diverse exercise sets. There are many routine computational exercises that develop basic skills. There are also many exercises and blocks of exercises that develop new concepts. Challenging exercises are included in large numbers and are so marked, with one star (*) indicating a difficult exercise or two stars (**) indicating an extremely difficult exercise. Exercises containing results used in the text are marked with a hand pointing toward them (☞). These exercises should be assigned whenever possible.

- **Answers to Exercises**

  The back of the book contains a section with answers to all odd-numbered exercises in the text. These answers are meant to be checks for students' work, and should not be used as a crutch.

- **Computer Projects**

  Each section of the text is followed by a selection of computer projects involving concepts or algorithms from that section. These projects are open-ended exercises that give students the opportunity to tie together the mathematics in the text with their computer skills. Each set of computer projects is split into two sections. The first section includes *programming projects*, while the second section includes *computations and explorations*. To do these computations and explorations, students will need to use either the programs they have written or computational programs, such as Maple, Mathematica, or another similar program.

- **Biographies**

Contributors to the development of number theory include mathematicians of ancient times, such as Euclid and Eratosthenes, mathematicians of the middle ages, such as Fibonacci, mathematicians of the 17th, 18th, and 19th centuries such as Fermat, Euler, and Gauss, and mathematicians of this century. Biographical footnotes describing the lives of more than 25 of these important contributors have been interspersed in the text to convey the human aspect of the subject.

- **Historical Notes**

There are numerous historical remarks included in the text to show the chronology of the development of the subject. These notes emphasize that number theory has an old and rich history as well as a modern vitality.

- **Latest Discoveries**

Important discoveries are made in number theory with surprising frequency. This edition incorporates the latest of these, including the newest discovery of a Mersenne prime with more than 200,000 digits.

- **Unsolved Questions**

I have made it a point to mention unsolved questions in number theory in both the text and exercise sets. This shows students that the subject offers many opportunities for future discoveries. Many of these problems have eluded solution for centuries. Students should be forewarned that attempts to settle such problems are often time-consuming and futile. Often people think they have solved such problems, only to discover some subtle flaw in their reasoning.

- **Bibliography**

An extensive bibliography, split into separate sections for books and articles, is included at the end of the text. Each section is subdivided by subject area. The first section contains lists of number theory texts and references, texts tying together computer science and number theory, books on the history of number theory, books on aspects of computer science dealt with in the text, books on cryptography, and general references. In the articles section there are lists of pertinent expository and research papers in number theory and in cryptography.

- **Appendix**

A set of five tables is included in the appendix to help students with their computations and experimentation. Students may want to compile tables different than those found in the text and in the appendix; compiling such tables would provide additional computer projects.

- **List of Symbols**

   A list of the symbols used in the text and where they are defined is included.

NEW FEATURES OF THE THIRD EDITION

The following list summarizes the major improvements in this third edition.

- Many routine computational exercises have been added.

- Many challenging exercises have been added.

- Answers or solutions for all odd-numbered exercises are now included at the end of the text.

- Additional examples have been selectively added to further clarify key concepts.

- Computations and explorations have been added to the computer projects for each section of the text.

- Two additional factorization methods, the Pollard $p - 1$ method and the Pollard rho method, have been added.

- New sections on check digits, factoring using the Pollard rho method, and zero-knowledge proofs have been added.

- Misprints and errors in the second edition have been corrected.

- The latest discoveries of large primes and factorizations of special numbers have been included.

- More than 10 biographies of mathematicians have been added.

- The bibliography has been expanded.

- Solutions to all exercises are now included in the Instructor's Manual.

HOW TO USE THIS BOOK

   This text is designed to be extremely flexible. The core material for a course in number theory is presented in Chapters 1, 2, and 5, and in Sections 3.1-3.3 and 6.1.

   Section 3.4 requires a background in linear algebra; the material in this section is used in Section 7.2; these two sections can be omitted if desired. Section 3.5 deals with a particular factorization method, and can be omitted.

   Each section in Chapter 4 is optional; Sections 4.1, 4.2, and 4.3 present traditional applications of number theory, Section 4.4 presents an application to computer science, Section 4.5 presents check digits, an application to coding theory.

Sections 6.2 and 6.3 discuss arithmetic functions, Mersenne primes, and perfect numbers; some of this material is used in Chapter 8.

Chapter 7 covers the applications of number theory to cryptology. Sections 7.1, 7.3, and 7.4, which contain discussions of classical and public-key cryptography, are highly recommended.

Chapter 8 deals with primitive roots; Sections 8.1-8.4 should be covered if possible, with Sections 8.5 and 8.6 optional. Section 8.7 will interest computer science students; it deals with pseudo-random numbers.

Sections 9.1 and 9.2 are about quadratic residues and reciprocity, a fundamental topic which should be covered if possible; Sections 9.3 and 9.4, which deal with Jacobi symbols and Euler pseudoprimes, are optional. Section 9.5 covers zero-knowledge proofs and is also optional.

Section 10.1, which covers decimal fractions, and Sections 11.1 and 11.2 which discuss Pythagorean triples and Fermat's last theorem are covered in most number theory courses. Sections 10.2-10.5 and 11.4 involve continued fractions; these sections are optional. Also optional is Section 11.3 which deals with sums of squares.

The dependence of chapters on earlier chapters is as follows. Each of Chapters 2,3, and 4 depends on all previous chapters; Chapter 5 depends on Chapters 1-3; Chapter 6 depends on Chapters 1-3 and 5; Chapter 7 depends on Chapters 1-3 and 5-6; Chapters 8 and 9 both depend on Chapters 1-3 and 5-6; Chapter 10 depends only on Chapters 1-3; and Chapter 11 depends only on Chapters 1-3 with the exception of Section 11.4 which depends on Chapter 10.

INSTRUCTOR'S MANUAL

An instructor's manual is available to adopters of this book. This manual contains the solutions of *all* exercises in the text. Also included are suggestions for teaching, sample exams, and comments on computer projects.

ACKNOWLEDGMENTS FOR THE THIRD EDITION

I thank my management at AT&T Bell Laboratories, including Dick Rosinski, Ken Huber, Randy Pilc, Jim Day, and Joe Timko for their support in the preparation of this edition.

Thanks go to Michael Payne, the editor of this edition, for his support and interest, and to Barbara Holland, who as the original editor of this edition, helped me start this project. I also thank Wayne Yuhasz, Jeff Pepper, and Tom Taylor, my editors for the first and second editions, for their enthusiasm and vision. I also appreciate the work done at Addison Wesley by Laurie Rosatone, who as editorial assistant coordinated this project, Peggy McMahon who managed the production of this book, and Andy Fisher, who as product manager was concerned with many business issues relating to this text. Special thanks go to Bart Goddard, along with Kevin O'Bryant, for their work on the solutions of exercises in this book. The high degree of accuracy in this edition was ensured by careful proofreading by Jerry Grossman, Bart Goddard, and Ayshhyah Eli Khazad.

I have appreciated all the suggestions, corrections, and comments from users of previous editions of this book. Many of their ideas have been incorporated in this third edition. My thanks go to the following reviewers of various editions of this text.

David Bressoud	*Pennsylvania State University*
Sydney Bulman-Fleming	*Wilfrid Laurier University*
Charles Cook	*University of South Carolina, Sumter*
Christopher Cotter	*University of Northern Colorado*
Euda Dean	*Tarleton State University*
Daniel Drucker	*Wayne State University*
Bob Gold	*Ohio State University*
Jerrold Grossman	*Oakland University*
Roy Jordan	*Monmouth College*
Herbert Kasube	*Bradley University*
Neil Koblitz	*University of Washington*
Charles Lewis	*Monmouth College*
James McKay	*Oakland University*
John Mairhuber	*University of Maine-Orono*
Rudolf Najar	*California State University, Fresno*
Carl Pomerance	*University of Georgia*
Tom Shemanske	*Dartmouth College*
Evelyn Bender Vaskas	*Clark University*
Samuel Wagstaff	*Purdue University*
Edward Wang	*Wilfrid Laurier University*
Paul Zwier	*Calvin College*

Finally, I thank in advance all those who send me suggestions and corrections in the future. Keep those cards and letters coming!

Holmdel, New Jersey   *Kenneth H. Rosen*
April, 1992

Contents

Introduction

Number theory is a subject that has interested people for thousands of years. In this introduction we give a brief overview of number theory, describing why the subject is both fascinating and important.

Number theory, in the most general sense, is the study of numbers and their properties. In this book, we primarily deal with the integers, 0, ±1, ±2, ... We will discuss many interesting properties of integers and relationships between integers. In addition, we will study the applications of number theory, particularly those directed towards computer science and cryptology. The development of modern science and technology depended on the development of adequate methods for representing integers and doing integer arithmetic. Starting as long as 5000 years ago different methods, using different integers as bases, have been devised to denote integers. The ancient Babylonians used 60 as the base for their number system and the ancient Mayans used 20. Our method of expressing integers, the decimal system using 10 as the base, was first developed in India approximately six centuries ago. Today, the binary system, which takes 2 as its base, is used extensively by computing machines.

Once a method for representing integers has been devised, procedures for doing arithmetic using these representations are needed. In fact, the term *algorithm*, which now refers to any procedure for solving a general problem, originally referred specifically to such procedures. The development of efficient procedures for arithmetic has become extremely important as computers work with progressively larger integers. Number theory has been used in many ways to devise efficient algorithms for efficient computer arithmetic. Substantial new results in this area have been discovered in the past decade.

The ancient Greeks in the school of Pythagoras, 2500 years ago, made the distinction between *primes* and *composites*. A *prime* is a positive integer greater than one with no positive factors other than one and the integer itself. Questions concerning primes have interested mathematicians from antiquity to modern times. Perhaps the first question about primes that comes to mind is whether there are infinitely many. A demonstration that there are infinitely many primes can be found in the *Elements* of Euclid, an ancient Greek mathematician. We will prove this result in the text. Another question that arises is: Are there formulae that generate primes? Pierre de Fermat, the great French number theorist of the seventeenth century, thought that all integers of the form $2^{2^n} + 1$ are prime; that this is false was shown, a century after Fermat made this claim, by the renowned Swiss mathematician Leonard Euler, who demonstrated that 641 is a factor of $2^{2^5} + 1$. We will discuss these, and other related results, in the sequel.

The problem of distinguishing primes from composites has been extensively studied. Although this problem was originally studied only out of intellectual curiousity, it is now an extremely important problem in cryptology. The ancient Greek scholar Eratosthenes devised a method, now called the *sieve of Eratosthenes*, that finds all primes less than a specified limit. It first removes all even integers, then removes remaining integers divisible by three, then removes remaining integers divisible by five, and so on, stopping when integers divisible by the largest prime not exceeding the square root of the upper limit are removed. All integers left, other than 1, are prime. We will provide the details of this procedure, and justify its validity, in the text.

Sometimes we need to determine whether a particular integer is prime. Unfortunately, it is inefficient to use the sieve of Eratosthenes to determine whether a particular integer is prime. So mathematicians have sought other methods for establishing the primality of an integer. Ancient Chinese mathematicians thought that the primes were precisely those positive integers n such that n divides $2^n - 2$. Fermat showed that if n is prime, then n does divide $2^n - 2$. However, by the early nineteenth century, it was known that there are composite integers n such that n divides $2^n - 2$, such as $n = 341$. These composite integers are called *pseudoprimes*. Because most composite integers are not pseudoprimes, it is possible to develop primality tests based on the original Chinese idea, together with extra observations. It is now possible to efficiently find primes; in fact, primes with as many as 200 decimal digits can be found in minutes of computer time. There is a tremendous amount of current research devoted to showing integers are prime. We will discuss methods for showing integers are prime throughout the text.

The *fundamental theorem of arithmetic*, known to the ancient Greeks, says that every positive integer can be written uniquely as the product of primes. This factorization can be found by trial division of the integer by primes less than its square-root; unfortunately, this method is very time-consuming. Fermat, Euler, and many other mathematicians have produced imaginative factorization techniques. However, using the most efficient technique yet devised, billions of years of computer time may be required to factor an integer with 200 decimal digits. There are large scale efforts underway today to find new ways to quickly factor large integers.

The German mathematician Carl Friedrich Gauss, considered to be one of the greatest mathematicians of all time, developed the language of *congruences* in the early nineteenth century. When doing certain computations, integers may be replaced by their remainders when divided by a specific integer, using the language of congruences. Many questions can be phrased using the notion of a congruence that can only be awkwardly stated without this terminology. Congruences have diverse applications to computer science, including applications to computer file storage, arithmetic with large integers, and the generation of pseudo-random numbers.

One of the most important applications of number theory to computer science is in the area of cryptography. Congruences can be used to develop various types of ciphers. Recently, a new type of cipher system, called a *public-key cipher system*, has been devised. When a public-key cipher is used, each individual has a public enciphering key and a private deciphering key. Messages are enciphered using the public key of the receiver. Moreover, only the receiver can decipher the message, since an overwhelming amount of computer time is required to decipher when just the enciphering key is known. The most widely used public-key cipher system relies on the disparity in computer time required to find large primes and to factor large integers. In particular, to produce an enciphering key requires that two large primes be found and then multiplied; this can be done in minutes on a computer. When these large primes are known, the deciphering key can be quickly found. To find the deciphering key from the enciphering key requires that a large integer, namely the product of the large primes, be factored. This may take billions of years.

In the following chapters, we discuss these and other topics of elementary number theory and its applications.

1

The Integers

1.1 Basic Properties

In this section we state a collection of fundamental properties for the set of *integers* $\{...,-2,-1,0,1,2,...\}$ which we will take as axioms. These properties provide the foundations for proving results in number theory. We begin with properties dealing with addition and multiplication. As usual, we denote the sum and product of a and b by $a+b$ and $a \cdot b$, respectively. Following convention, we will also write ab for $a \cdot b$.

- **Closure**: $a+b$ and $a \cdot b$ are integers whenever a and b are integers.

- **Commutative laws**: $a+b = b+a$ and $a \cdot b = b \cdot a$ for all integers a and b.

- **Associative laws**: $(a+b)+c = a+(b+c)$ and $(a \cdot b) \cdot c = a \cdot (b \cdot c)$ for all integers a, b, and c.

- **Distributive law**: $(a+b) \cdot c = a \cdot c + b \cdot c$ for all integers a, b, and c.

- **Identity elements**: $a+0 = a$ and $a \cdot 1 = a$ for all integers a.

- **Additive inverse**: For every integer a there is an integer solution x to the equation $a + x = 0$; this integer x is called the *additive inverse* of a and is denoted by $-a$. By $b-a$ we mean $b+(-a)$.

- **Cancellation law**: If a, b, and c are integers with $a \cdot c = b \cdot c$ with $c \neq 0$, then $a = b$.

We can use these axioms and the usual properties of equality to establish additional properties of integers. An example illustrating how this is done follows. In subsequent sections of the book results that are easily proved from these axioms will be used without comment.

Example 1.1. To show that $0 \cdot a = 0$ begin with the equation $0+0=0$; this holds since 0 is an identity element for addition. Next multiply both sides by a to obtain $(0+0) \cdot a = 0 \cdot a$. By the distributive law the left-hand side of this equation equals $(0+0) \cdot a = 0 \cdot a + 0 \cdot a$. Hence $0 \cdot a + 0 \cdot a = 0 \cdot a$. Next subtract $0 \cdot a$ from both sides (which is the same as adding the inverse of $0 \cdot a$). Using the associative law for addition and the fact that 0 is an additive identity element, the left-hand side becomes $0 \cdot a + (0 \cdot a - 0 \cdot a) = 0 \cdot a + 0 = 0 \cdot a$. The right-hand side becomes $0 \cdot a - 0 \cdot a = 0$. We conclude that $0 \cdot a = 0$. □

Ordering of integers is defined using the set of *positive integers* $\{1,2,3,...\}$. We have the following definition.

Definition. If a and b are integers, then $a < b$ if $b - a$ is a positive integer. If $a < b$, we also write $b > a$.

Note that a is a positive integer if and only if $a > 0$.

The fundamental properties of ordering of integers follow.

- **Closure for the Positive Integers**: $a + b$ and $a \cdot b$ are positive integers whenever a and b are positive integers.

- **Trichotomy law**: For every integer a, either $a > 0$, $a = 0$, or $a < 0$.

The set of integers is said to be an *ordered set* because it has a subset that is closed under addition and multiplication and because the trichotomy law holds for every integer.

Basic properties of ordering of integers can now be proved using our axioms, as the following example shows. We will use without proof properties of ordering that easily follow from our axioms.

Example 1.2. Suppose that a,b, and c integers with $a < b$ and $c > 0$. We can show that $ac < bc$. First, note that by the definition of $a < b$ we have $b - a > 0$. Since the set of positive integers is closed under multiplication, $c(b - a) > 0$. Since $c(b - a) = cb - ca$, it follows that $ca < cb$. □

We need one more property to complete our set of axioms.

- **The Well-Ordering Property.** Every nonempty set of positive integers has a least element.

We say that the set of positive integers is *well ordered*. On the other hand, the set of all integers is not well ordered since there are sets of integers that do not have a smallest element (as the reader should verify).

The set of integers less than or equal to a given integer has a largest element (the reader should supply the details of the proof of this statement, using the Well-Ordering Property). This leads to the following definition.

Definition. The *greatest integer* in a real number x, denoted by $[x]$, is the largest integer less than or equal to x. That is, $[x]$ is the integer satisfying

$$[x] \leq x < [x] + 1.$$

Example 1.3. We have $[5/2] = 2$, $[-5/2] = -3$, $[\pi] = 3$, $[-2] = -2$, and $[0] = 0$. □

Remark: The greatest integer function is also known as the *floor function*. Instead of using the notation $[x]$ for this function, computer scientists usually use the notation $\lfloor x \rfloor$. The *ceiling function* is a related function often used by computer scientists. The ceiling function of a real number x, denoted by $\lceil x \rceil$, is the smallest integer greater than or equal to x. For example, $\lceil 5/2 \rceil = 3$ and $\lceil -5/2 \rceil = -2$.

We will also deal with the set of rational numbers in this book. Recall that a real number x is *rational* if and only if there are integers a and b, $b \neq 0$, with $x = a/b$. A number that is not rational is called *irrational*. Examples of irrational numbers are π, $\sqrt{2}$, and e. We will now use the well-ordering property to show that $\sqrt{2}$ is irrational; we will prove this result in a different way in Chapter 3. (We refer the reader to Hardy and Wright [21] for proofs that π and e are irrational.)

Theorem 1.1. $\sqrt{2}$ is irrational.

Proof. Suppose that $\sqrt{2}$ were rational. Then there would exist positive integers a and b with $\sqrt{2} = a/b$. Consequently the set $S = \{k\sqrt{2} \mid k \text{ and } k\sqrt{2} \text{ are positive integers}\}$ is a nonempty set of positive integers (it is nonempty since $a = b\sqrt{2}$ is a member of S). Therefore by the well-ordering property S has a smallest element, say $s = t\sqrt{2}$. We have $s\sqrt{2} - s = s\sqrt{2} - t\sqrt{2} = (s-t)\sqrt{2}$. Since $s\sqrt{2} = 2t$ and s are both integers, $s\sqrt{2} - s = (s-t)\sqrt{2}$ must also be an integer. Furthermore, it is positive since $s\sqrt{2} - s = s(\sqrt{2} - 1)$ and $\sqrt{2} > 1$. It is less than s since $s = t\sqrt{2}$, $s\sqrt{2} = 2t$, and $\sqrt{2} < 2$. This contradicts the choice of s as the smallest positive integer in S. It follows that $\sqrt{2}$ is irrational. ∎

The set of integers, the set of positive integers, the set of rational numbers, and the set of real numbers, are traditionally denoted by $\mathbf{Z}$, $\mathbf{Z}^+$, $\mathbf{Q}$, and $\mathbf{R}$, respectively. Also, we write $x \in S$ to indicate that x belongs to the set S. Notation such as this will be used occasionally in this book.

___

1.1 Exercises

1. Use the axioms for the integers to prove the following statements for all integers a, b, and c.
   a) $a \cdot (b + c) = a \cdot b + a \cdot c$
   b) $(a+b)^2 = a^2 + 2ab + b^2$
   c) $a + (b + c) = (c + a) + b$
   d) $(b - a) + (c - b) + (a - c) = 0$

2. Use the axioms for the integers to prove the following statements for all integers a and b.
   a) $(-1) \cdot a = -a$
   b) $-(a \cdot b) = a \cdot (-b)$
   c) $(-a) \cdot (-b) = ab$
   d) $-(a+b) = (-a) + (-b)$

3. What is the value of -0? Give a reason for your answer.

4. Use the axioms for the integers to show that if a and b are integers with $ab = 0$, then $a = 0$ or $b = 0$.

5. Show that an integer a is positive if and only if $a > 0$.

6. Use the definition of ordering of integers and properties of the set of positive integers to prove the following statements for integers a, b, and c with $a < b$ and $c < 0$.
   a) $a + c < b + c$
   b) $a^2 \geq 0$
   c) $ac > bc$
   d) $c^3 < 0$

7. Show that if a, b, and c are integers with $a > b$ and $b > c$, then $a > c$.

☞ 8. Show that there is no positive integer that is less than 1.

9. Determine whether each of the following sets is well-ordered. Either give a proof using the well-ordering property of the set of positive integers or give an example of a subset of the set with no smallest element.

   a) the set of negative integers
   b) the set of integers greater than 3
   c) the set of even positive integers
   d) the set of positive rational numbers
   e) the set of positive rational numbers that can be written in the form $a/2$ where a is a positive integer
   f) the set of nonnegative rational numbers

10. Find the following values of the greatest integer function.

a) $[-3/4]$ c) $[22/7]$ e) $[3]$

b) $[1/4]$ d) $[-2]$ f) $[-\pi]$

11. Show that a nonempty set of negative integers has a largest element.

☞ **12.** Let k be an integer. Show that $[x+k] = [x] + k$ for every real number x.

13. What is the value of $[x] + [-x]$ where x is a real number?

14. Show that $[x] + [x+\frac{1}{2}] = [2x]$ whenever x is a real number.

15. Show that $[x+y] \geq [x] + [y]$ for all real numbers x and y.

16. Show that $[2x] + [2y] \geq [x] + [y] + [x+y]$ whenever x and y are real numbers.

17. Show that if x and y are positive real numbers then $[xy] \geq [x][y]$. What is the situation when both x and y are negative? When one of x and y is negative and the other positive?

18. Show that $-[-x]$ is the least integer greater than or equal to x when x is a real number.

19. Show that $[x + \frac{1}{2}]$ is the integer nearest to x (when there are two integers equidistant from x it is the larger of the two).

20. Show that $[[x] /n] = [x/n]$ whenever x is a real number and n is a positive integer.

21. Show that if a and b are positive integers, then there is a smallest positive integer of the form $a - bk, k \in \mathbf{Z}$.

☞ **22.** Prove that both the sum and the product of two rational numbers are rational.

23. Prove or disprove each of the following statements.

a) The sum of a rational and an irrational number is irrational.

b) The sum of two irrational numbers is irrational.

c) The product of a rational number and an irrational number is irrational.

d) The product of two irrational numbers is irrational.

* **24.** Use the well-ordering property to show that $\sqrt{3}$ is irrational.

** **25.** Suppose that a and b are irrational numbers such that $1/a + 1/b = 1$. Show that every nonnegative integer can be uniquely expressed as either $[ka]$ or $[kb]$ for some integer k.

The *Ulam numbers* u_n, $n=1,2,3,...$ are defined as follows. We specify that $u_1 = 1$ and $u_2 = 2$. For each successive integer m, $m>2$, this integer is an Ulam number if and only if it can be written uniquely as the sum of two distinct Ulam numbers. These numbers are named after Stanislaw Ulam* who first described them in 1964

26. Find the first ten Ulam numbers.

* 27. Show that there are infinitely many Ulam numbers.

1.1 Computer Projects

Programming Projects

Write programs to do the following:

1. Find the first n Ulam numbers where n is a positive integer.

Computations and Explorations

Using the programs you have written or a computation program, carry out the following computations and explorations:

1. Find the first 1000 Ulam numbers.

2. How many pairs of consecutive integers can you find that are both Ulam numbers?

3. Can the sum of any two consecutive Ulam numbers, other than 1 and 2, be another Ulam number? If so, how many examples can you find?

4. How large are the gaps between consecutive Ulam numbers? Do you think that these gaps can be arbitrarily long?

5. What conjectures can you make evidence about the number of Ulam numbers less than an integer n? Do your computations support these conjectures?

1.2 Summations and Products

In this section we introduce notation for summations and products. The following notation represents the sum of the numbers $a_1, a_2, ..., a_n$.

---

*   **STANISLAW M. ULAM (1909-1984)** was born in Lwow, Poland. He received his masters and doctoral degrees from the Polytechnic Institute of Poland, and, as a young mathematician, spent time at the Institute for Advanced Study in Princeton, New Jersey, Harvard College, and the University of Wisconsin. Ulam was an important member of the Manhattan Project, based in Los Alamos, New Mexico, which led to the development of the first atomic bomb. Ulam also used the Monte Carlo approximation methods to help develop the hydrogen bomb. He made contributions to many areas of mathematics, including real analysis, probability theory, number theory, and mathematical biology. Besides his work at Los Alamos, Ulam served on the faculties of the University of Southern California, the University of Colorado, and the University of Florida.

$$\sum_{k=1}^{n} a_k = a_1 + a_2 + \cdots + a_n.$$

The letter k, the *index of summation*, is a "dummy variable" and can be replaced by any letter, so that

$$\sum_{k=1}^{n} a_k = \sum_{j=1}^{n} a_j = \sum_{i=1}^{n} a_i \text{ , and so forth.}$$

Example 1.4. We see that

$$\sum_{j=1}^{5} j = 1 + 2 + 3 + 4 + 5 = 15,$$

$$\sum_{j=1}^{5} 2 = 2 + 2 + 2 + 2 + 2 = 10,$$

and

$$\sum_{j=1}^{5} 2^j = 2 + 2^2 + 2^3 + 2^4 + 2^5 = 62.$$

We also note that in summation notation, the index of summation may range between any two integers, as long as the lower limit does not exceed the upper limit. If m and n are integers such that $m \leq n$, then

$$\sum_{k=m}^{n} a_k = a_m + a_{m+1} + \cdots + a_n.$$

For instance, we have

$$\sum_{k=3}^{5} k^2 = 3^2 + 4^2 + 5^2 = 50,$$

$$\sum_{k=0}^{2} 3^k = 3^0 + 3^1 + 3^2 = 13,$$

and

$$\sum_{k=-2}^{1} k^3 = (-2)^3 + (-1)^3 + 0^3 + 1^3 = -8.$$

We will use several properties of summation notation in the sequel. We first consider sums where every term is multiplied by a constant. Using the distributive law, we have

$$\sum_{j=m}^{n} ka_j = ka_m + ka_{m+1} + \cdots + ka_n$$

$$= k(a_m + a_{m+1} + \cdots + a_n)$$

$$= k\sum_{j=m}^{n} a_j.$$

Next we consider sums of sums of terms from two sequences.

$$\sum_{j=m}^{n} (a_j + b_j) = (a_m + b_m) + (a_{m+1} + b_{m+1}) + \cdots + (a_n + b_n)$$

$$= (a_m + a_{m+1} + \cdots + a_n) + (b_m + b_{m+1} + \cdots + b_n)$$

$$= \sum_{j=m}^{n} a_j + \sum_{j=m}^{n} b_j.$$

Finally, we show that we can change the order of summation in a double sum. Using the first property we have established about summations twice we find that

$$\sum_{i=m}^{n} \sum_{j=p}^{q} a_i b_j = \sum_{i=m}^{n} [a_i(\sum_{j=p}^{q} b_j)]$$

$$= \sum_{i=m}^{n} a_i(b_p + \cdots + b_q)$$

$$= a_m(b_p + \cdots + b_q) + \cdots + a_n(b_p + \cdots + b_q)$$

$$= a_m b_p + \cdots + a_m b_q + \cdots + a_n b_p + \cdots + a_n b_q$$

$$= (a_m + a_{m+1} + \cdots + a_n)(b_p + b_{p+1} + \cdots + b_q)$$

$$= \left[\sum_{i=m}^{n} a_i\right]\left[\sum_{j=p}^{q} b_j\right].$$

Similarly, we find that

$$\sum_{j=p}^{q} \sum_{i=m}^{n} a_i b_j = \left(\sum_{j=p}^{q} b_j \right) \left(\sum_{i=m}^{n} a_i \right).$$

It follows that

$$\sum_{i=m}^{n} \sum_{j=p}^{q} a_i b_j = \sum_{j=p}^{q} \sum_{i=m}^{n} a_i b_j.$$

Summarizing, we have the following three properties for summations.

(1.1)
$$\sum_{j=m}^{n} k a_j = k \sum_{j=m}^{n} a_j.$$

(1.2)
$$\sum_{j=m}^{n} (a_j + b_j) = \sum_{j=m}^{n} a_j + \sum_{j=m}^{n} b_j.$$

(1.3)
$$\sum_{i=m}^{n} \sum_{j=p}^{q} a_i b_j = \left(\sum_{i=m}^{n} a_i \right) \left(\sum_{j=p}^{q} b_j \right) = \sum_{j=p}^{q} \sum_{i=m}^{n} a_i b_j.$$

We also define a notation for products, analogous to that for summations. The product of the numbers $a_1, a_2, ..., a_n$ is denoted by

$$\prod_{j=1}^{n} a_j = a_1 a_2 \cdots a_n.$$

The letter j above is a "dummy variable", and can be replaced arbitrarily.

Example 1.5. To illustrate the notation for products we have

$$\prod_{j=1}^{5} j = 1 \cdot 2 \cdot 3 \cdot 4 \cdot 5 = 120,$$

$$\prod_{j=1}^{5} 2 = 2 \cdot 2 \cdot 2 \cdot 2 \cdot 2 = 2^5 = 32, \text{ and}$$

$$\prod_{j=1}^{5} 2^j = 2 \cdot 2^2 \cdot 2^3 \cdot 2^4 \cdot 2^5 = 2^{15}. \qquad \square$$

The *factorial function* arises throughout number theory.

Definition. Let n be a positive integer. Then $n!$, read as n factorial, is the product of the integers $1, 2, ..., n$. We also specify that $0! = 1$. In terms of product notation, we have $n! = \prod_{j=1}^{n} j$.

Example 1.6. We have $1! = 1$, $4! = 1 \cdot 2 \cdot 3 \cdot 4 = 24$, and $12! = 1 \cdot 2 \cdot 3 \cdot 4 \cdot 5 \cdot 6 \cdot 7 \cdot 8 \cdot 9 \cdot 10 \cdot 11 \cdot 12 = 479{,}001{,}600$. $\square$

1.2 Exercises

1. Find the value of each of the following sums.

   a) $\displaystyle\sum_{j=1}^{10} 2$

   b) $\displaystyle\sum_{j=1}^{10} j$

   c) $\displaystyle\sum_{j=1}^{10} j^2$

   d) $\displaystyle\sum_{j=1}^{10} 2^j$

2. Find the value of each of the following sums.

   a) $\displaystyle\sum_{j=3}^{6} (j-1)$

   b) $\displaystyle\sum_{j=0}^{5} j!$

   c) $\displaystyle\sum_{j=-2}^{2} j^2$

   d) $\displaystyle\sum_{j=-8}^{-2} j(j+1)$

3. Find the value of each of the following products.

   a) $\displaystyle\prod_{j=1}^{5} 3$

   b) $\displaystyle\prod_{j=1}^{5} (j+1)$

   c) $\displaystyle\prod_{j=1}^{5} j^2$

   d) $\displaystyle\prod_{j=1}^{6} (-2)^j$

4. Find the value of each of the following products.

   a) $\displaystyle\prod_{j=3}^{7} 5$

   b) $\displaystyle\prod_{j=0}^{5} j!$

   c) $\displaystyle\prod_{j=-1}^{4} (j+1)$

   d) $\displaystyle\prod_{j=-5}^{-1} j^2$

5. Express each of the following sums in terms of $\displaystyle\sum_{i=m}^{n} a_i$ and $\displaystyle\sum_{i=m}^{n} b_i$.

   a) $\displaystyle\sum_{i=m}^{n} 5a_i$

   b) $\displaystyle\sum_{i=m}^{n} (-b_i)$

   c) $\displaystyle\sum_{i=m}^{n} (a_i-b_i)$

   d) $\displaystyle\sum_{i=m}^{n} (3\,a_i+4\,b_i)$

6. Find $n!$ for n equal to each of the first ten positive integers.

7. List the integers $100!$, 100^{100}, 2^{100}, and $(50!)^2$ in order of increasing size. Justify your answer.

8. Express each of the following products in terms of $\prod_{i=1}^{n} a_i$ where k is a constant.

   a) $\prod_{i=1}^{n} ka_i$ b) $\prod_{i=1}^{n} ia_i$ c) $\prod_{i=1}^{n} a_i^k$

9. Find the value of each of the following double sums.

   a) $\sum_{i=1}^{5} \sum_{j=1}^{6} ij$

   c) $\sum_{i=1}^{5} \sum_{j=1}^{6} i$

   b) $\sum_{i=1}^{5} \sum_{j=1}^{6} (i+1)(j+1)$

   d) $\sum_{i=1}^{5} \sum_{j=1}^{6} (i+j)$

10. Find the value of each of the following double products.

   a) $\prod_{i=1}^{4} \prod_{j=1}^{3} ij$

   c) $\prod_{i=1}^{4} \prod_{j=1}^{3} i$

   b) $\prod_{i=1}^{4} \prod_{j=1}^{3} (i+j)$

   d) $\prod_{i=1}^{4} \prod_{j=1}^{3} (i+1)(j-1)$

11. Find the value of each of the following quantities.

   a) $\sum_{i=1}^{3} \prod_{j=1}^{4} ij$

   c) $\sum_{i=1}^{3} \prod_{j=1}^{4} (i+j)$

   b) $\prod_{i=1}^{3} \sum_{j=1}^{4} ij$

   d) $\prod_{i=1}^{3} \sum_{j=1}^{4} (i+j)$

12. Show that $\sum_{j=1}^{n} (a_j - a_{j-1}) = a_n - a_0$ where $a_0, a_1, ..., a_n$ is a sequence of numbers. Such a sum is said to be *telescoping*.

13. Use the identity $\dfrac{1}{k(k+1)} = \dfrac{1}{k} - \dfrac{1}{k+1}$ and Exercise 12 to evaluate $\sum_{k=1}^{n} \dfrac{1}{k(k+1)}$.

14. Use the identity $\dfrac{1}{k^2 - 1} = \dfrac{1}{2}\left[\dfrac{1}{k-1} - \dfrac{1}{k+1}\right]$ and Exercise 12 to evaluate $\sum_{k=2}^{n} \dfrac{1}{k^2 - 1}$.

15. Sum both sides of the identity $(k+1)^2 - k^2 = 2k + 1$ from $k=1$ to $k=n$, and then use Exercise 12 to find a formula for $\sum_{k=1}^{n} k$.

16. Find a formula for $\sum_{k=1}^{n} k^2$ using a technique analogous to that outlined in Exercise 15 and the result of that exercise.

17. Find a formula for $\sum_{k=1}^{n} k^3$ using a technique analogous to that outlined in Exercise 15 and the results of Exercises 15 and 16.

18. Find the values of the following products.

    a) $\prod_{j=2}^{n} (1 - 1/j)$ b) $\prod_{j=2}^{n} (1 - 1/j^2)$.

19. Without multiplying all the terms, show that

    a) $10! = 6!\ 7!$ c) $16! = 14!\ 5!\ 2!$
    b) $10! = 7!\ 5!\ 3!$ d) $9! = 7!\ 3!\ 3!\ 2!$.

20. Let $a_1, a_2, \ldots, a_n$ be positive integers. Let $b = (a_1!\ a_2! \cdots a_n!) - 1$, and $c = a_1!\ a_2! \cdots a_n!$. Show that $c! = a_1!\ a_2! \cdots a_n!\ b!$.

21. Find all positive integers x, y, and z such that $x! + y! = z!$.

* 22. Show that $\sum_{j=0}^{n-1} [x + j/n] = [nx]$ whenever x is a real number and n is a positive integer.

1.2 Computer Projects

Programming Projects

Write programs to do the following:

1. Find $n!$ for a given nonnegative integer n.

Computations and Explorations

Using the programs you have written or a computation program, carry out the following computations and explorations:

1. Compute $n!$ for integers n as large as you can. Use your computations to estimate the number of decimal digits in $n!$.

1.3 Mathematical Induction

The *principle of mathematical induction* is a valuable tool for proving results about the integers. We will first state this principle and then show how it is used.

Afterwards, we will use the well-ordering principle to show that mathematical induction is a valid technique. In our study of number theory, we will use both the well-ordering property and the principle of mathematical induction many times.

Theorem 1.2. The Principle of Mathematical Induction. If a set of positive integers contains 1, and for each positive integer n, it contains $n + 1$ if it contains n then it is the set of all positive integers.

We need to do two things to prove by mathematical induction that a result holds for every positive integer. First, we must show that the result is true for the integer 1. This is called the *basis step*. Second, we must show for each positive integer n that the result is true for the positive integer $n + 1$ if it is true for the positive integer n. This is called the *inductive step*. Once these two steps are completed, by the principle of mathematical induction we can conclude that the statement is true for the set of all positive integers.

We illustrate how mathematical induction works with several examples. First we prove an inequality.

Example 1.7. We can show by mathematical induction that $n! \leq n^n$ for every positive integer n. The basis step, namely the case where $n = 1$, holds since $1! = 1 \leq 1^1 = 1$. Now assume that $n! \leq n^n$; this is the inductive hypothesis. To complete the proof we must show, under the assumption that the inductive hypothesis is true, that $(n+1)! \leq (n+1)^{n+1}$. Using the inductive hypothesis we have

$$
\begin{aligned}
(n+1)! &= (n+1) \cdot n! \\
&\leq (n+1) n^n \\
&< (n+1)(n+1)^n \\
&\leq (n+1)^{n+1}.
\end{aligned}
$$

This completes both the inductive step and the proof.   □

One of the most common applications of mathematical induction is in proofs of explicit formulae for sums. Often we can conjecture such a formula from the first few cases, and then use mathematical induction to confirm our conjecture, if it is indeed true. (Note that mathematical induction can only be used to prove formulae for sums and not find such formulae.) Consider the following example.

Example 1.8. By examining the sums of the first n odd positive integers for small values of n we can conjecture a formula for this sum. We have

$$1 = 1,$$
$$1+3 = 4,$$
$$1+3+5 = 9,$$
$$1+3+5+7 = 16,$$
$$1+3+5+7+9 = 25,$$
$$1+3+5+7+9+11 = 36.$$

From these values, we conjecture that $\sum_{j=1}^{n} (2j-1) = 1 + 3 + 5 + 7 + \cdots + 2n - 1 = n^2$ for every positive integer n.

We now attempt to prove this conjecture using mathematical induction. (Clearly this attempt will fail if our guess is wrong.) The basis step follows since $\sum_{j=1}^{1} (2j - 1) = 2 \cdot 1 - 1 = 1 = 1^2$. For the inductive step, we assume that the formula holds for n. We must show that under this assumption the formula holds for $n+1$. We assume that $\sum_{j=1}^{n} (2j - 1) = n^2$. Using the inductive hypothesis we have

$$\sum_{j=1}^{n+1} (2j - 1) = \sum_{j=1}^{n} (2j - 1) + 2(n+1) - 1$$
$$= n^2 + 2(n+1) - 1$$
$$= n^2 + 2n + 1$$
$$= (n+1)^2.$$

Since both the basis step and inductive step have been completed, we know that the result holds. $\square$

Geometric progressions are sequences of integers where each term is a constant multiple of the previous term. We will often need to sum terms of such progressions. After giving the formal definition of a geometric progression, we will use mathematical induction to establish a formula for the sum of the terms of such a sequence.

Definition. Given real numbers a and r, the real numbers

$$a, \; ar, \; ar^2, \; ar^3, \ldots$$

are said to form a *geometric progression.* Also, a is called the *initial term* and r is called the *common ratio.*

Example 1.9. The numbers $5, -15, 45, -135, \ldots$ form a geometric progression with initial term 5 and common ratio -3. $\square$

We give an explicit formula for the sum of the first n consecutive terms of a geometric progression in the following example.

Example 1.10. We will prove that if a and r are real numbers and $r \neq 1$, then for each positive integer n

(1.4) $$\sum_{j=0}^{n} ar^j = a + ar + ar^2 + \cdots + ar^n = \frac{ar^{n+1} - a}{r - 1}.$$

To prove that this formula is true we complete the basis step, that is, we must first show that the formula holds for $n = 1$. Then, we must complete the inductive step, that is, we must show that if the formula is valid for the positive integer n, it must also be true for the positive integer $n + 1$.

To start things off, let $n = 1$. Then, the left side of (1.4) is $a + ar$, while on the right side of (1.4) we have

$$\frac{ar^2 - a}{r - 1} = \frac{a(r^2 - 1)}{r - 1} = \frac{a(r+1)(r-1)}{r - 1} = a(r+1) = a + ar.$$

So the formula is valid when $n = 1$.

Now we assume that (1.4) holds for the positive integer n. That is, we assume that

(1.5) $$a + ar + ar^2 + \cdots + ar^n = \frac{ar^{n+1} - a}{r - 1}.$$

We must show that the formula also holds for the positive integer $n + 1$. What we must show is that

(1.6) $$a + ar + ar^2 + \cdots + ar^n + ar^{n+1} = \frac{ar^{(n+1)+1} - a}{r - 1} = \frac{ar^{n+2} - a}{r - 1}.$$

To show that (1.6) is valid, we add ar^{n+1} to both sides of (1.5), to obtain

(1.7) $$(a + ar + ar^2 + \cdots + ar^n) + ar^{n+1} = \frac{ar^{n+1} - a}{r - 1} + ar^{n+1}.$$

The left-hand side of (1.7) is identical to that of (1.6). To show that the right sides are equal, we note that

$$\frac{ar^{n+1} - a}{r - 1} + ar^{n+1} = \frac{ar^{n+1} - a}{r - 1} + \frac{ar^{n+1}(r - 1)}{r - 1}$$

$$= \frac{ar^{n+1} - a + ar^{n+2} - ar^{n+1}}{r - 1}$$

$$= \frac{ar^{n+2} - a}{r - 1}.$$

Since we have shown that (1.5) implies (1.6), we can conclude that (1.4) holds for all positive integers n.   □

The following example illustrates the usefulness of the formula proved in Example 1.10.

Example 1.11. Let n be a positive integer. To find the sum

$$\sum_{k=0}^{n} 2^k = 1 + 2 + 2^2 + \cdots + 2^n,$$

we use Example 1.10 with $a = 1$ and $r = 2$, to obtain

$$1 + 2 + 2^2 + \cdots + 2^n = \frac{2^{n+1} - 1}{2 - 1} = 2^{n+1} - 1.$$

Hence, the sum of consecutive nonnegative powers of 2 is one less than the next largest power of 2. $\square$

We now present the proof that the principle of mathematical induction follows from the well-ordering principle.

Proof. Let S be a set of positive integers containing the integer 1 and the integer $n + 1$ whenever it contains n. Assume that S is not the set of all positive integers. Therefore, there are some positive integers not contained in S. By the well-ordering property, since the set of positive integers not contained in S is nonempty, there is a least positive integer n which is not in S. Note that $n \neq 1$, since 1 is in S. Now since $n > 1$ (since there is no positive integer n with $n < 1$; see Exercise 8 in Section 1.1), the integer $n - 1$ is a positive integer smaller than n, and hence must be in S. But since S contains $n - 1$, it must also contain $(n - 1) + 1 = n$, which is a contradiction, since n is supposedly the smallest positive integer not in S. This shows that S must be the set of all positive integers. ∎

A slight variant of the principle of mathematical induction is also sometimes useful in proofs.

Theorem 1.3. The Second Principle of Mathematical Induction. A set of positive integers which contains the integer 1, and which has the property that for every positive integer n if it contains all the positive integers $1, 2,..., n$, then it also contains the integer $n + 1$, must be the set of all positive integers.

Before proving that the second principle of mathematical induction is valid, we will give an example to illustrate its use.

Example 1.12. We will show that any postage more than one cent can be formed using just two-cent and three-cent stamps. For the basis step, note that postage of two cents can be formed using one two-cent stamp and postage of three cents can be formed using one three-cent stamp. For the inductive step, assume that every postage not exceeding n cents, $n \geq 3$, can be formed using two-cent and three-cent stamps. Then a postage of $n + 1$ cents can be formed by forming a postage of $n - 1$ cents together with a two-cent stamp. This completes the proof. (This result can also be proved using the principle of mathematical induction, as the reader can verify.) $\square$

We will now prove that the second principle of mathematical induction is a valid technique.

Proof. Let T be a set of integers containing 1 and such that for every positive integer n if it contains 1, 2, ..., n, it also contains $n+1$. Let S be the set of all positive integers n such that all the positive integers less than or equal to n are in T. Then 1 is in S, and by the hypotheses, we see that if n is in S, then $n + 1$ is in S. Hence, by the principle of mathematical induction, S must be the set of all positive integers, so clearly T is also the set of all positive integers, since S is a subset of T. ■

Recursive Definitions

The principle of mathematical induction provides a method for defining the values of functions at positive integers. Instead of explicitly specifying the value of the function at n we give the value of the function at 1 and give a rule for finding for each positive integer n the value of the function at $n+1$ from the value of the function at n.

Definition. We say that the function f is *defined recursively* if the value of f at 1 is specified and if for each positive integer n a rule is provided for determining $f(n+1)$ from $f(n)$.

The principle of mathematical induction can be used to show that a function that is defined recursively is defined uniquely at each positive integer. (See Exercise 29 at the end of this section.) We illustrate how to define a function recursively with the following definition.

Example 1.13. We will recursively define the *factorial function* $f(n) = n!$. First, we specify that

$$f(1) = 1.$$

Then we give a rule for finding $f(n+1)$ from $f(n)$ for each positive integer, namely

$$f(n+1) = (n+1) \cdot f(n).$$

These two statements uniquely define $n!$.

To find the value of $f(6) = 6!$ from the recursive definition, use the second property successively, as follows:

$$f(6) = 6 \cdot f(5) = 6 \cdot 5 \cdot f(4) = 6 \cdot 5 \cdot 4 \cdot f(3) = 6 \cdot 5 \cdot 4 \cdot 3 \cdot f(2) = 6 \cdot 5 \cdot 4 \cdot 3 \cdot 2 \cdot f(1).$$

Then use the first statement of the definition to replace $f(1)$ by its stated value 1, to conclude that

$$6! = 6 \cdot 5 \cdot 4 \cdot 3 \cdot 2 \cdot 1 = 720.$$

□

The second principle of mathematical induction also serves as a basis for recursive definitions. We can define a function whose domain is the set of positive integers by specifying its value at 1 and giving a rule for each positive integer n for finding $f(n)$ from the values $f(j)$ for $1 \le j \le n-1$. (Exercise 32 at the end of this section asks for a proof that such a function is well defined.)

The Fibonacci Numbers. We illustrate how such an inductive definition works with the definition of the *Fibonacci sequence,* named after Fibonacci*, an Italian mathematician of the thirteenth century who used this sequence to solve a problem about rabbit populations (see Exercise 46 at the end of this section.)

Example 1.14. The Fibonacci sequence $f_1, f_2, \ldots, f_n$ is defined recursively by $f_1 = 1, f_2 = 1,$ and $f_n = f_{n-1} + f_{n-2}$ for $n \ge 3$. From this definition we see that

$$
\begin{aligned}
f_3 &= f_2 + f_1 = 1+1 = 2, \\
f_4 &= f_3 + f_2 = 2+1 = 3, \\
f_5 &= f_4 + f_3 = 3+2 = 5, \\
f_6 &= f_5 + f_4 = 5+3 = 8,
\end{aligned}
$$

and so on. (*Note*: A sequence is just a function from the set of positive integers to the set of real numbers. Consequently it makes sense to recursively define sequences.)

We also define the value of the Fibonacci sequence at $n=0$ by setting $f_0 = 0$. With this definition, we find that $f_2 = f_1 + f_0$. We can define f_n where n is a negative number so that the equality in the recursive definition is satisfied. (See Exercise 55.) $\square$

The Fibonacci numbers satisfy many identities. We prove one of these in the following example and give several others as exercises at the end of the section.

Example 1.15. We can conjecture a formula for $\sum_{j=1}^{n} f_j$ by examining this sum for small values of n. We have

$$
\begin{aligned}
f_1 &= 1, \\
f_1 + f_2 &= 1+1 = 2, \\
f_1 + f_2 + f_3 &= 1+1+2 = 4, \\
f_1 + f_2 + f_3 + f_4 &= 1+1+2+3 = 7, \\
f_1 + f_2 + f_3 + f_4 + f_5 &= 1+1+2+3+5 = 12.
\end{aligned}
$$

These sums are all one less than the Fibonacci number that is two terms out in the

---

* **FIBONACCI (c. 1180-1228),** (short for filus Bonacci, son of Bonacci), also known as Leonardo of Pisa, was born in the Italian commercial center of Pisa. Fibonacci was a merchant who traveled extensively throughout the Mideast, where he came into contact with mathematical works from the Arabic world. In his book *Liber Abaci* Fibonacci introduced Arabic notation for numerals and their algorithms for arithmetic into the European world. It was in this book that his famous rabbit problem appeared. Fibonacci also wrote *Practica geometriae,* a treatise on geometry and trigonometry, and *Liber quadratorum,* a book on Diophantine equations.

sequence. Hence it is reasonable to guess that

$$\sum_{j=1}^{n} f_j = f_{n+2} - 1.$$

We now attempt to prove this conjecture by mathematical induction. The basis step follows since $\sum_{j=1}^{1} f_j = 1$ and this is the same as $f_{1+2} - 1 = f_3 - 1 = 2 - 1 = 1$. The inductive hypothesis is $\sum_{j=1}^{n} f_j = f_{n+2} - 1$. We must show that under this assumption that $\sum_{j=1}^{n+1} f_j = f_{n+3} - 1$. To prove this, by the inductive hypothesis it follows that

$$\begin{aligned}
\sum_{j=1}^{n+1} f_j &= \left[\sum_{j=1}^{n} f_j\right] + f_{n+1} \\
&= (f_{n+2} - 1) + f_{n+1} \\
&= (f_{n+1} + f_{n+2}) - 1 \\
&= f_{n+3} - 1.
\end{aligned}$$

□

The following inequality for the Fibonacci numbers will be used in Chapter 2.

Example 1.16. We can use the second principle of mathematical induction to prove that $f_n > \alpha^{n-2}$ for $n \geq 3$ where $\alpha = (1+\sqrt{5})/2$. The basis step consists of verifying this inequality for $n=3$ and $n=4$. We have $\alpha < 2 = f_3$, so the theorem is true for $n=3$. Since $\alpha^2 = (3 + \sqrt{5})/2 < 3 = f_4$, the theorem is true for $n=4$.

The inductive hypothesis consists of assuming that $\alpha^{k-2} < f_k$ for all integers k with $k \leq n$. Since $\alpha = (1+\sqrt{5})/2$ is a solution of $x^2 - x - 1 = 0$, we have $\alpha^2 = \alpha + 1$. Hence

$$\alpha^{n-1} = \alpha^2 \cdot \alpha^{n-3} = (\alpha+1)\cdot\alpha^{n-3} = \alpha^{n-2} + \alpha^{n-3}.$$

By the induction hypothesis, we have the inequalities

$$\alpha^{n-2} < f_n, \quad \alpha^{n-3} < f_{n-1}.$$

By adding these two inequalities, we conclude that

$$\alpha^{n-1} < f_n + f_{n-1} = f_{n+1}.$$

This finishes the proof.

□

We have presented a few important results involving the Fibonacci numbers. There is a vast literature concerning these numbers and their many applications to botany, computer science, geography, physics, and other areas. There is even a scholarly journal, *The Fibonacci Quarterly*, devoted to the study of these numbers.

1.3 Exercises

1. Use mathematical induction to prove that $n < 2^n$ whenever n is a positive integer.

2. Conjecture a formula for the sum of the first n even positive integers. Prove your result using mathematical induction.

3. Use mathematical induction to prove that $\sum_{k=1}^{n} \frac{1}{k^2} = \frac{1}{1^2} + \frac{1}{2^2} + \cdots + \frac{1}{n^2} \le 2 - \frac{1}{n}$ whenever n is a positive integer.

4. Conjecture a formula for $\sum_{k=1}^{n} \frac{1}{k(k+1)} = \frac{1}{1 \cdot 2} + \frac{1}{2 \cdot 3} + \dots + \frac{1}{n(n+1)}$ from the value of this sum for small integers n. Prove that your conjecture is correct using mathematical induction. (Compare this to Exercise 13 in Section 1.2.)

5. Conjecture a formula for $\mathbf{A}^n$ where $\mathbf{A} = \begin{pmatrix} 1 & 1 \\ 0 & 1 \end{pmatrix}$. Prove your conjecture using mathematical induction.

6. Use mathematical induction to prove that $\sum_{j=1}^{n} j = 1 + 2 + 3 + \cdots + n = n(n+1)/2$ for every positive integer n.

7. Use mathematical induction to prove that $\sum_{j=1}^{n} j^2 = 1^2 + 2^2 + 3^2 + \cdots + n^2 = n(n+1)(2n+1)/6$ for every positive integer n.

8. Use mathematical induction to prove that $\sum_{j=1}^{n} j^3 = 1^3 + 2^3 + 3^3 + \cdots + n^3 = [n(n+1)/2]^2$ for every positive integer n.

9. Use mathematical induction to prove that $\sum_{j=1}^{n} j(j+1) = 1 \cdot 2 + 2 \cdot 3 + \cdots + n \cdot (n+1) = n(n+1)(n+2)/3$ for every positive integer n.

10. Use mathematical induction to prove that $\sum_{j=1}^{n} (-1)^{j-1} j^2 = 1^2 - 2^2 + 3^2 - \cdots + (-1)^{n-1} n^2 = (-1)^{n-1} n(n+1)/2$ for every positive integer n.

11. Find a formula for $\displaystyle\prod_{j=1}^{n} 2^{j}$.

12. Find $1 - 2 + 2^{2} - 2^{3} + \cdots + 2^{100}$.

13. Find $3 + 3 \cdot 5^{2} + 3 \cdot 5^{4} + \cdots + 3 \cdot 5^{1000}$.

14. Find $7^{3} \cdot 11^{2} + 7^{5} \cdot 11^{6} + \cdots + 7^{99} \cdot 11^{194}$.

15. Find $1 + \dfrac{1}{2} + \dfrac{1}{2^{2}} + \cdots + \dfrac{1}{2^{100}}$.

16. Show that $\displaystyle\sum_{j=1}^{n} j \cdot j! = 1 \cdot 1! + 2 \cdot 2! + \cdots + n \cdot n! = (n+1)! - 1$ for every positive integer n.

17. Show that any postage that is an integer number of cents greater than 11 cents can be formed using just 4-cent and 5-cent stamps.

18. Show that any postage that is an integer number of cents greater than 53 cents can be formed using just 7-cent and 10-cent stamps.

Let H_n be the nth partial sum of the harmonic series, that is, $H_n = \displaystyle\sum_{j=1}^{n} 1/j$.

* 19. Use mathematical induction to show that $H_{2^{n}} \geq 1 + n/2$.

* 20. Use mathematical induction to show that $H_{2^{n}} \leq 1 + n$.

21. Show by mathematical induction that if n is a positive integer, then $(2n)! < 2^{2n}(n!)^{2}$.

22. Use mathematical induction to prove that $x - y$ is a factor of $x^{n} - y^{n}$ where x and y are variables.

☞ 23. Use the principle of mathematical induction to show that a set of integers that contains the integer k and such that this set contains $n+1$ whenever it contains n contains the set of integers that are greater than or equal to k.

24. Use mathematical induction to prove that $2^{n} < n!$ for $n \geq 4$.

25. Use mathematical induction to prove that $n^{2} < n!$ for $n \geq 4$.

26. Show by mathematical induction that if $h \geq -1$ then $1 + nh \leq (1+h)^{n}$ for all nonnegative integers n.

27. A jigsaw puzzle is solved by putting its pieces together in the correct way. Show that exactly $n-1$ moves are required to solve a jigsaw puzzle with n pieces, where a move consists of putting together two blocks of pieces, with a block consisting of one or more assembled pieces. (*Hint:* Use the second principle of mathematical induction.)

28. What is wrong with the following proof by mathematical induction that all horses are the same color? Clearly all horses in any set of 1 horse are all the same color. This completes the basis step. Now assume that all horses in any set of n horses are the same color. Consider a set of $n+1$ horses, labeled with the integers $1,2,...,n+1$. By the induction hypothesis horses $1,2,...,n$ are all the same color, as are horses $2,3,...,n,n+1$. Since these two sets of horses have common members, namely horses $2,3,4,...,n$ all $n+1$ horses must be the same color. This completes the induction argument.

29. Use the principle of mathematical induction to show that the value at each positive integer of a function defined recursively is uniquely determined.

30. What function $f(n)$ is defined recursively by $f(1) = 2$ and $f(n+1) = 2f(n)$ for $n \geq 1$? Prove your answer using mathematical induction.

31. If g is defined recursively by $g(1) = 2$ and $g(n) = 2^{g(n-1)}$ for $n \geq 2$, what is $g(4)$?

32. Use the second principle of mathematical induction to show that if $f(1)$ is specified and a rule for finding $f(n+1)$ from the values of f at the first n positive integers is given, then $f(n)$ is uniquely determined for every positive integer n.

33. We define a function recursively for all positive integers n by $f(1) = 1$, $f(2) = 5$, and for $n > 2$, $f(n+1) = f(n) + 2f(n-1)$. Show that $f(n) = 2^n + (-1)^n$, using the second principle of mathematical induction.

The *triangular number* t_n is the number of dots in the triangular array with n rows that has j dots in the jth row. For instance $t_1 = 1, t_2 = 3, t_3 = 6$, and $t_4 = 10$.

$t_1 = 1 \qquad t_2 = 3 \qquad t_3 = 6 \qquad t_4 = 10$

Exercises 34 - 36 deal with triangular numbers.

34. Show that t_n satisfies the recursive defintion $t_1 = 1$ and $t_{n+1} = t_n + (n+1)$ for $n \geq 1$. From this, show that $t_n = \sum_{j=1}^{n} j$.

35. By putting together two triangular arrays of this kind, one with n rows and one with $n-1$ rows (as illustrated for $n=4$) to form a square, show that $t_{n-1} + t_n = n^2$.

36. By putting together two triangular arrays of this kind, each with n rows, to form a rectangular array of dots of size n by $n+1$ (as illustrated for $n=4$) show that $2t_n = n(n+1)$. From this conclude that $t_n = n(n+1)/2$.

In Exercises 37 - 50 f_n denotes the nth Fibonacci number.

37. Find the following Fibonacci numbers.

    a) f_{10} c) f_{15} e) f_{20}
    b) f_{13} d) f_{18} f) f_{25}

38. Use the fact that $f_k = f_{k+2} - f_{k+1}$ for every positive integer k to find $\sum_{k=1}^{n} f_k$.

39. Determine a formula for $\sum_{j=1}^{n} f_{2j-1} = f_1 + f_3 + \cdots + f_{2n-1}$ where n is a positive integer by examining the value of this sum for small positive integers. Prove this formula using mathematical induction.

40. Prove that $\sum_{j=1}^{n} f_j^2 = f_1^2 + f_2^2 + \cdots + f_n^2 = f_n f_{n+1}$ for every positive integer n.

41. Prove that $f_{n+1} f_{n-1} - f_n^2 = (-1)^n$ for every positive integer n.

42. Prove that $f_{n+1} f_n - f_{n-1} f_{n-2} = f_{2n-1}$ for every positive integer n, $n > 2$.

43. Prove that $f_1 f_2 + f_2 f_3 + \cdots + f_{2n-1} f_{2n} = f_{2n}^2$ if n is a positive integer.

44. Let $\alpha = (1+\sqrt{5})/2$ and $\beta = (1-\sqrt{5})/2$. Prove that $f_n = (\alpha^n - \beta^n)/\sqrt{5}$ for every positive integer n.

45. Show that $f_n \le \alpha^{n-1}$ for every integer n with $n \ge 2$ where $\alpha = (1+\sqrt{5})/2$.

46. Suppose that on January 1 a pair of baby rabbits is left on an island. These rabbits take two months to mature, and on March 1 they produce another pair of rabbits. They continually produce a new pair the first of every succeeding month. Each newborn pair takes two months to mature and produces a new pair on the first day of the third month of its life, and on the first day of every succeeding month. Show that the number of pairs of rabbits on the island after

n months is f_n, assuming that no rabbits ever die.

* **47.** Show that every positive integer can be written as the sum of distinct Fibonacci numbers.

48. Let $\mathbf{F} = \begin{bmatrix} 1 & 1 \\ 1 & 0 \end{bmatrix}$. Show that $\mathbf{F}^n = \begin{bmatrix} f_{n+1} & f_n \\ f_n & f_{n-1} \end{bmatrix}$ when $n \in \mathbf{Z}^+$.

49. By taking determinants of both sides of the result of Exercise 48 prove the identity in Exercise 41.

50. Define the *generalized Fibonacci numbers* recursively by $g_1 = a$, $g_2 = b$, and $g_n = g_{n-1} + g_{n-2}$ for $n \geq 3$. Show that $g_n = af_{n-2} + bf_{n-1}$ for $n \geq 3$.

51. Suppose that $a_0 = 1$, $a_1 = 3$, $a_2 = 9$, and $a_n = a_{n-1} + a_{n-2} + a_{n-3}$ for $n \geq 3$. Show that $a_n \leq 3^n$ for every nonnegative integer n.

52. The tower of Hanoi was a popular puzzle of the late nineteenth century. The puzzle includes three pegs and eight rings of different sizes placed in order of size, with the largest on the bottom, on one of the pegs. The goal of the puzzle is to move all the rings, one at a time without ever placing a larger ring on top of a smaller ring, from the first peg to the second, using the third peg as an auxiliary peg.

     a) Use mathematical induction to show that the minimum number of moves to transfer n rings, with the rules we have described, from one peg to another is $2^n - 1$.

     b) An ancient legend tells of the monks in a tower with 64 gold rings and 3 diamond pegs. They started moving the rings, one move per second, when the world was created. When they finish transferring the rings to the second peg, the world ends. How long will the world last?

** **53.** The *arithmetic mean* of the positive real numbers $a_1, a_2, \ldots, a_n$ is $A = (a_1 + a_2 + \cdots + a_n)/n$ and the *geometric mean* of these number is $G = (a_1 a_2 \cdots a_n)^{1/n}$. Use mathematical induction to prove that $A \geq G$ for every finite sequence of positive real numbers. When does equality hold?

* **54.** Use mathematical induction to show that a $2^n \times 2^n$ chessboard with one square missing can be covered with L-shaped pieces, where each L-shaped piece covers three squares.

55. Give a recursive definition of the Fibonacci number f_n when n is a negative integer. Use your definition to find f_n for $n = -1, -2, -3, \ldots, -10$.

56. Use the results of Exercise 55 to formulate a conjecture that relates the values of f_{-n} and f_n when n is a positive integer. Prove this conjecture using mathematical induction.

1.3 Computer Projects

Programming Projects

Write programs to do the following:

    1. Find the sum of the terms of a geometric series.

    2. Given a positive integer n find the first n terms of the Fibonacci sequence.

   * **3.** List the moves in the tower of Hanoi puzzle (see Exercise 52).

  ** **4.** Cover a $2^n \times 2^n$ chessboard missing one square with L-shaped pieces (see Exercise 54).

Computations and Explorations

Using the programs you have written or a computation program, carry out the following computations and explorations:

    1. Construct a list of the first 1000 Fibonacci numbers and use the list of Fibonacci numbers you constructed to conjecture estimates about the size of the nth Fibonacci number.

1.4 Binomial Coefficients

Sums of two terms are called *binomial expressions*. Powers of binomial expressions are used throughout number theory and thoroughout mathematics. In this section we will define the *binomial coefficients* and show that these are precisely the coefficients that arise in expansions of powers of binomial expressions.

Definition. Let m and k be nonnegative integers with $k \leq m$. The *binomial coefficient* $\binom{m}{k}$ is defined by

$$\binom{m}{k} = \frac{m!}{k!(m-k)!}.$$

When k and m are positive integers with $k > m$ we define $\binom{m}{k} = 0$.

In computing $\binom{m}{k}$, we see that there is a good deal of cancellation, because

$$\binom{m}{k} = \frac{m!}{k!(m-k)!} = \frac{1 \cdot 2 \cdot 3 \ \cdots \ (m-k)(m-k+1) \ \cdots \ (m-1)m}{k! \quad 1 \cdot 2 \cdot 3 \ \cdots \ (m-k)}$$

$$= \frac{(m-k+1) \ \cdots \ (m-1)m}{k!} .$$

Example 1.17. To evaluate the binomial coefficient $\binom{7}{3}$, we note that

$$\binom{7}{3} = \frac{7!}{3!4!} = \frac{1 \cdot 2 \cdot 3 \cdot 4 \cdot 5 \cdot 6 \cdot 7}{1 \cdot 2 \cdot 3 \cdot 1 \cdot 2 \cdot 3 \cdot 4} = \frac{5 \cdot 6 \cdot 7}{1 \cdot 2 \cdot 3} = 35. \qquad \square$$

We now prove some simple properties of binomial coefficients.

Theorem 1.4. Let n and k be nonnegative integers with $k \le n$. Then

$$(i) \qquad \binom{n}{0} = \binom{n}{n} = 1$$

$$(ii) \qquad \binom{n}{k} = \binom{n}{n-k} .$$

Proof. To see that (i) is true, note that

$$\binom{n}{0} = \frac{n!}{0!n!} = \frac{n!}{n!} = 1$$

and

$$\binom{n}{n} = \frac{n!}{n!0!} = \frac{n!}{n!} = 1.$$

To verify (ii), we see that

$$\binom{n}{k} = \frac{n!}{k!(n-k)!} = \frac{n!}{(n-k)!(n-(n-k))!} = \binom{n}{n-k} . \qquad \blacksquare$$

An important property of binomial coefficients is the following identity.

Theorem 1.5. Let n and k be positive integers with $n \ge k$. Then

$$\binom{n}{k} + \binom{n}{k-1} = \binom{n+1}{k} .$$

Proof. We perform the addition

$$\begin{bmatrix} n \\ k \end{bmatrix} + \begin{bmatrix} n \\ k-1 \end{bmatrix} = \frac{n!}{k!(n-k)!} + \frac{n!}{(k-1)!(n-k+1)!}$$

by using the common denominator $k!(n-k+1)!$. This gives

$$\begin{bmatrix} n \\ k \end{bmatrix} + \begin{bmatrix} n \\ k-1 \end{bmatrix} = \frac{n!(n-k+1)}{k!(n-k+1)!} + \frac{n!k}{k!(n-k+1)!}$$

$$= \frac{n!((n-k+1)+k)}{k!(n-k+1)!}$$

$$= \frac{n!(n+1)}{k!(n-k+1)!}$$

$$= \frac{(n+1)!}{k!(n-k+1)!}$$

$$= \begin{bmatrix} n+1 \\ k \end{bmatrix}. \qquad\qquad\blacksquare$$

Using Theorem 1.5 we can construct *Pascal's triangle*, named after the French mathematician Blaise Pascal* who used the binomial coefficients in his analysis of gambling games. In Pascal's triangle, the binomial coefficient $\begin{bmatrix} n \\ k \end{bmatrix}$ is the $(k+1)$st number in the $(n+1)$st row. The first nine rows of Pascal's triangle are displayed in Figure 1.1.

We see that the exterior numbers in the triangle are all 1. To find an interior number, we simply add the two numbers in the positions above, and to either side, of the position being filled. From Theorem 1.5, this yields the correct integer.

---

* **BLAISE PASCAL** (1623-1662) exhibited his mathematical talents early even though his father, who had made discoveries in analytic geometry, kept mathematical books from him to encourage his other interests. At 16 Pascal discovered an important result concerning conic sections. At 18 he designed a calculating machine, which he had built and successfully sold. Later Pascal made substantial contributions to hydrostatics. Pascal, together with Fermat, laid the foundations for the modern theory of probability. It was in his work on probability where Pascal made new discoveries concerning what is now called Pascal's triangle. In this work he gave what is considered the first lucid description of the principle of mathematical induction. In 1654, catalyzed by an intense religious experience, Pascal abandoned his mathematical and scientific pursuits to devote himself to theology. After this he returned to mathematics only once. One night he had insomnia caused by the discomfort of a toothache and, as a distraction, he studied the mathematical properties of the cycloid. Miraculously, his pain subsided, which he took as a signal of divine approval of the study of mathematics.

$$1$$
$$1 \; 1$$
$$1 \; 2 \; 1$$
$$1 \; 3 \; 3 \; 1$$
$$1 \; 4 \; 6 \; 4 \; 1$$
$$1 \; 5 \; 10 \; 10 \; 5 \; 1$$
$$1 \; 6 \; 15 \; 20 \; 15 \; 6 \; 1$$
$$1 \; 7 \; 21 \; 35 \; 35 \; 21 \; 7 \; 1$$
$$1 \; 8 \; 28 \; 56 \; 70 \; 56 \; 28 \; 8 \; 1$$

$$\cdots$$

Figure 1.1. Pascal's triangle.

Binomial coefficients occur in the expansions of powers of sums. Exactly how they occur is described by the *binomial theorem.*

Theorem 1.6 The Binomial Theorem. Let x and y be variables and n a positive integer. Then

$$(x+y)^n = \binom{n}{0} x^n + \binom{n}{1} x^{n-1} y + \binom{n}{2} x^{n-2} y^2 + \cdots$$

$$+ \binom{n}{n-2} x^2 y^{n-2} + \binom{n}{n-1} xy^{n-1} + \binom{n}{n} y^n,$$

or using summation notation,

$$(x+y)^n = \sum_{j=0}^{n} \binom{n}{j} x^{n-j} y^j.$$

Proof. We use mathematical induction. When $n = 1$, according to the binomial theorem, the formula becomes

$$(x+y)^1 = \binom{1}{0} x^1 y^0 + \binom{1}{1} x^0 y^1.$$

But because $\binom{1}{0} = \binom{1}{1} = 1$, this states that $(x+y)^1 = x + y$, which is obviously true.

We now assume that the theorem is true for the positive integer n, that is, we assume that

$$(x+y)^n = \sum_{j=0}^{n} \binom{n}{j} x^{n-j} y^j.$$

We must now verify that the corresponding formula holds with n replaced by $n + 1$, assuming the result holds for n. Hence, we have

$$(x+y)^{n+1} = (x+y)^n (x+y)$$

$$= \left[\sum_{j=0}^{n} \binom{n}{j} x^{n-j} y^j \right] (x+y)$$

$$= \sum_{j=0}^{n} \binom{n}{j} x^{n-j+1} y^j + \sum_{j=0}^{n} \binom{n}{j} x^{n-j} y^{j+1}.$$

We see by removing terms from the sums and subsequently shifting indices, that

$$\sum_{j=0}^{n} \binom{n}{j} x^{n-j+1} y^j = x^{n+1} + \sum_{j=1}^{n} \binom{n}{j} x^{n-j+1} y^j$$

and

$$\sum_{j=0}^{n} \binom{n}{j} x^{n-j} y^{j+1} = \sum_{j=0}^{n-1} \binom{n}{j} x^{n-j} y^{j+1} + y^{n+1}$$

$$= \sum_{j=1}^{n} \binom{n}{j-1} x^{n-j+1} y^j + y^{n+1}.$$

Hence, we find that

$$(x+y)^{n+1} = x^{n+1} + \sum_{j=1}^{n} \left[\binom{n}{j} + \binom{n}{j-1} \right] x^{n-j+1} y^j + y^{n+1}.$$

By Pascal's Identity we have

$$\binom{n}{j} + \binom{n}{j-1} = \binom{n+1}{j},$$

so we conclude that

$$(x+y)^{n+1} = x^{n+1} + \sum_{j=1}^{n} \binom{n+1}{j} x^{n-j+1} y^j + y^{n+1}$$

$$= \sum_{j=0}^{n+1} \binom{n+1}{j} x^{n+1-j} y^j.$$

This establishes the theorem.   □

The binomial theorem shows that the coefficients of $(x+y)^n$ are the numbers in the $(n+1)$st row of Pascal's triangle.

We now illustrate one use of the binomial theorem.

Corollary 1.6.1. Let n be a nonnegative integer. Then

$$2^n = (1+1)^n = \sum_{j=0}^{n} \binom{n}{j} 1^{n-j} 1^j = \sum_{j=0}^{n} \binom{n}{j}.$$

Proof. Let $x=1$ and $y=1$ in the binomial theorem. $\square$

Corollary 1.6.1 shows that if we add all elements of the $(n+1)$th row of Pascal's triangle, we get 2^n. For instance, for the fifth row, we find that

$$\binom{4}{0} + \binom{4}{1} + \binom{4}{2} + \binom{4}{3} + \binom{4}{4} = 1 + 4 + 6 + 4 + 1 = 16 = 2^4.$$

1.4 Exercises

1. Find the value of each of the following binomial coefficients.

    a) $\binom{100}{0}$ d) $\binom{11}{5}$

    b) $\binom{50}{1}$ e) $\binom{10}{7}$

    c) $\binom{20}{3}$ f) $\binom{70}{70}$

2. Find the binomial coefficients $\binom{9}{3}$, $\binom{9}{4}$, and $\binom{10}{4}$, and verify that

    $$\binom{9}{3} + \binom{9}{4} = \binom{10}{4}.$$

3. Use the binomial theoreom to write out all terms in the expansions of the following expressions.

    a) $(a+b)^5$ d) $(2a + 3b)^4$
    b) $(x+y)^{10}$ e) $(3x - 4y)^5$
    c) $(m-n)^7$ f) $(5x + 7)^8$

4. What is the coefficient of $x^{99} y^{101}$ in $(2x + 3y)^{200}$?

5. Let n be a positive integer. Using the binomial theorem to expand $(1+(-1))^n$ show that

$$\sum_{k=0}^{n} (-1)^k \binom{n}{k} = 0.$$

6. Use Corollary 1.6.1 and Exercise 5 to find

$$\binom{n}{0} + \binom{n}{2} + \binom{n}{4} + \cdots$$

and

$$\binom{n}{1} + \binom{n}{3} + \binom{n}{5} + \cdots.$$

7. Show that if n, r, and k are integers with $0 \le k \le r \le n$ then

$$\binom{n}{r}\binom{r}{k} = \binom{n}{k}\binom{n-k}{r-k}.$$

* 8. What is the largest value of $\binom{m}{n}$ where m is a positive integer and n is an integer such that $0 \le n \le m$? Justify your answer.

9. Show that

$$\binom{r}{r} + \binom{r+1}{r} + \cdots + \binom{n}{r} = \binom{n+1}{r+1}$$

where n and r are integers with $1 \le r \le n$.

10. Show that

$$\binom{n}{0} + \binom{n-1}{1} + \binom{n-2}{2} + \cdots = f_{n+1}$$

where n is a nonnegative integer and f_{n+1} is the $(n+1)$st Fibonacci number. (Here the sum ends with the term $\binom{1}{n-1}$.)

11. Prove that whenever n is a nonnegative integer $\sum_{j=1}^{n} \binom{n}{j} f_j = f_{2n}$ where f_j is the jth Fibonacci number.

The binomial coefficients $\binom{x}{n}$, where x is a real number and n is a positive integer, can be defined recursively by the equations $\binom{x}{1} = x$ and

$$\binom{x}{n+1} = \frac{x-n}{n+1}\binom{x}{n}.$$

12. Show from the recursive definition that if x is a positive integer, then
$$\binom{x}{k} = \frac{x!}{k!(x-k)!}, \text{ where } k \text{ is an integer with } 1 \le k \le x.$$

13. Show from the recursive definition that if x is a positive integer, then $\binom{x}{n} +$
$$\binom{x}{n+1} = \binom{x+1}{n+1}, \text{ whenever } n \text{ is a positive integer.}$$

14. Show that the binomial coefficient $\binom{n}{k}$, where n and k are integers with $0 \le k \le n$, gives the number of subsets with k elements of a set with n elements.

15. Use Exercise 14 to give an alternate proof of the binomial theorem.

16. Let S be a set with n elements and let P_1 and P_2 be two properties that an element of S may have. Show that the number of elements of S possessing neither property P_1 nor property P_2 is

$$n - [n(P_1) + n(P_2) - n(P_1,P_2)]$$

where $n(P_1)$, $n(P_2)$, and $n(P_1,P_2)$ are the number of elements of S with property P_1, with property P_2, and both properties P_1 and P_2.

17. Let S be a set with n elements and let P_1, P_2, and P_3 be three properties that an element S may have. Show that the number of elements of S possessing none of the properties P_1, P_2, and P_3 is

$$n - [n(P_1) + n(P_2) + n(P_3)$$
$$- n(P_1,P_2) - n(P_1,P_3) - n(P_2,P_3) + n(P_1,P_2,P_3)]$$

where $n(P_{i_1} \cdots P_{i_k})$ is the number of elements of S with properties $P_{i_1}, \ldots, P_{i_k}$.

* 18. In this exercise we develop the *principle of inclusion-exclusion*. Suppose that S is a set with n elements and let $P_1, P_2, \ldots, P_t$ be t different properties that an element of S may have. Show that the number of elements of S possessing *none* of the t properties is

$$n - [n(P_1) + n(P_2) + \cdots + n(P_t)]$$
$$+ [n(P_1,P_2) + n(P_1,P_3) + \cdots + n(P_{t-1},P_t)]$$
$$- [n(P_1,P_2,P_3) + n(P_1,P_2,P_4) + \cdots + n(P_{t-2},P_{t-1},P_t)]$$
$$+ \cdots + (-1)^t n(P_1,P_2,\ldots,P_t),$$

where $n(P_{i_1},P_{i_2},\ldots,P_{i_j})$ is the number of elements of S possessing all of the properties $P_{i_1},P_{i_2},\ldots,P_{i_j}$. The first expression in brackets contains a term for each property, the second expression in brackets contains terms for all combinations of two properties, the third expression contains terms for all combinations of three properties, and so forth. (*Hint:* For each element of S

determine the number of times it is counted in the above expression. If an element has k of the properties, show that it is counted $1 - \begin{pmatrix} k \\ 1 \end{pmatrix} + \begin{pmatrix} k \\ 2 \end{pmatrix} - \cdots + (-1)^k \begin{pmatrix} k \\ k \end{pmatrix}$ times. This is 0 when $k > 0$ by Exercise 5.)

* **19.** What are the coefficients of $(x_1 + x_2 + \cdots + x_m)^n$? These coefficients are called *multinomial coefficients*.

20. Write out all terms in the expansion of $(x + y + z)^7$.

21. What is the coefficient of $x^3 y^4 z^5$ in the expansion of $(2x - 3y + 5z)^{12}$?

1.4 Computer Projects

Programming Projects

Write programs to do the following:

1. Evaluate binomial coefficients.

2. Given a positive integer n, print out the first n rows of Pascal's triangle.

3. Expand $(x+y)^n$, given a positive integer n, using the binomial theorem.

Computations and Explorations

Using the programs you have written or a computation program, carry out the following computations and explorations:

1. Find the least integer n such that there is a binomial coefficient $\begin{pmatrix} n \\ k \end{pmatrix}$, where k is a positive integer, greater than 1000000.

1.5 Divisibility

When an integer is divided by a second nonzero integer, the quotient may or may not be an integer. For instance, $24/8 = 3$ is an integer, whereas $17/5 = 3.4$ is not. This observation leads to the following definition.

Definition. If a and b are integers with $a \neq 0$, we say that a *divides* b if there is an integer c such that $b = ac$. If a divides b, we also say that a is a *divisor* or *factor* of b.

If a divides b we write $a \mid b$ and if a does not divide b, we write $a \nmid b$.

Example 1.18. The following statements illustrate the concept of divisibility of integers: $13 \mid 182$, $-5 \mid 30$, $17 \mid 289$, $6 \nmid 44$, $7 \nmid 50$, $-3 \mid 33$, and $17 \mid 0$. □

Example 1.19. The divisors of 6 are ± 1, ± 2, ± 3, and ± 6. The divisors of 17 are ± 1 and ± 17. The divisors of 100 are ± 1, ± 2, ± 4, ± 5, ± 10, ± 20, ± 25, ± 50, and ± 100. □

In subsequent sections, we will need some simple properties of divisibility. We now state and prove these properties.

Theorem 1.7. If a, b, and c are integers with $a \mid b$ and $b \mid c$, then $a \mid c$.

Proof. Since $a \mid b$ and $b \mid c$, there are integers e and f with $ae = b$ and $bf = c$. Hence, $c = bf = (ae)f = a(ef)$, and we conclude that $a \mid c$. ∎

Example 1.20. Since $11 \mid 66$ and $66 \mid 198$, Theorem 1.7 tells us that $11 \mid 198$.

□

Theorem 1.8. If a, b, m, and n are integers, and if $c \mid a$ and $c \mid b$, then $c \mid (ma + nb)$.

Proof. Since $c \mid a$ and $c \mid b$, there are integers e and f such that $a = ce$ and $b = cf$. Hence, $ma + nb = mce + ncf = c(me + nf)$. Consequently, we see that $c \mid (ma + nb)$. ∎

Example 1.21. Since $3 \mid 21$ and $3 \mid 33$, Theorem 1.8 tells us that 3 divides

$$5 \cdot 21 - 3 \cdot 33 = 105 - 99 = 6.$$

□

The following theorem states an important fact about division.

Theorem 1.9. The Division Algorithm. If a and b are integers such that $b > 0$, then there are unique integers q and r such that $a = bq + r$ with $0 \leq r < b$.

In the equation given in the division algorithm, we call q the *quotient* and r the *remainder*. We also call a the *dividend* and b the *divisor*. (*Note*: We use the traditional name for this theorem even though the division algorithm is not an algorithm. We discuss algorithms, which are procedures for solving general problems, in Section 7 of this chapter.)

We note that a is divisible by b if and only if the remainder in the division algorithm is zero. Before we prove the division algorithm, consider the following examples.

Example 1.22. If $a = 133$ and $b = 21$, then $q = 6$ and $r = 7$, since $133 = 21 \cdot 6 + 7$. Likewise, if $a = -50$ and $b = 8$, then $q = -7$ and $r = 6$, since $-50 = 8(-7) + 6$. □

We now prove the division algorithm. We prove this theorem using the well-ordering property.

Proof. Consider the set S of all integers of the form $a - bk$ where k is an integer, that is, $S = \{ a - bk \mid k \in \mathbf{Z} \}$. Let T be the set of all nonnegative integers in S. T is nonempty, since $a - bk$ is positive whenever k is an integer with $k < a/b$. By the well-ordering property T has a least element $k = q$. We set $r = a - bq$. (These are the values for q and r specified in the theorem.) We know that $r \geq 0$ by construction, and it is easy to see that $r < b$. If $r \geq b$ then $r > r - b = a - bq - b = a - b(q+1) \geq 0$, which contradicts the choice of $r = a - bq$ as the least nonnegative integer of the form $a - bk$. Hence $0 \leq r < b$.

To show that these values for q and r are unique, assume that we have two equations $a = bq_1 + r_1$ and $a = bq_2 + r_2$, with $0 \le r_1 < b$ and $0 \le r_2 < b$. By subtracting the second of these from the first, we find that

$$0 = b(q_1 - q_2) + (r_1 - r_2).$$

Hence, we see that

$$r_2 - r_1 = b(q_1 - q_2).$$

This tells us that b divides $r_2 - r_1$. Since $0 \le r_1 < b$ and $0 \le r_2 < b$, we have $-b < r_2 - r_1 < b$. Hence b can divide $r_2 - r_1$ only if $r_2 - r_1 = 0$, or, in other words, if $r_1 = r_2$. Since $bq_1 + r_1 = bq_2 + r_2$ and $r_1 = r_2$ we also see that $q_1 = q_2$. This shows that the quotient q and the remainder r are unique. ■

We can use the greatest integer function to give explicit formulae for the quotient and remainder in the division algorithm. Since the quotient q is the largest integer such that $bq \le a$, and $r = a - bq$, it follows that

$$q = [a/b], \quad r = a - b[a/b].$$

The following examples display the quotient and remainder of a division.

Example 1.23. Let $a = 1028$ and $b = 34$. Then $a = bq + r$ with $0 \le r < b$, where $q = [1028/34] = 30$ and $r = 1028 - [1028/34] \cdot 34 = 1028 - 30 \cdot 34 = 8$. □

Example 1.24. Let $a = -380$ and $b = 75$. Then $a = bq + r$ with $0 \le r < b$, where $q = [-380/75] = -6$ and $r = -380 - [-380/75] \cdot 75 = -380 - (-6)75 = 70$. □

Given a positive integer d, we can classify integers according to their remainders when divided by d. For example, with $d = 2$, we see from the division algorithm that every integer when divided by 2 leaves a remainder of either 0 or 1. This leads to the following definition of some common terminology.

Definition. If the remainder when n is divided by 2 is 0, then $n = 2k$ for some positive integer k, and we say that n is *even*, while if the remainder when n is divided by 2 is 1, then $n = 2k + 1$ for some integer k, and we say that n is *odd*.

Similarly, when $d = 4$, we see from the division algorithm that when an integer n is divided by 4, the remainder is either $0, 1, 2,$ or 3. Hence, every integer is of the form $4k, 4k + 1, 4k + 2,$ or $4k + 3$, where k is a positive integer.

We will pursue these matters further in Chapter 3.

1.5 Exercises

1. Show that $3 \mid 99$, $5 \mid 145$, $7 \mid 343$, and $888 \mid 0$.

2. Show that 1001 is divisible by 7, by 11, and by 13.

3. Decide which of the following integers are divisible by 7.
   a) 0
   b) 707
   c) 1717
   d) 123321
   e) -285714
   f) -430597.

4. Decide which of the following integers are divisible by 22
   a) 0
   b) 444
   c) 1716
   d) 192544
   e) -32516
   f) -195518.

5. Find the quotient and remainder in the division algorithm with divisor 17 and dividend
   a) 100
   b) 289
   c) -44
   d) -100.

6. What can you conclude if a and b are nonzero integers such that $a \mid b$ and $b \mid a$?

7. Show that if a, b, c, and d are integers with a and c nonzero such that $a \mid b$ and $c \mid d$, then $ac \mid bd$.

8. Are there integers a, b, and c such that $a \mid bc$, but $a \nmid b$ and $a \nmid c$?

9. Show that if a, b, and $c \neq 0$ are integers, then $a \mid b$ if and only if $ac \mid bc$.

10. Show that if a and b are positive integers and $a \mid b$, then $a \leq b$.

☞ 11. Show that if a and b are integers such that $a \mid b$ then $a^k \mid b^k$ for every positive integer k.

12. Show that the sum of two even or of two odd integers is even, while the sum of an odd and an even integer is odd.

13. Show that the product of two odd integers is odd, while the product of two integers is even if either of the integers is even.

14. Show that if a and b are odd positive integers and b does not divide a, then there are integers s and t such that $a = bs + t$, where t is odd and $|t| < b$.

15. When the integer a is divided by the integer b where $b > 0$, the division algorithm gives a quotient of q and a remainder of r. Show that if $b \nmid a$, when $-a$ is divided by b, the division algorithm gives a quotient of $-(q+1)$ and a remainder of $b - r$, while if $b \mid a$, the quotient is $-q$ and the remainder is zero.

16. Show that if a, b, and c are integers with $b > 0$ and $c > 0$, such that when a is divided by b the quotient is q and the remainder is r, and when q is divided by c the quotient is t and the remainder is s, then when a is divided by bc, the quotient is t and the remainder is $bs + r$.

17. a) Extend the division algorithm by allowing negative divisors. In particular, show that whenever a and $b \neq 0$ are integers, there are unique integers q and r such that $a = bq + r$, where $0 \leq r < |b|$.

   b) Find the remainder when 17 is divided by -7.

18. Show that if a and b are positive integers, then there are unique integers q, r and $e = \pm 1$ such that $a = bq + er$ where $-b/2 < r \leq b/2$.

☞ 19. Show that if m and $n > 0$ are integers, then

$$\left[\frac{m+1}{n}\right] = \begin{cases} \left[\dfrac{m}{n}\right] & \text{if } m \neq kn - 1 \text{ for some integer } k \\ \left[\dfrac{m}{n}\right] + 1 & \text{if } m = kn - 1 \text{ for some integer } k. \end{cases}$$

20. Show that the integer n is even if and only if $n - 2[n/2] = 0$.

21. Show that the number of positive integers less than or equal to x, where x is a positive real number, that are divisible by the positive integer d equals $[x/d]$.

22. Find the number of positive integers not exceeding 1000 that are divisible by 5, by 25, by 125, and by 625.

23. How many integers between 100 and 1000 are divisible by 7? by 49?

24. Find the number of positive integers not exceeding 1000 that are not divisible by 3 or 5.

25. Find the number of positive integers not exceeding 1000 that are not divisible by 3, 5, or 7.

26. Find the number of positive integers not exceeding 1000 that are divisible by 3 but not by 4.

27. In 1992 to mail a first class letter in the U.S.A. it costs 29 cents for the first ounce and 23 cents for each additional ounce or fraction thereof. Find a formula involving the greatest integer function for the cost of mailing a letter. Could it possibly cost $1.45 or $2.13 to mail a first class letter in the U.S.A.?

28. Show that if a is an integer, then 3 divides $a^3 - a$.

29. Show that the product of two integers of the form $4k + 1$ is again of this form, while the product of two integers of the form $4k + 3$ is of the form $4k + 1$.

30. Show that the square of every odd integer is of the form $8k + 1$.

31. Show that the fourth power of every odd integer is of the form $16k + 1$.

32. Show that the product of two integers of the form $6k + 5$ is of the form $6k + 1$.

33. Show that the product of any three consecutive integers is divisible by 6.

34. Use mathematical induction to show that $n^5 - n$ is divisible by 5 for every positive integer n.

35. Use mathematical induction to show that the sum of the cubes of three consecutive integers is divisible by 9.

In Exercises 36 - 40 let f_n denote the nth Fibonacci number.

36. Show that f_n is even if and only if n is divisible by 3.

37. Show that f_n is divisible by 3 if and only if n is divisible by 4.

38. Show that f_n is divisible by 4 if and only if n is divisible by 6.

39. Show that $f_n = 5f_{n-4} + 3f_{n-5}$ whenever n is a positive integer with $n > 5$. Use this result to show that f_n is divisible by 5 whenever n is divisible by 5.

* 40. Show that $f_{n+m} = f_m f_{n+1} + f_{m-1} f_n$ whenever m and n are positive integers with $m > 1$. Use this result to show that $f_n \mid f_m$ when m and n are positive integers with $n \mid m$.

Let n be a positive integer. We define

$$T(n) = \begin{cases} n/2 & \text{if } n \text{ is even} \\ (3n+1)/2 & \text{if } n \text{ is odd.} \end{cases}$$

We then form the sequence obtained by iterating T: $n, T(n), T(T(n)), T(T(T(n))),\dots$. For instance, starting with $n = 7$ we have $7,11,17,26,13,20,10,5,8,4,2,1,2,1,2,1,\dots$. A well-known conjecture, sometimes called the *Collatz conjecture*, asserts that the sequence obtained by iterating T always reaches the integer 1 no matter which positive integer n begins the sequence.

41. Find the sequence obtained by iterating T starting with $n = 39$.

42. Show that the sequence obtained by iterating T starting with $n = (2^{2k} - 1)/3$, where k is a positive integer greater than 1, always reaches the integer 1.

43. Show that the Collatz conjecture is true if it can be shown that for every positive integer n with $n \geq 2$ there is a term in the sequence obtained by iterating T that is less than n.

44. Verify that there is a term in the sequence obtained by iterating T, starting with the positive integer n, that is less than n for all positive integers n with $2 \leq n \leq 100$. (*Hint*: Begin by considering sets of positive integers for which it is easy to show this is true.)

* 45. Show that $[(2+\sqrt{3})^n]$ is odd whenever n is a nonnegative integer.

1.5 Computer Projects

Programming Projects

Write programs to do the following:

1. Decide whether an integer is divisible by a given integer.

2. Find the quotient and remainder in the division algorithm.

3. Find the quotient, remainder, and sign in the modified division algorithm given in Exercise 18.

4. Compute the terms of the sequence n, $T(n)$, $T(T(n))$, $T(T(T(n)))$,... for a given positive integer n defined in the preamble to Exercise 41.

Computations and Explorations

Using the programs you have written or a computation program, carry out the following computations and explorations:

1. Verify the Collatz conjecture described in the preamble to Exercise 41 for all integers n not exceeding 10000.

2. Using numerical evidence what sort of conjectures can you make concerning the number of iterations needed before the sequence of iterations $T(n)$ reaches 1 where n is a given positive integer.

3. Using numerical evidence make conjectures about the divisibility of Fibonacci numbers by 7, by 8, by 9, by 11, by 13, and so on.

1.6 Representations of Integers

The conventional manner of expressing numbers is by decimal notation. We write out numbers using digits to represent multiples of powers of ten. For instance, when we write the integer 34765 we mean

$$3 \cdot 10^4 + 4 \cdot 10^3 + 7 \cdot 10^2 + 6 \cdot 10^1 + 5 \cdot 10^0.$$

There is no particular reason for the use of ten as the base of notation, other than the fact that we have ten fingers. Other civilizations have used different bases, including the Babylonians, who used base sixty, and the Mayans, who used base twenty. Electronic computers use two as a base for internal representation of integers, and either eight or sixteen for display purposes.

We now show that every positive integer greater than one may be used as a base.

Theorem 1.10. Let b be a positive integer with $b > 1$. Then every positive integer n can be written uniquely in the form

$$n = a_k b^k + a_{k-1} b^{k-1} + \cdots + a_1 b + a_0,$$

where k is a nonnegative integer, a_j is an integer with $0 \le a_j \le b-1$ for $j = 0,1,\ldots,k$ and the initial coefficient $a_k \ne 0$.

Proof. We obtain an expression of the desired type by successively applying the division algorithm in the following way. We first divide n by b to obtain

$$n = bq_0 + a_0, \quad 0 \le a_0 \le b-1.$$

If $q_0 \ne 0$, we continue by dividing q_0 by b to find that

$$q_0 = bq_1 + a_1, \quad 0 \le a_1 \le b-1.$$

We continue this process to obtain

$$q_1 = bq_2 + a_2, \quad 0 \le a_2 \le b-1,$$
$$q_2 = bq_3 + a_3, \quad 0 \le a_3 \le b-1,$$
$$\cdot$$
$$\cdot$$
$$\cdot$$
$$q_{k-2} = bq_{k-1} + a_{k-1}, \quad 0 \le a_{k-1} \le b-1,$$
$$q_{k-1} = b \cdot 0 + a_k, \quad 0 \le a_k \le b-1.$$

The last step of the process occurs when a quotient of 0 is obtained. To see this, first note that the sequence of quotients satisfies

$$n > q_0 > q_1 > q_2 > \cdots \ge 0.$$

Since the sequence $q_0, q_1, q_2, \ldots$ is a decreasing sequence of nonnegative integers which continues as long as its terms are positive, there are at most q_0 terms in this sequence and the last term equals 0.

From the first equation above we find that

$$n = bq_0 + a_0.$$

We next replace q_0 using the second equation, to obtain

$$n = b(bq_1 + a_1) + a_0 = b^2 q_1 + a_1 b + a_0,$$

Successively substituting for $q_1, q_2, \ldots, q_{k-1}$, we have

$$n = b^3 q_2 + a_2 b^2 + a_1 b + a_0,$$
$$\cdot$$
$$\cdot$$
$$\cdot$$
$$n = b^{k-1} q_{k-2} + a_{k-2} b^{k-2} + \cdots + a_1 b + a_0,$$
$$n = b^k q_{k-1} + a_{k-1} b^{k-1} + \cdots + a_1 b + a_0$$
$$= a_k b^k + a_{k-1} b^{k-1} + \cdots + a_1 b + a_0,$$

where $0 \le a_j \le b-1$ for $j = 0,1,\ldots,k$ and $a_k \ne 0$, since $a_k = q_{k-1}$ is the last nonzero quotient. Consequently, we have found an expansion of the desired type.

To see that the expansion is unique, assume that we have two such expansions equal to n, $i.e.$,

$$n = a_k b^k + a_{k-1} b^{k-1} + \cdots + a_1 b + a_0$$
$$= c_k b^k + c_{k-1} b^{k-1} + \cdots + c_1 b + c_0,$$

where $0 \le a_k < b$ and $0 \le c_k < b$ (and if necessary we add initial terms with zero coefficients to one of the expansions to have the number of terms agree). Subtracting one expansion from the other, we have

$$(a_k - c_k) b^k + (a_{k-1} - c_{k-1}) b^{k-1} + \cdots + (a_1 - c_1) b + (a_0 - c_0) = 0.$$

If the two expansions are different, there is a smallest integer j, $0 \le j \le k$, such that $a_j \ne c_j$. Hence,

$$b^j \left[(a_k - c_k) b^{k-j} + \cdots + (a_{j+1} - c_{j+1}) b + (a_j - c_j) \right] = 0,$$

so that

$$(a_k - c_k) b^{k-j} + \cdots + (a_{j+1} - c_{j+1}) b + (a_j - c_j) = 0.$$

Solving for $a_j - c_j$ we obtain

$$a_j - c_j = (c_k - a_k) b^{k-j} + \cdots + (c_{j+1} - a_{j+1}) b$$
$$= b \left[(c_k - a_k) b^{k-j-1} + \cdots + (c_{j+1} - a_{j+1}) \right].$$

Hence, we see that

$$b \mid (a_j - c_j).$$

But since $0 \le a_j < b$ and $0 \le c_j < b$, we know that $-b < a_j - c_j < b$. Consequently, $b \mid (a_j - c_j)$ implies that $a_j = c_j$. This contradicts the assumption that the two expansions are different. We conclude that our base b expansion of n is unique. ∎

For $b = 2$, we see by Theorem 1.10 that the following corollary holds.

Corollary 1.10.1. Every positive integer may be represented as the sum of distinct powers of two.

Proof. Let n be a positive integer. From Theorem 1.10 with $b = 2$, we know that $n = a_k 2^k + a_{k-1} 2^{k-1} + \cdots + a_1 2 + a_0$ where each a_j is either 0 or 1. Hence, every positive integer is the sum of distinct powers of 2. ∎

In the expansions described in Theorem 1.10, b is called the *base* or *radix* of the expansion. We call base 10 notation, our conventional way of writing integers, *decimal* notation. Base 2 expansions are called *binary* expansions, base 8 expansions are called *octal* expansions, and base 16 expansions are called *hexadecimal*, or *hex* for short, expansions. The coefficients a_j are called the *digits* of the expansion. Binary digits are called *bits* (*bi*nary dig*its*) in computer terminology.

To distinguish representations of integers with different bases, we use a special notation. We write $(a_k a_{k-1}...a_1 a_0)_b$ to represent the number $a_k b^k + a_{k-1} b^{k-1} + \cdots + a_1 b + a_0$.

Example 1.25. To illustrate base b notation, note that $(236)_7 = 2 \cdot 7^2 + 3 \cdot 7 + 6 = 135$ and $(10010011)_2 = 1 \cdot 2^7 + 1 \cdot 2^4 + 1 \cdot 2^1 + 1 = 147$.   □

Note that the proof of Theorem 1.10 gives us a method of finding the base b expansion of a given positive integer. We simply perform the division algorithm successively, replacing the dividend each time with the quotient, and stop when we come to a quotient which is zero. We then read up the list of remainders to find the base b expansion.

Example 1.26. To find the base 2 expansion of 1864, we use the division algorithm successively:

$$
\begin{aligned}
1864 &= 2 \cdot 932 + 0, \\
932 &= 2 \cdot 466 + 0, \\
466 &= 2 \cdot 233 + 0 \\
233 &= 2 \cdot 116 + 1, \\
116 &= 2 \cdot 58 + 0, \\
58 &= 2 \cdot 29 + 0, \\
29 &= 2 \cdot 14 + 1, \\
14 &= 2 \cdot 7 + 0, \\
7 &= 2 \cdot 3 + 1, \\
3 &= 2 \cdot 1 + 1, \\
1 &= 2 \cdot 0 + 1.
\end{aligned}
$$

To obtain the base 2 expansion of 1864, we simply take the remainders of these divisions. This shows that $(1864)_{10} = (11101001000)_2$.   □

Computers represent numbers internally by using a series of "switches" which may be either "on" or "off". (This may be done mechanically, using magnetic tape, electrical switches, or by other means.) Hence, we have two possible states for each switch. We can use "on" to represent the digit 1 and "off" to represent the digit 0. This is why computers use binary expansions to represent integers internally.

Computers use base 8 or base 16 for display purposes. In base 16, or hexadecimal, notation there are 16 digits, usually denoted by $0,1,2,3,4,5,6,7,8,9,A,B,C,D,E,F$. The letters A, B, C, D, E, and F are used to represent the digits that correspond to 10, 11, 12, 13, 14, and 15 (written in decimal notation). We give the following example to show how to convert from hexadecimal notation to decimal notation.

Example 1.27. To convert $(A35B0F)_{16}$ from hexadecimal notation to decimal notation we write

$$(A35B0F)_{16} = 10 \cdot 16^5 + 3 \cdot 16^4 + 5 \cdot 16^3 + 11 \cdot 16^2 + 0 \cdot 16 + 15$$
$$= (10705679)_{10}. \qquad \qquad \square$$

A simple conversion is possible between binary and hexadecimal notation. We can write each hex digit as a block of four binary digits according to the correspondence given in Table 1.1.

Hex Digit	Binary Digits	Hex Digit	Binary Digits
0	0000	8	1000
1	0001	9	1001
2	0010	A	1010
3	0011	B	1011
4	0100	C	1100
5	0101	D	1101
6	0110	E	1110
7	0111	F	1111

Table 1.1. Conversion from hex digits to blocks of binary digits.

Example 1.28. An example of conversion from hex to binary is $(2FB3)_{16} = (10111110110011)_2$. Each hex digit is converted to a block of four binary digits (the initial zeros in the initial block $(0010)_2$ corresponding to the digit $(2)_{16}$ are omitted).

To convert from binary to hex, consider $(11110111101001)_2$. We break this into blocks of four starting from the right. The blocks are, from right to left, 1001, 1110, 1101, and 0011 (we add the initial zeros). Translating each block to hex, we obtain $(3DE9)_{16}$. $\qquad \square$

We note that a conversion between two different bases is as easy as binary hex conversion, whenever one of the bases is a power of the other.

1.6 Exercises

1. Convert $(1999)_{10}$ from decimal to base 7 notation. Convert $(6105)_7$ from base 7 to decimal notation.

2. Convert $(89156)_{10}$ from decimal to base 8 notation. Convert $(706113)_8$ from base 8 to decimal notation.

3. Convert $(10101111)_2$ from binary to decimal notation and $(999)_{10}$ from decimal to binary notation.

4. Convert $(101001000)_2$ from binary to decimal notation and $(1984)_{10}$ from decimal to binary notation.

5. Convert $(100011110101)_2$ and $(11101001110)_2$ from binary to hexadecimal.

6. Convert $(ABCDEF)_{16}$, $(DEFACED)_{16}$, and $(9A0B)_{16}$ from hexadecimal to binary.

7. Explain why we really are using base 1000 notation when we break large decimal integers into blocks of three digits, separated by commas.

8. Show that if b is a negative integer less than -1, then every nonzero integer n can be uniquely written in the form

$$n = a_k b^k + a_{k-1} b^{k-1} + \cdots + a_1 b + a_0,$$

where $a_k \neq 0$ and $0 \leq a_j < |b|$ for $j = 0,1,2,...,k$. We write $n = (a_k a_{k-1}...a_1 a_0)_b$, just as we do for positive bases.

9. Find the decimal representation of $(101001)_{-2}$ and $(12012)_{-3}$.

10. Find the base -2 representations of the decimal numbers $-7, -17$, and 61.

11. Show that any weight not exceeding $2^k - 1$ may be measured using weights of $1, 2, 2^2, ..., 2^{k-1}$, when all the weights are placed in one pan.

12. Show that every nonzero integer can be uniquely represented in the form

$$e_k 3^k + e_{k-1} 3^{k-1} + \cdots + e_1 3 + e_0$$

where $e_j = -1, 0$, or 1 for $j = 0,1,2,...,k$ and $e_k \neq 0$. This expansion is called a *balanced ternary expansion*.

13. Use Exercise 12 to show that any weight not exceeding $(3^k - 1)/2$ may be measured using weights of $1, 3, 3^2, ..., 3^{k-1}$, when the weights may be placed in either pan.

14. Explain how to convert from base 3 to base 9 notation, and from base 9 to base 3 notation.

15. Explain how to convert from base r to base r^n notation, and from base r^n notation to base r notation, when $r > 1$ and n are positive integers.

16. Show that if $n = (a_k a_{k-1}...a_1 a_0)_b$, then the quotient and remainder when n is divided by b^j are $q = (a_k a_{k-1}...a_j)_b$ and $r = (a_{j-1}...a_1 a_0)_b$, respectively.

17. If the base b expansion of n is $n = (a_k a_{k-1}...a_1 a_0)_b$, what is the base b expansion of $b^m n$?

One's complement representations of integers are used to simplify computer arithmetic. To represent positive and negative integers with absolute value less than 2^n, a total of $n+1$ bits is used.

The leftmost bit is used to represent the sign. A zero bit in this position is used for positive integers and a one bit in this position is used for negative integers.

For positive integers the remaining bits are identical to the binary expansion of the integer. For negative integers, the remaining bits are obtained by first finding the binary expansion of the absolute value of the integer, and then taking the complement of each of these bits, where the complement of a 1 is a 0 and the complement of a 0 is a 1.

18. Find the one's complement representations, using bit strings of length six, of the following integers

a) 22   c) -7
b) 31   d) -19

19. What integer do each of the following one's complement representations of length five represent?

a) 11001   c) 10001
b) 01101   d) 11111

20. How is the one's complement representation of $-m$ obtained from the one's complement of m, when bit strings of length n are used?

21. Sometimes integers are encoded by using four digit binary expansions to represent each decimal digit. This produces the *binary coded decimal* form of the integer. For instance, 791 is encoded in this way by 011110010001. How many bits are required to represent a number with n decimal digits using this type of encoding?

A *Cantor expansion* of a positive integer n is a sum

$$n = a_m m! + a_{m-1}(m-1)! + \cdots + a_2 2! + a_1 1!$$

where each a_j is an integer with $0 \le a_j \le j$ and $a_m = 0$.

22. Find Cantor expansions of 14, 56, and 384.

* 23. Show that every positive integer has a unique Cantor expansion. (*Hint*: For each positive integer n there is a positive integer m such that $m! \le n < (m+1)!$. For a_m take the quotient from the division algorithm when n is divided by $m!$. Then iterate.)

The Chinese game of *nim* is played as follows. There are several piles of matches, each containing an arbitrary number of matches at the start of the game. To make a move a player removes one or more matches from one of the piles. The players take turns, with the player removing the last match winning the game.

A *winning position* is an arrangement of matches in piles so that if a player can move to this position, then, no matter what the second player does, the first player can continue to play in a way that will win the game. An example is the position where there are two piles each containing one match; this is a winning position, because the second player must remove a match leaving the first player the opportunity to win by removing the last match.

24. Show that the position in nim where there are two piles, each with two matches, is a winning position.

25. For each arrangement of matches into piles, write the number of matches in each pile in binary notation, and then line up the digits of these numbers into columns (adding initial zeroes if necessary to some of the numbers). Show that a position is a winning one if and only if the number of ones in each column is even (*Example*: Three piles of 3, 4, and 7 give

$$0\ 1\ 1$$
$$1\ 0\ 0$$
$$1\ 1\ 1$$

where each column has exactly two ones.) (*Hint*: Show that any move from a winning position produces a non-winning one. Show that there is a move from a non-winning position to a winning one.)

Let a be an integer with a four-digit decimal expansion, with not all digits the same. Let a' be the integer with a decimal expansion obtained by writing the digits of a in descending order, and let a'' be the integer with a decimal expansion obtained by writing the digits of a in ascending order. Define $T(a) = a' - a''$. For instance, $T(7318) = 8731 - 1378 = 7353$.

* **26.** Show that the only integer with a four-digit decimal expansion with not all digits the same such that $T(a) = a$ is $a = 6174$. The integer 6174 is called *Kaprekar's constant*, after the Indian mathematician D. R. Kaprekar*, because it is the only integer with this property.

** **27.** a) Show that if a is a positive integer with a four-digit decimal expansion with not all digits the same, then the sequence a, $T(a)$, $T(T(a))$, $T(T(T(a)))$,...., obtained by iterating T, eventually reaches the integer 6174.

   b) Determine the maximum number of steps required for the sequence defined in part (a) to reach 6174.

Let b be a positive integer and let a be an integer with a four-digit base b expansion, with not all digits the same. Define $T_b(a) = a' - a''$, where a' is the integer with base b expansion obtained by writing the base b digits of a in descending order, and let a'' is the integer with base b expansion obtained by writing the base b digits of a in ascending order.

** **28.** Let $b = 5$. Find the unique integer a_0 with a four-digit base 5 expansion such that $T_5(a_0) = a_0$. Show that this integer a_0 is a Kaprekar constant for the base 5, *i.e.*, a, $T(a)$, $T(T(a))$, $T(T(T(a)))$,... eventually reaches a_0, whenever a is an integer which a four-digit base 5 expansion with not all digits the same.

* **D.R. KAPREKAR (1905-1986)** was born in Dahanu, India and was interested in numbers even as a small child. He received his secondary school education in Thana and studied at Ferguson College in Poona. Kaprekar attended the University of Bombay, receiving his bachelors degree from this school in 1929. From 1930 until his retirement in 1962 he worked as a school teacher in Devlali, India. Kaprekar discovered many interesting properties in recreational number theory. He published extensively, writing about such topics as recurring decimals, magic squares, and integers with special properties.

* **29.** Show that no Kaprekar constant exists for four-digit numbers to the base 6.

* **30.** Determine whether there is a Kaprekar constant for three-digit integers to the base 10. Prove that your answer is correct.

1.6 Computer Projects

Programming Projects

Write programs to do the following:

1. Find the binary expansion of an integer from the decimal expansion of this integer and *vice versa.*

2. Convert from base b_1 notation to base b_2 notation, where b_1 and b_2 are arbitrary positive integers greater than one.

3. Convert from binary notation to hexadecimal notation and *vice versa.*

4. Find the base (-2) notation of an integer from its decimal notation (see Exercise 8).

5. Find the balanced ternary expansion of an integer from its decimal expansion (see Exercise 12).

6. Find the Cantor expansion of an integer from its decimal expansion (see the preamble to Exercise 22).

7. Play a winning strategy in the game of nim (see the preamble to Exercise 24).

* **8.** Find the sequence $a, T(a), T(T(a)), T(T(T(a))), \dots$ (defined in the preamble to Exercise 26), where a is a positive integer, to discover how many iterations are needed to reach 6174.

Computations and Explorations

Using the programs you have written or a computation program, carry out the following computations and explorations:

1. Find the Cantor expansions of the integers 100000, 10000000, and 1000000000. (See the preamble to Exercise 22 for the definition of Cantor expansions.)

2. Verify the result described in Exercise 26 for several different four-digit integers, with not all digits the same, of your choice.

3. Use numerical evidence to make conjectures about the behavior of the sequence a, $T(a)$, $T(T(a)), \dots$ where a is a five-digit integer in base 10 notation with not all digits the same and $T(a)$ is defined as in the preamble to Exercise 26.

4. Explore the behavior of the sequence a, $T(a), T(T(a)), \dots$ where a is a three-digit integer in base b notation for different bases b. What conjectures can you make? Repeat your exploration using four-digit and then five-digit integers in base b notation.

1.7 Computer Operations with Integers

Before computers were invented mathematicians did computations either by hand or by using mechanical devices. Either way, they were only able to work with integers of limited size. Many number theoretic computations, such as factoring and primality testing, require computations with integers with as many as 50 or even 100 digits. In this section we will study some of the basic algorithms for doing computer arithmetic. Then we will study the number of basic computer operations required to carry out these algorithms in Section 8 of this chapter.

We have mentioned that computers internally represent numbers using bits, or binary digits. Computers have a built-in limit on the size of integers that can be used in machine arithmetic. This upper limit is called the *word size*, which we denote by w. The word size is usually a power of 2, such as 2^{35}, although sometimes the word size is a power of 10.

To do arithmetic with integers larger than the word size, it is necessary to devote more than one word to each integer. To store an integer $n > w$, we express n in base w notation, and for each digit of this expansion we use one computer word. For instance, if the word size is 2^{35}, using ten computer words we can store integers as large as $2^{350} - 1$, since integers less than 2^{350} have no more than ten digits in their base 2^{35} expansions. Also note that to find the base 2^{35} expansion of an integer, we need only group together blocks of 35 bits.

The first step in discussing computer arithmetic with large integers is to describe how the basic arithmetic operations are methodically performed.

We will describe the classical methods for performing the basic arithmetic operations with integers in base r notation where $r > 1$ is an integer. These methods are examples of *algorithms**.

Definition.   An *algorithm* is a specified set of rules for obtaining a desired result from a set of input.

We will describe algorithms for performing addition, subtraction, and multiplication of two n-digit integers $a = (a_{n-1} a_{n-2} ... a_1 a_0)_r$ and $b = (b_{n-1} b_{n-2} ... b_1 b_0)_r$, where initial digits of zero are added if necessary to make both expansions the same length. The algorithms described are used both for binary arithmetic with integers less than the word size of a computer, and for

---

* The word "algorithm" has an interesting history, as does the evolution of the concept of an algorithm. "Algorithm" is a corruption of the original term "algorism" which originally comes from the book *Kitab al jabr w'al-muqabala (Rules of Restoration and Reduction)* written by Abu Ja'far Mohammed ibn Musa al-Khowarizmi in the ninth century. The word "algorism" originally referred only to the rules of performing arithmetic using Arabic numerals. The word "algorism" evolved into "algorithm" by the eighteenth century. With growing interest in computing machines, the concept of an algorithm became more general, to include all definite procedures for solving problems, not just the procedures for performing arithmetic with integers expressed in Arabic notation.

multiple precision arithmetic with integers larger than the word size w, using w as the base.

We first discuss the algorithm for addition. When we add a and b, we obtain the sum

$$a + b = \sum_{j=0}^{n-1} a_j r^j + \sum_{j=0}^{n-1} b_j r^j = \sum_{j=0}^{n-1} (a_j + b_j) r^j.$$

To find the base r expansion of the $a + b$, first note that by the division algorithm, there are integers C_0 and s_0 such that

$$a_0 + b_0 = C_0 r + s_0, \ 0 \le s_0 < r.$$

Because a_0 and b_0 are positive integers not exceeding r, we know that $0 \le a_0 + b_0 \le 2r - 2$, so that $C_0 = 0$ or 1; here C_0 is the *carry* to the next place. Next, we find that there are integers C_1 and s_1 such that

$$a_1 + b_1 + C_0 = C_1 r + s_1, \ 0 \le s_1 < r.$$

Since $0 \le a_1 + b_1 + C_0 \le 2r - 1$, we know that $C_1 = 0$ or 1. Proceeding inductively, we find integers C_i and s_i for $1 \le i \le n - 1$ by

$$a_i + b_i + C_{i-1} = C_i r + s_i, \ 0 \le s_i < r,$$

with $C_i = 0$ or 1. Finally, we let $s_n = C_{n-1}$, since the sum of two integers with n digits has $n + 1$ digits when there is a carry in the nth place. We conclude that the base r expansion for the sum is $a + b = (s_n s_{n-1} ... s_1 s_0)_r$.

When performing base r addition by hand, we can use the same familiar technique as is used in decimal addition.

Example 1.29. To add $(1101)_2$ and $(1001)_2$ we write

$$
\begin{array}{cccccc}
 & \mathit{1} & & & \mathit{1} & \\
 & & 1 & 1 & 0 & 1 \\
+ & & 1 & 0 & 0 & 1 \\
\hline
 & 1 & 0 & 1 & 1 & 0 \\
\end{array}
$$

where we have indicated carries by 1's in italics written above the appropriate column. We found the binary digits of the sum by noting that $1 + 1 = 1 \cdot 2 + 0$, $0 + 0 + 1 = 0 \cdot 2 + 1$, $1 + 0 + 0 = 0 \cdot 2 + 1$, and $1 + 1 + 0 = 1 \cdot 2 + 0$. □

We now turn our attention to subtraction. We consider

$$a - b = \sum_{j=0}^{n-1} a_j r^j - \sum_{j=0}^{n-1} b_j r^j = \sum_{j=0}^{n-1} (a_j - b_j) r^j,$$

where we assume that $a > b$. Note that by the division algorithm, there are integers B_0 and d_0 such that

$$a_0 - b_0 = B_0 r + d_0, \quad 0 \le d_0 < r,$$

and since a_0 and b_0 are positive integers less than r, we have

$$-(r - 1) \le a_0 - b_0 \le r - 1.$$

When $a_0 - b_0 \ge 0$, we have $B_0 = 0$. Otherwise, when $a_0 - b_0 < 0$, we have $B_0 = -1$; B_0 is the *borrow* from the next place of the base r expansion of a. We use the division algorithm again to find integers B_1 and d_1 such that

$$a_1 - b_1 + B_0 = B_1 r + d_1, \quad 0 \le d_1 < r.$$

From this equation, we see that the borrow $B_1 = 0$ as long as $a_1 - b_1 + B_0 \ge 0$, and $B_1 = -1$ otherwise, since $-r \le a_1 - b_1 + B_0 \le r - 1$. We proceed inductively to find integers B_i and d_i, such that

$$a_i - b_i + B_{i-1} = B_i r + d_i, \quad 0 \le d_i < r$$

with $B_i = 0$ or -1, for $1 \le i \le n - 1$. We see that $B_{n-1} = 0$, since $a > b$. We can conclude that

$$a - b = (d_{n-1} d_{n-2} ... d_1 d_0)_r.$$

When performing base r subtraction by hand, we use the same familiar technique as is used in decimal subtraction.

Example 1.30. To subtract $(10110)_2$ from $(11011)_2$, we have

$$
\begin{array}{r}
{\scriptstyle -1} \\
1\ 1\ 0\ 1\ 1 \\
-\ \ 1\ 0\ 1\ 1\ 0 \\
\hline
1\ 0\ 1
\end{array}
$$

where the -1 in italics above a column indicates a borrow. We found the binary digits of the difference by noting that $1 - 0 = 0 \cdot 2 + 1$, $1 - 1 + 0 = 0 \cdot 2 + 0$, $0 - 1 + 0 = -1 \cdot 2 + 1$, $1 - 0 - 1 = 0 \cdot 2 + 0$, and $1 - 1 + 0 = 0 \cdot 2 + 0$. □

Before discussing multiplication, we describe *shifting*. To multiply $(a_{n-1}...a_1 a_0)_r$ by r^m, we need only shift the expansion left m places, appending the expansion with m zero digits.

Example 1.31. To multiply $(101101)_2$ by 2^5, we shift the digits to the left five places and append the expansion with five zeros, obtaining $(10110100000)_2$. □

To deal with multiplication, we first discuss the multiplication of an n-place integer by a one-digit integer. To multiply $(a_{n-1}...a_1 a_0)_r$ by $(b)_r$, we first note that

$$a_0 b = q_0 r + p_0, \quad 0 \le p_0 < r,$$

and $0 \le q_0 \le r - 2$, since $0 \le a_0 b \le (r - 1)^2$. Next, we have

$$a_1 b + q_0 = q_1 r + p_1, \quad 0 \le p_1 < r,$$

and $0 \le q_1 \le r-1$. In general, we have

$$a_i b + q_{i-1} = q_i r + p_i, \quad 0 \le p_i < r$$

and $0 \le q_i \le r - 1$. Furthermore, we have $p_n = q_{n-1}$. This yields $(a_{n-1}...a_1 a_0)_r (b)_r = (p_n p_{n-1}...p_1 p_0)_r$.

To perform a multiplication of two n-place integers we write

$$ab = a \left[\sum_{j=0}^{n-1} b_j r^j \right] = \sum_{j=0}^{n-1} (ab_j) r^j.$$

For each j, we first multiply a by the digit b_j, then shift to the left j places, and finally add all of the n integers we have obtained to find the product.

When multiplying two integers with base r expansions, we use the familiar method of multiplying decimal integers by hand.

Example 1.32. To multiply $(1101)_2$ and $(1110)_2$ we write

$$
\begin{array}{r}
1\ 1\ 0\ 1 \\
\times\ 1\ 1\ 1\ 0 \\
\hline
0\ 0\ 0\ 0 \\
1\ 1\ 0\ 1 \\
1\ 1\ 0\ 1 \\
1\ 1\ 0\ 1 \\
\hline
1\ 0\ 1\ 1\ 0\ 1\ 1\ 0
\end{array}
$$

Note that we first multiplied $(1101)_2$ by each digit of $(1110)_2$, shifting each time by the appropriate number of places, and then we added the appropriate integers to find our product. □

We now discuss integer division. We wish to find the quotient q in the division algorithm

$$a = bq + R, \quad 0 \le R < b.$$

If the base r expansion of q is $q = (q_{n-1} q_{n-2}...q_1 q_0)_r$, then we have

$$a = b \left[\sum_{j=0}^{n-1} q_j r^j \right] + R, \quad 0 \le R < b.$$

To determine the first digit q_{n-1} of q, notice that

$$a - b q_{n-1} r^{n-1} = b \left[\sum_{j=0}^{n-2} q_j r^j \right] + R.$$

The right-hand side of this equation is not only positive, but also less than br^{n-1}, since $\sum_{j=0}^{n-2} q_j r^j \le \sum_{j=0}^{n-2} (r-1) r^j = \sum_{j=1}^{n-1} r^j - \sum_{j=0}^{n-2} r^j = r^{n-1} - 1$. Therefore, we know that

$$0 \le a - bq_{n-1}r^{n-1} < br^{n-1}.$$

This tells us that

$$q_{n-1} = [a/(br^{n-1})].$$

We can obtain q_{n-1} by successively subtracting br^{n-1} from a until a negative result is obtained, and then q_{n-1} is one less than the number of subtractions.

To find the other digits of q, we define the sequence of *partial remainders* R_i by

$$R_0 = a$$

and

$$R_i = R_{i-1} - bq_{n-i}r^{n-i}$$

for $i = 1, 2, ..., n$. By mathematical induction, we show that

(1.8)
$$R_i = \left[\sum_{j=0}^{n-i-1} q_j r^j \right] b + R.$$

For $i = 0$, this is clearly correct, since $R_0 = a = qb + R$. Now assume that

$$R_k = \left[\sum_{j=0}^{n-k-1} q_j r^j \right] b + R.$$

Then

$$R_{k+1} = R_k - bq_{n-k-1}r^{n-k-1}$$

$$= \left[\sum_{j=0}^{n-k-1} q_j r^j \right] b + R - bq_{n-k-1}r^{n-k-1}$$

$$= \left[\sum_{j=0}^{n-(k+1)-1} q_j r^j \right] b + R,$$

establishing (1.8).

By (1.8) we see that $0 \le R_i < r^{n-i}b$, for $i = 1, 2, ..., n$, since $\sum_{j=0}^{n-i-1} q_j r^j \le r_{n-i} - 1$. Consequently, since $R_i = R_{i-1} - bq_{n-i}r^{n-i}$ and $0 \le R_i < r^{n-1}b$, we see that the digit q_{n-i} is given by $[R_{i-1}/(br^{n-i})]$ and can be obtained by successively subtracting br^{n-i} from R_{i-1} until a negative result is obtained, and then q_{n-i} is one less than the number of subtractions. This is how we find the digits of q.

Example 1.33. To divide $(11101)_2$ by $(111)_2$, we let $q = (q_2q_1q_0)_2$. We subtract $2^2(111)_2 = (11100)_2$ once from $(11101)_2$ to obtain $(1)_2$, and once more to obtain a negative result, so that $q_2 = 1$. Now $R_1 = (11101)_2 - (11100)_2 = (1)_2$. We find that $q_1 = 0$, since $R_1 - 2(111)_2$ is less than zero, and likewise $q_0 = 0$. Hence, the quotient of the division is $(100)_2$ and the remainder is $(1)_2$. □

1.7 Exercises

1. Add $(101111011)_2$ and $(1100111011)_2$.

2. Add $(10001000111101)_2$ and $(11111101011111)_2$.

3. Subtract $(11010111)_2$ from $(1111000011)_2$.

4. Subtract $(101110101)_2$ from $(1101101100)_2$.

5. Multiply $(11101)_2$ and $(110001)_2$.

6. Multiply $(1110111)_2$ and $(10011011)_2$.

7. Find the quotient and remainder when $(110011111)_2$ is divided by $(1101)_2$.

8. Find the quotient and remainder when $(110100111)_2$ is divided by $(11101)_2$.

9. Add $(1234321)_5$ and $(2030104)_5$.

10. Subtract $(434421)_5$ from $(4434201)_5$.

11. Multiply $(1234)_5$ and $(3002)_5$.

12. Find the quotient and remainder when $(14321)_5$ is divided by $(334)_5$.

13. Add $(ABAB)_{16}$ and $(BABA)_{16}$.

14. Subtract $(CAFE)_{16}$ from $(FEED)_{16}$.

15. Multiply $(FACE)_{16}$ and $(BAD)_{16}$.

16. Find the quotient and remainder when $(BEADED)_{16}$ is divided by $(ABBA)_{16}$.

17. Explain how to add, subtract, and multiply the integers 18235187 and 22135674 on a computer with word size 1000.

18. Write algorithms for the basic operations with integers in base (-2) notation (see Exercise 8 of Section 1.6).

19. How is the one's complement representation of the sum of two integers obtained from the one's complement representations of these integers?

20. How is the one's complement representation of the difference of two integers obtained from the one's complement representations of these integers?

21. Give an algorithm for adding and an algorithm for subtracting Cantor expansions (see preamble to Exercise 22 of Section 1.6).

22. A *dozen* equals 12 and a *gross* equals 12^2. Using base 12, or *duodecimal*, arithmetic answer the following questions.

    a) If 3 gross, 7 dozen, and 4 eggs are removed from a total of 11 gross and 3 dozen eggs, how many eggs are left?

b) If 5 truckloads of 2 gross, 3 dozen, and 7 eggs each are delivered to the supermarket, how many eggs were delivered?

c) If 11 gross, 10 dozen and 6 eggs are divided in 3 groups of equal size, how many eggs are in each group?

23. A well-known rule used to find the square of an integer with decimal expansion $(a_n a_{n-1} \ldots a_1 a_0)_{10}$ with final digit $a_0 = 5$ is to find the decimal expansion of the product $(a_n a_{n-1} \ldots a_1)_{10} [(a_n a_{n-1} \ldots a_1)_{10} + 1]$ and append this with the digits $(25)_{10}$. For instance, we see that the decimal expansion of $(165)^2$ begins with $16 \cdot 17 = 272$, so that $(165)^2 = 27225$. Show that the rule just described is valid.

24. In this exercise we generalize the rule given in Exercise 19 to find the squares of integers with final base $2B$ digit B, where B is a positive integer. Show that the base $2B$ expansion of the integer $(a_n a_{n-1} \ldots a_1 a_0)_{2B}$ starts with the digits of the base $2B$ expansion of the integer $(a_n a_{n-1} \ldots a_1)_{2B}$ $[(a_n a_{n-1} \ldots a_1)_{2B} + 1]$ and ends with the digits $B/2$ and 0 when B is even, and the digits $(B-1)/2$ and B when B is odd.

1.7 Computer Projects

Programming Projects

Write programs to do the following:

1. Perform addition with arbitrarily large integers.

2. Perform subtraction with arbitrarily large integers.

3. Multiply two arbitrarily large integers using the conventional algorithm.

4. Divide arbitrarily large integers, finding the quotient and remainder.

Computations and Explorations

Using the programs you have written or a computation program, carry out the following computations and explorations:

1. Verify the rules given in Exercises 23 and 24 for examples of your choice.

1.8 Complexity of Integer Operations

Once an algorithm is specified for an operation we can consider the amount of time required to perform this algorithm on a computer. We will measure the amount of time needed in terms of *bit operations*. By a bit operation we mean the addition, subtraction, or multiplication of two binary digits, the division of a two-bit integer by a one bit (obtaining a quotient and a remainder), or the shifting of a binary integer one place. (The actual amount of time required to carry out a bit operation on a computer varies depending on the computer architecture and capacity.) When we describe the number of bit operations needed to perform an algorithm, we are describing the *computational complexity* of this algorithm.

In describing the number of bit operations needed to perform calculations we will use *big-O* notation. To motivate the definition of this notation, consider the following situation. Suppose that performing a specified operation on an integer n requires at most $n^3 + 8n^2 \log n$ bit operations. Since $8n^2 \log n < 8n^3$ for every positive integer, less than $9n^3$ bit operations are required for this operation for every integer n. Since the number of bit operations required is always less than a constant times n^3, namely $9n^3$, we say that $O(n^3)$ bit operations are needed. In general we have the following definition.

Definition. If f and g are functions taking positive values, defined for all $x \in S$, where S is a specified set of real numbers, then f is $O(g)$ on S if there is a positive constant K such that $f(x) < Kg(x)$ for all $x \in S$. (Normally we take S to be the set of positive integers and we drop all reference to S.)

We illustrate this concept with several examples.

Example 1.34. We can show on the set of positive integers that $n^4 + 2n^3 + 5$ is $O(n^4)$. To do this, note that $n^4 + 2n^3 + 5 \le n^4 + 2n^4 + 5n^4 = 8n^4$ for all positive integers. (We take $K = 8$ in the definition.) The reader should also note that n^4 is $O(n^4 + 2n^3 + 5)$. □

Example 1.35. We can easily give a big-O estimate for $\sum_{j=1}^{n} j$. Noting that each summand is less than n tells us that $\sum_{j=1}^{n} j \le \sum_{j=1}^{n} n = n \cdot n = n^2$. Note that we could also easily derive this estimate from the formula $\sum_{j=1}^{n} j = n(n+1)/2$. □

We now will give some useful results for working with big-O estimates for combinations of functions.

Theorem 1.11. If f is $O(g)$ and c is a positive constant, then cf is $O(g)$.

Proof. If f is $O(g)$, then there is a constant K with $f(x) < Kg(x)$ for all x under consideration. Hence $cf(x) < (cK)g(x)$, so cf is $O(g)$. ∎

Theorem 1.12. If f_1 is $O(g_1)$ and f_2 is $O(g_2)$, then $f_1 + f_2$ is $O(g_1 + g_2)$ and $f_1 f_2$ is $O(g_1 g_2)$.

Proof. If f is $O(g_1)$ and f_2 is $O(g_2)$, then there are constants K_1 and K_2 such that $f_1(x) < K_1 g_1(x)$ and $f_2(x) < K_2 g_2(x)$ for all x under consideration. Hence

$$f_1(x) + f_2(x) < K_1 g_1(x) + K_2 g_2(x)$$
$$\le K(g_1(x) + g_2(x))$$

where K is the maximum of K_1 and K_2. Hence $f_1 + f_2$ is $O(g_1 + g_2)$.

Also

$$f_1(x)f_2(x) < K_1g_1(x) \ K_2g_2(x)$$
$$= (K_1K_2)(g_1(x)g_2(x)),$$

so f_1f_2 is $O(g_1g_2)$. ∎

Corollary 1.12.1. If f_1 and f_2 are $O(g)$, then $f_1 + f_2$ is $O(g)$.

Proof. Theorem 1.12 tells us that $f_1 + f_2$ is $O(2g)$. But if $f_1 + f_2 < K(2g)$, then $f_1 + f_2 < (2K)g$, so $f_1 + f_2$ is $O(g)$. ∎

We illustrate how to use these theorems with the following example.

Example 1.36. To give a big-O estimate for $(n + 8 \log n) (10n \log n + 17n^2)$, first note that $n + 8 \log n$ is $O(n)$ and $10n \log n + 17n^2$ is $O(n^2)$ (since $\log n$ is $O(n)$ and $n \log n$ is $O(n^2)$) by Theorems 1.11 and 1.12 and Corollary 1.12.1. By Theorem 1.12 we see that $(n + 8 \log n)(10n \log n + 17n^2)$ is $O(n^3)$. □

Using big-O notation we can see that to add or subtract two n-bit integers takes $O(n)$ bit operations, while to multiply two n-bit integers in the conventional way takes $O(n^2)$ bit operations (see Exercises 12 and 13 at the end of this section). Surprisingly, there are faster algorithms for multiplying large integers. To develop one such algorithm, we first consider the multiplication of two $2n$-bit integers, say $a = (a_{2n-1}a_{2n-2}...a_1a_0)_2$ and $b = (b_{2n-1}b_{2n-2}...b_1b_0)_2$. We write

$$a = 2^nA_1 + A_0 \quad b = 2^nB_1 + B_0,$$

where

$$A_1 = (a_{2n-1}a_{2n-2}...a_{n+1}a_n)_2 \quad A_0 = (a_{n-1}a_{n-2}...a_1a_0)_2$$
$$B_1 = (b_{2n-1}b_{2n-2}...b_{n+1}b_n)_2 \quad B_0 = (b_{n-1}b_{n-2}...b_1b_0)_2.$$

We will use the identity

(1.9) $\quad ab = (2^{2n}+2^n)A_1B_1 + 2^n(A_1-A_0)(B_0-B_1) + (2^n+1)A_0B_0.$

To find the product of a and b using (1.9), requires that we perform three multiplications of n-bit integers (namely A_1B_1, $(A_1 - A_0)(B_0 - B_1)$, and A_0B_0), as well as a number of additions and shifts. This is illustrated by the following example.

Example 1.37. We can use (1.9) to multiply $(1101)_2$ and $(1011)_2$. We have $(1101)_2 = 2^2(11)_2 + (01)_2$ and $(1011)_2 = 2^2(10)_2 + (11)_2$. Using (1.9) we find that

$$(1101)_2(1011)_2 = (2^4 + 2^2)(11)_2(10)_2 + 2^2((11)_2 - (01)_2) \cdot$$
$$((11)_2 - (10)_2) + (2^2 + 1)(01)_2(11)_2$$
$$= (2^4 + 2^2)(110)_2 + 2^2(10)_2(01)_2 + (2^2 + 1)(11)_2$$
$$= (1100000)_2 + (11000)_2 + (1000)_2 + (1100)_2 + (11)_2$$
$$= (10001111)_2.$$

$\square$

We will now estimate the number of bit operations required to multiply two n-bit integers using (1.9) repeatedly. If we let $M(n)$ denote the number of bit operations needed to multiply two n-bit integers, we find from (1.9) that

(1.10) $$M(2n) \le 3M(n) + Cn,$$

where C is a constant, since each of the three multiplications of n-bit integers takes $M(n)$ bit operations, while the number of additions and shifts needed to compute ab via (1.9) does not depend on n, and each of these operations takes $O(n)$ bit operations.

From (1.10), using mathematical induction, we can show that

(1.11) $$M(2^k) \le c(3^k - 2^k),$$

where c is the maximum of the quantities $M(2)$ and C (the constant in (1.10)). To carry out the induction argument, we first note that with $k = 1$, we have $M(2) \le c(3^1 - 2^1) = c$, since c is the maximum of $M(2)$ and C.

As the induction hypothesis, we assume that

$$M(2^k) \le c(3^k - 2^k).$$

Then, using (1.10), we have

$$M(2^{k+1}) \le 3M(2^k) + C2^k$$
$$\le 3c(3^k - 2^k) + C2^k$$
$$\le c3^{k+1} - c \cdot 3 \cdot 2^k + c2^k$$
$$\le c(3^{k+1} - 2^{k+1}).$$

This establishes that (1.11) is valid for all positive integers k.

Using inequality (1.11), we can prove the following theorem.

Theorem 1.13. Multiplication of two n-bit integers can be performed using $O(n^{\log_2 3})$ bit operations. (*Note:* $\log_2 3$ is approximately 1.585, which is considerably less than the exponent 2 that occurs in the estimate of the number of bit operations needed for the conventional multiplication algorithm.)

Proof. From (1.11) we have

$$M(n) = M(2^{\log_2 n}) \le M(2^{\lceil \log_2 n \rceil + 1})$$
$$\le c(3^{\lceil \log_2 n \rceil + 1} - 2^{\lceil \log_2 n \rceil + 1})$$
$$\le 3c \cdot 3^{\lceil \log_2 n \rceil} \le 3c \cdot 3^{\log_2 n} = 3cn^{\log_2 3}$$
$$(\text{since } 3^{\log_2 n} = n^{\log_2 3}).$$

Hence, $M(n)$ is $O(n^{\log_2 3})$. ∎

We now state, without proof, two pertinent theorems. Proofs may be found in Knuth [90] or Kronsjö [92].

Theorem 1.14. Given a positive number $\varepsilon > 0$, there is an algorithm for multiplication of two n-bit integers using $O(n^{1+\varepsilon})$ bit operations.

Note that Theorem 1.13 is a special case of Theorem 1.14 with $\varepsilon = \log_2 3 - 1$, which is approximately 0.585.

Theorem 1.15. There is an algorithm to multiply two n-bit integers using $O(n \log_2 n \log_2 \log_2 n)$ bit operations.

Since $\log_2 n$ and $\log_2 \log_2 n$ are much smaller than n^ε for large numbers n, Theorem 1.15 is an improvement over Theorem 1.14. Although we know that $M(n)$ is $O(n \log_2 n \log_2 \log_2 n)$, for simplicity we will use the obvious fact that $M(n)$ is $O(n^2)$ in our subsequent discussions.

The conventional algorithm described in Section 1.7 performs a division of a $2n$-bit integer by an n-bit integer with $O(n^2)$ bit operations. However, the number of bit operations needed for integer division can be related to the number of bit operations needed for integer multiplication. We state the following theorem, which is based on an algorithm which is discussed in Knuth [90].

Theorem 1.16. There is an algorithm to find the quotient $q = \lfloor a/b \rfloor$, when the $2n$-bit integer a is divided by the integer b having no more than n bits, using $O(M(n))$ bit operations, where $M(n)$ is the number of bit operations needed to multiply two n-bit integers.

1.8 Exercises

1. Determine whether each of the following functions is $O(n)$ on the set of positive integers.

   a) $2n + 7$ d) $\log(n^2 + 1)$
   b) $n^2/3$ e) $\sqrt{n^2 + 1}$
   c) 10 f) $(n^2 + 1)/(n + 1)$

2. Show that $2n^4 + 3n^3 + 17$ is $O(n^4)$ on the set of positive integers.

3. Show that $(n^3 + 4n^2 \log n + 101n^2)(14n \log n + 8n)$ is $O(n^4 \log n)$.

4. Show that $n!$ is $O(n^n)$ on the set of positive integers.

5. Show that $(n! + 1)(n + \log n) + (n^3 + n^n)((\log n)^3 + n + 7)$ is $O(n^{n+1})$.

6. Suppose that m is a positive real number. Show that $\sum\limits_{j=1}^{n} j^m$ is $O(n^{m+1})$.

* 7. Show that $n \log n$ is $O(\log n!)$ on the set of positive integers.

8. Show that if f_1 and f_2 are $O(g_1)$ and $O(g_2)$, respectively, and c_1 and c_2 are constants, then $c_1 f_1 + c_2 f_2$ is $O(g_1 + g_2)$.

9. Show that if f is $O(g)$, then f^k is $O(g^k)$ for all positive integers k.

10. Let r be a positive real number greater than 1. Show that a function f is $O(\log_2 n)$ if and only if f is $O(\log_r n)$. (*Hint:* Recall that $\log_a n / \log_b n = \log_a b$.)

11. Show that the base b expansion of a positive integer n has $[\log_b n] + 1$ digits.

12. Analyzing the conventional algorithms for subtraction and addition, show that with n-bit integers these operations require $O(n)$ bit operations.

13. Show that to multiply an n-bit and an m-bit integer in the conventional manner requires $O(nm)$ bit operations.

14. Estimate the number of bit operations needed to find $1 + 2 + \cdots + n$

    a) by performing all the additions.

    b) by using the identity $1 + 2 + \cdots + n = n(n+1)/2$, and multiplying and shifting.

15. Give an estimate for the number of bit operations needed to find each of the following quantities.

    a) $n!$ b) $\dbinom{n}{k}$

16. Give an estimate of the number of bit operations needed to find the binary expansion of an integer from its decimal expansion.

17. Use identity (1.9) with $n = 2$ to multiply $(1001)_2$ and $(1011)_2$.

18. Use identity (1.9) with $n = 4$ and then with $n = 2$ to multiply $(10010011)_2$ and $(11001001)_2$.

19. a) Show there is an identity analogous to (1.9) for decimal expansions.

b) Using part (a), multiply 73 and 87 performing only three multiplications of one-digit integers, plus shifts and additions.

c) Using part (a), reduce the multiplication of 4216 and 2733 to three multiplications of two-digit integers, plus shifts and additions, and then using part (a) again, reduce each of the multiplications of two-digit integers into three multiplications of one-digit integers, plus shifts and additions. Complete the multiplication using only nine multiplications of one-digit integers, and shifts and additions.

20. If $\mathbf{A}$ and $\mathbf{B}$ are $n \times n$ matrices, with entries a_{ij} and b_{ij} for $1 \le i \le n$, $1 \le j \le n$, then $\mathbf{AB}$ is the $n \times n$ matrix with entries $c_{ij} = \sum_{k=1}^{n} a_{ik} b_{kj}$. Show that n^3 multiplications of integers are used to find $\mathbf{AB}$ directly from its definition.

21. Show that it is possible to multiply two 2×2 matrices using only seven multiplications of integers by using the identity

$$
\begin{bmatrix} a_{11} & a_{12} \\ a_{21} & a_{22} \end{bmatrix} \begin{bmatrix} b_{11} & b_{12} \\ b_{21} & b_{22} \end{bmatrix} =
$$

$$
\begin{bmatrix}
a_{11}b_{11} + a_{12}b_{21} & x + (a_{21} + a_{22})(b_{12} - b_{11}) + \\
& (a_{11} + a_{12} - a_{21} - a_{22})b_{22} \\
& \\
x + (a_{11} - a_{21})(b_{22} - b_{12}) - & x + (a_{11} - a_{21})(b_{22} - b_{12}) + \\
a_{22}(b_{11} - b_{21} - b_{12} + b_{22}) & (a_{21} + a_{22})(b_{12} - b_{11})
\end{bmatrix}
$$

where $x = a_{11}b_{11} - (a_{11} - a_{21} - a_{22})(b_{11} - b_{12} + b_{22})$.

* 22. Using an inductive argument, and splitting $(2n) \times (2n)$ matrices into four $n \times n$ matrices, use Exercise 21 to show that it is possible to multiply two $2^k \times 2^k$ matrices using only 7^k multiplications, and less than 7^{k+1} additions.

23. Conclude from Exercise 22 that two $n \times n$ matrices can be multiplied using $O(n^{\log_2 7})$ bit operations when all entries of the matrices have less than c bits, where c is a constant.

1.8 Computer Projects

Programming Projects

Write programs to do the following:

* 1. Multiply two arbitrarily large integers using the identity (1.9).

** 2. Multiply two $n \times n$ matrices using the algorithm discussed in Exercise 21-23.

Computations and Explorations

Using the programs you have written or a computation program, carry out the following computations and explorations:

1. Multiply 81873569 and 41458892 by using identity (1.9) with these eight-digit integers, with the resulting four-digit integers, and with the resulting two-digit integers.

2. Multiply two 8×8 matrices of your choice using the identity in Exercise 21 with these matrices and then again for the multiplication of the resulting 4×4 matrices.

1.9 Prime Numbers

The positive integer 1 has just one positive divisor. Every other positive integer has at least two positive divisors, because it is divisible by 1 and by itself. Integers with exactly two positive divisors are of great importance in number theory; they are called *primes*.

Definition. A *prime* is a positive integer greater than 1 that is divisible by no positive integers other than 1 and itself.

Example 1.38. The integers $2, 3, 5, 13, 101$ and 163 are primes. □

Definition. A positive integer greater than 1 that is not prime, is called *composite*.

Example 1.39. The integers $4 = 2 \cdot 2$, $8 = 4 \cdot 2$, $33 = 3 \cdot 11$, $111 = 3 \cdot 37$, and $1001 = 7 \cdot 11 \cdot 13$ are composite. □

The primes are the building blocks of the integers. Later, we will show that every positive integer can be written uniquely as the product of primes.

Here, we briefly discuss the distribution of primes and mention some conjectures about primes. We start by showing that there are infinitely many primes. The following lemma is needed.

Lemma 1.1. Every positive integer greater than one has a prime divisor.

Proof. We prove the lemma by contradiction; we assume that there is a positive integer greater than 1 having no prime divisors. Then, since the set of positive integers greater than 1 with no prime divisors is non-empty, the well-ordering property tells us that there is a least positive integer n greater than 1 with no prime divisors. Since n has no prime divisors and n divides n, we see that n is not prime. Hence, we can write $n = ab$ with $1 < a < n$ and $1 < b < n$. Because $a < n$, a must have a prime divisor. By Theorem 1.7 any divisor of a is also a divisor of n, so n must have a prime divisor, contradicting the fact that n has no prime divisors. We can conclude that every positive integer greater than 1 has at least one prime divisor. ∎

We now show that the number of primes is infinite. This result was known to the ancient Greek mathematician Euclid. We give a proof similar to that given by Euclid (the proof from his *Elements* is outlined in Exercise 8).

Theorem 1.17. There are infinitely many primes.

Proof. Consider the integer

$$Q_n = n! + 1, \qquad n \geq 1.$$

Lemma 1.1. tells us that Q_n has at least one prime divisor, which we denote by q_n. Thus, q_n must be larger than n; for if $q_n \leq n$, it would follow that $q_n \mid n!$, and then, by Thereom 1.8, $q_n \mid (Q_n - n!) = 1$, which is impossible.

Since we have found a prime larger than n, for every positive integer n, there must be infinitely many primes. ∎

Later we will be interested in finding, and using, extremely large primes. We will be concerned throughout this book with the problem of determining whether a given integer is prime. We first deal with this question by showing that by trial divisions of n by primes not exceeding the square root of n, we can determine whether n is prime.

Theorem 1.18. If n is a composite integer, then n has a prime factor not exceeding $\sqrt{n}$.

Proof. Since n is composite, we can write $n = ab$, where a and b are integers with $1 < a \leq b < n$. We must have $a \leq \sqrt{n}$, since otherwise $b \geq a > \sqrt{n}$ and $ab > \sqrt{n} \cdot \sqrt{n} = n$. Now, by Lemma 1.1 a must have a prime divisor, which by Theorem 1.7 is also a divisor of n and which is clearly less than or equal to $\sqrt{n}$. ∎

We can use Theorem 1.18 to find all the primes less than or equal to a given positive integer n. This procedure is called the *sieve of Eratosthenes*, since it was invented by the ancient Greek mathematician Eratosthenes*. We illustrate its use in Figure 1.2 by finding all primes less than 100. We first note that every composite integer less than 100 must have a prime factor less than $\sqrt{100} = 10$. Since the only primes less than 10 are 2,3,5, and 7, we only need to check each integer less than 100 for divisibility by these primes. We first cross out, below by a horizontal slash

---

* **ERATOSTHENES** (276 - 194 B.C.E.) was born in Cyrene, which was a Greek colony west of Egypt. It is known that he spent some time studying at Plato's school in Athens. King Ptolemy II invited Eratosthenes to Alexandria to tutor his son. Later Eratosthenes became the chief librarian of the famous library at Alexandria which was a central repository of ancient works of literature, art, and science. He was an extremely versatile scholar, having written on mathematics, geography, astronomy, history, philosophy, and literary criticism. Besides his work in mathematics, Eratosthenes was most noted for his chronology of ancient history and for his geographical measurements, including his famous measurement of the size of the earth.

—, all multiples of 2 greater than 2. Next we cross out with a slash / those integers remaining that are multiples of 3, other than 3 itself. Then all multiples of 5, other than 5, that remain are crossed out, below by a backslash \. Finally, all multiples of 7, other than 7, that are left are crossed out, below with a vertical slash |. All remaining integers (other than 1) must be prime.

1	2	3	4	5	6	7	8	9	10
11	12	13	14	15	16	17	18	19	20
21	22	23	24	25	26	27	28	29	30
31	32	33	34	35	36	37	38	39	40
41	42	43	44	45	46	47	48	49	50
51	52	53	54	55	56	57	58	59	60
61	62	63	64	65	66	67	68	69	70
71	72	73	74	75	76	77	78	79	80
81	82	83	84	85	86	87	88	89	90
91	92	93	94	95	96	97	98	99	100

Figure 1.2. Finding the Primes Less Than 100 Using the Sieve of Eratosthenes.

Although the sieve of Eratosthenes produces all primes less than or equal to a fixed integer, to determine whether a particular integer n is prime in this manner, it is necessary to check n for divisibility by all primes not exceeding $\sqrt{n}$. This is quite inefficient; later on we will have better methods for deciding whether or not an integer is prime.

We know that there are infinitely many primes, but can we estimate how many primes there are less than a positive real number x? One of the most famous theorems of number theory, and of all mathematics, is the *prime number theorem* which answers this question. To state this theorem, we introduce some notation.

Definition. The function $\pi(x)$, where x is a positive real number, denotes the number of primes not exceeding x.

Example 1.40. From our example illustrating the sieve of Eratosthenes, we see that $\pi(10) = 4$ and $\pi(100) = 25$.   □

We now state the prime number theorem.

Theorem 1.19. The Prime Number Theorem. The ratio of $\pi(x)$ to $x/\log x$ approaches one as x grows without bound. (Here $\log x$ denotes the natural logarithm of x. In the language of limits, we have $\lim\limits_{x \to \infty} \pi(x)/\dfrac{x}{\log x} = 1$).

The prime number theorem was conjectured by Gauss in 1793, but it was not proved until 1896, when a French mathematician J. Hadamard* and a Belgian mathematician C. J. de la Vallée-Poussin** produced independent proofs. We will not prove the prime number theorem here; the various proofs known are either quite complicated or rely on advanced mathematics. In Table 1.1 we give some numerical evidence to indicate the validity of the theorem.

x	$\pi(x)$	$x/\log x$	$\pi(x)/\dfrac{x}{\log x}$	$li(x)$	$\pi(x)/li(x)$
10^3	168	144.8	1.160	178	0.9438202
10^4	1229	1085.7	1.132	1246	0.9863563
10^5	9592	8685.9	1.104	9630	0.9960540
10^6	78498	72382.4	1.085	78628	0.9983466
10^7	664579	620420.7	1.071	664918	0.9998944
10^8	5761455	5428681.0	1.061	5762209	0.9998691
10^9	50847534	48254942.4	1.054	50849235	0.9999665
10^{10}	455052512	434294481.9	1.048	455055614	0.9999932
10^{11}	4118054813	3948131663.7	1.043	4118165401	0.9999731
10^{12}	37607912018	36191206825.3	1.039	37607950281	0.9999990
10^{13}	346065536839	334072678387.1	1.036	346065645810	0.9999997
10^{14}	3204941750802	3102103442166	1.033	3204942065692	0.9999999

Table 1.1. Approximations to $\pi(x)$.

---

* **JACQUES HADAMARD (1865-1963)** was born in Versailles, France. His father was a Latin teacher and his mother a distinguished piano teacher. After completing his undergraduate studies he taught at a Paris secondary school. After receiving his doctorate in 1892 he became lecturer at the Faculté des Sciences de Bordeaux. He subsequently served on the faculties of the Sorbonne, the College de France, the Ecole Polytechnique, and the Ecole Centrale des Arts et Manufactures. Hadamard made important contributions to complex analysis, functional analysis, and mathematical physics. His proof of the prime number was based on his work in complex analysis. Hadamard was a famous teacher. His wrote numerous articles about elementary mathematics that were used in French schools and his text on elementary geometry was used for many years.

** **CHARLES-JEAN-GUSTAVE-NICHOLAS DE LA VALLEE-POUSSIN (1866-1962),** the son of a geology professor, was born at Louvain, Belgium. He studied at the Jesuit College at Mons, first studying philosophy, later turning to engineering. After receiving his degree, instead of pursuing a career in engineering, he devoted himself to mathematics. De La Vallée-Poussin's most significant contribution to mathematics was his proof of the prime number theorem. Extending this work, he established results about the distribution of primes in arithmetic progression and the distribution of primes represented by quadratic forms. Furthermore, he refined the prime number theorem to include error estimates. He made important contributions to differential equations, approximation theory, and analysis. His textbook *Cours d'analyse* had a strong impact on mathematical thought in the first half of the 20th century.

The prime number theorem tells us the ratio between $x/\log x$ and $\pi(x)$ is close to 1 when x is large. However, there are functions for which the ratio between these functions and $\pi(x)$ approaches 1 more rapidly than it does for $x/\log x$. In particular, it has been shown that an even better approximation is given by

$$li(x) = \int_2^x \frac{dt}{\log t}$$

(where $\int_2^x \frac{dt}{\log t}$ represents the area under the curve $y = 1/\log t$ and above the t-axis from $t = 2$ to $t = x$). In Table 1.1 we see evidence that $li(x)$ is an excellent approximation of $\pi(x)$. (Note that the values of $li(x)$ have been rounded to the nearest integer.)

We can now estimate the number of bit operations needed to show that an integer n is prime by trial divisions of n by all primes not exceeding $\sqrt{n}$. The prime number theorem tells us that there are approximately $\sqrt{n}/\log\sqrt{n} = 2\sqrt{n}/\log n$ primes not exceeding $\sqrt{n}$. To divide n by an integer m takes $O(\log_2 n \cdot \log_2 m)$ bit operations. Therefore, the number of bit operations needed to show that n is prime by this method is at least $(2\sqrt{n}/\log n)(c \log_2 n) = c\sqrt{n}$ (where we have ignored the $\log_2 m$ term since it is at least 1, even though it sometimes is as large as $(\log_2 n)/2$). This method of showing that an integer n is prime is very inefficient, for not only is it necessary to know all the primes not larger than $\sqrt{n}$, but it is also necessary to do at least a constant multiple of $\sqrt{n}$ bit operations. Later on we will have more efficient methods of showing that an integer is prime.

We remark here that it is not necessary to find all primes not exceeding x in order to compute $\pi(x)$. One way that $\pi(x)$ can be evaluated without finding all the primes less then x is to use a counting argument based on the sieve of Eratosthenes (see Exercise 18). (Recently, very efficient ways of finding $\pi(x)$ using $O(x^{3/5+\varepsilon})$ bit operations have been devised by Lagarias and Odlyzko [112].)

We have shown that there are infinitely many primes and we have discussed the abundance of primes below a given bound x, but we have yet to discuss how regularly primes are distributed throughout the positive integers. We first give a result that shows that there are arbitrarily long runs of integers containing no primes.

Theorem 1.20. For any positive integer n, there are at least n consecutive composite positive integers.

Proof. Consider the n consecutive positive integers

$$(n + 1)! + 2, (n + 1)! + 3, \ldots, (n + 1)! + n + 1.$$

When $2 \leq j \leq n + 1$, we know that $j \mid (n + 1)!$. By Theorem 1.8 it follows that $j \mid (n + 1)! + j$. Hence, these n consecutive integers are all composite. ∎

Example 1.41. The seven consecutive integers beginning with $8! + 2 = 40322$ are all composite. (However, these are much larger than the smallest seven consecutive composites, 90, 91, 92, 93, 94, 95, and 96.) □

Theorem 1.20 shows that the gap between consecutive primes is arbitrarily long. On the other hand, primes may often be close together. The only consecutive primes are 2 and 3, because 2 is the only even prime. However, many pairs of primes differ by two; these pairs of primes are called *twin primes*. Examples are the primes 5 and 7, 11 and 13, 101 and 103, and 4967 and 4969.* A famous unsettled conjecture asserts that there are infinitely many twin primes.

There are many conjectures concerning the number of primes of various forms. For instance, it is unknown whether there are infinitely many primes of the form $n^2 + 1$ where n is a positive integer. Questions such as this may be easy to state, but are sometimes extremely difficult to resolve.

We conclude this section by discussing perhaps the most notorious conjecture about primes.

Goldbach's Conjecture. Every even positive integer greater than two can be written as the sum of two primes.

This conjecture was stated by Christian Goldbach** in a letter to Leonhard Euler in 1742. It has been verified for all even integers less than a million. One sees by experimentation, as the following example illustrates, that usually there are many sums of two primes equal to a particular even integer, but a proof that there always is at least one such sum has not yet been found.

Example 1.42. The integers 10, 24, and 100 can be written as the sum of two primes in the following ways:

$$
\begin{aligned}
10 &= 3 + 7 = 5 + 5, \\
24 &= 5 + 19 = 7 + 17 = 11 + 13, \\
100 &= 3 + 97 = 11 + 89 = 17 + 83 \\
&= 29 + 71 = 41 + 59 = 47 + 53.
\end{aligned}
$$

□

---

* The largest twin primes currently known are $1706595 \cdot 2^{11235} \pm 1$.

** **CHRISTIAN GOLDBACH** (1690-1764) was born in Königsberg, Prussia (the city noted in mathematical circles for its famous bridge problem). He became professor of mathematics at the Imperial Academy of St. Petersburg in 1725. In 1728 Goldbach went to Moscow to tutor Tsarevich Peter II. In 1742 he entered the Russian Ministry of Foreign Affairs as a staff member. Goldbach is most noted for his correspondence with eminent mathematicians, in particular Leonhard Euler and Daniel Bernoulli. Besides his well known conjectures that every even positive integer greater than 2 in the sum of two primes and that every odd positive integer greater than five is the sum of three primes, Goldbach made several notable contributions to analysis.

Interested readers can find many unsolved conjectures about primes in the references listed at the end of the text. These conjectures are excellent fodder for computational explorations.

1.9 Exercises

1. Determine which of the following integers are primes.

    a) 101   c) 107   e) 113
    b) 103   d) 111   f) 121

2. Determine which of the following integers are primes.

    a) 201   c) 207   e) 213
    b) 203   d) 211   f) 221

3. Use the sieve of Eratosthenes to find all primes less than 150.

4. Use the sieve of Eratosthenes to find all primes less than 200.

5. Find all primes that are the difference of the fourth powers of two integers.

6. Show that no integer of the form $n^3 + 1$ is a prime, other than $2 = 1^3 + 1$.

7. Show that if a and n are positive integers such that $a^n - 1$ is prime, then $a = 2$ and n is prime. (*Hint:* Use the identity $a^{kl} - 1 = (a^k - 1)$ $(a^{k(l-1)} + a^{k(l-2)} + \cdots + a^k + 1.)$

8. This exercise constructs another proof of the infinitude of primes. Assume that there are only finitely many primes $p_1, p_2, ..., p_n$. Form the integer $Q = p_1 p_2 \cdots p_n + 1$. Show that Q has a prime factor not in the above list. Conclude that there are infinitely many primes.

9. Let $Q_n = p_1 p_2 \cdots p_n + 1$ where $p_1, p_2, ..., p_n$ are the n smallest primes. Determine the smallest prime factor of Q_n for $n = 1, 2, 3, 4, 5$, and 6. Do you think Q_n is prime infinitely often? (*Note:* This is an unresolved question.)

10. Let $p_1, p_2, ..., p_n$ be the first n primes and let m be an integer with $1 < m < n$. Let Q be the product of a set of m primes in the list and let R be the product of the remaining primes. Show that $Q + R$ is not divisible by any primes in the list, and hence must have a prime factor not in the list. Conclude that there are infinitely many primes.

11. Show that if the smallest prime factor p of the positive integer n exceeds $\sqrt[3]{n}$ then n/p must be prime or 1.

12. a) Find the smallest five consecutive composite integers.
    b) Find one million consecutive composite integers.

13. Show that there are no "prime triplets", *i.e.* primes p, $p + 2$, and $p + 4$, other than $3, 5$, and 7.

14. Verify Goldbach's conjecture for each of the following values of n.

a) 50   c) 102   e) 200
b) 98   d) 144   f) 222

15. Goldbach also conjectured that every odd positive integer greater than five is the sum of three primes. Verify this conjecture for each of the following odd integers.

a) 7   c) 27   e) 101
b) 17   d) 97   f) 199

16. Show that every integer greater than 11 is the sum of two composite integers.

17. Use the second principle of mathematical induction to prove that every integer greater than one is either prime or the product of two or more primes.

* 18. Use the principle of inclusion-exclusion (Exercise 18 of Section 1.4) to show that

$$\pi(n) = (\pi(\sqrt{n}) - 1) + n - \left(\left\lfloor \frac{n}{p_1} \right\rfloor + \left\lfloor \frac{n}{p_2} \right\rfloor + \cdots + \left\lfloor \frac{n}{p_r} \right\rfloor \right)$$

$$+ \left(\left\lfloor \frac{n}{p_1 p_2} \right\rfloor + \left\lfloor \frac{n}{p_1 p_3} \right\rfloor + \cdots + \left\lfloor \frac{n}{p_{r-1} p_r} \right\rfloor \right)$$

$$- \left(\left\lfloor \frac{n}{p_1 p_2 p_3} \right\rfloor + \left\lfloor \frac{n}{p_1 p_2 p_4} \right\rfloor + \cdots + \left\lfloor \frac{n}{p_{r-2} p_{r-1} p_r} \right\rfloor \right) + \cdots,$$

where $p_1, p_2, ..., p_r$ are the primes less than or equal to $\sqrt{n}$ (with $r = \pi(\sqrt{n})$). (*Hint:* Let property P_i be the property that an integer is divisible by p_i.)

19. Use Exercise 18 to find $\pi(250)$.

20. Show that $x^2 - x + 41$ is prime for all integers x with $0 \leq x \leq 40$. Show, however, that it is composite for $x = 41$.

* 21. Show that if $f(x) = a_n x^n + a_{n-1} x^{n-1} + \cdots + a_1 x + a_0$ where the coefficients are integers, then there is an integer y such that $f(y)$ is composite. (*Hint:* Assume that $f(x) = p$ is prime, and show p divides $f(x + kp)$ for all integers k. Conclude from the fact that a polynomial of degree n, $n > 1$ takes on each value at most n times, that there is an integer y such that $f(y)$ is composite.)

The *lucky numbers* are generated by the following sieving process. Start with the positive integers. Begin the process by crossing out every second integer in the list,

starting your count with the integer 1. Other than 1 the smallest integer left is 3, so
we continue by crossing out every third integer left, starting the count with the integer
1. The next integer left is 7, so we cross out every seventh integer left. Continue this
process, where at each stage we cross out every kth integer left where k is the smallest
integer left other than one. The integers that remain are the lucky numbers.

22. Find all lucky numbers less than 100.

23. Show that there are infinitely many lucky numbers.

24. Show that if p is prime and $1 \leq k < p$, then the binomial coefficient $\binom{p}{k}$ is
divisible by p.

* 25. A *prime power* is an integer of the form p^n where p is prime and n is a positive
integer greater than 1. Find all pairs of prime powers that differ by 1. Prove
that your answer is correct.

26. Let n be a positive integer greater than 1 and let $p_1, p_2,...,p_t$ be the primes not
exceeding n. Show that $p_1 p_2 \cdots p_t < 4^n$.

* 27. Let n be a positive integer greater than 3 and let p be a prime such that
$2n/3 < p \leq n$. Show that p does not divide the binomial coefficient $\binom{2n}{n}$.

** 28. Use Exercises 26 and 27 to show that if n is a positive integer then there exists
a prime p such that $n < p < 2n$. (This is know as *Bertrand's Conjecture*,
after Joseph Bertrand who conjectured this result in the 1840s. It was proved
by the Russian mathematician P. Chebyshev in the 1850s.)

29. Use Exercise 28 to show that if p_n is the nth prime, then $p_n \leq 2^n$.

1.9 Computer Projects

Programming Projects

Write programs to do the following:

1. Decide whether a given positive integer is prime using trial division of the
integer by all primes not exceeding its square root.

* 2. Use the sieve of Eratosthenes to find all primes less than n where n is a given
positive integer.

** 3. Find $\pi(n)$, the number of primes less than or equal to n, using Exercise 18.

4. Verify Goldbach's conjecture for all even integers less than n where n is a
given positive integer.

5. Find all twin primes less than n where n is a given positive integer.

6. Find the first m primes of the form $n^2 + 1$ where n is a positive integer and m
is a given positive integer.

* 7. Find the lucky numbers less than n where n is a given integer. (see the
preamble to Exercise 22).

Computations and Explorations

Using the programs you have written or a computation program, carry out the following computations and explorations:

1. Find the smallest prime factor of $n! + 1$ for all positive integers n not exceeding 20.

2. Find the smallest prime factor of $p_1 p_2 \cdots p_k + 1$ where $p_1, p_2, \ldots, p_k$ are the kth smallest primes for all positive integers k not exceeding 50.

3. Verify as much of the information given in Table 1.1 as you can.

4. Use the sieve of Eratosthenes to find all primes less than 10000.

5. Use the result given in Exercise 18 to find the $\pi(10000)$, the number of primes not exceeding 1000.

6. Find all lucky numbers (defined in the preamble to Exercise 22) not exceeding 10000.

7. Explore the conjecture that every even integer is the sum of two, not necessarily distinct, lucky numbers. Continue by exploring the conjecture that given a positive integer k there is a positive integer n that can be expressed as the sum of two lucky numbers in exactly k ways.

8. Verify Goldbach's conjecture for all even positive integers less than 10000.

9. Find all twin primes less than 10000. Can you make any conjectures about the number of twin primes less than a given positive integer n?

2

Greatest Common Divisors and Prime Factorization

2.1 Greatest Common Divisors

If a and b are integers, not both zero, then the set of common divisors of a and b is a finite set of integers, always containing the integers $+1$ and -1. We are interested in the largest integer among the common divisors of the two integers.

Definition. The *greatest common divisor* of two integers a and b, that are not both zero, is the largest integer that divides both a and b.

The greatest common divisor of a and b is written as (a,b). We also define $(0,0) = 0$.

Example 2.1. The common divisors of 24 and 84 are ± 1, ± 2, ± 3, ± 4, ± 6, and ± 12. Hence $(24,84) = 12$. Similarly, looking at sets of common divisors, we find that $(15,81) = 3$, $(100,5) = 5$, $(17,25) = 1$, $(0,44) = 44$, $(-6,-15) = 3$, and $(-17,289) = 17$. $\square$

We are particularly interested in pairs of integers sharing no common divisors greater than 1. Such pairs of integers are called *relatively prime*.

Definition. The integers a and b are called *relatively prime* if a and b have greatest common divisor $(a,b) = 1$.

Example 2.2. Since $(25,42) = 1$, 25 and 42 are relatively prime. $\square$

Note that since the divisors of $-a$ are the same as the divisors of a, it follows that $(a,b) = (|a|,|b|)$ (where $|a|$ denotes the absolute value of a which equals a if $a \geq 0$

and equals $-a$ if $a < 0$). Hence, we can restrict our attention to greatest common divisors of pairs of positive integers.

We now prove some properties of greatest common divisors.

Theorem 2.1. Let a, b, and c be integers with $(a,b) = d$. Then

$$(i) \quad (a/d, b/d) = 1$$
$$(ii) \quad (a+cb, b) = (a,b).$$

Proof. (i) Let a and b be integers with $(a,b) = d$. We will show that a/d and b/d have no common positive divisors other than 1. Assume that e is a positive integer such that $e \mid (a/d)$ and $e \mid (b/d)$. Then, there are integers k and l with $a/d = ke$ and $b/d = le$, so that $a = dek$ and $b = del$. Hence, de is a common divisor of a and b. Since d is the greatest common divisor of a and b, $de \le d$, so that e must be 1. Consequently, $(a/d, b/d) = 1$.

(ii) Let a, b, and c be integers. We will show that the common divisors of a and b are exactly the same as the common divisors of $a + cb$ and b. This will show that $(a+cb, b) = (a,b)$. Let e be a common divisor of a and b. By Theorem 1.8 we see that $e \mid (a+cb)$, so that e is a common divisor of $a + cb$ and b. If f is a common divisor of $a + cb$ and b, then by Theorem 1.8, we see that f divides $(a+cb) - cb = a$, so that f is a common divisor of a and b. Hence $(a+cb, b) = (a,b)$. ∎

We will show that the greatest common divisor of the integers a and b, not both zero, can be written as a sum of multiples of a and b. To phrase this more succinctly, we use the following definition.

Definition. If a and b are integers, then a *linear combination* of a and b is a sum of the form $ma + nb$, where both m and n are integers.

We can now state and prove the following theorem about greatest common divisors.

Theorem 2.2. The greatest common divisor of the integers a and b, not both zero, is the least positive integer that is a linear combination of a and b.

Proof. Let d be the least positive integer that is a linear combination of a and b. (There is a *least* such positive integer, using the well-ordering property, since at least one of two linear combinations $1 \cdot a + 0 \cdot b$ and $(-1)a + 0 \cdot b$, where $a \ne 0$, is positive.) We write

(2.1) $$d = ma + nb,$$

where m and n are integers. We will show that $d \mid a$ and $d \mid b$.

By the division algorithm, we have

$$a = dq + r, \quad 0 \le r < d.$$

From this equation and (2.1), we see that

$$r = a - dq = a - q(ma + nb) = (1 - qm)a - qnb.$$

This shows that the integer r is a linear combination of a and b. Since $0 \le r < d$, and d is the least positive linear combination of a and b, we conclude that $r = 0$, and hence $d \mid a$. In a similar manner, we can show that $d \mid b$.

We now demonstrate that d is the *greatest* common divisor of a and b. To show this, all we need to show is that any common divisor c of a and b must divide d. Since $d = ma + nb$, if $c \mid a$ and $c \mid b$, Theorem 1.8 tells us that $c \mid d$. ∎

We have shown that the greatest common divisor of the integers a and b, that are not both zero, is a linear combination of a and b. How to find a particular linear combination of a and b equal to (a,b) will be discussed in the next section.

We can also define the greatest common divisor of more than two integers.

Definition.   Let $a_1, a_2, ..., a_n$ be   integers,   not   all   zero.   The *greatest common divisor* of these integers is the largest integer which is a divisor of all of the integers in the set. The greatest common divisor of $a_1, a_2, ..., a_n$ is denoted by $(a_1, a_2, ..., a_n)$.

Example 2.3. We easily see that $(12, 18, 30) = 6$ and $(10, 15, 25) = 5$. □

We can use the following lemma to find the greatest common divisor of a set of more than two integers.

Lemma 2.1. If $a_1, a_2, ..., a_n$ are integers, not all zero, then $(a_1, a_2, ..., a_{n-1}, a_n) = (a_1, a_2, ..., a_{n-2}, (a_{n-1}, a_n))$.

Proof. Any common divisor of the n integers $a_1, a_2, ..., a_{n-1}, a_n$ is, in particular, a divisor of a_{n-1} and a_n, and therefore a divisor of (a_{n-1}, a_n). Also, any common divisor of the $n-1$ integers $a_1, a_2, ..., a_{n-2}$, and (a_{n-1}, a_n) must be a common divisor of all n integers, for if it divides (a_{n-1}, a_n), it must divide both a_{n-1} and a_n. Since the set of n integers and the set of the first $n-2$ integers together with the greatest common divisor of the last two integers have exactly the same divisors, their greatest common divisors are equal. ∎

Example 2.4. To find the greatest common divisor of the three integers 105, 140, and 350, we use Lemma 2.1 to see that $(105, 140, 350) = (105, (140, 350)) = (105, 70) = 35$. □

Definition. We say that the integers $a_1, a_2, ..., a_n$ are *mutually relatively prime* if $(a_1, a_2, ..., a_n) = 1$. These integers are called *pairwise relatively prime* if for each pair of integers a_i and a_j from the set, $(a_i, a_j) = 1$, that is, if each pair of integers from the set is relatively prime.

It is easy to see that if integers are pairwise relatively prime, they must be mutually relatively prime. However, the converse is false as the following example shows.

Example 2.5. Consider the integers $15, 21$, and 35. Since

$$(15,21,35) = (15,(21,35)) = (15,7) = 1,$$

we see that the three integers are mutually relatively prime. However, any two of these integers are not relatively prime because $(15,21) = 3$, $(15,35) = 5$, and $(21,35) = 7$. □

2.1 Exercises

1. Find the greatest common divisor of each of the following pairs of integers.

   a) $15, 35$ d) $99, 100$
   b) $0, 111$ e) $11, 121$
   c) $-12, 18$ f) $100, 102$

2. Find the greatest common divisor of each of the following pairs of integers.

   a) $5, 15$ d) $-90, 100$
   b) $0, 100$ e) $100, 121$
   c) $-27, -45$ f) $1001, 289$

3. Let a be a positive integer. What is the greatest common divisor of a and $2a$?

4. Let a be a positive integer. What is the greatest common divisor of a and a^2?

5. Let a be a positive integer. What is the greatest common divisor of a and $a+1$?

6. Let a be a positive integer. What is the greatest common divisor of a and $a+2$?

7. Show that if a and b are integers that are not both zero, and c is a nonzero integer, then $(ca,cb) = |c| (a,b)$.

8. Show that if a and b are integers with $(a,b) = 1$, then $(a+b,a-b) = 1$ or 2.

9. What is $(a^2+b^2,a+b)$, where a and b are relatively prime integers that are not both zero?

10. Show that if a and b are both even integers that are not both zero, then $(a,b) = 2(a/2,b/2)$.

11. Show that if a is an even integer and b is an odd integer, then $(a,b) = (a/2,b)$.

12. Show that if a,b, and c are integers such that $(a,b) = 1$ and $c \mid (a+b)$, then $(c,a) = (c,b) = 1$.

13. Show that if a, b, and c are mutually relatively prime nonzero integers, then $(a,bc) = (a,b)(a,c)$.

☞ 14. a) Show that if a, b, and c are integers with $(a,b) = (a,c) = 1$, then $(a,bc) = 1$.

   b) Use mathematical induction to show that if $a_1,a_2,...,a_n$ are integers, and b is another integer such that $(a_1,b) = (a_2,b) = \cdots = (a_n,b) = 1$, then $(a_1 a_2 \cdots a_n,b) = 1$.

15. Find a set of three integers that are mutually relatively prime, but any two of which are not relatively prime. Do not use examples from the text.

16. Find four integers that are mutually relatively prime such that any three of these integers are not mutually relatively prime.

17. Find the greatest common divisor of each of the following sets of integers

   a) 8, 10, 12   d) 6, 15, 21
   b) 5, 25, 75   e) -7, 28, -35
   c) 99, 9999, 0   f) 0, 0, 1001

18. Find three mutually relatively prime integers from among the integers 66, 105, 42, 70, and 165.

19. Show that if a_1, $a_2,...$, a_n are integers that are not all zero and c is a positive integer, then $(ca_1,ca_2,...,ca_n) = c(a_1,a_2...,a_n)$.

20. Show that the greatest common divisor of the integers $a_1,a_2,...,a_n$ not all zero is the least positive integer that is a linear combination of $a_1,a_2,...,a_n$.

21. Show that if k is an integer, then the integers $6k-1$, $6k+1$, $6k+2$, $6k+3$, and $6k+5$ are pairwise relatively prime.

22. Show that if k is a positive integer, then $3k+2$ and $5k+3$ are relatively prime.

23. Show that every positive integer greater than six is the sum of two relatively prime integers greater than one.

The *Farey series* $\mathcal{F}_n$ *of order* n is the set of fractions h/k where h and k are integers, $0 \le h \le k \le n$, and $(h,k) = 1$, in ascending order. We include 0 and 1 in the forms $\frac{0}{1}$ and $\frac{1}{1}$, respectively. For instance, the Farey series of order 4 is

$$\frac{0}{1}, \frac{1}{4}, \frac{1}{3}, \frac{1}{2}, \frac{2}{3}, \frac{3}{4}, \frac{1}{1}.$$

Exercises 24 - 27 deal with the Farey series.

24. Find the Farey series of order 7.

* 25. Show that if a/b, c/d, and e/f are successive terms of a Farey series, then

$$\frac{c}{d} = \frac{a+e}{b+f}.$$

* 26. Show that if a/b and c/d are successive terms of a Farey series, then $ad - bc = -1$.

* 27. Show that if a/b and c/d are successive terms of the Farey series of order n, then $b+d > n$.

* 28. a) Show that if a and b are positive integers, then $((a^n - b^n)/(a-b), a-b) = (n(a,b)^{n-1}, a-b)$.

b) Show that if a and b are relatively prime positive integers, then $((a^n - b^n)/(a-b), a-b) = (n, a-b)$.

29. Show that if a, b, c, and d are integers such that b and d are positive, $(a,b) = (c,d) = 1$, and $\frac{a}{b} + \frac{c}{d}$ is an integer, then $b = d$.

30. What can you conclude if a, b, and c are positive integers such that $(a,b) = (b,c) = 1$ and $\frac{1}{a} + \frac{1}{b} + \frac{1}{c}$ is an integer?

☞ **31.** Show that if a and b are positive integers then the set of linear combinations $ma + nb$, where m and n are integers, is the set of integer multiples of (a,b).

2.1 Computer Projects

Programming Projects

Write programs to do the following things.

1. Find the greatest common divisor of two integers from the lists of their divisors.

2. Print out the Farey series of order n for a given positive integer n.

Computations and Explorations

Using the programs you have written or a computation program, carry out the following computations and explorations:

1. Construct the Farey sequence of order 100.

2. Verify the properties of the Farey sequence given in Exercises 25, 26, and 27 for successive terms of your choice in the Farey series of order 100.

2.2 The Euclidean Algorithm

We are going to develop a systematic method, or *algorithm*, to find the greatest common divisor of two positive integers. This method is called the *Euclidean algorithm*. It is named after the ancient Greek mathematician Euclid* who describes this algorithm in his book *The Elements*.

Before we discuss the algorithm in general, we demonstrate its use with an example. We find the greatest common divisor of 30 and 72. First, we use the division algorithm to write $72 = 30 \cdot 2 + 12$, and we use Theorem 2.1 to note that $(30,72) = (30,72 - 2 \cdot 30) = (30,12)$. Another way to see that $(30,72) = (30,12)$ is to notice that any common divisor of 30 and 72 must also divide 12 because $12 = 72 - 30 \cdot 2$, and conversely, any common divisor of 12 and 30 must also divide 72, since $72 = 30 \cdot 2 + 12$. Note we have replaced 72 by the smaller number 12 in our computations since $(72,30) = (30,12)$. Next, we use the division algorithm again to write $30 = 2 \cdot 12 + 6$. Using the same reasoning as before, we see that $(30,12) = (12,6)$. Because $12 = 6 \cdot 2 + 0$, we now see that $(12,6) = (6,0) = 6$. Consequently, we can conclude that $(72,30) = 6$, without finding all the common divisors of 30 and 72.

We now set up the general format of the Euclidean algorithm for computing the greatest common divisor of two positive integers.

Theorem 2.3. The Euclidean Algorithm. Let $r_0 = a$ and $r_1 = b$ be integers such that $a \geq b > 0$. If the division algorithm is successively applied to obtain $r_j = r_{j+1} q_{j+1} + r_{j+2}$ with $0 < r_{j+2} < r_{j+1}$ for $j = 0,1,2,...,n-2$ and $r_{n+1} = 0$, then $(a,b) = r_n$, the last nonzero remainder.

From this theorem, we see that the greatest common divisor of a and b is the last nonzero remainder in the sequence of equations generated by successively using the division algorithm continuing until a remainder is 0, where at each step, the dividend and divisor are replaced by smaller numbers, namely the divisor and remainder.

To prove that the Euclidean algorithm produces greatest common divisors, the following lemma will be helpful.

---

* **EUCLID (c. 350 B.C.E)** was the author of the most successful mathematics textbook ever written, namely his *Elements*, which has appeared in over a thousand different editions from ancient to modern times. Very little is known about Euclid's life other than that he taught at the famed academy at Alexandria. Evidently he did not stress the applications of mathematics for it is reputed that when asked by a student for the use of geometry, Euclid had his slave give the student some coins, "since he must needs make gain of what he learns." Euclid's *Elements* provides an introduction to plane and solid geometry and to number theory. The Euclidean algorithm is found in Book VII of the 13 books in the *Elements* and his proof of the infinitude of primes is found in Book IX. Euclid also wrote books on a variety of other topics, including astronomy, optics, music, and mechanics.

Lemma 2.2. If c and d are integers and $c = dq + r$ where q and r are integers, then $(c,d) = (d,r)$.

Proof. If an integer e divides both c and d, then since $r = c - dq$, Theorem 1.8 shows that $e \mid r$. If $e \mid d$ and $e \mid r$, then since $c = dq + r$, from Theorem 1.8 we see that $e \mid c$. Since the common divisors of c and d are the same as the common divisors of d and r, we see that $(c,d) = (d,r)$. ∎

We now prove that the Euclidean algorithm works.

Proof. Let $r_0 = a$ and $r_1 = b$ be positive integers with $a \geq b$. By successively applying the division algorithm, we find that

$$
\begin{aligned}
r_0 &= r_1 q_1 + r_2 & 0 &\leq r_2 < r_1, \\
r_1 &= r_2 q_2 + r_3 & 0 &\leq r_3 < r_2, \\
&\quad\vdots \\
r_{j-2} &= r_{j-1} q_{j-1} + r_j & 0 &\leq r_j < r_{j-1}, \\
&\quad\vdots \\
r_{n-4} &= r_{n-3} q_{n-3} + r_{n-2} & 0 &\leq r_{n-2} < r_{n-3}, \\
r_{n-3} &= r_{n-2} q_{n-2} + r_{n-1} & 0 &\leq r_{n-1} < r_{n-2}, \\
r_{n-2} &= r_{n-1} q_{n-1} + r_n & 0 &\leq r_n < r_{n-1}, \\
r_{n-1} &= r_n q_n.
\end{aligned}
$$

We can assume that we eventually obtain a remainder of zero since the sequence of remainders $a = r_0 > r_1 > r_2 > \cdots \geq 0$ cannot contain more than a terms. By Lemma 2.2 we see that $(a,b) = (r_0, r_1) = (r_1, r_2) = (r_2, r_3) = \cdots = (r_{n-3}, r_{n-2}) = (r_{n-2}, r_{n-1}) = (r_{n-1}, r_n) = (r_n, 0) = r_n$. Hence $(a,b) = r_n$, the last nonzero remainder. ∎

We illustrate the use of the Euclidean algorithm with the following example.

Example 2.6. To find $(252, 198)$, we use the division algorithm successively to obtain

$$
\begin{aligned}
252 &= 1 \cdot 198 + 54 \\
198 &= 3 \cdot 54 + 36 \\
54 &= 1 \cdot 36 + 18 \\
36 &= 2 \cdot 18.
\end{aligned}
$$

Hence $(252, 198) = 18$. □

Later in this section, we give estimates for the maximum number of divisions used by the Euclidean algorithm to find the greatest common divisor of two positive integers. However, we first show that given any positive integer n, there are integers a and b such that exactly n divisions are required to find (a,b) using the

Euclidean algorithm. We can find such numbers by taking successive terms of the Fibonacci sequence.

The reason that the Euclidean algorithm operates most slowly when it finds the greatest common divisor of successive Fibonacci numbers is that the quotient in all but the last step is 1, as illustrated in the following example.

Example 2.7. We apply the Euclidean algorithm to find $(34,55)$. Note that $f_9 = 34$ and $f_{10} = 55$. We have

$$55 = 34 \cdot 1 + 21$$
$$34 = 21 \cdot 1 + 13$$
$$21 = 13 \cdot 1 + 8$$
$$13 = 8 \cdot 1 + 5$$
$$8 = 5 \cdot 1 + 3$$
$$5 = 3 \cdot 1 + 2$$
$$3 = 2 \cdot 1 + 1$$
$$2 = 1 \cdot 2.$$

Observe that when the Euclidean algorithm is used to find the greatest common divisor of $f_9 = 34$ and $f_{10} = 55$, a total of eight divisions are required. Furthermore, $(34,55) = 1$. □

The following theorem tells us how many divisions are needed to find the greatest common divisor of successive Fibonacci numbers.

Theorem 2.4. Let f_{n+1} and f_{n+2} be successive terms of the Fibonacci sequence, with $n > 1$. Then the Euclidean algorithm takes exactly n divisions to show that $(f_{n+1}, f_{n+2}) = 1$.

Proof. Applying the Euclidean algorithm, and using the defining relation for the Fibonacci numbers $f_j = f_{j-1} + f_{j-2}$ in each step, we see that

$$f_{n+2} = f_{n+1} \cdot 1 + f_n,$$
$$f_{n+1} = f_n \cdot 1 + f_{n-1},$$

$$\cdot$$
$$\cdot$$
$$\cdot$$

$$f_4 = f_3 \cdot 1 + f_2,$$
$$f_3 = f_2 \cdot 2.$$

Hence, the Euclidean algorithm takes exactly n divisions, to show that $(f_{n+2}, f_{n+1}) = f_2 = 1$. ∎

We can now prove a theorem first proved by Gabriel Lamé*, a French mathematician of the nineteenth century, which gives an estimate for the number of divisions needed to find the greatest common divisor using the Euclidean algorithm.

Theorem 2.5. Lamé's Theorem. The number of divisions needed to find the greatest common divisor of two positive integers using the Euclidean algorithm does not exceed five times the number of decimal digits in the smaller of the two integers.

Proof. When we apply the Euclidean algorithm to find the greatest common divisor of $a = r_0$ and $b = r_1$ with $a > b$, we obtain the following sequence of equations:

$$
\begin{aligned}
r_0 &= r_1 q_1 + r_2, & 0 \le r_2 < r_1, \\
r_1 &= r_2 q_2 + r_3, & 0 \le r_3 < r_2, \\
&\quad\vdots \\
r_{n-2} &= r_{n-1} q_{n-1} + r_n, & 0 \le r_n < r_{n-1}, \\
r_{n-1} &= r_n q_n.
\end{aligned}
$$

We have used n divisions. We note that each of the quotients $q_1, q_2, \ldots, q_{n-1}$ is greater than or equal to 1, and $q_n \ge 2$, since $r_n < r_{n-1}$. Therefore,

$$
\begin{aligned}
r_n &\ge 1 = f_2, \\
r_{n-1} &\ge 2r_n \ge 2f_2 = f_3, \\
r_{n-2} &\ge r_{n-1} + r_n \ge f_3 + f_2 = f_4, \\
r_{n-3} &\ge r_{n-2} + r_{n-1} \ge f_4 + f_3 = f_5, \\
&\quad\vdots \\
r_2 &\ge r_3 + r_4 \ge f_{n-1} + f_{n-2} = f_n, \\
b = r_1 &\ge r_2 + r_3 \ge f_n + f_{n-1} = f_{n+1}.
\end{aligned}
$$

Thus, for there to be n divisions used in the Euclidean algorithm we must have $b \ge f_{n+1}$. By Example 1.16, we know that $f_{n+1} > \alpha^{n-1}$ for $n > 2$ where $\alpha = (1+\sqrt{5})/2$. Hence, $b > \alpha^{n-1}$. Now, since $\log_{10} \alpha > 1/5$, we see that

---

* **GABRIEL LAMÉ** (1795 -1870) was a graduate of the École Polytechnique. Lamé was a civil and railway engineer. He advanced the mathematical theory of elasticity and invented curvilinear coordinates. Although his main contributions were to mathematical physics, he made several discoveries in number theory, including the estimate of the number of steps required by the Euclidean algorithm, and the proof that Fermat's last theorem holds for $n = 7$ (see Section 11.2). It is interesting to note that Gauss considered Lamé to be the foremost French mathematician of his time.

$$\log_{10} b > (n-1)\log_{10}\alpha > (n-1)/5.$$

Consequently,

$$n - 1 < 5 \cdot \log_{10} b.$$

Let b have k decimal digits, so that $b < 10^k$ and $\log_{10} b < k$. Hence, we see that $n - 1 < 5k$ and since k is an integer, we can conclude that $n \le 5k$. This establishes Lamé's theorem.   ∎

The following result is a consequence of Lamé's theorem.

Corollary 2.5.1. The greatest common divisor of two positive integers a and b with $a > b$ can be found using $O((\log_2 a)^3)$ bit operations.

Proof. We know from Lamé's theorem that $O(\log_2 a)$ divisions, each taking $O((\log_2 a)^2)$ bit operations, are needed to find (a, b). Hence, by Theorem 1.12 (a,b) may be found using a total of $O((\log_2 a)^3)$ bit operations.   ∎

The Euclidean algorithm can be used to express the greatest common divisor of two integers as a linear combination of these integers. We illustrate this by expressing $(252,198) = 18$ as a linear combination of 252 and 198. Referring to the steps of the Euclidean algorithm used to find $(252,198)$, by the next to the last step we see that

$$18 = 54 - 1 \cdot 36.$$

By the preceding step it follows that

$$36 = 198 - 3 \cdot 54,$$

which implies that

$$18 = 54 - 1 \cdot (198 - 3 \cdot 54) = 4 \cdot 54 - 1 \cdot 198.$$

Likewise, by the first step we have

$$54 = 252 - 1 \cdot 198,$$

so that

$$18 = 4(252 - 1 \cdot 198) - 1 \cdot 198 = 4 \cdot 252 - 5 \cdot 198.$$

This last equation exhibits $18 = (252,198)$ as a linear combination of 252 and 198.

In general, to see how $d = (a,b)$ may be expressed as a linear combination of a and b, refer to the series of equations that is generated by use of the Euclidean algorithm. By the penultimate equation we have

$$r_n = (a,b) = r_{n-2} - r_{n-1}q_{n-1}.$$

This expresses (a,b) as a linear combination of r_{n-2} and r_{n-1}. The second to the last equation can be used to express r_{n-1} as $r_{n-3} - r_{n-2}q_{n-2}$. Using this last equation to eliminate r_{n-1} in the previous expression for (a,b), we find that

$$r_n = r_{n-3} - r_{n-2}q_{n-2},$$

so that

$$(a, b) = r_{n-2} - (r_{n-3} - r_{n-2}q_{n-2})q_{n-1}$$
$$= (1 + q_{n-1}q_{n-2})r_{n-2} - q_{n-1}r_{n-3},$$

which expresses (a,b) as a linear combination of r_{n-2} and r_{n-3}. We continue working backwards through the steps of the Euclidean algorithm to express (a, b) as a linear combination of each preceding pair of remainders until we have found (a,b) as a linear combination of $r_0 = a$ and $r_1 = b$. Specifically, if we have found at a particular stage that

$$(a,b) = sr_j + tr_{j-1},$$

then, since

$$r_j = r_{j-2} - r_{j-1}q_{j-1},$$

we have

$$(a,b) = s(r_{j-2} - r_{j-1}q_{j-1}) + tr_{j-1}$$
$$= (t - sq_{j-1})r_{j-1} + sr_{j-2}.$$

This shows how to move up through the equations that are generated by the Euclidean algorithm so that, at each step, the greatest common divisor of a and b may be expressed as a linear combination of a and b.

This method for expressing (a,b) as a linear combination of a and b is somewhat inconvenient for calculation, because it is necessary to work out the steps of the Euclidean algorithm, save all these steps, and then proceed backwards through the steps to write (a,b) as a linear combination of each successive pair of remainders. There is another method for finding (a,b) which requires working through the steps of the Euclidean algorithm only once. The following theorem gives this method.

Theorem 2.6. Let a and b be positive integers. Then

$$(a,b) = s_n a + t_n b,$$

for $n = 0,1,2,...,$ where s_n and t_n are the nth terms of the sequences defined recursively by

$$s_0 = 1, t_0 = 0,$$
$$s_1 = 0, t_1 = 1,$$

and

$$s_j = s_{j-2} - q_{j-1}s_{j-1}, \quad t_j = t_{j-2} - q_{j-1}t_{j-1}$$

for $j = 2, 3, ..., n$, where the q_j's are the quotients in the divisions of the Euclidean algorithm when it is used to find (a,b).

Proof. We will prove that

(2.2) $$r_j = s_j a + t_j b$$

for $j = 0, 1, \ldots, n$. Since $(a, b) = r_n$, once we have established (2.2), we will know that

$$(a, b) = s_n a + t_n b.$$

We prove (2.2) using the second principle of mathematical induction. For $j = 0$, we have $a = r_0 = 1 \cdot a + 0 \cdot b = s_0 a + t_0 b$. Hence, (2.2) is valid for $j = 0$. Likewise, $b = r_1 = 0 \cdot a + 1 \cdot b = s_1 a + t_1 b$, so that (2.2) is valid for $j = 1$.

Now, assume that

$$r_j = s_j a + t_j b$$

for $j = 1, 2, \ldots, k - 1$. Then, from the kth step of the Euclidean algorithm, we have

$$r_k = r_{k-2} - r_{k-1} q_{k-1}.$$

Using the induction hypothesis, we find that

$$\begin{aligned} r_k &= (s_{k-2} a + t_{k-2} b) - (s_{k-1} a + t_{k-1} b) q_{k-1} \\ &= (s_{k-2} - s_{k-1} q_{k-1}) a + (t_{k-2} - t_{k-1} q_{k-1}) b \\ &= s_k a + t_k b. \end{aligned}$$

This finishes the proof. ∎

The following example illustrates the use of this algorithm for expressing (a, b) as a linear combination of a and b.

Example 2.8. Let $a = 252$ and $b = 198$. Then

$s_0 = 1,$	$t_0 = 0,$
$s_1 = 0,$	$t_1 = 1,$
$s_2 = s_0 - s_1 q_1 = 1 - 0 \cdot 1 = 1,$	$t_2 = t_0 - t_1 q_1 = 0 - 1 \cdot 1 = -1,$
$s_3 = s_1 - s_2 q_2 = 0 - 1 \cdot 3 = -3,$	$t_3 = t_1 - t_2 q_2 = 1 - (-1)3 = 4,$
$s_4 = s_2 - s_3 q_3 = 1 - (-3) \cdot 1 = 4,$	$t_4 = t_2 - t_3 q_3 = -1 - 4 \cdot 1 = -5.$

Since $r_4 = 18 = (252, 198)$ and $r_4 = s_4 a + t_4 b$, we have

$$18 = (252, 198) = 4 \cdot 252 - 5 \cdot 198. \qquad \square$$

Note that the greatest common divisor of two integers may be expressed as a linear combination of these integers in an infinite number of different ways. To see this, let $d = (a, b)$ and let $d = sa + tb$ be one way to write d as a linear combination of a and b, guaranteed to exist by the previous discussion. Then

$$d = (s + k(b/d)) a + (t - k(a/d)) b$$

for all integers k.

Example 2.9. With $a = 252$ and $b = 198$, we have $18 = (252, 198) = (4 + 11k)252 + (-5 - 14k)198$ whenever k is an integer. $\qquad \square$

2.2 Exercises

1. Use the Euclidean algorithm to find each of the following greatest common divisors.

   a) $(45, 75)$ c) $(666, 1414)$
   b) $(102, 222)$ d) $(20785, 44350)$

2. Use the Euclidean algorithm to find each of the following greatest common divisors.

   a) $(51, 87)$ c) $(981, 1234)$
   b) $(105, 300)$ d) $(34709, 100313)$

3. For each pair of integers in Exercise 1, express the greatest common divisor of the integers as a linear combination of these integers.

4. For each pair of integers in Exercise 2, express the greatest common divisor of the integers as a linear combination of these integers.

5. Find the greatest common divisor of each of the following sets of integers.

   a) 6, 10, 15
   b) 70, 98, 105
   c) 280, 330, 405, 490

6. Find the greatest common divisor of each of the following sets of integers.

   a) 15, 35, 90
   b) 300, 2160, 5040
   c) 1240, 6660, 15540, 19980

The greatest common divisor of the n integers $a_1, a_2,..., a_n$ can be expressed as a linear combination of these integers. To do this, first express (a_1,a_2) as a linear combination of a_1 and a_2. Then express $(a_1,a_2,a_3) = ((a_1,a_2),a_3)$ as a linear combination of a_1, a_2, and a_3. Repeat this until $(a_1,a_2, \ldots, a_n)$ is expressed as a linear combination of $a_1, a_2,...,a_n$. Use this procedure in Exercises 7 and 8.

7. Express the greatest common divisor of each set of numbers in Exercise 5 as a linear combination of the numbers in that set.

8. Express the greatest common divisor of each set of numbers in Exercise 6 as a linear combination of the numbers in that set.

The greatest common divisor of two integers can be found using only subtractions, parity checks, and shifts of binary expansions, without using any divisions. The algorithm proceeds recursively using the following reduction

$$(a,b) = \begin{cases} a & \text{if } a = b \\ 2(a/2,b/2) & \text{if } a \text{ and } b \text{ are even} \\ (a/2,b) & \text{if } a \text{ is even and } b \text{ is odd} \\ (a-b,b) & \text{if } a \text{ and } b \text{ are odd where } a > b. \end{cases}$$

(*Note:* Reverse the roles of a and b when necessary.) Exercises 9-13 refer to this algorithm.

9. Find $(2106,8318)$ using this algorithm.

10. Show that this algorithm always produces the greatest common divisor of a pair of positive integers.

* 11. How many steps does this algorithm use to find (a,b) if $a = (2^n - (-1)^n)/3$ and $b = 2(2^{n-1} - (-1)^{n-1})/3$ when n is a positive integer?

* 12. Show that to find (a,b) this algorithm uses the subtraction step in the reduction no more than $1 + [\log_2 \max(a,b)]$ times.

* 13. Devise an algorithm for finding the greatest common divisor of two positive integers using their balanced ternary expansions.

In Exercise 18 of Section 1.5, a modified division algorithm is given which states that if a and $b > 0$ are integers, then there exist unique integers q, r, and e such that $a = bq + er$, where $e = \pm1$, $r \geq 0$, and $-b/2 < er \leq b/2$. We can set up an algorithm, analogous to the Euclidean algorithm, based on this modified division algorithm, called the *least-remainder algorithm*. It works as follows. Let $r_0 = a$ and $r_1 = b$, where $a > b > 0$. Using the modified division algorithm repeatedly, obtain the greatest common divisor of a and b as the last nonzero remainder r_n in the sequence of divisions

$$r_0 = r_1 q_1 + e_2 r_2, \qquad -r_1/2 < e_2 r_2 \leq r_1/2$$

.

.

.

$$r_{n-2} = r_{n-1} q_{n-1} + e_n r_n, \qquad -r_{n-1}/2 < e_n r_n \leq r_{n-1}/2$$
$$r_{n-1} = r_n q_n.$$

14. Use the least-remainder algorithm to find $(384,226)$.

15. Show that the least-remainder algorithm always produces the greatest common divisor of two integers.

** 16. Show that the least-remainder algorithm is always faster, or as fast, as the Euclidean algorithm. (*Hint:* First show that if a and b are positive integers with $2b < a$, then the least-remainder algorithm can find (a,b) with no more steps than it uses to find $(a,a-b)$.)

* 17. Find a sequence of integers $v_0, v_1, v_2,...$ such that the least-remainder algorithm takes exactly n divisions to find (v_{n+1},v_{n+2}).

* **18.** Show that the number of divisions needed to find the greatest common divisor of two positive integers using the least-remainder algorithm is less than 8/3 times the number of digits in the smaller of the two numbers, plus 4/3.

* **19.** Let m and n be positive integers and let a be an integer greater than one. Show that $(a^m - 1, a^n - 1) = a^{(m,n)} - 1$.

* **20.** Show that if m and n are positive integers then $(f_m, f_n) = f_{(m,n)}$.

The next two exercises deal with the *game of Euclid*. Two players begin with a pair of positive integers and take turns making moves of the following type. A player can move from the pair of positive integers $\{x,y\}$ with $x \geq y$, to any of the pairs $\{x-ty,y\}$, where t is a positive integer and $x-ty \geq 0$. A *winning move* consists of moving to a pair with one element equal to 0.

21. Show that every sequence of moves starting with the pair $\{a,b\}$ must eventually end with the pair $\{0, (a,b)\}$.

* **22.** Show that in a game beginning with the pair $\{a,b\}$, the first player may play a winning strategy if $a = b$ or if $a > b(1+ \sqrt{5})/2$; otherwise the second player may play a winning strategy. (*Hint*: First show that if $y < x \leq y(1+\sqrt{5})/2$ then there is a unique move from $\{x,y\}$ that goes to a pair $\{z,y\}$ with $y > z(1+\sqrt{5})/2$.)

* **23.** Show that the number of bit operations needed to find the greatest common divisor of two positive integers a and b with $a > b$ is $O((\log_2 a)^2)$. (*Hint*: First show that the complexity of division of the positive intger q by the positive integer d is $O(\log d \log q)$.)

* **24.** Let a and b be positive integers and let r_j and q_j, $j = 1,2,...,n$ be the remainders and quotients of the steps of the Euclidean algorithm as defined in this section.

   a) Find the value of $\sum_{j=1}^{n} r_j q_j$.

   b) Find the value of $\sum_{j=1}^{n} r_j^2 q_j$.

2.2 Computer Projects

Programming Projects

Write programs to do the following:

1. Find the greatest common divisor of two integers using the Euclidean algorithm.

2. Find the greatest common divisor of two integers using the modified Euclidean algorithm given in the preamble to Exercise 14.

3. Find the greatest common divisor of two integers using no divisions (see the preamble to Exercise 9).

   4. Find the greatest common divisor of a set of more than two integers.

   5. Express the greatest common divisor of two integers as a linear combination of these integers.

   6. Express the greatest common divisor of a set of more than two integers as a linear combination of these integers.

* **7.** Play the game of Euclid described in the preamble to Exercise 21.

Computations and Explorations

   Using the programs you have written or a computation program, carry out the following computations and explorations:

   1. Verify Lamé's Theorem for several different pairs of large positive integers of your choice.

   2. Compare the number of steps required to find the greatest common divisor of different pairs of large positive integers of your choice using the Euclidean algorithm, the algorithm described in the preamble of Exercise 9 and the least-remainder algorithm described in the preamble to Exercise 14.

   3. Estimate the proportion of pairs of positive integers (a,b) which are relatively prime, where a and b are positive integers not exceeding 1000, not exceeding 10000, not exceeding 100000, and not exceeding 1000000. To do so, you may want to test a random selection of some relatively small number of such pairs. (See Section 8.7 for material on pseudorandom numbers.) Can you make any conjectures from this evidence?

2.3 The Fundamental Theorem of Arithmetic

   The fundamental theorem of arithmetic is an important result that shows that the primes are the building blocks of the integers. Here is what the theorem says.

Theorem 2.7. The Fundamental Theorem of Arithmetic. Every positive integer greater than one can be written uniquely as a product of primes, with the prime factors in the product written in order of nondecreasing size.

Example 2.10. The factorizations of some positive integers are given by

$$240 = 2 \cdot 2 \cdot 2 \cdot 2 \cdot 3 \cdot 5 = 2^4 \cdot 3 \cdot 5, \ 289 = 17 \cdot 17 = 17^2, \ 1001 = 7 \cdot 11 \cdot 13. \qquad \square$$

   Note that it is convenient to combine all the factors of a particular prime into a power of this prime, such as in the previous example. There, for the factorization of 240, all the factors of 2 were combined to form 2^4. Factorizations of integers in which the factors of primes are combined to form powers are called *prime-power factorizations*.

To prove the fundamental theorem of arithmetic, we need the following lemma concerning divisibility.

Lemma 2.3. If a, b, and c are positive integers such that $(a,b) = 1$ and $a \mid bc$, then $a \mid c$.

Proof. Since $(a,b) = 1$, there are integers x and y such that $ax + by = 1$. Multiplying both sides of this equation by c, we have $acx + bcy = c$. By Theorem 1.8, a divides $acx + bcy$, since this is a linear combination of a and bc, both of which are divisible by a. Hence $a \mid c$. ∎

The following consequence of this lemma will be needed in the proof of the Fundamental Theorem of Arithmetic.

Lemma 2.4. If p divides $a_1 a_2 \cdots a_n$ where p is a prime and $a_1, a_2, ..., a_n$ are positive integers, then there is an integer i with $1 \leq i \leq n$ such that p divides a_i.

Proof. We prove this result by induction. The case where $n = 1$ is trivial. Assume that the result is true for n. Consider a product of $n + 1$ integers $a_1 a_2 \cdots a_{n+1}$ that is divisible by the prime p. Since $p \mid a_1 a_2 \cdots a_{n+1} = (a_1 a_2 \cdots a_n)a_{n+1}$, we know from Lemma 2.3 that $p \mid a_1 a_2 \cdots a_n$ or $p \mid a_{n+1}$. Now, if $p \mid a_1 a_2 \cdots a_n$, from the induction hypothesis there is an integer i with $1 \leq i \leq n$ such that $p \mid a_i$. Consequently $p \mid a_i$ for some i with $1 \leq i \leq n + 1$. This establishes the result. ∎

We now begin the proof of the fundamental theorem of arithmetic. First, we will show that every positive integer greater than one can be written as the product of primes in at least one way. Then we will show that this product is unique up to the order of primes that appear.

Proof. We use proof by contradiction. Assume that some positive integer cannot be written as the product of primes. Let n be the smallest such integer (such an integer must exist from the well-ordering property). If n is prime, it is obviously the product of a set of primes, namely the one prime n. So n must be composite. Let $n = ab$, with $1 < a < n$ and $1 < b < n$. But since a and b are smaller than n they must be the product of primes. Then, since $n = ab$, we conclude that n is also a product of primes. This contradiction shows that every positive integer can be written as the product of primes.

We now finish the proof of the fundmental theorem of arithmetic by showing that the factorization is unique. Suppose that there is an integer n that has two different factorizations into primes:

$$n = p_1 p_2 \cdots p_s = q_1 q_2 \cdots q_t,$$

where $p_1, p_2, ..., p_s, q_1, ..., q_t$ are all primes, with $p_1 \leq p_2 \leq \cdots \leq p_s$ and $q_1 \leq q_2 \leq \cdots \leq q_t$.

Remove all common primes from the two factorizations to obtain

$$p_{i_1} p_{i_2} \cdots p_{i_u} = q_{j_1} q_{j_2} \cdots q_{j_v},$$

where the primes on the left-hand side of this equation differ from those on the right-hand side, $u \geq 1$, and $v \geq 1$ (since the two original factorizations were presumed to differ). However, this leads to a contradiction of Lemma 2.4; by this lemma p_{i_1} must divide q_{i_j} for some j, which is impossible, since each q_{i_j} is prime and is different from p_{i_1}. Hence the prime factorization of a positive integer n is unique. ∎

The prime factorization of an integer is often useful. As an example, let us find all the divisors of an integer from its prime factorization.

Example 2.11. The positive divisors of $120 = 2^3 \cdot 3 \cdot 5$ are those positive integers with prime power factorizations containing only the primes 2, 3, and 5, to powers less than or equal to 3, 1, and 1, respectively. These divisors are

1	3	5	$3 \cdot 5 = 15$
2	$2 \cdot 3 = 6$	$2 \cdot 5 = 10$	$2 \cdot 3 \cdot 5 = 30$
$2^2 = 4$	$2^2 \cdot 3 = 12$	$2^2 \cdot 5 = 20$	$2^2 \cdot 3 \cdot 5 = 60$
$2^3 = 8$	$2^3 \cdot 3 = 24$	$2^3 \cdot 5 = 40$	$2^3 \cdot 3 \cdot 5 = 120.$

□

Another way in which we can use prime factorizations is to find greatest common divisors, as illustrated in the following example.

Example 2.12. To be a common divisor of $720 = 2^4 \cdot 3^2 \cdot 5$ and $2100 = 2^2 \cdot 3 \cdot 5^2 \cdot 7$, a positive integer can contain only the primes 2, 3, and 5 in its prime-power factorization, and the power to which one of these primes appears cannot be larger than either of the powers of that prime in the factorizations of 720 and 2100. Consequently, to be a common divisor of 720 and 2100, a positive integer can contain only the primes 2, 3, and 5 to powers no larger than 2, 1, and 1, respectively. Therefore, the greatest common divisor of 720 and 2100 is $2^2 \cdot 3 \cdot 5 = 60$.

□

To describe, in general, how prime factorizations can be used to find greatest common divisors, let $\min(a,b)$ denote the smaller or minimum, of the two numbers a and b. Now let the prime factorizations of a and b be

$$a = p_1^{a_1} p_2^{a_2} \cdots p_n^{a_n}, \quad b = p_1^{b_1} p_2^{b_2} \cdots p_n^{b_n},$$

where each exponent is a nonnegative integer and where all primes occurring in the prime factorizations of a and of b are included in both products, perhaps with zero exponents. We note that

$$(a,b) = p_1^{\min(a_1,b_1)} p_2^{\min(a_2,b_2)} \cdots p_n^{\min(a_n,b_n)},$$

since for each prime p_i, a and b share exactly $\min(a_i,b_i)$ factors of p_i.

Prime factorizations can also be used to find the smallest integer that is a multiple of two positive integers. The problem of finding this integer arises when fractions are added.

Definition. The *least common multiple* of two positive integers a and b is the smallest positive integer that is divisible by a and b.

The least common multiple of a and b is denoted by $[a,b]$.

Example 2.13. We have the following least common multiples: $[15,21] = 105$, $[24,36] = 72$, $[2,20] = 20$, and $[7,11] = 77$. □

Once the prime factorizations of a and b are known, it is easy to find $[a,b]$. If $a = p_1^{a_1} p_2^{a_2} \cdots p_n^{a_n}$ and $b = p_1^{b_1} p_2^{b_2} \cdots p_n^{b_n}$, where $p_1, p_2, \ldots, p_n$ are the primes occurring in the prime-power factorizations of a and b, then for an integer to be divisible by both a and b, it is necessary that in the factorization of the integer, each p_j occurs with a power at least as large as a_j and b_j. Hence, $[a,b]$, the smallest positive integer divisible by both a and b is

$$[a,b] = p_1^{\max(a_1,b_1)} p_2^{\max(a_2,b_2)} \cdots p_n^{\max(a_n,b_n)}$$

where $\max(x,y)$ denotes the larger, or maximum, of x and y.

Finding the prime factorization of large integers is time-consuming. Therefore, we would prefer a method for finding the least common multiple of two integers without using the prime factorizations of these integers. We will show that we can find the least common multiple of two positive integers once we know the greatest common divisor of these integers. The latter can be found via the Euclidean algorithm. First, we prove the following lemma.

Lemma 2.5. If x and y are real numbers, then $\max(x,y) + \min(x,y) = x + y$.

Proof. If $x \geq y$, then $\min(x,y) = y$ and $\max(x,y) = x$, so that $\max(x,y) + \min(x,y) = x + y$. If $x < y$, then $\min(x,y) = x$ and $\max(x,y) = y$, and again we find that $\max(x,y) + \min(x,y) = x + y$. ∎

We use the following theorem to find $[a,b]$ once (a,b) is known.

Theorem 2.8. If a and b are positive integers, then $[a,b] = ab/(a,b)$, where $[a,b]$ and (a,b) are the least common multiple and greatest common divisor of a and b, respectively.

Proof. Let a and b have prime-power factorizations $a = p_1^{a_1} p_2^{a_2} \cdots p_n^{a_n}$ and $b = p_1^{b_1} p_2^{b_2} \cdots p_n^{b_n}$, where the exponents are nonnegative integers and all primes occurring in either factorization occur in both, perhaps with zero exponents. Now let $M_j = \max(a_j,b_j)$ and $m_j = \min(a_j,b_j)$. Then, we have

$$[a,b](a,b) = p_1^{M_1} p_2^{M_2} \cdots p_n^{M_n} p_1^{m_1} p_2^{m_2} \cdots p_n^{m_n}$$
$$= p_1^{M_1+m_1} p_2^{M_2+m_2} \cdots p_n^{M_n+m_n}$$
$$= p_1^{a_1+b_1} p_2^{a_2+b_2} \cdots p_n^{a_n+b_n}$$
$$= p_1^{a_1} p_2^{a_2} \cdots p_n^{a_n} p_1^{b_1} \cdots p_n^{b_n}$$
$$= ab,$$

since $M_j + m_j = \max(a_j,b_j) + \min(a_j,b_j) = a_j + b_j$ by Lemma 2.5.   ∎

The following consequence of the fundamental theorem of arithmetic will be needed later.

Lemma 2.6. Let m and n be relatively prime positive integers. Then if d is a positive divisor of mn, there is a unique pair of positive divisors d_1 of m and d_2 of n such that $d = d_1 d_2$. Conversely, if d_1 and d_2 are positive divisors of m and n, respectively, then $d = d_1 d_2$ is a positive divisor of mn.

Proof. Let the prime-power factorizations of m and n be $m = p_1^{m_1} p_2^{m_2} \cdots p_s^{m_s}$ and $n = q_1^{n_1} q_2^{n_2} \cdots q_t^{n_t}$. Since $(m,n) = 1$, the set of primes $p_1, p_2,..., p_s$ and the set of primes $q_1, q_2,..., q_t$ have no common elements. Therefore, the prime-power factorization of mn is

$$mn = p_1^{m_1} p_2^{m_2} \cdots p_s^{m_s} q_1^{n_1} q_2^{n_2} \cdots q_t^{n_t}.$$

Hence, if d is a positive divisor of mn, then

$$d = p_1^{e_1} p_2^{e_2} \cdots p_s^{e_s} q_1^{f_1} q_2^{f_2} \cdots q_t^{f_t}$$

where $0 \le e_i \le m_i$ for $i = 1,2,...,s$ and $0 \le f_j \le n_j$ for $j = 1,2,...,t$. Now let $d_1 = (d,m)$ and $d_2 = (d,n)$, so that

$$d_1 = p_1^{e_1} p_2^{e_2} \cdots p_s^{e_s}$$

and

$$d_2 = q_1^{f_1} q_2^{f_2} \cdots q_t^{f_t}.$$

Clearly $d = d_1 d_2$ and $(d_1,d_2) = 1$. This is the decomposition of d we desire. Furthermore this decomposition is unique. To see this, note that every prime power in the factorization of d must occur in either d_1 or d_2, and prime powers in the factorization of d that are powers of primes dividing m must appear in d_1, while prime powers in the factorization of d that are powers of primes dividing n must appear in d_2. It follows that d_1 must be (d,m) and d_2 must be (d,n).

Conversely, let d_1 and d_2 be positive divisors of m and n, respectively. Then

$$d_1 = p_1^{e_1} p_2^{e_2} \cdots p_s^{e_s}$$

where $0 \le e_i \le m_i$ for $i = 1, 2,..., s$, and

$$d_2 = q_1^{f_1} q_2^{f_2} \cdots q_t^{f_t}$$

where $0 \le f_j \le n_j$ for $j = 1, 2,..., t$. The integer

$$d = d_1 d_2 = p_1^{e_1} p_2^{e_2} \cdots p_s^{e_s} q_1^{f_1} q_2^{f_2} \cdots q_t^{f_t}$$

is clearly a divisor of

$$mn = p_1^{m_1} p_2^{m_2} \cdots p_s^{m_s} q_1^{n_1} q_2^{n_2} \cdots q_t^{n_t},$$

since the power of such prime occurring in the prime-power factorization of d is less than or equal to the power of that prime in the prime-power factorization of mn. ∎

A famous result of number theory deals with primes in arithmetic progressions.

Theorem 2.9. Dirichlet's Theorem on Primes in Arithmetic Progressions. Let a and b be relatively prime positive integers. Then the arithmetic progression $an + b$, $n = 1, 2, 3,...$, contains infinitely many primes.

G. Lejeune Dirichlet*, proved this theorem in 1837. Since proofs of Dirichlet's Theorem are complicated and rely on advanced techniques, we do not present a proof here. However, it is not difficult to prove special cases of Dirichlet's theorem, as the following theorem illustrates.

Theorem 2.10. There are infinitely many primes of the form $4n + 3$, where n is a positive integer.

Before we prove this result, we first prove a useful lemma.

Lemma 2.7. If a and b are integers both of the form $4n + 1$, then the product ab is also of this form.

Proof. Since a and b are both of the form $4n + 1$, there exist integers r and s such that $a = 4r + 1$ and $b = 4s + 1$. Hence,

$$ab = (4r + 1)(4s + 1) = 16rs + 4r + 4s + 1 = 4(4rs + r + s) + 1,$$

which is again of the form $4n + 1$. ∎

We now prove the desired result.

Proof. Let us assume that there are only a finite number of primes of the form

---

* **G. LEJEUNE DIRICHLET** (1805-1859) was born into a French family living in the vicinity of Cologne, Germany. He studied at the University of Paris when this was an important world center of mathematics. He held positions at the University of Breslau and the University of Berlin, and in 1855 was chosen to succeed Gauss at the University of Göttingen. Dirichlet is said to be the first person to master Gauss' *Disquisitiones Arithmeticae,* which had appeared 20 years earlier. He is said to have kept a copy of this book at his side even when he traveled. His book on number theory, *Vorlesungen über Zahlentheorie,* helped make Gauss' discoveries accessible to other mathematicians. Besides his fundamental work in number theory, Dirichlet made many important contributions to analysis. His famous "drawer principle", also called the pigeonhole principle, is used extensively in combinatorics and in number theory.

$4n + 3$, say $p_0 = 3, p_1, p_2, ..., p_r$. Let

$$Q = 4p_1 p_2 \cdots p_r + 3.$$

Then, there is at least one prime in the factorization of Q of the form $4n + 3$. Otherwise, all of these primes would be of the form $4n + 1$, and by Lemma 2.7, this would imply that Q would also be of this form, which is a contradiction. However, none of the primes $p_0, p_1,..., p_n$ divides Q. The prime 3 does not divide Q, for if $3 \mid Q$, then $3 \mid (Q-3) = 4p_1 p_2 \cdots p_r$, which is a contradiction. Likewise, none of the primes p_j can divide Q, because $p_j \mid Q$ implies $p_j \mid (Q-4p_1 p_2 \cdots p_r) = 3$ which is absurd. Hence, there are infinitely many primes of the form $4n + 3$. ■

We conclude this section by proving some results about irrational numbers. If α is a rational number then we may write α as the quotient of two integers in infinitely many ways, for if $\alpha = a/b$, where a and b are integers with $b \neq 0$, then $\alpha = ka/kb$ whenever k is a nonzero integer. It is easy to see that a positive rational number may be written uniquely as the quotient of two relatively prime positive integers; when this is done we say that the rational number is in *lowest terms*. We note that the rational number $11/21$ is in lowest terms. We also see that

$$\cdots = -33/-63 = -22/-42 = -11/-21 = 11/21 = 22/42 = 33/63 = \cdots .$$

The next two results show that certain numbers are irrational. We start by giving another proof that $\sqrt{2}$ is irrational (we proved this originally in Section 1.1).

Example 2.14. Suppose that $\sqrt{2}$ is rational. Then $\sqrt{2} = a/b$, where a and b are relatively prime integers with $b \neq 0$. It follows that $2 = a^2/b^2$, so that $2b^2 = a^2$. Since $2 \mid a^2$, it follows from the Fundamental Theorem of Arithmetic (see Exercise 32 at the end of this section) that $2 \mid a$. Let $a = 2c$, so that $b^2 = 2c^2$. Hence, $2 \mid b^2$, and by Exercise 32, 2 also divides b. However, since $(a,b) = 1$, we know that 2 cannot divide both a and b. This contradiction shows that $\sqrt{2}$ is irrational. □

We can also use the following more general result to show that $\sqrt{2}$ is irrational.

Theorem 2.11. Let α be a root of the polynomial $x^n + c_{n-1}x^{n-1} + \cdots + c_1 x + c_0$ where the coefficients $c_0, c_1, ..., c_{n-1}$ are integers. Then α is either an integer or an irrational number.

Proof. Suppose that α is rational. Then we can write $\alpha = a/b$ where a and b are relatively prime integers with $b \neq 0$. Since α is a root of $x^n + c_{n-1}x^{n-1} + \cdots + c_1 x + c_0$, we have

$$(a/b)^n + c_{n-1}(a/b)^{n-1} + \cdots + c_1(a/b) + c_0 = 0.$$

Multiplying by b^n, we find that

$$a^n + c_{n-1}a^{n-1}b + \cdots + c_1 ab^{n-1} + c_0 b^n = 0.$$

Since

$$a^n = b(-c_{n-1}a^{n-1} - \cdots - c_1 ab^{n-2} - c_0 b^{n-1}),$$

we see that $b \mid a^n$. Assume that $b \neq \pm 1$. Then, b has a prime divisor p. Since $p \mid b$ and $b \mid a^n$, we know that $p \mid a^n$. Hence, by Exercise 33 we see that $p \mid a$. However, since $(a, b) = 1$, this is a contradiction which shows that $b = \pm 1$. Consequently, if α is rational then $\alpha = \pm a$, so that α must be an integer. ∎

We illustrate the use of Theorem 2.11 with the following example.

Example 2.15. Let a be a positive integer that is not the mth power of an integer, so that $\sqrt[m]{a}$ is not an integer. Then $\sqrt[m]{a}$ is irrational by Theorem 2.11, since $\sqrt[m]{a}$ is a root of $x^m - a$. Consequently, such numbers as $\sqrt{2}, \sqrt[3]{5}, \sqrt[10]{17}$, *etc* are irrational. □

2.3 Exercises

1. Find the prime factorizations of each of the following integers.

   a) 36   e) 222   i) 5040
   b) 39   f) 256   j) 8000
   c) 100   g) 515   k) 9555
   d) 289   h) 989   l) 9999

2. Find the prime factorization of 111111.

3. Find the prime factorization of 4849845.

4. Show that all the powers in the prime-power factorization of an integer n are even if and only if n is a perfect square.

5. Which positive integers have exactly three positive divisors? Which have exactly four positive divisors?

6. Show that every positive integer can be written as the product of possibly a square and a square-free integer. A *square-free integer* is an integer that is not divisible by any perfect squares other than 1.

7. An integer n is called *powerful* if whenever a prime p divides n, p^2 divides n. Show that every powerful number can be written as the product of a perfect square and a perfect cube.

8. Show that if a and b are positive integers and $a^3 \mid b^2$, then $a \mid b$.

Let p be a prime and n a positive integer. If $p^a \mid n$, but $p^{a+1} \nmid n$, we say that p^a *exactly divides* n, and we write $p^a \parallel n$.

9. Show that if $p^a \parallel m$ and $p^b \parallel n$, then $p^{a+b} \parallel mn$.

10. Show that if $p^a \parallel m$, then $p^{ka} \parallel m^k$.

11. Show that if $p^a \parallel m$ and $p^b \parallel n$ with $a \neq b$, then $p^{\min(a,b)} \parallel (m+n)$.

12. Let n be a positive integer. Show that the power of the prime p occurring in the prime power factorization of $n!$ is

$$[n/p] + [n/p^2] + [n/p^3] + \cdots.$$

13. Use Exercise 10 to find the prime-power factorization of 20!.

14. How many zeros are there at the end of 1000! in decimal notation? How many in base eight notation?

15. Find all positive integers n such that $n!$ ends with exactly 74 zeros in decimal notation.

16. Show that if n is a positive integer it is impossible for $n!$ to end with exactly 153, 154, or 155 zeros when it is written in decimal notation.

The next four exercises present an example of a system where unique factorization into primes fails. Let H be the set of all positive integers of the form $4k + 1$, where k is a nonnegative integer.

17. Show that the product of two elements of H is also in H.

18. An element $h \neq 1$ in H is called a *Hilbert prime* (named after the famous German mathematician David Hilbert if the only way it can be written as the product of two integers in H is $h = h \cdot 1 = 1 \cdot h$. Find the 20 smallest Hilbert primes.

19. Show that every element of H can be factored into Hilbert primes.

20. Show that factorization of elements of H into Hilbert primes is not necessarily unique by finding two different factorizations of 693 into Hilbert primes.

21. Which positive integers n are divisible by all integers not exceeding $\sqrt{n}$?

---

* **DAVID HILBERT** (1862-1943), born in Königsburg, the city famous in mathematics for its seven bridges, was the son of a judge. During his tenure at Göttingen University from 1892 to 1930, Hilbert made many fundamental contributions to a wide range of mathematical subjects. He almost always worked on one area of mathematics at a time, making important contributions, then moving to a new mathematical subject. Some areas in which Hilbert worked are the calculus of variations, geometry, algebra, number theory, logic, and mathematical physics. Besides his many outstanding original contributions, Hilbert is remembered for his famous list of 23 difficult problems. He described these problems at the 1900 International Congress of Mathematicians as a challenge to mathematicians at the birth of the twentieth century. Since that time they have spurred a tremendous amount and variety of research. Although many of these problems have now been solved, several remain open. Hilbert was also the author of several important textbooks in number theory and in geometry.

22. Find the least common multiple of each of the following pairs of integers.

a) 8, 12   d) 111, 303
b) 14, 15   e) 256, 5040
c) 28, 35   f) 343, 999

23. Find the least common multiple of each of the following pairs of integers.

a) 7, 11   d) 101, 333
b) 12, 18   e) 1331, 5005
c) 25, 30   f) 5040, 7700

24. Find the greatest common divisor and least common multiple of the following pairs of integers

a) $23^2 5^3 , 2^2 3^3 7^2$
b) $2 \cdot 3 \cdot 5 \cdot 7, 7 \cdot 11 \cdot 13$
c) $2^8 3^6 5^4 11^{13} , 2 \cdot 3 \cdot 5 \cdot 11 \cdot 13$
d) $41^{101} 47^{43} 103^{1001} , 41^{11} 43^{47} 83^{111}$.

25. Find the greatest common divisor and least common multiple of the following pairs of integers

a) $2^2 3^3 5^5 7^7 , 2^7 3^5 5^3 7^2$
b) $2 \cdot 3 \cdot 5 \cdot 7 \cdot 11 \cdot 13, 17 \cdot 19 \cdot 23 \cdot 29$
c) $2^3 5^7 11^{13} , 2 \cdot 3 \cdot 5 \cdot 7 \cdot 11 \cdot 13$
d) $47^{11} 79^{111} 101^{1001} , 41^{11} 83^{111} 101^{1000}$.

26. Show that every common multiple of the positive integers a and b is divisible by the least common multiple of a and b.

27. Periodical cicadas are insects with very long larval periods and brief adult lives. For each species of periodical cicada with larval period of 17 years, there is a similar species with a larval period of 13 years. If both the 17-year and 13-year species emerged in a particular location in 1900, when will they next both emerge in that location?

28. Which pairs of integers a and b have greatest common divisor 18 and least common multiple 540?

29. Show that if a and b are positive integers, then $(a,b) \mid [a,b]$. When does $(a,b) = [a,b]$?

30. Show that if a and b are positive integers, then there are divisors c of a and d of b with $(c,d) = 1$ and $cd = [a,b]$.

☞ **31.** Show that if a, b, and c are integers, then $[a,b] \mid c$ if and only if $a \mid c$ and $b \mid c$.

☞ **32.** Show that if p is a prime and a is an integer with $p \mid a^2$, then $p \mid a$.

☞ **33.** Show that if p is a prime, a is an integer, and n is a positive integer such that $p \mid a^n$, then $p \mid a$.

34. Show that if a, b, and c are integers with $c \mid ab$, then $c \mid (a,c)(b,c)$.

35. a) Show that if a and b are positive integers with $(a,b) = 1$, then $(a^n, b^n) = 1$ for all positive integers n.

    b) Use part (a) to prove that if a and b are integers such that $a^n \mid b^n$ where n is a positive integer, then $a \mid b$.

36. Show that $\sqrt[3]{5}$ is irrational

    a) by an argument similar to that given in Example 2.14.
    b) using Theorem 2.11.

37. Show that $\sqrt{2} + \sqrt{3}$ is irrational.

38. Show that $\log_2 3$ is irrational.

39. Show that $\log_p b$ is irrational, where p is a prime and b is a positive integer that is not a power of p.

* 40. Let n be a positive integer greater than 1. Show that $1 + \dfrac{1}{2} + \dfrac{1}{3} + \cdots + \dfrac{1}{n}$ is not an integer.

41. Show that if a and b are positive integers then $(a,b) = (a+b,[a,b])$.

42. Find the two positive integers with sum 798 and least common multiple 10780. (*Hint*: Use Exercise 4.)

43. Show that if a, b, and c are positive integers, then $([a,b],c) = [(a,c),(b,c)]$ and $[(a,b),c] = ([a,c],[b,c])$.

The *least common multiple* of the integers $a_1, a_2, \ldots, a_n$, that are not all zero, is the smallest positive integer that is divisible by all the integers $a_1, a_2, \ldots, a_n$; it is denoted by $[a_1, a_2, \ldots, a_n]$.

44. Find $[6,10,15]$ and $[7,11,13]$.

45. Show that $[a_1, a_2, \ldots, a_{n-1}, a_n] = [[a_1, a_2, \ldots, a_{n-1}], a_n]$.

46. Let n be a positive integer. How many pairs of positive integers satisfy $[a,b] = n$? (*Hint*: Consider the prime factorization of n.)

47. a) Show that if a, b, and c are positive integers, then

    $$\max(a,b,c) = a + b + c - \min(a,b) - \min(a,c) - \min(b,c) + \min(a,b,c).$$

    b) Use part (a) to show that

    $$[a,b,c] = \frac{abc(a,b,c)}{(a,b)(a,c)(b,c)}.$$

48. Generalize Exercise 47 to find a formula relating $(a_1, a_2, \ldots, a_n)$ and $[a_1, a_2, \ldots, a_n]$ where $a_1, a_2, \ldots, a_n$ are positive integers.

49. Show that if a, b, and c are positive integers then $(a,b,c)[ab,ac,bc] = abc$.

50. Show that if a, b, and c are positive integers then $[a,b,c](ab,ac,bc) = abc$.

51. Show that if a, b, and c are positive integers then $([a,b],[a,c],[b,c]) = [(a,b),(a,c),(b,c)]$.

52. Prove that there are infinitely many primes of the form $6k + 5$, where k is a positive integer.

* 53. Show that if a and b are integers, then the arithmetic progression $a, a+b, a+2b, \ldots$ contains an arbitrary number of consecutive composite terms.

54. Find the prime factorizations of each of the following integers.

a) $10^6 - 1$ d) $2^{24} - 1$
b) $10^8 - 1$ e) $2^{30} - 1$
c) $2^{15} - 1$ f) $2^{36} - 1$

55. A discount store sells a camera at a price less than its usual retail price of $99. If they sell $8137 worth of this camera and the discounted dollar price is an integer, how many cameras did they sell?

56. Show that if a and b are positive integers, then $a^2 \mid b^2$ implies that $a \mid b$.

57. Show that if a, b, and c are positive integers with $(a,b) = 1$ and $ab = c^n$, then there are positive integers d and e such that $a = d^n$ and $b = e^n$.

☞ 58. Show that if $a_1, a_2, \ldots, a_n$ are pairwise relatively prime integers, then $[a_1, a_2, \ldots, a_n] = a_1 a_2 \cdots a_n$.

59. Show that among any set of $n + 1$ positive integers not exceeding $2n$ there is an integer that divides a different integer in the set.

60. Show that $\dfrac{(m+n)!}{m!n!}$ is an integer whenever m and n are positive integers.

* 61. Find all solutions of the equation $m^n = n^m$ where m and n are integers.

The next six exercises establish some estimates for the size of $\pi(x)$, the number of primes less than or equal to x. These results were originally proved in the 19th century by the Russian mathematician Chebyshev.*

62. Let p be a prime and let n be a positive integer. Show that p divides $\dbinom{2n}{n}$

---

* **PAFNUTY LVOVICH CHEBYSHEV (1821-1894)** was born on the estate of his parents in Okatovo, Russia. His father was a retired army officer. In 1832 Chebyshev's family moved to Moscow where he completed his secondary education with study at home. In 1837 Chebyshev entered Moscow University, graduating in 1841. While still an undergraduate he made his first original contribution, a new method for approximating roots of equations. Chebyshev joined the faculty of Petersburg University in 1843, where he remained until 1882. His doctoral thesis, written in 1849, was long used as a number theory textbook at Russian universities. Chebyshev made contributions to many areas of mathematics besides number theory, including probability theory, numerical analysis, real analysis. He worked in theoretical and applied mechanics and had a bent for constructing mechanisms, including linkages and hinges. He was a popular teacher and had a strong influence on the development of Russian mathematics.

exactly

$$([2n/p] - 2[n/p]) + ([2n/p^2] - 2[n/p^2]) + \cdots + ([2n/p'] - 2[n/p'])$$

times, where $t = [\log_p 2n]$. Conclude that if p^r divides $\binom{2n}{n}$ then $p^r \leq 2n$.

63. Use Exercise 62 to show that

$$\binom{2n}{n} \leq (2n)^{\pi(2n)}.$$

64. Show that the product of all primes between n and $2n$ is between $\binom{2n}{n}$ and $n^{\pi(2n) - \pi(n)}$. (*Hint*: Use the fact that every prime between n and $2n$ divides $(2n)!$ but not $(n!)^2$.)

65. Use Exercises 63 and 64 to show that

$$\pi(2n) - \pi(n) < n \log 4/\log n$$

* 66. Use Exercise 65 to show that

$$\pi(2n) = (\pi(2n) - \pi(n)) + (\pi(n) - \pi(n/2)) + (\pi(n/2) - \pi(n/4))$$
$$+ \cdots \leq n \log 64/\log n.$$

* 67. Use Exercises 63 and 66 to show that there are positive constants c_1 and c_2 such that

$$c_1 x/\log x < \pi(x) < c_2 x/\log x$$

for all $x \geq 2$. (Compare this to the strong statement given in the Prime Number Theorem, stated as Theorem 1.19 in Section 1.9.)

2.3 Computer Projects

Programming Projects

Write programs to do the following:

1. Find all positive divisors of a positive integer from its prime factorization.

2. Find the greatest common divisor of two positive integers from their prime factorizations.

3. Find the least common multiple of two positive integers from their prime factorizations.

4. Find the number of zeros at the end of the decimal expansion of $n!$ where n is a positive integer.

5. Find the prime factorization of $n!$ where n is a positive integer.

Computations and Explorations

Using the programs you have written or a computation program, carry out the following computations and explorations:

1. Compare the number of primes of the form $4n + 1$ and the number of primes of the form $4n + 3$ for a range of values of n. Can you make any conjectures above the relationship between these numbers?

2. Find the smallest prime of the form $an + b$ given integers a and b for a range of values of a and b. Can you make any conjectures about such primes?

2.4 The Fermat Numbers and Factorization Methods

From the fundamental theorem of arithmetic, we know that every positive integer can be written uniquely as the product of primes. In this section, we discuss the problem of determining this factorization. The most direct way to find the factorization of the positive integer n is as follows. Recall from Theorem 1.18 that n either is prime, or else has a prime factor not exceeding $\sqrt{n}$. Consequently, when we divide n by the primes $2,3,5,\ldots$ successively not exceeding $\sqrt{n}$, either we find a prime factor p_1 of n or else we conclude that n is prime. If we have located a prime factor p_1 of n, we next look for a prime factor of $n_1 = n/p_1$, beginning our search with the prime p_1, since n_1 has no prime factor less than p_1, and any factor of n_1 is also a factor of n. We continue, if necessary, determining whether any of the primes not exceeding $\sqrt{n_1}$ divide n_1. We continue in this manner, proceeding iteratively, to find the prime factorization of n.

Example 2.16. Let $n = 42833$. We note that n is not divisible by 2, 3 or 5, but that $7 \mid n$. We have

$$42833 = 7 \cdot 6119.$$

Trial divisions show that 6119 is not divisible by any of the primes $7,11,13,17,19,$ or 23. However, we see that

$$6119 = 29 \cdot 211.$$

Since $29 > \sqrt{211}$, we know that 211 is prime. We conclude that the prime factorization of 42833 is $42833 = 7 \cdot 29 \cdot 211$. $\qquad\square$

Unfortunately, this method for finding the prime factorization of an integer is quite inefficient. To factor an integer N, it may be necessary to perform as many as $\pi(\sqrt{N})$ divisions, altogether requiring on the order of $\sqrt{N} \log N$ bit operations, since from the prime number theorem $\pi(\sqrt{N})$ is approximately $\sqrt{N}/\log\sqrt{N} = 2\sqrt{N}/\log N$, and from Theorem 1.16, these divisions take $O(\log^2 N)$ bit operations each. More efficient algorithms for factorization have been developed, requiring fewer bit operations than the direct method of factorization previously described. In general, these algorithms are complicated and rely on ideas that we have not yet discussed.

For information about these algorithms we refer the reader to Bressoud [55], Dixon [105], Guy [108], Knuth [90], Niven, Zuckerman, and Montgomery [30], Pomerance [58] and [83], Riesel [123], Rumely [124], and Wagstaff and Smith [128]. We note that the quickest method yet devised (the so-called "number field sieve") can factor an integer N in approximately

$$\exp(c(\log N)^{1/3}(\log \log N)^{2/3})$$

bit operations, where the constant c is a positive real number and where exp stands for the exponential function.

In Table 2.1, we give the time required to factor integers of various sizes using the most efficient algorithm known in 1992, where the time for each bit operation has been estimated as one nanosecond (one nanosecond is 10^{-9} seconds). Research currently underway will certainly reduce these times by discovering more efficient factoring algorithms as well as increasing the speed of computer operations.

Number of decimal digits	Number of bit operations	Time
50	1.4×10^{10}	14 seconds
75	9.0×10^{12}	3 hours
100	2.3×10^{15}	26 days
200	1.2×10^{23}	3.8×10^{15} years
300	1.5×10^{29}	4.9×10^{21} years
500	1.3×10^{39}	4.2×10^{32} years

Table 2.1. Time Required For Factorization of Large Integers.

Later we will show that it is far easier to decide whether an integer is prime, than it is to factor the integer. This difference is the basis of a cryptographic system discussed in Chapter 7.

We now describe a factorization technique which is interesting, although it is not always efficient. This technique is known as *Fermat factorization*, since it was discovered by the French mathematician Pierre de Fermat* in the 17th century. This factorization method is based on the following lemma.

---

* **PIERRE DE FERMAT** (1601-1665), was a lawyer by profession. He was a noted jurist at the provincial parliament in the French city of Toulouse. Fermat was probably the most famous amateur mathematician in history. He published almost none of his mathematical discoveries, but did correspond with contemporary mathematicians about these discoveries. From his correspondents, especially the French monk Mersenne (discussed in Chapter 6), the world learned about his many contributions to mathematics. Fermat was one of the inventors of analytic geometry. Furthermore, he laid the foundations of calculus. Fermat, along with Pascal, gave a mathematical basis to the concept of probability. Some of Fermat's discoveries come to us only because he made notes in the margins of his copy of the work of Diophantus. His son found his copy with these notes and published this so that other mathematicians would be aware of Fermat's results and claims.

Lemma 2.7. If n is an odd positive integer, then there is a one-to-one correspondence between factorizations of n into two positive integers and differences of two squares that equal n.

Proof. Let n be an odd positive integer and let $n = ab$ be a factorization of n into two positive integers. Then n can be written as the difference of two squares, since

$$n = ab = s^2 - t^2 ,$$

where $s = (a+b)/2$ and $t = (a-b)/2$ are both integers since a and b are both odd.

Conversely, if n is the difference of two squares, say $n = s^2 - t^2$, then we can factor n by noting that $n = (s-t)(s+t)$.

We leave it to the reader to show this is a one-to-one correspondence. ∎

To carry out the method of Fermat factorization, we look for solutions of the equation $n = x^2 - y^2$ by searching for perfect squares of the form $x^2 - n$. Hence, to find factorizations of n, we search for a square among the sequence of integers

$$t^2 - n, \ (t+1)^2 - n, \ (t+2)^2 - n,...$$

where t is the smallest integer greater than $\sqrt{n}$. This procedure is guaranteed to terminate, since the trivial factorization $n = n \cdot 1$ leads to the equation

$$n = \left[\frac{n+1}{2} \right]^2 - \left[\frac{n-1}{2} \right]^2 .$$

Example 2.17. We factor 6077 using the method of Fermat factorization. Since $77 < \sqrt{6077} < 78$, we look for a perfect square in the sequence

$$78^2 - 6077 = 7$$
$$79^2 - 6077 = 164$$
$$80^2 - 6077 = 323$$
$$81^2 - 6077 = 484 = 22^2.$$

Since $6077 = 81^2 - 22^2$, we see that $6077 = (81-22)(81+22) = 59 \cdot 103$. □

Unfortunately, Fermat factorization can be very inefficient. To factor n using this technique, it may be necessary to check as many as $(n+1)/2 - \sqrt{n}$ integers to determine whether they are perfect squares. Fermat factorization works best when it is used to factor integers having two factors of similar size. Although Fermat factorization is rarely used to factor large integers, it is the basis for many more powerful factorization algorithms used extensively in computer calculations.

Fermat factorization is one of several elementary factorization techniques that we will discuss in this text. Among the other techniques of this type that we will describe are two methods invented by J.M. Pollard, the Pollard rho method, discussed in Section 3.5, the Pollard $p-1$ method, discussed in Section 5.1, and factorization using continued fractions, discussed in Section 10.5.

The Fermat Numbers

The integers $F_n = 2^{2^n} + 1$ are called the *Fermat numbers*. Fermat conjectured that these integers are all primes. Indeed, the first few are primes, namely $F_0 = 3$, $F_1 = 5$, $F_2 = 17$, $F_3 = 257$, and $F_4 = 65537$. Unfortunately, $F_5 = 2^{2^5} + 1$ is composite as we will now demonstrate.

Example 2.18. The Fermat number $F_5 = 2^{2^5} + 1$ is divisible by 641. We can show that $641 \mid F_5$ without actually performing the division. Note that

$$641 = 5 \cdot 2^7 + 1 = 2^4 + 5^4.$$

Hence,

$$
\begin{aligned}
2^{2^5} + 1 &= 2^{32} + 1 = 2^4 \cdot 2^{28} + 1 = (641 - 5^4)2^{28} + 1 \\
&= 641 \cdot 2^{28} - (5 \cdot 2^7)^4 + 1 = 641 \cdot 2^{28} - (641 - 1)^4 + 1 \\
&= 641(2^{28} - 641^3 + 4 \cdot 641^2 - 6 \cdot 641 + 4).
\end{aligned}
$$

Therefore, we see that $641 \mid F_5$.   □

The following result is a valuable aid in the factorization of Fermat numbers.

Theorem 2.12. Every prime divisor of the Fermat number $F_n = 2^{2^n} + 1$ is of the form $2^{n+2}k + 1$.

The proof of Theorem 2.12 is left until later in the book. It is presented as an exercise in Chapter 9. Here, we indicate how Theorem 2.12 is useful in determining the factorization of Fermat numbers.

Example 2.19. From Theorem 2.11, we know that every prime divisor of $F_3 = 2^{2^3} + 1 = 257$ must be of the form $2^5 k + 1 = 32 \cdot k + 1$. Since there are no primes of this form less than or equal to $\sqrt{257}$, we can conclude that $F_3 = 257$ is prime.   □

Example 2.20. When factoring $F_6 = 2^{2^6} + 1$, we use Theorem 2.11 to see that all its prime factors are of the form $2^8 k + 1 = 256 \cdot k + 1$. Hence, we need only perform trial divisions of F_6 by primes of the form $256 \cdot k + 1$ that do not exceed $\sqrt{F_6}$. After considerable computation, we find that a prime divisor is obtained with $k = 1071$, *i.e.* $274177 = (256 \cdot 1071 + 1) \mid F_6$.   □

A great deal of effort has been devoted to the factorization of Fermat numbers. As yet, no new Fermat primes have been found, and many people believe that no additional Fermat primes exist. We will develop a primality test for Fermat numbers in Chapter 9 which has been used to show that many Fermat numbers are composite.

At the present time (early 1992), the factorizations of only six composite Fermat numbers, F_5, F_6, F_7, F_8, F_9, and F_{11}, are known. The Fermat number F_9, a number with 155 decimal digits, was factored in 1990 by Mark Manasse and Arjen Lenstra, using an algorithm known as the number field sieve. This algorithm breaks the problem of factoring an integer into a large number of smaller factoring problems which can be done in parallel. Even though Manasse and Lenstra farmed out computations for the factorization of F_9 to hundreds of mathematicians and computer scientists, it still took about two months to complete the required computations. (For more details about the factorization of F_9 see Cipra [102]).) The prime factorization of the Fermat number F_{11} was discovered by Brent in 1989 using a factoring algorithm known as the elliptic curve method (described in detail in [54]). There are 617 decimal digits in F_{11} and $F_{11} = 319489 \cdot 974849 \cdot p_{21} \cdot p_{22} p_{564}$ where p_{21}, p_{22}, and p_{564} are primes with 21, 22, and 564 digits, respectively. Note that at present time the factorization of F_{10} is not known.

Although the complete factorization of F_n is known only when $n \leq 9$ and $n = 11$, at least one prime factor of F_n is known when $n = 10, 12, 13, 15, 16, 17, 18, 19, 21, 23, 25, 26, 27, 30$. These results were established using results such as Theorem 2.12. It is known that F_n is composite for F_{14} and F_{20}, but no factors of these numbers have yet been found. Moreover, it is currently known that F_n is composite for many integers n, with the largest such integer being 23473. F_{22} is the smallest Fermat number which has not been shown to be composite, if it is indeed composite. Because of advances in computer performance, we can expect new results on the nature of Fermat numbers and their factorization in the next few years.

The factorization of Fermat numbers is part of the *Cunnigham Project*, sponsored by the American Mathematical Society. This project, devoted to building a table of all the known factors of integers of the form $b^n \pm 1$, is called the Cunnigham Project because A.J. Cunnigham, a colonel in the British army, compiled a table of factors of integers of this sort in the early years of the 20th century. (See [56] for the latest set of tables of the Cunnigham Project.) Numbers of the form $b^n \pm 1$ are of special interest because of their importance in generating pseudorandom numbers, their importance in abstract algebra, and their significance in number theory.

In conjunction with the Cunnigham project, a list of the "ten most wanted" integers to be factored is kept by Samuel Wagstaff of Purdue University. For example, until it was factored in 1990, F_9 was on this most wanted list. With advances in factoring techniques and in computer power, the numbers on the list at any given time are growing in size. In the early 1980s they had between 50 and 70 decimal digits. But now, they contain more than 100 decimal digits.

It is possible to prove that there are infinitely many primes using Fermat numbers. We begin by showing that any two distinct Fermat numbers are relatively prime. The following lemma will be used.

Lemma 2.8. Let $F_k = 2^{2^k} + 1$ denote the kth Fermat number, where k is a nonnegative integer. Then for all positive integers n, we have

$$F_0 F_1 F_2 \cdots F_{n-1} = F_n - 2.$$

Proof. We will prove the lemma using mathematical induction. For $n = 1$, the identity reads

$$F_0 = F_1 - 2.$$

This is obviously true since $F_0 = 3$ and $F_1 = 5$. Now let us assume that the identity holds for the positive integer n, so that

$$F_0 F_1 F_2 \cdots F_{n-1} = F_n - 2.$$

With this assumption we can easily show that the identity holds for the integer $n + 1$, since

$$\begin{aligned}
F_0 F_1 F_2 \cdots F_{n-1} F_n &= (F_0 F_1 F_2 \cdots F_{n-1}) F_n \\
&= (F_n - 2) F_n = (2^{2^n} - 1)(2^{2^n} + 1) \\
&= (2^{2^n})^2 - 1 = 2^{2^{n+1}} - 1 = F_{n+1} - 2.
\end{aligned}$$
$\square$

This leads to the following theorem.

Theorem 2.12. Let m and n be distinct nonnegative integers. Then the Fermat numbers F_m and F_n are relatively prime.

Proof. Let us assume that $m < n$. By Lemma 2.8 we know that

$$F_0 F_1 F_2 \cdots F_m \cdots F_{n-1} = F_n - 2.$$

Assume that d is a common divisor of F_m and F_n. Then, Thereom 1.8 tells us that

$$d \mid (F_n - F_0 F_1 F_2 \cdots F_m \cdots F_{n-1}) = 2.$$

Hence, either $d = 1$ or $d = 2$. However, since F_m and F_n are odd, d cannot be 2. Consequently, $d = 1$ and $(F_m, F_n) = 1$.
$\square$

Using Fermat numbers we can give another proof that there are infinitely many primes. First, we note that from Lemma 1.1 from Section 1.9, every Fermat number F_n has a prime divisor p_n. Since $(F_m, F_n) = 1$, we know that $p_m \neq p_n$ whenever $m \neq n$. Hence, we can conclude that there are infinitely many primes.

The Fermat primes are important in geometry. The proof of the following famous theorem of Gauss may be found in Ore [32]. (For a biography of Gauss see Section 3.1.)

Theorem 2.13. A regular polygon of n sides can be constructed using a ruler and compass if and only if n is of the form $n = 2^a p_1 \cdots p_t$ where p_i, $i = 1,2,...,t$ are distinct Fermat primes and a is a nonnegative integer.

2.4 Exercises

1. Find the prime factorization of each of the following positive integers.

   a) 33776925   b) 210733237   c) 1359170111.

2. Find the prime factorization of each of the following positive integers.

   a) 33108075   b) 7300977607   c) 4165073376607.

3. Using the Fermat's factorization method factor each of the following positive integers.

   a)   143   c)   43
   b)   2279   d)   11413

4. Using the Fermat's factorization method factor each of the following positive integers.

   a)   8051   d)   11021
   b)   73   e)   3200399
   c)   10897   f)   24681023

5. Show that the last two decimal digits of a perfect square must be one of the following pairs: 00, $e1$, $e4$, 25, $o6$, $e9$, where e stands for any even digit and o stands for any odd digit. (*Hint:* Show that n^2, $(50+n)^2$, and $(50-n)^2$ all have the same final decimal digits, and then consider those integers n with $0 \leq n \leq 25$.)

6. Explain how the result of Exercise 5 can be used to speed up Fermat's factorization method.

7. Show that if the smallest prime factor of n is p, then $x^2 - n$ will not be a perfect square for $x > (n+p^2)/(2p)$ with the single exception $x = (n+1)/2$.

Exercises 8-10 involve the method of *Draim factorization*. To use this technique to search for a factor of the positive integer $n = n_1$, we start by using the division algorithm, to obtain

$$n_1 = 3q_1 + r_1, \quad 0 \leq r_1 < 3.$$

Setting $m_1 = n_1$, we let

$$m_2 = m_1 - 2q_1, \quad n_2 = m_2 + r_1.$$

We use the division algorithm again, to obtain

$$n_2 = 5q_2 + r_2, \quad 0 \le r_2 < 5,$$

and we let

$$m_3 = m_2 - 2q_2, \quad n_3 = m_3 + r_2.$$

We proceed recursively, using the division algorithm, to write

$$n_k = (2k+1)q_k + r_k, \quad 0 \le r_k < 2k+1,$$

and we define

$$m_k = m_{k-1} - 2q_{k-1}, \quad n_k = m_k + r_{k-1}.$$

We stop when we obtain a remainder $r_k = 0$.

8. Show that $n_k = kn_1 - (2k+1)(q_1 + q_2 + \cdots + q_{k-1})$ and $m_k = n_1 - 2 \cdot (q_1 + q_2 + \cdots + q_{k-1})$.

9. Show that if $(2k+1) \mid n$, then $(2k+1) \mid n_k$ and $n = (2k+1)m_{k+1}$.

10. Factor 5899 using the method of Draim factorization.

In Exercises 11-13 we develop a factorization technique known as *Euler's method*. It is applicable when the integer being factored is odd and can be written as the sum of two squares in two different ways. Let n be odd and let $n = a^2 + b^2 = c^2 + d^2$, where a and c are odd positive integers, and b and d are even positive integers.

11. Let $u = (a-c, b-d)$. Show that u is even and that if $r = (a-c)/u$ and $s = (d-b)/u$, then $(r,s) = 1$, $r(a+c) = s(d+b)$, and $s \mid (a+c)$.

12. Let $sv = a+c$. Show that $rv = d + b$, $v = (a+c, d+b)$, and v is even.

13. Conclude that n may be factored as $n = [(u/2)^2 + (v/2)^2](r^2 + s^2)$.

14. Use Euler's method to factor each of the following integers.

   a) $221 = 10^2 + 11^2 = 5^2 + 14^2$
   b) $2501 = 50^2 + 1^2 = 49^2 + 10^2$

      c) $1000009 = 1000^2 + 3^2 = 972^2 + 235^2$

15. Show that any number of the form $2^{4n+2} + 1$ can be easily factored by the use of the identity $4x^4 + 1 = (2x^2 + 2x + 1)(2x^2 - 2x + 1)$. Factor $2^{18} + 1$ using this identity.

16. Show that if a is a positive integer and $a^m + 1$ is a prime, then $m = 2^n$ for some positive integer n. (*Hint*: Recall the identity $a^m + 1 = (a^k + 1)$ $(a^{k(l-1)} - a^{k(l-2)} + \cdots - a^k + 1)$ where $m = kl$ and l is odd).

17. Show that the last digit in the decimal expansion of $F_n = 2^{2^n} + 1$ is 7 if $n \geq 2$. (*Hint*: Using mathematical induction, show that the last decimal digit of 2^{2^n} is 6.)

18. Use the fact that every prime divisor of $F_4 = 2^{2^4} + 1 = 65537$ is of the form $2^6 k + 1 = 64k + 1$ to verify that F_4 is prime. (You should need only one trial division.)

19. Use the fact that every prime divisor of $F_5 = 2^{2^5} + 1$ is of the form $2^7 k + 1 = 128k + 1$ to demonstrate that the prime factorization of F_5 is $F_5 = 641 \cdot 6700417$.

20. Find all primes of the form $2^{2^n} + 5$, where n is a nonnegative integer.

21. Estimate the number of decimal digits in the Fermat number F_n.

* 22. What is the greatest common divisor of n and F_n where n is a positive integer? Prove that your answer is correct.

23. Show that the only integer of the form $2^m + 1$ where m is a positive integer that is a power of a positive integer (*i.e.* of the form n^k where n and k are positive integers with $k \geq 2$) occurs when $m = 3$.

24. Factoring kn by the Fermat factorization method, where k is a small positive integer, is sometimes easier than factoring n by this method. Show that to factor 901 by the Fermat factorization method, it is easier to factor $3 \cdot 901 = 2703$ than to factor 901.

2.4 Computer Projects

Programming Projects

Write programs to do the following:

    1. Given a positive integer n, find the prime factorization of n.

    2. Given a positive integer n, perform the Fermat factorization method on n.

    3. Given a positive integer n, perform Draim factorization (see the preamble to Exercise 11) on n.

    4. Check a Fermat number F_n where n is a positive integer for prime factors, using Theorem 2.11.

Computations and Explorations

Using the programs you have written or a computation program, carry out the following computations and explorations:

1. Using trial division find the prime factorization of several integers exceeding 10000 of your choice.

2. Factor several integers of your choice exceeding 10000 using Fermat factorization.

3. Factor the Fermat numbers F_6 and F_7 using Theorem 2.11.

2.5 Linear Diophantine Equations

Consider the following problem. A man wishes to purchase $510 of travelers checks. The checks are available only in denominations of $20 and $50. How many of each denomination should he buy? If we let x denote the number of $20 checks and y the number of $50 checks that he should buy, then the equation $20x + 50y = 510$ must be satisfied. To solve this problem, we need to find all solutions of this equation, where both x and y are nonnegative integers.

A related problem arises when a woman wishes to mail a package. The postal clerk determines the cost of postage to be 83 cents but only 6-cent and 15-cent stamps are available. Can some combination of these stamps be used to mail the package? To answer this, we first let x denote the number of 6-cent stamps and y the number of 15-cent stamps to be used. Then we must have $6x + 15y = 83$, where both x and y are nonnegative integers.

When we require that solutions of a particular equation come from the set of integers, we have a *diophantine equation*. Diophantine equations get their name from the ancient Greek mathematician Diophantus*, who wrote extensively on such equations. The type of diophantine equation $ax + by = c$, where a, b, and c are integers is called a *linear diophantine equations in two variables*. We now develop the theory for solving such equations. The following theorem tells us when such an equation has solutions, and when there are solutions, explicitly describes them.

---

* **DIOPHANTUS (c. 250 C.E.)** wrote the *Arithmetica* which is the earliest book we know of on algebra. This book contains the first systematic use of mathematical notation to represent unknowns in equations and powers of these unknowns. Almost nothing is known about Diophantus' life, other than that he lived in Alexandria around 250 C.E. The only source of details about his life comes from an epigram found in a collection called the *Greek Anthology:* "Diophantus passed one sixth of his life in childhood, one twelfth in youth, and one seventh as a bachelor. Five years after his marriage was born a son who died four years before his father, at half his father's age." From this the reader can infer that Diophantus lived to the age of 84.

Theorem 2.14. Let a and b be integers with $d = (a,b)$. The equation $ax + by = c$ has no integral solutions if $d \nmid c$. If $d \mid c$, then there are infinitely many integral solutions. Moreover, if $x = x_0$, $y = y_0$ is a particular solution of the equation, then all solutions are given by

$$x = x_0 + (b/d)n, \quad y = y_0 - (a/d)n,$$

where n is an integer.

Proof. Assume that x and y are integers such that $ax + by = c$. Then, since $d \mid a$ and $d \mid b$, by Theorem 1.8 $d \mid c$ as well. Hence, if $d \nmid c$, there are no integral solutions of the equation.

Now assume that $d \mid c$. From Theorem 2.2, there are integers s and t with

(2.3) $$d = as + bt.$$

Since $d \mid c$, there is an integer e with $de = c$. Multiplying both sides of (2.3) by e, we have

$$c = de = (as + bt)e = a(se) + b(te).$$

Hence, one solution of the equation is given by $x = x_0$ and $y = y_0$, where $x_0 = se$ and $y_0 = te$.

To show that there are infinitely many solutions, let $x = x_0 + (b/d)n$ and $y = y_0 - (a/d)n$, where n is an integer. We see that this pair (x,y) is a solution, since

$$ax + by = ax_0 + a(b/d)n + by_0 - b(a/d)n = ax_0 + by_0 = c.$$

We now show that every solution of the equation $ax + by = c$ must be of the form described in the theorem. Suppose that x and y are integers with $ax + by = c$. Since

$$ax_0 + by_0 = c,$$

by subtraction we find that

$$(ax + by) - (ax_0 + by_0) = 0,$$

which implies that

$$a(x - x_0) + b(y - y_0) = 0.$$

Hence,

$$a(x - x_0) = b(y_0 - y).$$

Dividing both sides of this last equality by d, we see that

$$(a/d)(x - x_0) = (b/d)(y_0 - y).$$

By Theorem 2.1 we know that $(a/d, b/d) = 1$. Using Lemma 2.3, it follows that $(a/d) \mid (y_0 - y)$. Hence, there is an integer n with $(a/d)n = y_0 - y$; this means that $y = y_0 - (a/d)n$. Now putting this value of y into the equation $a(x - x_0) = b(y_0 - y)$, we find that $a(x - x_0) = b(a/d)n$, which implies that

$x = x_0 + (b/d)n.$ □

The following examples illustrate the use of Theorem 2.14.

Example 2.21. By Theorem 2.14 there are no integral solutions of the diophantine equation $15x + 6y = 7$ since $(15,6) = 3$ but $3 \nmid 7$ □

Example 2.22. By Theorem 2.14 there are infinitely many solutions of the diophantine equation $21x + 14y = 70$ since $(21,14) = 7$ and $7 \mid 70$. To find these solutions, note that by the Euclidean algorithm, $1 \cdot 21 + (-1) \cdot 14 = 7$, so that $10 \cdot 21 + (-10) \cdot 14 = 70$. Hence $x_0 = 10$, $y_0 = -10$ is a particular solution. All solutions are given by $x = 10 + 2n, y = -10 - 3n$ where n is an integer.   □

We will now use Theorem 2.14 to solve the two problems described at the beginning of the section.

Example 2.23. Consider the problem of forming 83 cents postage using only 6-cent and 15-cent stamps. If x denotes the number of 6-cent stamps and y denotes the number of 15-cent stamps, we have $6x + 15y = 83$. Since $(6,15) = 3$ does not divide 83, by Theorem 2.14 we know that there are no integral solutions. Hence, no combination of 6- and 15-cent stamps gives the correct postage.   □

Example 2.24. Consider the problem of purchasing $510 of travelers checks using only $20 checks and $50 checks. How many of each type of check should be used? Let x be the number of $20 checks and let y be the number of $50 checks. We have the equation $20x + 50y = 510$. Note that the greatest common divisor of 20 and 50 is $(20,50) = 10$. Since $10 \mid 510$, there are infinitely many integral solutions of this linear diophantine equation. Using the Euclidean algorithm, we find that $20(-2) + 50 = 10$. Multiplying both sides by 51, we obtain $20(-102) + 50(51) = 510$. Hence, a particular solution is given by $x_0 = -102$ and $y_0 = 51$. Theorem 2.14 tells us that all integral solutions are of the form $x = -102 + 5n$ and $y = 51 - 2n$. Since we want both x and y to be nonnegative, we must have $-102 + 5n \geq 0$ and $51 - 2n \geq 0$; thus, $n \geq 20 \, 2/5$ and $n \leq 25 \, 1/2$. Since n is an integer, it follows that $n = 21, 22, 23, 24,$ or 25. Hence, we have the following 5 solutions: $(x,y) = (3,9), (8,7), (13,5), (18,3),$ and $(23,1)$. So the teller can give the customer 3 $20 checks and 9 $50 checks, 8 $20 checks and 7 $50 checks, 13 $20 checks and 5 $50 checks, 18 $20 checks and 3 $50 checks, or 23 $20 checks and 1 $50 check.   □

We can extend Theorem 2.14 to cover linear diophantine equations with more than two variables. We have the following theorem.

Theorem 2.15. If $a_1, a_2, \ldots, a_n$ are nonzero positive integers, then the equation $a_1 x_1 + a_2 x_2 + \cdots + a_n x_n = c$ has an integral solution if and only if $d = (a_1, a_2, \ldots, a_n)$ divides c. Furthermore, when there is a solution there are infinitely many different solutions.

Proof. If there are integers $x_1, x_2, \ldots, x_n$ such that $a_1 x_1 + a_2 x_2 + \cdots + a_n x_n = c$, then since d divides a_i for $i = 1, 2, \ldots, n$, by Theorem 1.8, d also divides c. Hence, if $d \nmid c$ there are no integral solutions of the equation.

We will use mathematical induction to prove there are infinitely many integral solutions when $d \mid c$. Note that by Theorem 2.14 this is true when $n = 2$.

Now suppose that there are infinitely many solutions for all equations in n variables satisfying the hypotheses. By Exercise 31 of Section 3.1, the set of linear combinations $a_{n-1} x_{n-1} + a_{n+1} x_{n+1}$ is the same as the set of multiples of (a_n, a_{n+1}). Hence, for every integer y there are infinitely many solutions of the linear diophantine equation $a_n x_n + a_{n+1} x_{n+1} = (a_n, a_{n+1}) y$. It follows that the original equation in $n+1$ variables can be reduced to a linear diophantine equation in n variables:

$$a_1 x_1 + a_2 x_2 + \cdots + a_{n-1} x_{n-1} + (a_n, a_{n+1}) y = c.$$

Note that c is divisible by $(a_1, a_2, \ldots, a_{n-1}, (a_n, a_{n+1}))$ since by Lemma 2.1, this greatest common divisor equals $(a_1, a_2, \ldots, a_n, a_{n+1})$. By the inductive hypothesis this equation has infinitely many integer solutions, since it is a linear diophantine equation in n variables where the greatest common divisor of the coefficients divides the constant c. It follows that there are infinitely many solutions to the original equation. □

The solutions of linear diophantine equations in more than two variables can be found using the reduction in the proof of Theorem 2.15. We leave an application of Theorem 2.15 to the exercises.

2.5 Exercises

1. For each of the following linear diophantine equations, either find all solutions, or show that there are no integral solutions.

   a) $2x + 5y = 11$
   b) $17x + 13y = 100$
   c) $21x + 14y = 147$
   d) $60x + 18y = 97$
   e) $1402x + 1969y = 1$

2. For each of the following linear diophantine equations, either find all solutions, or show that there are no integral solutions.

   a) $3x + 4y = 7$
   b) $12x + 18y = 50$
   c) $30x + 47y = -11$
   d) $25x + 95y = 970$
   e) $102x + 1001y = 1$

3. A Japanese businessman returning home from a trip to North America exchanges his U.S. dollars and his Canadian dollars into yen. If he receives 15,286 yen and has received 122 yen for each U.S. dollar and 112 yen for each Canadian dollar, how many of each type of currency did he exchange?

4. A student returning from Europe changes his French francs and Swiss francs into U.S. money. If he receives $17.06 and has received 19¢ for each French franc and 59¢ for each Swiss franc, how much of each type of currency did he exchange?

5. A grocer orders apples and oranges at a total cost of $8.39. If apples cost him 25¢ each and oranges cost him 18¢ each, how many of each type of fruit did he order?

6. A shopper spends a total of $5.49 for oranges, which cost 18¢ each, and grapefruits, which cost 33¢ each. What is the minimum number of pieces of fruit the shopper could have bought?

7. A postal clerk has only 14-cent and 21-cent stamps to sell. What combinations of these may be used to mail a package requiring postage of exactly each of the following amounts?

    a) $3.50   b) $4.00   c) $7.77

8. At a clambake, the total cost of a lobster dinner is $11 and of a chicken dinner is $8. What can you conclude if the total bill is each of the following amounts?

    a) $777   b) $96   c) $69

* 9. Find all integer solutions of each of the following linear diophantine equations.

    a) $2x + 3y + 4z = 5$
    b) $7x + 21y + 35z = 8$
    c) $101x + 102y + 103z = 1$

* 10. Find all integer solutions of each of the following linear diophantine equations.

    a) $2x_1 + 5x_2 + 4x_3 + 3x_4 = 5$
    b) $12x_1 + 21x_2 + 9x_3 + 15x_4 = 9$
    c) $15x_1 + 6x_2 + 10x_3 + 21x_4 + 35x_5 = 1$

11. Which combinations of pennies, dimes, and quarters have a total value 99¢?

12. How many ways can change be made for one dollar using each of the following coins?

    a)   dimes and quarters
    b)   nickels, dimes, and quarters
    c)   pennies, nickels, dimes, and quarters

In Exercises 13-16 we consider simultaneous linear diophantine equations. To solve these, first eliminate all but two variables and then solve the resulting equation in two variables.

13. Find all integer solutions of the following systems of linear diophantine equations.

a) $x + y + z = 100$
$x + 8y + 50z = 156$

b) $x + y + z = 100$
$x + 6y + 21z = 121$

c) $x + y + z + w = 100$
$x + 2y + 3z + 4w = 300$
$x + 4y + 9z + 16w = 1000$

14. A piggy bank contains 24 coins, all nickels, dimes, and quarters. If the total value of the coins is two dollars, what combinations of coins are possible?

15. Nadir Airways offers three types of tickets on their Boston to New York flights. First-class tickets are $140, second-class tickets are $110, and stand-by tickets are $78. If 69 passengers pay a total of 6548 for their tickets on a particular flight, how many of each type of tickets were sold?

16. Is it possible to have 50 coins, all pennies, dimes, and quarters worth $3?

Let a and b be relatively prime positive integers and let n be a positive integer. We call a solution (x,y) of the linear diophantine equation $ax + by = n$ *nonnegative* when both x and y are nonnegative.

* 17. Show that whenever $n \geq (a-1)(b-1)$ there is a nonnegative solution of $ax + by = n$.

* 18. Show that if $n = ab - a - b$, then there are no nonnegative solutions of $ax + by = n$.

* 19. Show that there are exactly $(a-1)(b-1)/2$ nonnegative integers $n < ab - a - b$ such that the equation has a nonnegative solution.

20. The post office in a small Maine town is left with stamps of only two values. They discover that there are exactly 33 postage amounts that cannot be made up using these stamps, including 46¢. What are the values of the remaining stamps?

* 21. An ancient Chinese puzzle found in the sixth century work of the mathematician Chang Ch'iu-chien, called the "hundred fowls" problem, asks: If a cock is worth five coins, a hen three coins, and three chickens together are worth 1 coin, how many cocks, hens, and chickens, totaling 100, can be bought for 100 coins? Solve this problem.

* **22.**   Find all solutions where x and y are integers to the diophantine equation

$$\frac{1}{x} + \frac{1}{y} = \frac{1}{14}.$$

2.5 Computer Projects

Programming Projects

Write programs to do the following:

1. Find the solutions of a linear diophantine equation in two variables.

2. Find the positive solutions of a linear diophantine equation in two variables.

3. Find the solutions of a linear diophantine equation in three variables.

* **4.**   Find all positive integers n for which the linear diophantine equation $ax + by = n$ has no positive solutions (see the preamble to Exercise 17).

Computations and Explorations

Using the programs you have written or a computation program, carry out the following computations and explorations:

1. Determine which positive integers are of the form $ax + by$ where x and y are nonnegative integers and a and b are relatively prime positive integers of your choice. Use your evidence to confirm the results of Exercises 17-19.

3

Congruences

3.1 Introduction to Congruences

The special language of congruences that we introduce in this chapter is extremely useful in number theory. This language of congruences was developed at the beginning of the nineteenth century by Karl Friedrich Gauss*, one of the most famous mathematicians in history.

---

* **KARL FRIEDRICH GAUSS** (1777-1855) was the son of a bricklayer. It was quickly apparent that he was a child prodigy. In fact, at the age of three he corrected an error in his father's payroll. In his first arithmetic class the teacher gave an assignment designed to keep the class busy, namely to find the sum of the first 100 positive integers. Gauss, who was eight at the time, realized that this sum is $50 \cdot 101 = 5050$ because the terms can be grouped as 1+100=101, 2+99=101, ... ,49+52=101, and 50+51=101. In 1796 Gauss made an important discovery in an area of geometry that had not progressed since ancient times. In particular, he showed that a regular heptadecagon (17-sided polygon) could be drawn using just a ruler and a compass. In 1799 he presented the first rigorous proof of the Fundamental Theorem of Algebra, which states that a polynomial of degree n with real coefficients has exactly n roots. Gauss made fundamental contributions to astronomy including calculating the orbit of the asteroid Ceres. On the basis of this calculation, Gauss was appointed director of the Göttingen Observatory. He laid the foundations of modern number theory with his book *Disquistiones Arithmeticae* in 1801. Gauss was called Princeps Mathematicorum (the Prince of Mathematicians) by his contemporary mathematicians. Although Gauss is noted for his many discoveries in geometry, algebra, analysis, astronomy, and mathematical physics, he had a special interest in number theory. This can be seen from his statement: "Mathematics is the queen of sciences, and the theory of numbers is the queen of mathematics." Gauss made most of his important discoveries early in his life, and spent his later years refining them. Gauss made several fundamental discoveries that he did not reveal. Mathematicians making the same discoveries were often surprised to find that Gauss had described the same results years earlier in his unpublished notes.

Definition. Let m be a positive integer. If a and b are integers, we say that a is *congruent to b modulo m* if $m \mid (a-b)$.

If a is congruent to b modulo m, we write $a \equiv b \pmod{m}$. If $m \nmid (a-b)$, we write $a \not\equiv b \pmod{m}$, and say that a and b are *incongruent modulo m*.

Example 3.1. We have $22 \equiv 4 \pmod 9$, since $9 \mid (22-4) = 18$. Likewise $3 \equiv -6 \pmod 9$ and $200 \equiv 2 \pmod 9$. On the other hand, $13 \not\equiv 5 \pmod 9$ since $9 \nmid (13-5) = 8$. □

Congruences often arise in everyday life. For instance, clocks work either modulo 12 or 24 for hours, and modulo 60 for minutes and seconds, calendars work modulo 7 for days of the week and modulo 12 for months. Utility meters often operate modulo 1000, and odometers usually work modulo 100000.

In working with congruences, it is often useful to translate them into equalities. To do this, the following theorem is needed.

Theorem 3.1. If a and b are integers, then $a \equiv b \pmod{m}$ if and only if there is an integer k such that $a = b + km$.

Proof. If $a \equiv b \pmod{m}$, then $m \mid (a-b)$. This means that there is an integer k with $km = a - b$, so that $a = b + km$.

Conversely, if there is an integer k with $a = b + km$, then $km = a - b$. Hence $m \mid (a-b)$, and consequently, $a \equiv b \pmod{m}$. ∎

Example 3.2. We have $19 \equiv -2 \pmod 7$ and $19 = -2 + 3 \cdot 7$. □

The following proposition establishes some important properties of congruences.

Theorem 3.2. Let m be a positive integer. Congruences modulo m satisfy the following properties:

    (i)   *Reflexive property.* If a is an integer, then $a \equiv a \pmod{m}$.

    (ii) *Symmetric property.* If a and b are integers such that $a \equiv b \pmod{m}$, then $b \equiv a \pmod{m}$.

    (iii) *Transitive property.* If a, b, and c are integers with $a \equiv b \pmod{m}$ and $b \equiv c \pmod{m}$, then $a \equiv c \pmod{m}$.

Proof.

    (i)   We see that $a \equiv a \pmod{m}$, since $m \mid (a-a) = 0$.

    (ii) If $a \equiv b \pmod{m}$, then $m \mid (a-b)$. Hence, there is an integer k with $km = a - b$. This shows that $(-k)m = b - a$, so that $m \mid (b-a)$. Consequently, $b \equiv a \pmod{m}$.

   (iii) If $a \equiv b \pmod{m}$ and $b \equiv c \pmod{m}$, then $m \mid (a-b)$ and $m \mid (b-c)$. Hence, there are integers k and l with $km = a - b$ and $lm = b - c$. Therefore, $a - c = (a-b) + (b-c) = km + lm = (k+l)m$. It follows that $m \mid (a-c)$ and $a \equiv c \pmod{m}$. ∎

By Theorem 3.2 we see that the set of integers is divided into m different sets called *congruence classes modulo m*, each containing integers that are mutually congruent modulo m.

Example 3.3. The four congruence classes modulo 4 are given by

$$\cdots \equiv -8 \equiv -4 \equiv 0 \equiv 4 \equiv 8 \equiv \cdots \pmod{4}$$
$$\cdots \equiv -7 \equiv -3 \equiv 1 \equiv 5 \equiv 9 \equiv \cdots \pmod{4}$$
$$\cdots \equiv -6 \equiv -2 \equiv 2 \equiv 6 \equiv 10 \equiv \cdots \pmod{4}$$
$$\cdots \equiv -5 \equiv -1 \equiv 3 \equiv 7 \equiv 11 \equiv \cdots \pmod{4}.$$

□

Suppose that m is a positive integer. Given an integer a, by the division algorithm we have $a = bm + r$ where $0 \le r \le m - 1$. We call r the *least nonnegative residue* of a modulo m. We say that r is the result of *reducing a modulo m*. We also use the notation $a \mathbf{\ mod\ } m = r$ to denote that r is the remainder obtained when a is divided by m. For example, $17 \mathbf{\ mod\ } 5 = 2$ and $-8 \mathbf{\ mod\ } 7 = 6$.

Now note that from the equation $a = bm + r$, it follows that $a \equiv r \pmod{m}$. Hence, every integer is congruent modulo m to one of the integers of the set $0, 1, ..., m - 1$, namely the remainder when it is divided by m. Since no two of the integers $0, 1, ..., m - 1$ are congruent modulo m, we have m integers such that every integer is congruent to exactly one of these m integers.

Definition. A *complete system of residues modulo m* is a set of integers such that every integer is congruent modulo m to exactly one integer of the set.

Example 3.4. The division algorithm shows that the set of integers $0, 1, 2, ..., m - 1$ is a complete system of residues modulo m. This is called the set of *least nonnegative residues modulo m*. □

Example 3.5. Let m be an odd positive integer. Then the set of integers

$$-\frac{m-1}{2}, \ -\frac{m-3}{2}, ..., -1, \ 0, \ 1, ..., \ \frac{m-3}{2}, \ \frac{m-1}{2},$$

the set of *absolute least residues modulo m*, is a complete system of residues.   □

We will often do arithmetic with congruences. Congruences have many of the same properties that equalities do. First, we show that an addition, subtraction, or multiplication to both sides of a congruence preserves the congruence.

Theorem 3.3. If a, b, c, and m are integers with $m > 0$ such that $a \equiv b \pmod{m}$, then

    (i) $a + c \equiv b + c \pmod{m}$,

    (ii) $a - c \equiv b - c \pmod{m}$,

    (iii) $ac \equiv bc \pmod{m}$.

Proof. Since $a \equiv b \pmod{m}$, we know that $m \mid (a-b)$. From the identity $(a+c) - (b+c) = a - b$, we see $m \mid ((a+c) - (b+c))$, so that (i) follows. Likewise, (ii) follows from the fact that $(a-c) - (b-c) = a - b$. To show that (iii) holds, note that $ac - bc = c(a-b)$. Since $m \mid (a-b)$, it follows that $m \mid c(a-b)$, and hence, $ac \equiv bc \pmod{m}$. ∎

Example 3.6. Since $19 \equiv 3 \pmod{8}$, it follows from Theorem 3.3 that $26 = 19 + 7 \equiv 3 + 7 = 10 \pmod{8}$, $15 = 19 - 4 \equiv 3 - 4 \equiv -1 \pmod{8}$, and $38 = 19 \cdot 2 \equiv 3 \cdot 2 = 6 \pmod{8}$. ∎

What happens when both sides of a congruence are divided by an integer? Consider the following example.

Example 3.7. We have $14 = 7 \cdot 2 \equiv 4 \cdot 2 = 8 \pmod{6}$. But we cannot cancel the common factor of 2 since $7 \not\equiv 4 \pmod{6}$.   □

This example shows that it is not necessarily true that we preserve a congruence when we divide both sides by an integer. However, the following theorem gives a valid congruence when both sides of a congruence are divided by the same integer.

Theorem 3.4. If a, b, c and m are integers such that $m > 0$, $d = (c,m)$, and $ac \equiv bc \pmod{m}$, then $a \equiv b \pmod{m/d}$.

Proof. If $ac \equiv bc \pmod{m}$, we know that $m \mid (ac - bc) = c(a-b)$. Hence, there is an integer k with $c(a-b) = km$. By dividing both sides by d, we have $(c/d)(a-b) = k(m/d)$. Since $(m/d,c/d) = 1$, by Lemma 2.3 it follows that $m/d \mid (a-b)$. Hence, $a \equiv b \pmod{m/d}$. ∎

Example 3.8. Since $50 \equiv 20 \pmod{15}$ and $(10,15) = 5$, we see that $50/10 \equiv 20/10 \pmod{15/5}$, or $5 \equiv 2 \pmod{3}$.   □

The following corollary, which is a special case of Theorem 3.4, is used often.

Corollary 3.4.1. If a, b, c, and m are integers such that $m > 0$, $(c,m) = 1$, and

$ac \equiv bc \pmod{m}$, then $a \equiv b \pmod{m}$.

Example 3.9. Since $42 \equiv 7 \pmod 5$ and $(5,7) = 1$, we can conclude that $42/7 \equiv 7/7 \pmod 5$, or that $6 \equiv 1 \pmod 5$. □

The following theorem which is more general than Theorem 3.3 is also useful.

Theorem 3.5. If a, b, c, d, and m are integers such that $m > 0$, $a \equiv b \pmod m$, and $c \equiv d \pmod m$, then

(i) $a + c \equiv b + d \pmod m$,

(ii) $a - c \equiv b - d \pmod m$,

(iii) $ac \equiv bd \pmod m$.

Proof. Since $a \equiv b \pmod m$ and $c \equiv d \pmod m$, we know that $m \mid (a-b)$ and $m \mid (c-d)$. Hence, there are integers k and l with $km = a - b$ and $lm = c - d$.

To prove (i), note that $(a+c) - (b+d) = (a-b) + (c-d) = km + lm = (k+l)m$. Hence, $m \mid [(a+c) - (b+d)]$. Therefore, $a + c \equiv b + d \pmod m$.

To prove (ii), note that $(a-c) - (b-d) = (a-b) - (c-d) = km - lm = (k-l)m$. Hence, $m \mid [(a-c)-(b-d)]$, so that $a - c \equiv b - d \pmod m$.

To prove (iii), note that $ac - bd = ac - bc + bc - bd = c(a-b) + b(c-d) = ckm + blm = m(ck+bl)$. Hence, $m \mid (ac - bd)$. Therefore, $ac \equiv bd \pmod m$. ∎

Example 3.10. Since $13 \equiv 3 \pmod 5$ and $7 \equiv 2 \pmod 5$, using Theorem 3.5 we see that $20 = 13 + 7 \equiv 3 + 2 = 5 \pmod 5$, $6 = 13 - 7 \equiv 3 - 2 = 1 \pmod 5$, and $91 = 13 \cdot 7 \equiv 3 \cdot 2 = 6 \pmod 5$. □

Theorem 3.6. If $r_1, r_2, ..., r_m$ is a complete system of residues modulo m, and if a is a positive integer with $(a,m) = 1$, then

$$ar_1 + b, \ ar_2 + b, \ ..., \ ar_m + b$$

is a complete system of residues modulo m for any integer b.

Proof. First, we show that no two of the integers

$$ar_1 + b, \ ar_2 + b, \ ..., \ ar_m + b$$

are congruent modulo m. To see this, note that if

$$ar_j + b \equiv ar_k + b \pmod m,$$

then, by (ii) of Theorem 3.3, we know that

$$ar_j \equiv ar_k \pmod m.$$

Because $(a,m) = 1$, Corollary 3.4.1 shows that

$$r_j \equiv r_k \pmod{m}.$$

Since $r_j \not\equiv r_k \pmod{m}$ if $j \neq k$, we conclude that $j = k$.

Since the set of integers in question consists of m incongruent integers modulo m and there are m congruences class modulo m, these integers form a complete system of residues modulo m.   ∎

The following theorem shows that a congruence is preserved when both sides are raised to the same positive integral power.

Theorem 3.7. If a, b, k, and m are integers such that $k > 0$, $m > 0$, and $a \equiv b \pmod{m}$, then $a^k \equiv b^k \pmod{m}$.

Proof. Because $a \equiv b \pmod{m}$, we have $m \mid (a - b)$. Since

$$a^k - b^k = (a-b)(a^{k-1}+a^{k-2}b+ \cdots +ab^{k-2}+b^{k-1}),$$

we see that $(a - b) \mid (a^k - b^k)$. Therefore, by Theorem 1.7 it follows that $m \mid (a^k - b^k)$. Hence, $a^k \equiv b^k \pmod{m}$.   ∎

Example 3.11. Since $7 \equiv 2 \pmod 5$, Theorem 3.7 tells us that $343 = 7^3 \equiv 2^3 \equiv 8 \pmod 5$.   □

The following result shows how to combine congruences of two numbers to different moduli.

Theorem 3.8. If $a \equiv b \pmod{m_1}, a \equiv b \pmod{m_2},...,a \equiv b \pmod{m_k}$ where $a,b,m_1,m_2,...,m_k$ are integers with $m_1,m_2,...,m_k$ positive, then

$$a \equiv b \pmod{[m_1,m_2,...,m_k]},$$

where $[m_1,m_2,...,m_k]$ is the least common multiple of $m_1,m_2,...,m_k$.

Proof. Since $a \equiv b \pmod{m_1}$, $a \equiv b \pmod{m_2}$,..., $a \equiv b \pmod{m_k}$, we know that $m_1 \mid (a - b)$, $m_2 \mid (a - b)$, ..., $m_k \mid (a-b)$. By Exercise 27 of Section 2.3 we see that

$$[m_1,m_2,...,m_k] \mid (a - b).$$

Consequently,

$$a \equiv b \pmod{[m_1,m_2,...,m_k]}.$$ ∎

An immediate and useful consequence of this theorem is the following result.

Corollary 3.8.1. If $a \equiv b \pmod{m_1}$, $a \equiv b \pmod{m_2}$,..., $a \equiv b \pmod{m_k}$ where a and b are integers and $m_1,m_2,...,m_k$ are pairwise relatively prime positive integers, then

$$a \equiv b \pmod{m_1 m_2 \cdots m_k}.$$

Proof. Since $m_1, m_2, \ldots, m_k$ are pairwise relatively prime, Exercise 52 of Section 2.3 tells us that

$$[m_1, m_2, \ldots, m_k] = m_1 m_2 \cdots m_k.$$

Hence, by Theorem 3.8 we know that

$$a \equiv b \pmod{m_1 m_2 \cdots m_k}. \qquad \blacksquare$$

In our subsequent studies, we will be working with congruences involving large powers of integers. For example, we will want to find the least positive residue of 2^{644} modulo 645. If we attempt to find this least positive residue by first computing 2^{644}, we would have an integer with 194 decimal digits, a most undesirable thought. Instead, to find 2^{644} modulo 645 we first express the exponent 644 in binary notation:

$$(644)_{10} = (1010000100)_2.$$

Next, we compute the least positive residues of $2, 2^2, 2^4, 2^8, \ldots, 2^{512}$ by successively squaring and reducing modulo 645. This gives us the congruences

$$
\begin{array}{rcll}
2 & \equiv & 2 & (\bmod\ 645), \\
2^2 & \equiv & 4 & (\bmod\ 645), \\
2^4 & \equiv & 16 & (\bmod\ 645), \\
2^8 & \equiv & 256 & (\bmod\ 645), \\
2^{16} & \equiv & 391 & (\bmod\ 645), \\
2^{32} & \equiv & 16 & (\bmod\ 645), \\
2^{64} & \equiv & 256 & (\bmod\ 645), \\
2^{128} & \equiv & 391 & (\bmod\ 645), \\
2^{256} & \equiv & 16 & (\bmod\ 645), \\
2^{512} & \equiv & 256 & (\bmod\ 645).
\end{array}
$$

We can now compute 2^{644} modulo 645 by multiplying the least positive residues of the appropriate powers of 2. This gives

$$2^{644} = 2^{512+128+4} = 2^{512} 2^{128} 2^4 \equiv 256 \cdot 391 \cdot 16 = 1601536 \equiv 1 \pmod{645}.$$

We have just illustrated a general procedure for *modular exponentiation*, that is, for computing b^N modulo m where b, m, and N are positive integers. We first express the exponent N in binary notation, as $N = (a_k a_{k-1} \ldots a_1 a_0)_2$. We then find the least positive residues of $b, b^2, b^4, \ldots, b^{2^k}$ modulo m, by successively squaring and reducing modulo m. Finally, we multiply the least positive residues modulo m of b^{2^j} for those j with $a_j = 1$, reducing modulo m after each multiplication.

In our subsequent discussions, we will need an estimate for the number of bit operations needed for modular exponentiation. This is provided by the following proposition.

Theorem 3.9. Let b, m, and N be positive integers with $b < m$. Then the least positive residue of b^N modulo m can be computed using $O((\log_2 m)^2 \log_2 N)$ bit operations.

Proof. To find the least positive residue of b^N modulo m, we can use the algorithm just described. First, we find the least positive residues of $b, b^2, b^4, ..., b^{2^k}$ modulo m, where $2^k \le N < 2^{k+1}$, by successively squaring and reducing modulo m. This requires a total of $O((\log_2 m)^2 \log_2 N)$ bit operations, because we perform $[\log_2 N]$ squarings modulo m, each requiring $O((\log_2 m)^2)$ bit operations. Next, we multiply together the least positive residues of the integers b^{2^j} corresponding to the binary digits of N which are equal to one, and we reduce modulo m after each multiplication. This also requires $O((\log_2 m)^2 \log_2 N)$ bit operations, because there are at most $\log_2 N$ multiplications, each requiring $O((\log_2 m)^2)$ bit operations. Therefore, a total of $O((\log_2 m)^2 \log_2 N)$ bit operations is needed. ∎

3.1 Exercises

1. Show that each of the following congruences holds.

   a) $13 \equiv 1 \pmod 2$ e) $-2 \equiv 1 \pmod 3$

   b) $22 \equiv 7 \pmod 5$ f) $-3 \equiv 30 \pmod{11}$

   c) $91 \equiv 0 \pmod{13}$ g) $111 \equiv -9 \pmod{40}$

   d) $69 \equiv 62 \pmod 7$ h) $666 \equiv 0 \pmod{37}$

2. Determine whether each of the following pairs of integers are congruent modulo 7.

   a) 1,15   d) −1,8

   b) 0,42   e) −9,5

   c) 2,99   f) −1,699

3. For which positive integers m are each of the following statements true?

   a) $27 \equiv 5 \pmod m$

   b) $1000 \equiv 1 \pmod m$

   c) $1331 \equiv 0 \pmod m$

4. Show that if a is an even integer, then $a^2 \equiv 0 \pmod 4$, and if a is an odd integer, then $a^2 \equiv 1 \pmod 4$.

☞ 5. Show that if a is an odd integer, then $a^2 \equiv 1 \pmod 8$.

6. Find the least nonnegative residue modulo 13 of each of the following integers.

   a) 22
   b) 100
   c) 1001

   d) -1
   e) -100
   f) -1000

7. Find the least positive residue of $1! + 2! + 3! + \cdots + 100!$ modulo each of the following integers.

   a) 2
   b) 7

   c) 12
   d) 25

8. Show that if a, b, m, and n are integers such that $m > 0$, $n > 0$, $n \mid m$, and $a \equiv b \pmod{m}$, then $a \equiv b \pmod{n}$.

9. Show that if a, b, c, and m are integers such that $c > 0$, $m > 0$, and $a \equiv b \pmod{m}$, then $ac \equiv bc \pmod{mc}$.

10. Show that if a, b, and c are integers with $c > 0$ such that $a \equiv b \pmod{c}$, then $(a,c) = (b,c)$.

11. Show that if $a_j \equiv b_j \pmod{m}$ for $j = 1,2,\ldots,n$, where m is a positive integer and $a_j, b_j, j = 1,2,\ldots,n$, are integers, then

    a) $\displaystyle\sum_{j=1}^{n} a_j \equiv \sum_{j=1}^{n} b_j \pmod{m}$

    b) $\displaystyle\prod_{j=1}^{n} a_j \equiv \prod_{j=1}^{n} b_j \pmod{m}$

In Exercises 12-14 construct tables for arithmetic modulo 6 using the least nonnegative residues modulo 6 to represent the congruence classes.

12. Construct a table for addition modulo 6.

13. Construct a table for subtraction modulo 6.

14. Construct a table for multiplication modulo 6.

15. What time does a clock read

    a) 29 hours after it reads 11 o'clock
    b) 100 hours after it reads 2 o'clock
    c) 50 hours before it reads 6 o'clock?

16. Which decimal digits occur as the final digit of a fourth power of an integer?

17. What can you conclude if $a^2 \equiv b^2 \pmod{p}$, where a and b are integers and p is prime?

18. Show that if $a^k \equiv b^k \pmod{m}$ and $a^{k+1} \equiv b^{k+1} \pmod{m}$, where a, b, k, and m are integers with $k > 0$ and $m > 0$ such that $(a,m) = 1$, then $a \equiv b \pmod{m}$. If the condition $(a,m) = 1$ is dropped, is the conclusion that $a \equiv b \pmod{m}$ still valid?

19. Show that if n is an odd positive integer, then

$$1 + 2 + 3 + \cdots + (n-1) \equiv 0 \pmod{n}.$$

Is this statement true if n is even?

20. Show that if n is an odd positive integer or if n is divisible by 4, then

$$1^3 + 2^3 + 3^3 + \cdots + (n-1)^3 \equiv 0 \pmod{n}.$$

Is this statement true if n is even but not divisible by 4?

21. For which positive integers n is it true that

$$1^2 + 2^2 + 3^2 + \cdots + (n-1)^2 \equiv 0 \pmod{n}?$$

22. Show by mathematical induction that if n is a positive integer then $4^n \equiv 1 + 3n \pmod 9$.

23. Show by mathematical induction that if n is a positive integer then $5^n \equiv 1 + 4n \pmod{16}$.

24. Give a complete system of residues modulo 13 consisting entirely of odd integers.

25. Show that if $n \equiv 3 \pmod 4$, then n cannot be the sum of the squares of two integers.

26. Show that if p is prime, then the only solutions of the congruence $x^2 \equiv x \pmod p$ are those integers x with $x \equiv 0$ or $1 \pmod p$.

27. Show that if p is prime and k is a positive integer, then the only solutions of $x^2 \equiv x \pmod{p^k}$ are those integers x such that $x \equiv 0$ or $1 \pmod{p^k}$.

28. Find the least positive residues modulo 47 of each of the following integers.

a) 2^{32} b) 2^{47} c) 2^{200}

29. Let $m_1, m_2, \ldots, m_k$ be pairwise relatively prime positive integers. Let $M = m_1 m_2 \cdots m_k$ and $M_j = M/m_j$ for $j = 1, 2, \ldots, k$. Show that

$$M_1 a_1 + M_2 a_2 + \cdots + M_k a_k$$

runs through a complete system of residues modulo M when $a_1, a_2, \ldots, a_k$ run through complete systems of residues modulo $m_1, m_2, \ldots, m_k$, respectively.

30. Explain how to find the sum $u + v$ from the least positive residue of $u + v$ modulo m, where u and v are positive integers less than m. (*Hint:* Assume that $u \leq v$ and consider separately the cases where the least positive residue of $u + v$ is less than u, and where it is greater than v.)

31. On a computer with word size w, multiplication modulo n, where $n < w/2$, can be performed as outlined. Let $T = [\sqrt{n} + \frac{1}{2}]$, and $t = T^2 - n$. For each computation, show that all the required computer arithmetic can be done without exceeding the word size. (This method was described by Head [96]).

a) Show that $|t| \leq T$.

b) Show that if x and y are nonnegative integers less than n, then

$$x = aT + b, \quad y = cT + d$$

where a, b, c, and d are integers such that $0 \le a \le T$, $0 \le b < T$, $0 \le c < T$, and $0 \le d < T$.

c) Let $z \equiv ad + bc \pmod{n}$, with $0 \le z < n$. Show that

$$xy \equiv act + zT + bd \pmod{n}.$$

d) Let $ac = eT + f$ where e and f are integers with $0 \le e < T$ and $0 \le f \le T$. Show that

$$xy \equiv (z+et)T + ft + bd \pmod{n}.$$

e) Let $v = z + et \pmod{n}$, with $0 \le v < n$. Show that we can write

$$v = gT + h,$$

where g and h are integers with $0 \le g \le T, 0 \le h < T$, and such that

$$xy \equiv hT + (f+g)t + bd \pmod{n}.$$

f) Show that the right-hand side of the congruence of part (e) can be computed without exceeding the word size by first finding j with

$$j \equiv (f+g)t \pmod{n}$$

and $0 \le j < n$, and then finding k with

$$k \equiv j + bd \pmod{n}$$

and $0 \le k < n$, so that

$$xy \equiv hT + k \pmod{n}.$$

This gives the desired result.

32. Develop an algorithm for modular exponentiation from the base three expansion of the exponent.

33. Find the least positive residue of

a) 3^{10} modulo 11
b) 2^{12} modulo 13
c) 5^{16} modulo 17
d) 3^{22} modulo 23.
e) Can you propose a theorem from the above congruences?

34. Find the least positive residues of

a) $6!$ modulo 7
b) $10!$ modulo 11
c) $12!$ modulo 13
d) $16!$ modulo 17.
e) Can you propose a theorem from the above congruences?

* **35.** Show that for every positive integer m there are infinitely many Fibonacci numbers f_n such that m divides f_n. (*Hint*: Show that the sequence of least positive residues modulo m of the Fibonacci numbers is a repeating sequence.)

36. Prove Theorem 3.7 using mathematical induction.

37. Show that the least nonnegative residue modulo m of the product of two positive integers less than m can be computed using $O(\log^2 m)$ bit operations.

* **38.** Five men and a monkey are shipwrecked on an island. The men have collected a pile of coconuts which they plan to divide equally among themselves the next morning. Not trusting the other men, one of the group wakes up during the night and divides the coconuts into five equal parts with one left over, which he gives to the monkey. He then hides his portion of the pile. During the night, each of the other four men does exactly the same thing by dividing the pile he finds into five equal parts leaving one coconut for the monkey and hiding his portion. In the morning, the men gather and split the remaining pile of coconuts into five parts and one is left over for the monkey. What is the minimum number of coconuts the men could have collected for their original pile?

* **39.** Answer the same question as in Exercise 38 if instead of five men and one monkey, there are n men and k monkeys, and at each stage the monkeys receive one coconut each.

3.1 Computer Projects

Programming Projects

Write computer programs to do the following:

1. Find the least nonnegative residue of an integer with respect to a fixed modulus.

2. Perform modular addition and subtraction when the modulus is less than half of the word size of the computer.

3. Perform modular multiplication when the modulus is less than half of the word size of the computer using Exercise 31.

4. Perform modular exponentiation using the algorithm described in the text.

Computations and Explorations

Using the programs you have written or a computation program, carry out the following computations and explorations:

1. Compute the least positive residue modulo 10403 of 7651^{891}.

2. Compute the least positive residue modulo 10403 of $7651^{20!}$.

3.2 Linear Congruences

A congruence of the form

$$ax \equiv b \pmod{m},$$

where x is an unknown integer, is called a *linear congruence in one variable*. In this section we will see that the study of such congruences is similar to the study of linear diophantine equations in two variables.

We first note that if $x = x_0$ is a solution of the congruence $ax \equiv b \pmod{m}$, and if $x_1 \equiv x_0 \pmod{m}$, then $ax_1 \equiv ax_0 \equiv b \pmod{m}$, so that x_1 is also a solution. Hence, if one member of a congruence class modulo m is a solution, then all members of this class are solutions. Therefore, we may ask how many of the m congruence classes modulo m give solutions; this is exactly the same as asking how many incongruent solutions there are modulo m. The following theorem tells us when a linear congruence in one variable has solutions, and if it does, tells exactly how many incongruent solutions there are modulo m.

Theorem 3.10. Let a, b, and m be integers with $m > 0$ and $(a,m) = d$. If $d \nmid b$, then $ax \equiv b \pmod{m}$ has no solutions. If $d \mid b$, then $ax \equiv b \pmod{m}$ has exactly d incongruent solutions modulo m.

Proof. By Theorem 3.1 the linear congruence $ax \equiv b \pmod{m}$ is equivalent to the linear diophantine equation in two variables $ax - my = b$. The integer x is a solution of $ax \equiv b \pmod{m}$ if and only if there is an integer y with $ax - my = b$. By Theorem 2.14 we know that if $d \nmid b$, there are no solutions, while if $d \mid b$, $ax - my = b$ has infinitely many solutions, given by

$$x = x_0 + (m/d)t, \quad y = y_0 + (a/d)t,$$

where $x = x_0$ and $y = y_0$ is a particular solution of the equation. The values of x given above,

$$x = x_0 + (m/d)t,$$

are the solutions of the linear congruence; there are infinitely many of these.

To determine how many incongruent solutions there are, we find the condition that describes when two of the solutions $x_1 = x_0 + (m/d)t_1$ and $x_2 = x_0 + (m/d)t_2$ are congruent modulo m. If these two solutions are congruent, then

$$x_0 + (m/d)t_1 \equiv x_0 + (m/d)t_2 \pmod{m}.$$

Subtracting x_0 from both sides of this congruence, we find that

$$(m/d)t_1 \equiv (m/d)t_2 \pmod{m}.$$

Now $(m,m/d) = m/d$ since $(m/d) \mid m$, so that by Theorem 3.4 we see that

$$t_1 \equiv t_2 \pmod{d}.$$

This shows that a complete set of incongruent solutions is obtained by taking

$x = x_0 + (m/d)t$, where t ranges through a complete system of residues modulo d. One such set is given by $x = x_0 + (m/d)t$ where $t = 0,1,2,...,d - 1$. ∎

We now illustrate the use of Theorem 3.10.

Example 3.12. To find all solutions of $9x \equiv 12 \pmod{15}$, we first note that since $(9,15) = 3$ and $3 \mid 12$, there are exactly three incongruent solutions. We can find these solutions by first finding a particular solution and then adding the appropriate multiples of $15/3 = 5$.

To find a particular solution, we consider the linear diophantine equation $9x - 15y = 12$. The Euclidean algorithm shows that

$$15 = 9 \cdot 1 + 6$$
$$9 = 6 \cdot 1 + 3$$
$$6 = 3 \cdot 2,$$

so that $3 = 9 - 6 \cdot 1 = 9 - (15 - 9 \cdot 1) = 9 \cdot 2 - 15$. Hence $9 \cdot 8 - 15 \cdot 4 = 12$, and a particular solution of $9x - 15y = 12$ is given by $x_0 = 8$ and $y_0 = 4$.

From the proof of Theorem 3.10, we see that a complete set of 3 incongruent solutions is given by $x = x_0 \equiv 8 \pmod{15}$, $x = x_0 + 5 \equiv 13 \pmod{15}$, and $x = x_0 + 5 \cdot 2 \equiv 18 \equiv 3 \pmod{15}$. □

We now consider congruences of the special form $ax \equiv 1 \pmod{m}$. By Theorem 3.10 there is a solution to this congruence if and only if $(a,m) = 1$, and then all solutions are congruent modulo m.

Definition. Given an integer a with $(a,m) = 1$, a solution of $ax \equiv 1 \pmod{m}$ is called an *inverse of* a modulo m.

Example 3.13. Since the solutions of $7x \equiv 1 \pmod{31}$ satisfy $x \equiv 9 \pmod{31}$, 9, and all integers congruent to 9 modulo 31, are inverses of 7 modulo 31. Analogously, since $9 \cdot 7 \equiv 1 \pmod{31}$, 7 is an inverse of 9 modulo 31. □

When we have an inverse of a modulo m, we can use it to solve any congruence of the form $ax \equiv b \pmod{m}$. To see this, let $\bar{a}$ be an inverse of a modulo m, so that $a\bar{a} \equiv 1 \pmod{m}$. Then, if $ax \equiv b \pmod{m}$, we can multiply both sides of this congruence by $\bar{a}$ to find that $\bar{a}(ax) \equiv \bar{a}b \pmod{m}$, so that $x \equiv \bar{a}b \pmod{m}$.

Example 3.14. To find the solutions of $7x \equiv 22 \pmod{31}$, we multiply both sides of this congruence by 9, an inverse of 7 modulo 31, to obtain $9 \cdot 7x \equiv 9 \cdot 22 \pmod{31}$. Hence, $x \equiv 198 \equiv 12 \pmod{31}$. □

We note here that if $(a,m) = 1$, then the linear congruence $ax \equiv b \pmod{m}$ has a unique solution modulo m.

Example 3.15. To find all solutions of $7x \equiv 4 \pmod{12}$, we note that since $(7,12) = 1$, there is a unique solution modulo 12. To find this, we need only obtain a solution of the linear diophantine equation $7x - 12y = 4$. The Euclidean algorithm gives

$$12 = 7 \cdot 1 + 5$$
$$7 = 5 \cdot 1 + 2$$
$$5 = 2 \cdot 2 + 1$$
$$2 = 1 \cdot 2.$$

Hence $\qquad 1 = 5 - 2 \cdot 2 = 5 - (7 - 5 \cdot 1) \cdot 2 = 5 \cdot 3 - 2 \cdot 7 = (12 - 7 \cdot 1) \cdot 3 - 2 \cdot 7 = 12 \cdot 3 - 5 \cdot 7$. Therefore, a particular solution to the linear diophantine equation is $x_0 = -20$ and $y_0 = 12$. Hence, all solutions of the linear congruences are given by $x \equiv -20 \equiv 4 \pmod{12}$. $\qquad\square$

Later we will want to know which integers are their own inverses modulo p where p is prime. The following theorem tells us which integers have this property.

Theorem 3.11. Let p be prime. The positive integer a is its own inverse modulo p if and only if $a \equiv 1 \pmod{p}$ or $a \equiv -1 \pmod{p}$.

Proof. If $a \equiv 1 \pmod{p}$ or $a \equiv -1 \pmod{p}$, then $a^2 \equiv 1 \pmod{p}$, so that a is its own inverse modulo p.

Conversely, if a is its own inverse modulo p, then $a^2 = a \cdot a \equiv 1 \pmod{p}$. Hence, $p \mid (a^2 - 1)$. Since $a^2 - 1 = (a-1)(a+1)$, either $p \mid (a-1)$ or $p \mid (a+1)$. Therefore, either $a \equiv 1 \pmod{p}$ or $a \equiv -1 \pmod{p}$. $\qquad\blacksquare$

3.2 Exercises

1. Find all solutions of each of the following linear congruences.

   a) $2x \equiv 5 \pmod{7}$
   b) $3x \equiv 6 \pmod{9}$
   c) $19x \equiv 30 \pmod{40}$

   d) $9x \equiv 5 \pmod{25}$
   e) $103x \equiv 444 \pmod{999}$
   f) $980x \equiv 1500 \pmod{1600}$

2. Find all solutions of each of the following linear congruences.

   a) $3x \equiv 2 \pmod{7}$
   b) $6x \equiv 3 \pmod{9}$
   c) $17x \equiv 14 \pmod{21}$

   d) $15x \equiv 9 \pmod{25}$
   e) $128x \equiv 833 \pmod{1001}$
   f) $987x \equiv 610 \pmod{1597}$

3. Find all solutions to the congruence $6789783x \equiv 2474010 \pmod{28927591}$.

4. Let a, b, and m be positive integers with $a > 0$, $m > 0$, and $(a,m) = 1$. The following method can be used to solve the linear congruence $ax \equiv b \pmod{m}$.

   a) Show that if the integer x is a solution of $ax \equiv b \pmod{m}$, then x is also a solution of the linear congruence

   $$a_1 x \equiv -b[m/a] \pmod{m}.$$

where a_1 is the least positive residue of m modulo a. Note that this congruence is of the same type as the original congruence, with a positive integer smaller than a as the coefficient of x.

b) When the procedure of part (a) is iterated, one obtains a sequence of linear congruences with coefficients of x equal to $a_0 = a > a_1 > a_2 > \cdots$. Show that there is a positive integer n with $a_n = 1$, so that at the nth stage, one obtains a linear congruence $x \equiv B \pmod{m}$.

c) Use the method described in part (b) to solve the linear congruence $6x \equiv 7 \pmod{23}$.

5. An astronomer knows that a satellite orbits the Earth in a period that is an exact multiple of 1 hour that is less than 1 day. If the astronomer notes that the satellite completes 11 orbits in an interval starting when a 24-hour clock reads 0 hours and ending when the clock reads 17 hours, how long is the orbital period of the satellite?

6. For which integers c with $0 \le c < 30$ does the congruence $12x \equiv c \pmod{30}$ have solutions? When there are solutions, how many incongruent solutions are there?

7. For which integers c with $0 \le c < 1001$ does the congruence $154x \equiv c \pmod{1001}$ have solutions? When there are solutions, how many incongruent solutions are there?

8. Find an inverse modulo 13 of each of the following integers.

    a) 2   c) 5
    b) 3   d) 11

9. Find an inverse modulo 17 of each of the following integers.

    a) 4   c) 7
    b) 5   d) 16

10. Show that if $\bar{a}$ is an inverse of a modulo m and $\bar{b}$ is an inverse of b modulo m, then $\bar{a}\,\bar{b}$ is an inverse of ab modulo m.

11. Show that the linear congruence in two variables $ax + by \equiv c \pmod{m}$, where $a, b, c,$ and m are integers, $m > 0$, with $d = (a,b,m)$, has exactly dm incongruent solutions if $d \mid c$, and no solutions otherwise.

12. Find all solutions of each of the following linear congruences in two variables.

    a) $2x + 3y \equiv 1 \pmod{7}$ c) $6x + 3y \equiv 0 \pmod{9}$
    b) $2x + 4y \equiv 6 \pmod{8}$ d) $10x + 5y \equiv 9 \pmod{15}$

13. Let p be an odd prime and k a positive integer. Show that the congruence $x^2 \equiv 1 \pmod{p^k}$ has exactly two incongruent solutions, namely $x \equiv \pm 1 \pmod{p^k}$.

14. Show that the congruence $x^2 \equiv 1 \pmod{2^k}$ has exactly four incongruent solutions, namely $x \equiv \pm 1$ or $\pm(1 + 2^{k-1}) \pmod{2^k}$, when $k > 2$. Show that when $k = 1$ there is one solution and when $k = 2$ there are two incongruent solutions.

15. Show that if a and m are relatively prime positive integers with $a < m$, then an inverse of a modulo m can be found using $O(\log^3 m)$ bit operations.

16. Show that if p is an odd prime and a is a positive integer not divisible by p, then the congruence $x^2 \equiv a \pmod{p}$ has either no solution or exactly two incongruent solutions.

3.2 Computer Projects

Programming Projects

Write programs to do the following:

1. Solve linear congruence using the method given in the text.

2. Solve linear congruences using the method given in Exercise 2.

3. Find inverses modulo m of integers relatively prime to m where m is a positive integer.

4. Solve linear congruences using inverses.

5. Solve linear congruences in two variables.

Computations and Explorations

Using the programs you have written or a computation program, carry out the following computations and explorations:

1. Find the inverses of 734342, 499999, and 1000001 modulo 1533331.

3.3 The Chinese Remainder Theorem

In this section and in the one following, we discuss systems of simultaneous congruences. We will study two types of such systems. In the first type, there are two or more linear congruences in one variable, with different moduli (*moduli* is the plural of *modulus*). The second type consists of more than one simultaneous congruence in more than one variable, where all congruences have the same modulus.

First, we consider systems of congruences that involve only one variable, but different moduli. Such systems arose in ancient Chinese puzzles such as the following: Find a number that leaves a remainder of 1 when divided by 3, a remainder of 2 when divided by 5, and a remainder of 3 when divided by 7. This puzzle leads to the following system of congruences:

$$x \equiv 1 \ (\text{mod } 3), \ x \equiv 2 \ (\text{mod } 5), \ x \equiv 3 \ (\text{mod } 7).$$

We now give a method for finding all solutions of systems of simultaneous congruences such as this. The theory behind the solution of systems of this type is provided by the following theorem, which derives its name from the ancient Chinese heritage of the problem*.

Theorem 3.12. The Chinese Remainder Theorem. Let $m_1, m_2, \ldots, m_r$ be pairwise relatively prime positive integers. Then the system of congruences

$$x \equiv a_1 (\text{mod } m_1),$$
$$x \equiv a_2 (\text{mod } m_2),$$

.

.

.

$$x \equiv a_r (\text{mod } m_t),$$

has a unique solution modulo $M = m_1 m_2 \cdots m_r$.

Proof. First, we construct a simultaneous solution to the system of congruences. To do this, let $M_k = M/m_k = m_1 m_2 \cdots m_{k-1} m_{k+1} \cdots m_r$. We know that $(M_k, m_k) = 1$ by Exercise 14 of Section 2.1, since $(m_j, m_k) = 1$ whenever $j \neq k$. Hence, by Theorem 3.10 we can find an inverse y_k of M_k modulo m_k, so that $M_k y_k \equiv 1 \ (\text{mod } m_k)$. We now form the sum

$$x = a_1 M_1 y_1 + a_2 M_2 y_2 + \cdots + a_r M_r y_r.$$

The integer x is a simultaneous solution of the r congruences. To demonstrate this, we must show that $x \equiv a_k \ (\text{mod } m_k)$ for $k = 1, 2, \ldots, r$. Since $m_k \mid M_j$ whenever $j \neq k$, we have $M_j \equiv 0 \ (\text{mod } m_k)$. Therefore, in the sum for x, all terms except the kth term are congruent to $0 \ (\text{mod } m_k)$. Hence, $x \equiv a_k M_k y_k \equiv a_k \ (\text{mod } m_k)$, since $M_k y_k \equiv 1 \ (\text{mod } m_k)$. We now show that any two solutions are congruent modulo M. Let x_0 and x_1 both be simultaneous solutions to the system of r congruences. Then, for each k, $x_0 \equiv x_1 \equiv a_k \ (\text{mod } m_k)$, so that $m_k \mid (x_0 - x_1)$. Using Theorem 3.8, we see that $M \mid (x_0 - x_1)$. Therefore, $x_0 \equiv x_1 \ (\text{mod } M)$. This shows that the simultaneous solution of the system of r congruences is unique modulo M.   ■

We illustrate the use of the Chinese remainder theorem by solving the system that arises from the ancient Chinese puzzle.

---

* Problems involving systems of linear congruences arose as early as the first century. Such problems can be found in the *Arithmetic* of the Chinese mathematician Sun-Tsu who lived at that time. Problems involving such systems and methods for solving them can be found throughout the works of ancient Chinese and Hindu mathematicians. For more information on this topic consult the references on the history of number theory listed in the Bibliography which appears at the end of the text.

Example 3.16. To solve the system

$$x \equiv 1 \ (\text{mod } 3)$$
$$x \equiv 2 \ (\text{mod } 5)$$
$$x \equiv 3 \ (\text{mod } 7),$$

we have $M = 3 \cdot 5 \cdot 7 = 105$, $M_1 = 105/3 = 35$, $M_2 = 105/5 = 21$, and $M_3 = 105/7 = 15$. To determine y_1, we solve $35y_1 \equiv 1 \ (\text{mod } 3)$, or equivalently, $2y_1 \equiv 1 \ (\text{mod } 3)$. This yields $y_1 \equiv 2 \ (\text{mod } 3)$. We find y_2 by solving $21y_2 \equiv 1 \ (\text{mod } 5)$; this immediately gives $y_2 \equiv 1 \ (\text{mod } 5)$. Finally, we find y_3 by solving $15y_3 \equiv 1 \ (\text{mod } 7)$. This gives $y_3 \equiv 1 \ (\text{mod } 7)$. Hence,

$$x \equiv 1 \cdot 35 \cdot 2 + 2 \cdot 21 \cdot 1 + 3 \cdot 15 \cdot 1$$
$$\equiv 157 \equiv 52 \ (\text{mod } 105).$$

We can check that x satisfies this system of congruences whenever $x \equiv 52 \ (\text{mod } 105)$ by noting that $52 \equiv 1 \ (\text{mod } 3)$, $52 \equiv 2 \ (\text{mod } 5)$, and $52 \equiv 1 \ (\text{mod } 7)$. □

There is also an iterative method for solving simultaneous systems of congruences. We illustrate this method with an example.

Example 3.17. Suppose we wish to solve the system

$$x \equiv 1 \ (\text{mod } 5)$$
$$x \equiv 2 \ (\text{mod } 6)$$
$$x \equiv 3 \ (\text{mod } 7).$$

We use Theorem 3.1 to rewrite the first congruence as an equality, namely $x = 5t + 1$, where t is an integer. Inserting this expression for x into the second congruence, we find that

$$5t + 1 \equiv 2 \ (\text{mod } 6),$$

which can easily be solved to show that $t \equiv 5 \ (\text{mod } 6)$. Using Theorem 3.1 again, we write $t = 6u + 5$ where u is an integer. Hence, $x = 5(6u+5) + 1 = 30u + 26$. When we insert this expression for x into the third congruence, we obtain

$$30u + 26 \equiv 3 \ (\text{mod } 7).$$

When this congruence is solved, we find that $u \equiv 6 \ (\text{mod } 7)$. Consequently,

Theorem 3.1 tells us that $u = 7v + 6$, where v is an integer. Hence,

$$x = 30(7v+6) + 26 = 210v + 206.$$

Translating this equality into a congruence, we find that

$$x \equiv 206 \ (\text{mod} \ 210),$$

and this is the simultaneous solution. $\qquad\qquad\qquad\qquad\qquad$ □

Note that the method we have just illustrated shows that a system of simultaneous questions can be solved by successively solving linear congruences. This can be done even when the moduli of the congruences are not relatively prime as long as congruences are consistent. (See Exercises 11-14 at the end of this section.)

The Chinese remainder theorem provides a way to perform computer arithmetic with large integers. To store very large integers and do arithmetic with them requires special techniques. The Chinese remainder theorem tells us that given pairwise relatively prime moduli $m_1, m_2, \ldots, m_r$, a positive integer n with $n < M = m_1 m_2 \cdots m_r$ is uniquely determined by its least positive residues moduli m_j for $j = 1, 2, \ldots, r$. Suppose that the word size of a computer is only 100, but that we wish to do arithmetic with integers as large as 10^6. First, we find pairwise relatively prime integers less than 100 with a product exceeding 10^6; for instance, we can take $m_1 = 99$, $m_2 = 98$, $m_3 = 97$, and $m_4 = 95$. We convert integers less than 10^6 into 4-tuples consisting of their least positive residues modulo m_1, m_2, m_3, and m_4. (To convert integers as large as 10^6 into their list of least positive residues, we need to work with large integers using multiprecision techniques. However, this is done only once for each integer in the input and once for the output.) Then, for instance, to add integers, we simply add their respective least positive residues modulo m_1, m_2, m_3, and m_4, making use of the fact that if $x \equiv x_i \ (\text{mod} \ m_i)$ and $y \equiv y_i \ (\text{mod} \ m_i)$, then $x + y \equiv x_i + y_i \ (\text{mod} \ m_i)$. We then use the Chinese remainder theorem to convert the set of four least positive residues for the sum back to an integer.

The following example illustrates this technique.

Example 3.18. We wish to add $x = 123684$ and $y = 413456$ on a computer of word size 100. We have

$x \equiv 33 \ (\text{mod} \ 99),$	$y \equiv 32 \ (\text{mod} \ 99),$
$x \equiv 8 \ (\text{mod} \ 98),$	$y \equiv 92 \ (\text{mod} \ 98),$
$x \equiv 9 \ (\text{mod} \ 97),$	$y \equiv 42 \ (\text{mod} \ 97),$
$x \equiv 89 \ (\text{mod} \ 95),$	$y \equiv 16 \ (\text{mod} \ 95),$

so that

$$x + y \equiv 65 \pmod{99}$$
$$x + y \equiv 2 \pmod{98}$$
$$x + y \equiv 51 \pmod{97}$$
$$x + y \equiv 10 \pmod{95}.$$

We now use the Chinese remainder theorem to find $x + y$ modulo $99 \cdot 98 \cdot 97 \cdot 95$. We have $M = 99 \cdot 98 \cdot 97 \cdot 95 = 89403930$, $M_1 = M/99 = 903070$, $M_2 = M/98 = 912285$, $M_3 = M/97 = 921690$, and $M_4 = M/95 = 941094$. We need to find the inverse of $M_i \pmod{y_i}$ for $i = 1, 2, 3, 4$. To do this, we solve the following congruences (using the Euclidean algorithm):

$$903070 y_1 \equiv 91 y_1 \equiv 1 \pmod{99},$$
$$912285 y_2 \equiv 3 y_2 \equiv 1 \pmod{98},$$
$$921690 y_3 \equiv 93 y_3 \equiv 1 \pmod{97},$$
$$941094 y_4 \equiv 24 y_4 \equiv 1 \pmod{95}.$$

We find that $y_1 \equiv 37 \pmod{99}$, $y_2 \equiv 35 \pmod{98}$, $y_3 \equiv 24 \pmod{97}$, and $y_4 \equiv 4 \pmod{95}$. Hence,

$$x + y \equiv 65 \cdot 903070 \cdot 37 + 2 \cdot 912285 \cdot 33 + 51 \cdot 921690 \cdot 24 + 10 \cdot 941094 \cdot 4$$
$$= 3397886480$$
$$\equiv 537140 \pmod{89403930}.$$

Since $0 < x + y < 89403930$, we conclude that $x + y = 537140$. □

On most computers the word size is a large power of 2, with 2^{35} a common value. Hence, to use modular arithmetic and the Chinese remainder theorem to do computer arithmetic, we need integers less than 2^{35} that are pairwise relatively prime which multiply together to give a large integer. To find such integers, we use numbers of the form $2^m - 1$, where m is a positive integer. Computer arithmetic with these numbers turns out to be relatively simple (see Knuth [90]). To produce a set of pairwise relatively prime numbers of this form, we first prove some lemmata (*lemmata* is the plural of *lemma*; some people use the plural *lemmas*).

Lemma 3.1. If a and b are positive integers, then the least positive residue of $2^a - 1$ modulo $2^b - 1$ is $2^r - 1$, where r is the least positive residue of a modulo b.

Proof. From the division algorithm, $a = bq + r$ where r is the least positive residue of a modulo b. We have $2^a - 1 = 2^{bq+r} - 1 = (2^b - 1)(2^{b(q-1)+r} + \cdots + 2^{b+r} + 2^r) + (2^r - 1)$, which shows that the remainder when $2^a - 1$ is divided by $2^b - 1$ is $2^r - 1$; this is the least positive residue of $2^a - 1$ modulo $2^b - 1$. ∎

We use Lemma 3.1 to prove the following result.

Lemma 3.2. If a and b are positive integers, then the greatest common divisor of $2^a - 1$ and $2^b - 1$ is $2^{(a,b)} - 1$.

Proof. When we perform the Euclidean algorithm with $a = r_0$ and $b = r_1$, we obtain

$$r_0 = r_1 q_1 + r_2 \qquad\qquad 0 \leq r_2 < r_1$$
$$r_1 = r_2 q_2 + r_3 \qquad\qquad 0 \leq r_3 < r_2$$

$$\cdot$$
$$\cdot$$
$$\cdot$$

$$r_{n-3} = r_{n-2} q_{n-2} + r_{n-1} \qquad\qquad 0 \leq r_{n-1} < r_{n-2}$$
$$r_{n-2} = r_{n-1} q_{n-1}$$

where the last remainder, r_{n-1}, is the greatest common divisor of a and b.

Using Lemma 3.1, and the steps of the Euclidean algorithm with $a = r_0$ and $b = r_1$, when we perform the Euclidean algorithm on the pair $2^a - 1 = R_0$ and $2^b - 1 = R_1$, we obtain

$$R_0 = R_1 Q_1 + R_2 \qquad\qquad R_2 = 2^{r_2} - 1$$
$$R_1 = R_2 Q_2 + R_3 \qquad\qquad R_3 = 2^{r_3} - 1$$

$$\cdot$$
$$\cdot$$
$$\cdot$$

$$R_{n-3} = R_{n-2} Q_{n-2} + R_{n-1} \qquad\qquad R_{n-1} = 2^{r_{n-1}} - 1$$
$$R_{n-2} = R_{n-1} Q_{n-1}.$$

Here the last non-zero remainder, $R_{n-1} = 2^{r_{n-1}} - 1 = 2^{(a,b)} - 1$, is the greatest common divisor of R_0 and R_1. ∎

Using Lemma 3.2 we have the following theorem.

Theorem 3.13. The positive integers $2^a - 1$ and $2^b - 1$ are relatively prime if and only if a and b are relatively prime.

We can now use Theorem 3.13 to produce a set of pairwise relatively prime integers, each of which is less than 2^{35}, with product greater than a specified integer. Suppose that we wish to do arithmetic with integers as large as 2^{184}. We pick $m_1 = 2^{35} - 1$, $m_2 = 2^{34} - 1$, $m_3 = 2^{33} - 1$, $m_4 = 2^{31} - 1$, $m_5 = 2^{29} - 1$, and $m_6 = 2^{23} - 1$. Since the exponents of 2 in the expressions for the m_j are pairwise relatively prime, by Theorem 3.13 the m_j's are pairwise relatively prime. Also, we have $M = m_1 m_2 m_3 m_4 m_5 m_6 > 2^{184}$. We can now use modular arithmetic and the Chinese remainder theorem to perform arithmetic with integers as large as 2^{184}.

Although it is somewhat awkward to do computer operations with large integers using modular arithmetic and the Chinese remainder theorem, there are some definite advantages to this approach. First, on many high-speed computers, operations can be performed simultaneously. So, reducing an operation involving two large integers to a set of operations involving smaller integers, namely the least positive residues of the large integers with respect to the various moduli, leads to simultaneous computations which may be performed more rapidly than one operation with large integers. Second, even without taking into account the advantages of simultaneous computations, multiplication of large integers may be done faster using these ideas than with many other multiprecision methods. The interested reader should consult Knuth [90].

3.3 Exercises

1. Which integers leave a remainder of one when divided by either 2 or 3?

2. Find an integer that leaves a remainder of one when divided by either 2 or 5, but that is divisible by 3.

3. Find an integer that leaves a remainder of 2 when divided by either 3 or 5 but that is divisible by 4.

4. Find all the solutions of each of the following systems of linear congruences.

    a) $x \equiv 4 \pmod{11}$ c) $x \equiv 0 \pmod 2$
        $x \equiv 3 \pmod{17}$ $x \equiv 0 \pmod 3$
                                           $x \equiv 1 \pmod 5$
    b) $x \equiv 1 \pmod 2$ $x \equiv 6 \pmod 7$
        $x \equiv 2 \pmod 3$
        $x \equiv 3 \pmod 5$ d) $x \equiv 2 \pmod{11}$
                                           $x \equiv 3 \pmod{12}$
                                           $x \equiv 4 \pmod{13}$
                                           $x \equiv 5 \pmod{17}$
                                           $x \equiv 6 \pmod{19}$

5. Find all the solutions to the system of linear congruences $x \equiv 1 \pmod 2$, $x \equiv 2 \pmod 3$, $x \equiv 3 \pmod 5$, $x \equiv 4 \pmod 7$, and $x \equiv 5 \pmod{11}$.

6. Find all the solutions to the system of linear congruences $x \equiv 1 \pmod{999}$, $x \equiv 2 \pmod{1001}$, $x \equiv 3 \pmod{1003}$, $x \equiv 4 \pmod{1004}$, and $x \equiv 5 \pmod{1007}$.

7. A troop of 17 monkeys store their bananas in eleven piles of equal size with a twelfth pile of six left over. When they divide the bananas into 17 equal groups none remain. What is the smallest number of bananas they can have?

8. As an odometer check, a special counter measures the miles a car travels modulo 7. Explain how this counter can be used to determine whether the car has been driven 49335, 149335, or 249335 miles when the odometer reads 49335 and works modulo 100000.

9. Chinese generals counted troops remaining after a battle by lining them up in rows of different lengths, counting the number left over each time, and calculating the total from these remainders. If a general had 1200 troops at the start of a battle and if there were 3 left over when they lined up 5 at a time, 3 left over when they lined up 6 at a time, 1 left over when they lined up 7 at a time, and none left over when they lined up 11 at a time, how many troops remained after the battle?

10. Find an integer that leaves a remainder of 9 when it is divided by either 10 or 11, but that is divisible by 13.

11. Find a multiple of 11 that leaves a remainder of 1 when divided by each of the integers 2,3,5, and 7.

12. Solve the following ancient Indian problem: If eggs are removed from a basket two, three, four, five, and six at a time, there remain, respectively, one, two, three, four, and five eggs. But if the eggs are removed seven at a time no eggs remain. What is the least number of eggs that could have been in the basket?

13. Show that there are arbitrarily long strings of integers each divisible by a perfect square. (*Hint*: Use the Chinese remainder theorem to show that there is a simultaneous solution to the system of congruences $x \equiv 0 \pmod 4$, $x \equiv -1 \pmod 9$, $x \equiv -2 \pmod{25}$, ..., $x \equiv -k+1 \pmod{p_k^2}$, where p_k is the kth prime.)

* 14. Show that if a, b, and c are integers with $(a,b) = 1$, then there is an integer n such that $(an+b,c) = 1$.

In Exercises 15-18 we will consider systems of congruences where the moduli of the congruences are not necessarily relatively prime.

15. Show that the system of congruences

$$x \equiv a_1 \pmod{m_1}$$
$$x \equiv a_2 \pmod{m_2}$$

has a solution if and only if $(m_1,m_2) \mid (a_1-a_2)$. Show that when there is a solution, it is unique modulo $[m_1,m_2]$. (*Hint*: Write the first congruence as $x = a_1 + km_1$ where k is an integer, and then insert this expression for x into the second congruence.)

16. Using Exercise 15 solve each of the following simultaneous system of congruences,

a) $x \equiv 4 \pmod 6$ b) $x \equiv 7 \pmod{10}$
     $x \equiv 13 \pmod{15}$ $x \equiv 4 \pmod{15}$

17. Using Exercise 15 solve each of the following simultaneous system of congruences,

a) $x \equiv 10 \pmod{60}$ b) $x \equiv 2 \pmod{910}$
     $x \equiv 80 \pmod{350}$ $x \equiv 93 \pmod{1001}$

18. Does the system of congruences $x \equiv 1 \pmod 8$, $x \equiv 3 \pmod 9$, and $x \equiv 2 \pmod{12}$ have any simultaneous solutions?

19. Show that the system of congruences

$$x \equiv a_1 \pmod{m_1}$$
$$x \equiv a_2 \pmod{m_2}$$
$$.$$
$$.$$
$$.$$
$$x \equiv a_r \pmod{m_r}$$

has a solution if and only if $(m_i, m_j) \mid (a_i - a_j)$ for all pairs of integers (i,j) with $1 \leq i < j \leq r$. Show that if a solution exists, then it is unique modulo $[m_1, m_2, ..., m_r]$. (*Hint*: Use Exercise 15 and mathematical induction.)

20. Using Exercise 19 solve each of the following systems of congruences.

a) $x \equiv 5 \pmod 6$
  $x \equiv 3 \pmod{10}$
  $x \equiv 8 \pmod{15}$

d) $x \equiv 2 \pmod 6$
  $x \equiv 4 \pmod 8$
  $x \equiv 2 \pmod{14}$
  $x \equiv 14 \pmod{15}$

b) $x \equiv 2 \pmod{14}$
  $x \equiv 16 \pmod{21}$
  $x \equiv 10 \pmod{30}$

e) $x \equiv 7 \pmod 9$
  $x \equiv 2 \pmod{10}$
  $x \equiv 3 \pmod{12}$

c) $x \equiv 2 \pmod 9$
  $x \equiv 8 \pmod{15}$
  $x \equiv 10 \pmod{25}$

  $x \equiv 6 \pmod{15}$

21. What is the smallest number of lobsters in a tank if one lobster is left over when they are removed 2, 3, 5, or 7 at a time, but no lobsters are left over when they are removed 11 at a time?

22. Using the Chinese remainder theorem, explain how to add and how to multiply 784 and 813 on a computer of word size 100.

A positive integer $x \neq 1$ with n base b digits is called an *automorph to the base b* if the last n base b digits of x^2 are the same as those of x.

* 23. Find the base 10 automorphs with four digits (with initial zeros allowed).

* 24. How many base b automorphs are there with n or fewer base b digits, if b has prime-power factorization $b = p_1^{b_1} p_2^{b_2} \cdots p_k^{b_k}$?

According to the theory of *biorhythms*, there are three cycles in your life that start the day you are born. These are the *physical, emotional,* and *intellectual cycles,* of lengths 23, 28, and 33 days, respectively. Each cycle follows a sine curve with period equal to the length of that cycle, starting with amplitude zero, climbing to amplitude one one quarter of the way through the cycle, dropping back to amplitude zero one half of the way through the cycle, dropping further to amplitude minus one three quarters of the way through the cycle, and climbing back to amplitude zero at the end of the cycle.

Answer the following questions about biorhythms, measuring time in quarter days (so that the units will be integers).

25. For which days of your life will you be at a triple peak, where all of your three cycles are at maximum amplitudes?

26. For which days of your life will you be at a triple nadir, where all three of your cycles have lowest amplitude?

27. When in your life will all three cycles be a neutral position (amplitude 0)?

A set of congruences to distinct moduli greater than one that has the property that every integer satisfies at least one of the congruences is called a *covering set of congruences*.

28. Show the set of congruences $x \equiv 0 \pmod{2}$, $x \equiv 0 \pmod{3}$, $x \equiv 1 \pmod{4}$, $x \equiv 1 \pmod{6}$, and $x \equiv 11 \pmod{12}$ is a covering set of congruences.

29. Show that the set of congruences $x \equiv 0 \pmod{2}$, $x \equiv 0 \pmod{3}$, $x \equiv 0 \pmod{5}$, $x \equiv 0 \pmod{7}$, $x \equiv 1 \pmod{6}$, $x \equiv 1 \pmod{10}$, $x \equiv 1 \pmod{14}$, $x \equiv 2 \pmod{15}$, $x \equiv 2 \pmod{21}$, $x \equiv 23 \pmod{30}$, $x \equiv 4 \pmod{35}$, $x \equiv 5 \pmod{42}$, $x \equiv 59 \pmod{70}$, and $x \equiv 104 \pmod{105}$ is a covering set of congruences.

* 30. Let m be a positive integer with prime-power factorization $m = 2^{a_0} p_1^{a_1} p_2^{a_2} \cdots p_r^{a_r}$. Show that the congruence $x^2 \equiv 1 \pmod{m}$ has exactly 2^{r+e} solutions where $e = 0$ if $a_0 = 0$ or 1, $e = 1$ if $a_0 = 2$, and $e = 2$ if $a_0 > 2$. (*Hint:* Use Exercises 13 and 14 of Section 3.2.)

31. The three children in a family have feet that are 5 inches, 7 inches, and 9 inches long. When they measure the length of the dining room of their house using their feet, they each find that there are 3 inches left over. How long is the dining room?

32. Find all solutions of the congruence $x^2 + 6x - 31 \equiv 0 \pmod{72}$. (*Hint:* First note that $72 = 2^3 3^2$. Find, by trial and error, the solutions of this congruence modulo 8 and modulo 9. Then apply the Chinese remainder theorem.)

33. Find all solutions of the congruence $x^2 + 18x - 823 \equiv 0 \pmod{1800}$ (*Hint:* First note that $1800 = 2^3 3^2 5^2$. Find, by trial and error, the solutions of this congruence modulo 8, modulo 9, and modulo 25. Then apply the Chinese remainder theorem.)

3.3 Computer Projects

Programming Projects

Write programs to do the following:

1. Solve systems of linear congruences of the type found in the Chinese remainder theorem.

2. Solve systems of linear congruences of the type given in Exercises 15-20.

3. Add large integers exceeding the word size of the computer using the Chinese remainder theorem.

4. Multiply large integers exceeding the word size of the computer using the Chinese remainder theorem.

5. Find automorphs to the base b, where b is a positive integer greater than one (see preamble to Exercise 23).

6. Plot biorhythm charts and find triple peaks and triple nadirs (see preamble to Exercise 25).

Computations and Explorations

Using the programs you have written or a computation program, carry out the following computations and explorations:

1. Using Exercise 13 find a string of 100 consecutive positive integers each divisible by a perfect square. Can you find such a set of smaller integers?

2. Find a covering set of congruences (described in the preamble to Exercise 28) where the smallest modulus of one of the congruences in the covering set is 3, where the smallest modulus of one of the congruences in the covering set is 6, and where the smallest modulus of one of the congruences in the covering set is 8.

3.4 Systems of Linear Congruences

We will consider systems of more than one congruence involving the same number of unknowns as congruences, where all congruences have the same modulus. We begin our study with an example.

Suppose we wish to find all integers x and y such that both of the congruences

$$3x + 4y \equiv 5 \pmod{13}$$
$$2x + 5y \equiv 7 \pmod{13}$$

are satisfied. To attempt to find the unknowns x and y, we multiply the first congruence by 5 and the second by 4, to obtain

$$15x + 20y \equiv 25 \pmod{13}$$
$$8x + 20y \equiv 28 \pmod{13}.$$

We subtract the first congruence from the second, to find that

$$7x \equiv -3 \text{ (mod } 13).$$

Since 2 is an inverse of 7 (mod 13), we multiply both sides of the above congruences by 2. This gives

$$2 \cdot 7 \, x \equiv -2 \cdot 3 \text{ (mod } 13),$$

which tells us that

$$x \equiv 7 \text{ (mod } 13).$$

Likewise, we can multiply the first congruence by 2 and the second by 3, to see that

$$6x + 8y \equiv 10 \text{ (mod } 13)$$
$$6x + 15y \equiv 21 \text{ (mod } 13).$$

When we subtract the first congruence from the second, we obtain

$$7y \equiv 11 \text{ (mod } 13).$$

To solve for y, we multiply both sides of this congruence by 2, an inverse of 7 modulo 13. We get

$$2 \cdot 7 \, y \equiv 2 \cdot 11 \text{ (mod } 13),$$

so that

$$y \equiv 9 \text{ (mod } 13).$$

What we have shown is that any solution (x,y) must satisfy

$$x \equiv 7 \text{ (mod } 13), \, y \equiv 9 \text{ (mod } 13).$$

When we insert these congruences for x and y into the original system, we see that these pairs actually are solutions, since

$$3x + 4y \equiv 3 \cdot 7 + 4 \cdot 9 = 57 \equiv 5 \text{ (mod } 13)$$
$$2x + 5y \equiv 2 \cdot 7 + 5 \cdot 9 = 59 \equiv 7 \text{ (mod } 13).$$

Hence, the solutions of this system of congruences are all pairs (x,y) with $x \equiv 7$ (mod 13) and $y \equiv 9$ (mod 13).

We now give a general result concerning certain systems of two congruences in two unknowns. (This result resembles Cramer's Rule for solving systems of linear equalities.)

Theorem 3.14. Let $a, b, c, d, e, f,$ and m be integers with $m > 0$, such that $(\Delta, m) = 1$, where $\Delta = ad - bc$. Then the system of congruences

$$ax + by \equiv e \text{ (mod } m)$$
$$cx + dy \equiv f \text{ (mod } m)$$

has a unique solution modulo m given by

$$x \equiv \bar{\Delta} \ (de - bf) \ (\text{mod } m)$$
$$y \equiv \bar{\Delta} \ (af - ce) \ (\text{mod } m),$$

where $\bar{\Delta}$ is an inverse of Δ modulo m.

Proof. We multiply the first congruence of the system by d and the second by b, to obtain

$$adx + bdy \equiv de \ (\text{mod } m)$$
$$bcx + bdy \equiv bf \ (\text{mod } m).$$

Then we subtract the second congruence from the first, to find that

$$(ad - bc)x \equiv de - bf \ (\text{mod } m),$$

or, since $\Delta = ad - bc$,

$$\Delta x \equiv de - bf \ (\text{mod } m).$$

Next, we multiply both sides of this congruence by $\bar{\Delta}$, an inverse of Δ modulo m, to conclude that

$$x \equiv \bar{\Delta} \ (de - bf) \ (\text{mod } m).$$

In a similar way, we multiply the first congruence by c and the second by a, to obtain

$$acx + bcy \equiv ce \ (\text{mod } m)$$
$$acx + ady \equiv af \ (\text{mod } m).$$

We subtract the first congruence from the second, to find that

$$(ad - bc)y \equiv af - ce \ (\text{mod } m)$$

or

$$\Delta y \equiv af - ce \ (\text{mod } m).$$

Finally, we multiply both sides of this congruence by $\bar{\Delta}$ to see that

$$y \equiv \bar{\Delta} \ (af - ce) \ (\text{mod } m).$$

We have shown that if (x, y) is a solution of the system of congruences, then

$$x \equiv \bar{\Delta} \ (de - bf) \ (\text{mod } m) \ , \quad y \equiv \bar{\Delta} \ (af - ce) \ (\text{mod } m).$$

We can easily check that any such pair (x, y) is a solution. When $x \equiv \bar{\Delta} \ (de - bf) \ (\text{mod } m)$ and $y \equiv \bar{\Delta} \ (af - ce) \ (\text{mod } m)$, we have

$$ax + by \equiv a\overline{\Delta} \ (de - bf) + b\overline{\Delta} \ (af - ce)$$
$$\equiv \overline{\Delta} \ (ade - abf - abf - bce)$$
$$\equiv \overline{\Delta} \ (ad - bc) \ e$$
$$\equiv \overline{\Delta}\Delta e$$
$$\equiv e \ (\mathrm{mod} \ m),$$

and

$$cx + dy \equiv c\overline{\Delta} \ (de - bf) + d\overline{\Delta} \ (af - ce)$$
$$\equiv \overline{\Delta} \ (cde - bcf + adf - cde)$$
$$\equiv \overline{\Delta} \ (ad - bc)f$$
$$\equiv \overline{\Delta}\Delta f$$
$$\equiv f \ (\mathrm{mod} \ m).$$

This establishes the theorem.   ∎

By similar methods, we may solve systems of n congruences involving n unknowns. However, we will develop the theory of solving such systems, as well as larger systems, by methods taken from linear algebra. Readers unfamiliar with linear algebra may wish to skip the remainder of this section.

Systems of n linear congruences involving n unknowns will arise in our subsequent cryptographic studies. To study these systems when n is large, it is helpful to use the language of matrices. We will use some of the basic notions of matrix arithmetic which are discussed in most linear algebra texts, such as Anton [96].

We need to define congruences of matrices before we proceed.

Definition. Let **A** and **B** be $n \times k$ matrices with integer entries, with (i,j)th entries a_{ij} and b_{ij}, respectively. We say that **A** is *congruent to* **B** *modulo* m if $a_{ij} \equiv b_{ij} \ (\mathrm{mod} \ m)$ for all pairs (i,j) with $1 \le i \le n$ and $1 \le j \le k$. We write $\mathbf{A} \equiv \mathbf{B} \ (\mathrm{mod} \ m)$ if **A** is congruent to **B** modulo m.

The matrix congruence $\mathbf{A} \equiv \mathbf{B} \ (\mathrm{mod} \ m)$ provides a succinct way of expressing the nk congruences $a_{ij} \equiv b_{ij} \ (\mathrm{mod} \ m)$ for $1 \le i \le n$ and $1 \le j \le k$.

Example 3.19. We easily see that

$$\begin{bmatrix} 15 & 3 \\ 8 & 12 \end{bmatrix} \equiv \begin{bmatrix} 4 & 3 \\ -3 & 1 \end{bmatrix} \ (\mathrm{mod} \ 11).$$

$\square$

The following proposition will be needed.

Theorem 3.15. If **A** and **B** are $n \times k$ matrices with $\mathbf{A} \equiv \mathbf{B} \ (\mathrm{mod} \ m)$, **C** is a $k \times p$ matrix and **D** is a $p \times n$ matrix, all with integer entries, then $\mathbf{AC} \equiv \mathbf{BC} \ (\mathrm{mod} \ m)$ and $\mathbf{DA} \equiv \mathbf{DB} \ (\mathrm{mod} \ m)$.

Proof. Let the entries of **A** and **B** be a_{ij} and b_{ij}, respectively, for $1 \le i \le n$ and

$1 \leq j \leq k$, and let the entries of $\mathbf{C}$ be c_{ij} for $1 \leq i \leq k$ and $1 \leq j \leq p$. The (i,j)th entries of $\mathbf{AC}$ and $\mathbf{B\,C}$ are $\sum_{t=1}^{k} a_{it} c_{tj}$ and $\sum_{t=1}^{k} b_{it} c_{tj}$, respectively, for $1 \leq i \leq n$ and $1 \leq j \leq p$. Since $\mathbf{A} \equiv \mathbf{B}$ (mod m), we know that $a_{it} \equiv b_{it}$ (mod m) for all i and k. Hence, by Theorem 3.3 we see that $\sum_{t=1}^{k} a_{it} c_{tj} \equiv \sum_{t=1}^{k} b_{it} c_{tj}$ (mod m). Consequently, $\mathbf{AC} \equiv \mathbf{BC}$ (mod m).

The proof that $\mathbf{DA} \equiv \mathbf{DB}$ (mod m) is similar and is omitted. ■

Now let us consider the system of congruences

$$
\begin{aligned}
a_{11}\,x_1 + a_{12}\,x_2 + &\cdots + a_{1n}\,x_n \equiv b_1 \ (\text{mod } m)\\
a_{21}\,x_1 + a_{22}\,x_2 + &\cdots + a_{2n}\,x_n \equiv b_2 \ (\text{mod } m)\\
&\ \ \vdots\\
a_{n1}\,x_1 + a_{n2}\,x_2 + &\cdots + a_{nn}\,x_n \equiv b_n \ (\text{mod } m).
\end{aligned}
$$

Using matrix notation, we see that this system of n congruences is equivalent to the matrix congruence $\mathbf{AX} \equiv \mathbf{B}$ (mod m),

$$
\text{where } \mathbf{A} =
\begin{bmatrix}
a_{11} & a_{12} & \cdots & a_{1n}\\
a_{21} & a_{22} & \cdots & a_{2n}\\
& & \vdots & \\
a_{n1} & a_{n2} & \cdots & a_{nn}
\end{bmatrix},
\ \mathbf{X} =
\begin{bmatrix}
x_1\\
x_2\\
\vdots\\
x_n
\end{bmatrix},
\ \text{and } \mathbf{B} =
\begin{bmatrix}
b_1\\
b_2\\
\vdots\\
b_n
\end{bmatrix}.
$$

Example 3.20. The system

$$
\begin{aligned}
3x + 4y &\equiv 5 \ (\text{mod } 13)\\
2x + 5y &\equiv 7 \ (\text{mod } 13)
\end{aligned}
$$

can be written as

$$
\begin{bmatrix} 3 & 4\\ 2 & 5 \end{bmatrix}
\begin{bmatrix} x\\ y \end{bmatrix}
\equiv
\begin{bmatrix} 5\\ 7 \end{bmatrix}
\ (\text{mod } 13).
$$

□

We now develop a method for solving congruences of the form $\mathbf{AX} \equiv \mathbf{B}$ (mod m). This method is based on finding a matrix $\overline{\mathbf{A}}$ such that $\overline{\mathbf{A}}\mathbf{A} \equiv \mathbf{I}$ (mod m), where $\mathbf{I}$ is the identity matrix.

Definition. If **A** and $\overline{\textbf{A}}$ are $n\times n$ matrices of integers and $\overline{\textbf{A}}\textbf{A} \equiv \textbf{A}\overline{\textbf{A}} \equiv \textbf{I} \pmod{m}$,

where $\textbf{I} = \begin{bmatrix} 1 & 0 & \cdots & 0 \\ 0 & 1 & \cdots & 0 \\ & & \cdots & \\ 0 & 0 & \cdots & 1 \end{bmatrix}$ is the identity matrix of order n, then $\overline{\textbf{A}}$ is said to be an

inverse of **A** *modulo m.*

If $\overline{\textbf{A}}$ is an inverse of **A** and $\textbf{B} \equiv \overline{\textbf{A}} \pmod{m}$, then **B** is also an inverse of **A**. This follows from Theorem 3.15, since $\textbf{BA} \equiv \overline{\textbf{A}}\textbf{A} \equiv \textbf{I} \pmod{m}$. Conversely, if $\textbf{B}_1$ and $\textbf{B}_2$ are both inverses of **A**, then $\textbf{B}_1 \equiv \textbf{B}_2 \pmod{m}$. To see this, using Theorem 3.15 and the congruence $\textbf{B}_1\textbf{A} \equiv \textbf{B}_2\textbf{A} \equiv \textbf{I} \pmod{m}$, we have $\textbf{B}_1\textbf{AB}_1 \equiv \textbf{B}_2\textbf{AB}_1 \pmod{m}$. Since $\textbf{AB}_1 \equiv \textbf{I} \pmod{m}$, we conclude that $\textbf{B}_1 \equiv \textbf{B}_2 \pmod{m}$.

Example 3.21. Since

$$\begin{bmatrix} 1 & 3 \\ 2 & 4 \end{bmatrix} \begin{bmatrix} 3 & 4 \\ 1 & 2 \end{bmatrix} = \begin{bmatrix} 6 & 10 \\ 10 & 16 \end{bmatrix} \equiv \begin{bmatrix} 1 & 0 \\ 0 & 1 \end{bmatrix} \pmod{5}$$

and

$$\begin{bmatrix} 3 & 4 \\ 1 & 2 \end{bmatrix} \begin{bmatrix} 1 & 3 \\ 2 & 4 \end{bmatrix} = \begin{bmatrix} 11 & 25 \\ 5 & 11 \end{bmatrix} \equiv \begin{bmatrix} 1 & 0 \\ 0 & 1 \end{bmatrix} \pmod{5},$$

we see that the matrix $\begin{bmatrix} 3 & 4 \\ 1 & 2 \end{bmatrix}$ is an inverse of $\begin{bmatrix} 1 & 3 \\ 2 & 4 \end{bmatrix}$ modulo 5.   □

The following proposition gives an easy method for finding inverses for 2×2 matrices.

Theorem 3.16. Let $\textbf{A} = \begin{bmatrix} a & b \\ c & d \end{bmatrix}$ be a matrix of integers, such that $\Delta = \det \textbf{A} = ad - bc$ is relatively prime to the positive integer m. Then, the matrix

$$\overline{\textbf{A}} = \overline{\Delta} \begin{bmatrix} d & -b \\ -c & a \end{bmatrix},$$

where $\overline{\Delta}$ is the inverse of Δ modulo m, is an inverse of **A** modulo m.

Proof. To verify that the matrix $\overline{\textbf{A}}$ is an inverse of **A** modulo m, we need only verify that $\textbf{A}\overline{\textbf{A}} \equiv \overline{\textbf{A}}\textbf{A} \equiv \textbf{I} \pmod{m}$.

To see this, note that

$$\mathbf{A}\overline{\mathbf{A}} \equiv \begin{bmatrix} a & b \\ c & d \end{bmatrix} \overline{\Delta} \begin{bmatrix} d & -b \\ -c & a \end{bmatrix} \equiv \overline{\Delta} \begin{bmatrix} ad - bc & 0 \\ 0 & -bc + ad \end{bmatrix}$$

$$\equiv \overline{\Delta} \begin{bmatrix} \Delta & 0 \\ 0 & \Delta \end{bmatrix} \equiv \begin{bmatrix} \overline{\Delta}\Delta & 0 \\ 0 & \overline{\Delta}\Delta \end{bmatrix} \equiv \begin{bmatrix} 1 & 0 \\ 0 & 1 \end{bmatrix} = \mathbf{I} \ (\text{mod } m)$$

and

$$\overline{\mathbf{A}}\mathbf{A} \equiv \overline{\Delta} \begin{bmatrix} d & -b \\ -c & a \end{bmatrix} \begin{bmatrix} a & b \\ c & d \end{bmatrix} \equiv \overline{\Delta} \begin{bmatrix} ad - bc & 0 \\ 0 & -bc + ad \end{bmatrix}$$

$$\equiv \overline{\Delta} \begin{bmatrix} \Delta & 0 \\ 0 & \Delta \end{bmatrix} \equiv \begin{bmatrix} \overline{\Delta}\Delta & 0 \\ 0 & \overline{\Delta}\Delta \end{bmatrix} \equiv \begin{bmatrix} 1 & 0 \\ 0 & 1 \end{bmatrix} = \mathbf{I} \ (\text{mod } m),$$

where $\overline{\Delta}$ is an inverse of Δ (mod m), which exists because $(\Delta, m) = 1$. ∎

Example 3.22. Let $\mathbf{A} = \begin{bmatrix} 3 & 4 \\ 2 & 5 \end{bmatrix}$. Since 2 is an inverse det $\mathbf{A} = 7$ modulo 13, we have

$$\overline{\mathbf{A}} \equiv 2 \begin{bmatrix} 5 & -4 \\ -2 & 3 \end{bmatrix} \equiv \begin{bmatrix} 10 & -8 \\ -4 & 6 \end{bmatrix} \equiv \begin{bmatrix} 10 & 5 \\ 9 & 6 \end{bmatrix} \ (\text{mod } 13).$$

□

To provide a formula for an inverse of an $n \times n$ matrix where n is a positive integer greater than 2, we need a result from linear algebra. This result may be found in Anton [96; page 79]. It involves the notion of the adjoint of a matrix, which is defined as follows.

Definition. The *adjoint* of an $n \times n$ matrix $\mathbf{A}$ is the $n \times n$ matrix with (i,j)th entry C_{ji}, where C_{ij} is $(-1)^{i+j}$ times the determinant of the matrix obtained by deleting the ith row and jth column from $\mathbf{A}$. The adjoint of $\mathbf{A}$ is denoted by adj($\mathbf{A}$).

Theorem 3.17. If $\mathbf{A}$ is an $n \times n$ matrix with det $\mathbf{A} \neq 0$, then $\mathbf{A}(\text{adj } \mathbf{A}) = (\det \mathbf{A})\mathbf{I}$, where adj $\mathbf{A}$ is the adjoint of $\mathbf{A}$.

Using this theorem, the following theorem follows readily.

Theorem 3.18. If $\mathbf{A}$ is an $n \times n$ matrix with integer entries and m is a positive integer such that $(\det \mathbf{A}, m) = 1$, then the matrix $\overline{\mathbf{A}} = \overline{\Delta} \ (\text{adj } \mathbf{A})$ is an inverse of $\mathbf{A}$ modulo m, where $\overline{\Delta}$ is an inverse of $\Delta = \det \mathbf{A}$ modulo m.

Proof. If $(\det \mathbf{A}, m) = 1$, then we know that det $\mathbf{A} \neq 0$. Hence, by Theorem 3.15 we have

$$\mathbf{A} \cdot \text{adj } \mathbf{A} = (\det \mathbf{A})\mathbf{I} = \Delta\mathbf{I}.$$

Since $(\det \mathbf{A}, m) = 1$, there is an inverse $\overline{\Delta}$ of $\Delta = \det \mathbf{A}$ modulo m. Hence,

$$\mathbf{A}\,(\overline{\Delta}\,\text{adj}\,\mathbf{A}) \equiv \mathbf{A}\cdot(\text{adj}\,\mathbf{A})\overline{\Delta} \equiv \Delta\overline{\Delta}\mathbf{I} \equiv \mathbf{I}\ (\text{mod}\ m),$$

and

$$\overline{\Delta}\,(\text{adj}\,\mathbf{A})\,\mathbf{A} \equiv \overline{\Delta}\,(\text{adj}\,\mathbf{A}\cdot\mathbf{A}) \equiv \Delta\overline{\Delta}\mathbf{I} \equiv \mathbf{I}\ (\text{mod}\ m).$$

This shows that $\overline{\mathbf{A}} = \overline{\Delta}\cdot(\text{adj}\,\mathbf{A})$ is an inverse of $\mathbf{A}$ modulo m.   ∎

Example 3.23. Let $\mathbf{A} = \begin{bmatrix} 2 & 5 & 6 \\ 2 & 0 & 1 \\ 1 & 2 & 3 \end{bmatrix}$. Then $\det \mathbf{A} = -5$. Since $(\det \mathbf{A}, 7) = 1$, and an inverse of $\det \mathbf{A} = -5$ is $4\ (\text{mod}\ 7)$, we find that

$$\overline{\mathbf{A}} = 4\,(\text{adj}\,\mathbf{A}) = 4\begin{bmatrix} -2 & -3 & 5 \\ -5 & 0 & 10 \\ 4 & 1 & -10 \end{bmatrix} = \begin{bmatrix} -8 & -12 & 20 \\ -20 & 0 & 40 \\ 16 & 4 & -40 \end{bmatrix} \equiv \begin{bmatrix} 6 & 2 & 6 \\ 1 & 0 & 5 \\ 2 & 4 & 2 \end{bmatrix}\ (\text{mod}\ 7).$$ □

We can use an inverse of $\mathbf{A}$ modulo m to solve the system

$$\mathbf{A}\mathbf{X} \equiv \mathbf{B}\ (\text{mod}\ m),$$

where $(\det \mathbf{A}, m) = 1$. By Theorem 3.15, when we multiply both sides of this congruence by an inverse $\overline{\mathbf{A}}$ of $\mathbf{A}$, we obtain

$$\overline{\mathbf{A}}(\mathbf{A}\mathbf{X}) \equiv \overline{\mathbf{A}}\mathbf{B}\ (\text{mod}\ m)$$
$$(\overline{\mathbf{A}}\,\mathbf{A})\mathbf{X} \equiv \overline{\mathbf{A}}\mathbf{B}\ (\text{mod}\ m)$$
$$\mathbf{X} \equiv \overline{\mathbf{A}}\mathbf{B}\ (\text{mod}\ m).$$

Hence, we find the solution $\mathbf{X}$ by forming $\overline{\mathbf{A}}\,\mathbf{B}\ (\text{mod}\ m)$.

Note that this method provides another proof of Theorem 3.14. To see this,

let $\mathbf{A}\mathbf{X} = \mathbf{B}$, where $\mathbf{A} = \begin{bmatrix} a & b \\ c & d \end{bmatrix}$, $\mathbf{X} = \begin{bmatrix} x \\ y \end{bmatrix}$, and $\mathbf{B} = \begin{bmatrix} e \\ f \end{bmatrix}$. If $\Delta = \det \mathbf{A} = ad - bc$ is relatively prime to m, then

$$\begin{bmatrix} x \\ y \end{bmatrix} = \mathbf{X} \equiv \overline{\mathbf{A}}\,\mathbf{B} \equiv \overline{\Delta}\begin{bmatrix} d & -b \\ -c & a \end{bmatrix}\begin{bmatrix} e \\ f \end{bmatrix} = \overline{\Delta}\begin{bmatrix} de - bf \\ af - ce \end{bmatrix}\ (\text{mod}\ m).$$

This demonstrates that (x, y) is a solution if and only if

$$x \equiv \overline{\Delta}\,(de - bf)\ (\text{mod}\ m),\quad y \equiv \overline{\Delta}\,(af - ce)\ (\text{mod}\ m).$$

Next, we give an example of the solution of a system of three congruences in three unknowns using matrices.

Example 3.24. We consider the system of three congruences

$$2x_1 + 5x_2 + 6x_3 \equiv 3 \pmod 7$$
$$2x_1 + x_3 \equiv 4 \pmod 7$$
$$x_1 + 2x_2 + 3x_3 \equiv 1 \pmod 7.$$

This is equivalent to the matrix congruence

$$\begin{bmatrix} 2 & 5 & 6 \\ 2 & 0 & 1 \\ 1 & 2 & 3 \end{bmatrix} \begin{bmatrix} x_1 \\ x_2 \\ x_3 \end{bmatrix} \equiv \begin{bmatrix} 3 \\ 4 \\ 1 \end{bmatrix} \pmod 7.$$

We have previously shown that the matrix $\begin{bmatrix} 6 & 2 & 6 \\ 1 & 0 & 5 \\ 2 & 4 & 2 \end{bmatrix}$ is an inverse of $\begin{bmatrix} 2 & 5 & 6 \\ 2 & 0 & 1 \\ 1 & 2 & 3 \end{bmatrix} \pmod 7$.

Hence, we have

$$\begin{bmatrix} x_1 \\ x_2 \\ x_3 \end{bmatrix} = \begin{bmatrix} 6 & 2 & 6 \\ 1 & 0 & 5 \\ 2 & 4 & 2 \end{bmatrix} \begin{bmatrix} 3 \\ 4 \\ 1 \end{bmatrix} = \begin{bmatrix} 32 \\ 8 \\ 24 \end{bmatrix} \equiv \begin{bmatrix} 4 \\ 1 \\ 3 \end{bmatrix} \pmod 7. \qquad \square$$

Before leaving this subject, we should mention that many methods used for solving systems of linear equations may be adapted to solve systems of congruences. For instance, Gaussian elimination may be adapted to solve systems of congruences where division is always replaced by multiplication by inverses modulo m. Also, there is a method for solving systems of congruences analogous to Cramer's rule. We leave the development of these methods as exercises for those readers familiar with linear algebra.

3.4 Exercises

1. Find the solutions of each of the following systems of linear congruences.

   a) $x + 2y \equiv 1 \pmod 5$
      $2x + y \equiv 1 \pmod 5$

   b) $x + 3y \equiv 1 \pmod 5$
      $3x + 4y \equiv 2 \pmod 5$

   c) $4x + y \equiv 2 \pmod 5$
      $2x + 3y \equiv 1 \pmod 5$

2. Find the solutions of each of the following systems of linear congruences.

   a) $2x + 3y \equiv 5 \pmod 7$
      $x + 5y \equiv 6 \pmod 7$

b) $4x + y \equiv 5 \pmod 7$
   $x + 2y \equiv 4 \pmod 7$

* 3.   What are the possibilities for the number of incongruent solutions of the system of linear congruences

$$ax + by \equiv c \pmod p$$
$$dx + ey \equiv f \pmod p,$$

where p is a prime and $a,b,c,d,e,$ and f are positive integers?

4.   Find the matrix $\mathbf{C}$ such that

$$\mathbf{C} \equiv \begin{bmatrix} 2 & 1 \\ 4 & 3 \end{bmatrix}\begin{bmatrix} 4 & 0 \\ 2 & 1 \end{bmatrix} \pmod 5$$

and all entries of $\mathbf{C}$ are nonnegative integers less than 5.

5.   Use mathematical induction to prove that if $\mathbf{A}$ and $\mathbf{B}$ are $n \times n$ matrices with integer entries such that $\mathbf{A} \equiv \mathbf{B} \pmod m$, then $\mathbf{A}^k \equiv \mathbf{B}^k \pmod m$ for all positive integers k.

A matrix $\mathbf{A} \neq \mathbf{I}$ is called *involutory modulo* m if $\mathbf{A}^2 \equiv \mathbf{I} \pmod m$.

6.   Show that $\begin{bmatrix} 4 & 11 \\ 1 & 22 \end{bmatrix}$ is involutory modulo 26.

7.   Prove or disprove that if $\mathbf{A}$ is a 2×2 involutory matrix modulo m, then $\det \mathbf{A} \equiv \pm 1 \pmod m$.

8.   Find an inverse modulo 5 of each of the following matrices.

a) $\begin{bmatrix} 0 & 1 \\ 1 & 0 \end{bmatrix}$ b) $\begin{bmatrix} 1 & 2 \\ 3 & 4 \end{bmatrix}$ c) $\begin{bmatrix} 2 & 2 \\ 1 & 2 \end{bmatrix}$

9.   Find an inverse modulo 7 of each of the following matrices.

a) $\begin{bmatrix} 1 & 1 & 0 \\ 1 & 0 & 1 \\ 0 & 1 & 1 \end{bmatrix}$ b) $\begin{bmatrix} 1 & 2 & 3 \\ 1 & 2 & 5 \\ 1 & 4 & 6 \end{bmatrix}$ c) $\begin{bmatrix} 1 & 1 & 1 & 0 \\ 1 & 1 & 0 & 1 \\ 1 & 0 & 1 & 1 \\ 0 & 1 & 1 & 1 \end{bmatrix}$

10.   Using Exercise 9 find all solutions of each of the following systems.

a) $x + y \equiv 1 \pmod 7$
   $x + z \equiv 2 \pmod 7$
   $y + z \equiv 3 \pmod 7$

b) $x + 2y + 3z \equiv 1 \pmod 7$
   $x + 2y + 5z \equiv 1 \pmod 7$
   $x + 4y + 6z \equiv 1 \pmod 7$

c) $x+y+z \equiv 1 \pmod 7$
   $x+y+w \equiv 1 \pmod 7$
   $x+z+w \equiv 1 \pmod 7$
   $y+z+w \equiv 1 \pmod 7$

11. How many incongruent solutions does each of the following systems of congruences have?

a) $x + y + z \equiv 1 \pmod 5$
   $2x + 4y + 3z \equiv 1 \pmod 5$

b) $2x + 3y + z \equiv 3 \pmod 5$
   $x + 2y + 3z \equiv 1 \pmod 5$
   $2x + z \equiv 1 \pmod 5$

c) $3x + y + 3z \equiv 1 \pmod 5$
   $x + 2y + 4z \equiv 2 \pmod 5$
   $4x + 3y + 2z \equiv 3 \pmod 5$

d) $2x + y + z \equiv 1 \pmod 5$
   $x + 2y + z \equiv 1 \pmod 5$
   $x + y + 2z \equiv 1 \pmod 5$

* 12. Develop an analogue of Cramer's rule for solving systems of n linear congruences in n unknowns.

* 13. Develop an analogue of Gaussian elimination to solve systems of n linear congruences in m unknowns (where m and n may be different).

A *magic square* is a square array of integers with the property that the sum of the integers in a row or in a column is always the same. In this exercise, we present a method for producing magic squares.

* 14. Show that the n^2 integers $0,1,...,n^2 - 1$ are put into the n^2 positions of an $n \times n$ square, without putting two integers in the same position, if the integer k is placed in the ith row and jth column, where

$$i \equiv a + ck + e[k/n] \pmod n,$$
$$j \equiv b + dk + f[k/n] \pmod n,$$

$1 \le i \le n, \ 1 \le j \le n,$ and $a, b, c, d, e,$ and f are integers with $(cf - de, n) = 1$.

* 15. Show that a magic square is produced in Exercise 14 if $(c,n) = (d,n) = (e,n) = (f,n) = 1$.

* 16. The *positive* and *negative diagonals* of an $n \times n$ square consist of the integers in positions (i,j), where $i + j \equiv k \pmod n$ and $i - j \equiv k \pmod n$, respectively, where k is a given integer. A square is called *diabolic* if the sum of the integers in a positive or negative diagonal is always the same. Show that a diabolic square is produced using the procedure given in Exercise 14 if $(c+d,n) = (c-d,n) = (e+f,n) = (e-f,n) = 1$.

3.4 Computer Projects

Programming Projects

Write programs to do the following:

1. Find the solutions of a system of two linear congruences in two unknowns using Theorem 3.14.

2. Find inverses of 2×2 matrices using Theorem 3.16.

3. Find inverses of $n\times n$ matrices using Theorem 3.18.

4. Solve systems of n linear congruences in n unknowns using inverses of matrices.

5. Solve systems of n linear congruences in n unknowns using an analogue of Cramer's rule (see Exercise 12).

6. Solve system of n linear congruences in m unknowns using an analogue of Gaussian elimination (see Exercise 13).

7. Given a posiitve integer produce an $n\times n$ magic square by the method given in Exercise 14.

Computations and Explorations

Using the programs you have written or a computation program, carry out the following computations and explorations:

1. Produce 4×4, 5×5 and 6×6 magic squares.

3.5 Factoring Using the Pollard Rho Method

In this section we will describe a factorization method based on congruences which was developed in 1974 by J.M. Pollard. Pollard called this technique the *Monte Carlo method* because it relies on generating certain integers so that they have properties of randomly chosen integers. Although Pollard called this method the Monte Carlo method, it is now commonly known as the *Pollard rho method*, for reasons which will be explained later.

Suppose n is a large composite integer and that p is its smallest prime divisor. Our goal is to choose integers $x_0, x_1,...,x_s$ so that these integers have distinct least nonnegative residues modulo n, but where their least nonnegative residues modulo p are not all distinct. As can be seen using probabilistic arguments (see [24], [30], or [55], for example), this is likely to be the case when s is large compared to $\sqrt{p}$ but small when compared to $\sqrt{n}$ and the numbers are chosen randomly.

Once we have found integers x_i and x_j where $0 \le i < j \le s$ such that $x_i \equiv x_j \pmod{p}$ but $x_i \not\equiv x_j \pmod{n}$, it follows that $(x_i - x_j, n)$ is a nontrivial divisor of n, since p divides $x_i - x_j$, but n does not. The number $(x_i - x_j, n)$ can be found quickly using the Euclidean algorithm. However, to find $(x_i - x_j, n)$ for each pair (i, j) with $0 \le i < j \le s$ requires that we find $O(s^2)$ greatest common divisors. We will show how reduce the number of times we need to use the Euclidean algorithm.

To find such integers x_i and x_j we use the following procedure. We start with a seed value x_0 which is chosen randomly and a polynomial function $f(x)$ with integer coefficients of degree greater than one. We compute the terms x_k, $k = 1, 2, 3, ...,$ using the recursive definition

$$x_{k+1} \equiv f(x_k) \pmod{n}, \quad 0 \le x_{k+1} < n.$$

The polynomial $f(x)$ should have the property that the sequence $x_0, x_1, ..., x_k,$ ... behaves much like a truly random sequence would. (Of course, it is not a random sequence since we have given a rule for computing terms. However, this sequence can have terms which have some properties that a random sequence would). The following example illustrates how this sequence is generated.

Example 3.25. Let $n = 8051$ and suppose that $x_0 = 2$ and $f(x) = x^2 + 1$. We find that $x_1 = 5$, $x_2 = 26$, $x_3 = 677$, $x_4 = 7474$, $x_5 = 2839$, $x_6 = 871$, and so on. □

Now note that by the recursive definition of x_k, it follows that if

$$x_i \equiv x_j \pmod{d},$$

where d is a positive integer, then

$$x_{i+1} \equiv f(x_i) \equiv f(x_j) \equiv x_{j+1} \pmod{d}.$$

It follows that if $x_i \equiv x_j \pmod{d}$ then the sequence x_k becomes periodic modulo d with a period $j - i$. That is, $x_q \equiv x_r \pmod{d}$ whenever $q \equiv r \pmod{j-i}$, and $q \ge i$ and $r \ge i$. It follows that if s is the smallest multiple of $j-i$ that is as least i, then $x_s \equiv x_{2s} \pmod{d}$.

It follows that to look for a factor of n, we find the greatest common divisor of $x_{2k} - x_k$ and n for $k = 1, 2, 3, ...$. We have found a factor of n when we have found a value k for which $1 < x_{2k} - x_k < n$. From the observations we have made, we see that it is likely that we will find such an integer k with k close to $\sqrt{p}$.

In practice, when the Pollard rho method is used, the polynomial $f(x) = x^2 + 1$ is often chosen to generate the sequence of integers $x_0, x_1, x_2, ..., x_k, ...$. Furthermore, the seed $x_0 = 2$ is often used. This choice of polynomial and seed produces a sequence that behaves much like a random sequence for the purposes of this factorization method.

Example 3.26. Use the Pollard rho method with seed $x_0 = 2$ and generator polynomial $f(x) = x^2 + 1$ to find a nontrivial factor of $n = 8051$. We find that

$x_1 = 5$, $x_2 = 26$, $x_3 = 677$, $x_4 = 7474$, $x_5 = 2839$, $x_6 = 871$. Using the Euclidean algorithm it follows that $(x_2 - x_1, 8051) = (26 - 5, 8051) = (21, 8051) = 1$ and $(x_4 - x_2, 8051) = (7474 - 26, 8051) = (7448, 8051) = 1$. However, we find a nontrivial factor of 8051 at the next step since $(x_6 - x_3, 8051) = (871 - 677, 8051) = (194, 8051) = 97$. We see that 97 is a factor of 8051. □

To see why this method is called the Pollard rho method look at Figure 3.1. This figure shows the periodic behavior of the sequence x_i where $x_0 = 2$ and $x_{i+1} = x_i^2 + 1 \pmod{97}$, $i \geq 1$. The part of this sequence that occurs before the periodicity is the tail of the rho and the loop is the periodic part.

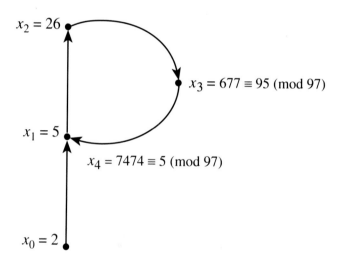

Figure 3.1 The Pollard Rho Method.

The Pollard rho method has proved to be practical for the factorization of integers with moderately large prime factors. In practice, the first attempt to factor a large integer is to do trial division by small primes, say by all primes less than 10000. Next, the Pollard rho method is used to look for prime factors of intermediate size (say up to 10^{15}, for instance.) Only after trial division by small primes and the Pollard rho method have been used are the really big guns brought in, such as the quadratic sieve or the elliptic curve method.

3.5 Exercises

1. Use the Pollard rho method with $x_0 = 2$ and $f(x) = x^2 + 1$ to find the prime factorization of each of the following integers.

    a) 133   c) 1927   e) 36287
    b) 1189   d) 8131   f) 48227

2. Use the Pollard rho method to factor the integer 1387 with

    a) $x_0 = 2, f(x) = x^2 + 1$
    b) $x_0 = 3, f(x) = x^2 + 1$
    c) $x_0 = 2, f(x) = x^2 - 1$
    d) $x_0 = 2, f(x) = x^3 + x + 1$

* 3. Explain why choosing $f(x)$ to be a linear polynomial, that is, a function of the
    form $f(x) = ax + b$ where a and b are integers, is a poor choice.

3.5 Computer Projects

Programming Projects

Write computer programs to do the following:

1. Given a positive integer n find a prime factor of this integer using the Pollard
    rho method.

Computations and Explorations

Using the programs you have written or a computation program, carry out the following
computations and explorations:

1. Use the Pollard rho method to factor ten different integers with between 15 and 20
    decimal digits.

2. Use the Pollard rho method to factor a large number of integers that are close to
    100000, keeping track of the number of steps required. Can you make any
    conjectures based on your data?

4

Applications of Congruences

4.1 Divisibility Tests

Using congruences we can develop divisibility tests for integers based on their expansions with respect to different bases.

We begin with tests that use decimal notation. In the following discussion let $n = (a_k a_{k-1} \ldots a_1 a_0)_{10}$. Then $n = a_k 10^k + a_{k-1} 10^{k-1} + \cdots + a_1 10 + a_0$, with $0 \le a_j \le 9$ for $j = 0, 1, 2, \ldots, k$.

First, we develop tests for divisibility by powers of 2.

Since $10 \equiv 0 \pmod 2$, it follows that $10^j \equiv 0 \pmod{2^j}$ for all positive integers j. Hence,

$$n \equiv (a_0)_{10} \pmod 2,$$
$$n \equiv (a_1 a_0)_{10} \pmod{2^2},$$
$$n \equiv (a_2 a_1 a_0)_{10} \pmod{2^3},$$

.

.

.

$$n \equiv (a_{k-1} a_{k-2} \ldots a_2 a_1 a_0)_{10} \pmod{2^k}.$$

These congruences tell us that to determine whether an integer n is divisible by 2, we only need to examine its last digit for divisibility by 2. Similarly, to determine whether n is divisible by 4, we only need to check the integer made up of the last two digits of n for divisibility by 4. In general, to test n for divisibility by 2^j, we only need to check the integer made up of the last j digits of n for divisibility by 2^j.

Example 4.1. Let $n = 32688048$. We see that $2 \mid n$ since $2 \mid 8$, $4 \mid n$ since

$4 \mid 48$, $8 \mid n$ since $8 \mid 48$, $16 \mid n$ since $16 \mid 8048$, but $32 \nmid n$ since $32 \nmid 88048$. $\qquad$ □

Next, we develop divisibility tests for powers of 5.

To develop tests for divisibility by powers of 5, first note that since $10 \equiv 0 \pmod{5}$ we have $10^j \equiv 0 \pmod{5^j}$. Hence, divisibility tests for powers of 5 are analogous to those for powers of 2. We only need to check the integer made up of the last j digits of n to determine whether n is divisible by 5^j.

Example 4.2. Let $n = 15535375$. Since $5 \mid 5$, $5 \mid n$, since $25 \mid 75$, $25 \mid n$, since $125 \mid 375$, $125 \mid n$, but since $625 \nmid 5375$, $625 \nmid n$. □

Next, we develop tests for divisibility by 3 and by 9.

Note that both the congruences $10 \equiv 1 \pmod{3}$ and $10 \equiv 1 \pmod{9}$ hold. Hence, $10^k \equiv 1 \pmod{3}$ and $10^k \equiv 1 \pmod{9}$. This gives us the useful congruences

$$
\begin{aligned}
(a_k a_{k-1} \cdots a_1 a_0) &= a_k 10^k + a_{k-1} 10^{k-1} + \cdots + a_1 10 + a_0 \\
&\equiv a_k + a_{k-1} + \cdots + a_1 + a_0 \pmod{3} \text{ and } \pmod{9}.
\end{aligned}
$$

Hence, we only need to check whether the sum of the digits of n is divisible by 3, or by 9, to see whether n is divisible by 3, or by 9.

Example 4.3. Let $n = 4127835$. Then, the sum of the digits of n is $4 + 1 + 2 + 7 + 8 + 3 + 5 = 30$. Since $3 \mid 30$ but $9 \nmid 30$, $3 \mid n$ but $9 \nmid n$. □

A rather simple test can be found for divisibility by 11.

Since $10 \equiv -1 \pmod{11}$, we have

$$
\begin{aligned}
(a_k a_{k-1} \ldots a_1 a_0)_{10} &= a_k 10^k + a_{k-1} 10^{k-1} + \cdots + a_1 10 + a_0 \\
&\equiv a_k(-1)^k + a_{k-1}(-1)^{k-1} + \cdots - a_1 + a_0 \pmod{11}.
\end{aligned}
$$

This shows that $(a_k a_{k-1} \ldots a_1 a_0)_{10}$ is divisible by 11 if and only if $a_0 - a_1 + a_2 - \cdots + (-1)^k a_k$, the integer formed by alternately adding and subtracting the digits, is divisible by 11.

Example 4.4. We see that 723160823 is divisible by 11, since alternately adding and subtracting its digits yields $3 - 2 + 8 - 0 + 6 - 1 + 3 - 2 + 7 = 22$ which is divisible 11. On the other hand, 33678924 is not divisible by 11, since $4 - 2 + 9 - 8 + 7 - 6 + 3 - 3 = 4$ is not divisible by 11. □

Next, we develop a test to simultaneously test for divisibility by the primes 7, 11, and 13.

Note that $7 \cdot 11 \cdot 13 = 1001$ and $10^3 = 1000 \equiv -1 \pmod{1001}$. Hence,

$$
\begin{aligned}
(a_k a_{k-1}...a_0)_{10} &= a_k 10^k + a_{k-1} 10^{k-1} + \cdots + a_1 10 + a_0 \\
&\equiv (a_0 + 10a_1 + 100a_2) + 1000(a_3 + 10a_4 + 100a_5) + \\
&\quad (1000)^2(a_6 + 10a_7 + 100a_8) + \cdots \\
&\equiv (100a_2 + 10a_1 + a_0) - (100a_5 + 10a_4 + a_3) + \\
&\quad (100a_8 + 10a_7 + a_6) - \cdots \\
&= (a_2 a_1 a_0)_{10} - (a_5 a_4 a_3)_{10} + (a_8 a_7 a_6)_{10} - \cdots \pmod{1001}.
\end{aligned}
$$

This congruence tells us that an integer is congruent modulo 1001 to the integer formed by successively adding and subtracting the three-digit integers with decimal expansions formed from successive blocks of three decimal digits of the original number, where digits are grouped starting with the rightmost digit. As a consequence, since 7, 11, and 13 are divisors of 1001, to determine whether an integer is divisible by 7, 11, or 13, we only need to check whether this alternating sum and difference of blocks of three digits is divisible by 7, 11, or 13.

Example 4.5. Let $n = 59358208$. Since the alternating sum and difference of the integers formed from blocks of three digits, $208 - 358 + 59 = -91$, is divisible by 7 and 13, but not by 11, we see that n is divisible by 7 and 13, but not by 11. ☐

All of the divisibility tests we have developed thus far are based on decimal representations. We now develop divisibility tests using base b representations, where b is a positive integer.

Theorem 4.1. If $d \mid b$ and j and k are positive integers with $j < k$, then $(a_k...a_1 a_0)_b$ is divisible by d^j if and only if $(a_{j-1}...a_1 a_0)_b$ is divisible by d^j.

Proof. Since $b \equiv 0 \pmod{d}$, it follows that $b^j \equiv 0 \pmod{d^j}$. Hence,

$$
\begin{aligned}
(a_k a_{k-1}...a_1 a_0)_b &= a_k b^k + \cdots + a_j b^j + a_{j-1} b^{j-1} + \cdots + a_1 b + a_0 \\
&\equiv a_{j-1} b^{j-1} + \cdots + a_1 b + a_0 \\
&= (a_{j-1}...a_1 a_0)_b \pmod{d^j}.
\end{aligned}
$$

Consequently, $d \mid (a_k a_{k-1}...a_1 a_0)_b$ if and only if $d \mid (a_{j-1}...a_1 a_0)_b$. ∎

Theorem 4.1 extends to other bases the divisibility tests of integers expressed in decimal notation by powers of 2 and by powers of 5.

Theorem 4.2. If $d \mid (b-1)$, then $n = (a_k...a_1 a_0)_b$ is divisible by d if and only if the sum of digits $a_k + \cdots + a_1 + a_0$ is divisible by d.

Proof. Since $d \mid (b-1)$, we have $b \equiv 1 \pmod{d}$, so that by Theorem 3.7 we have $b^j \equiv 1 \pmod{d}$ for all positive integers b. Hence, $n = (a_k...a_1 a_0)_b = a_k b^k + \cdots + a_1 b + a_0 \equiv a_k + \cdots + a_1 + a_0 \pmod{d}$. This shows that $d \mid n$ if and only if $d \mid (a_k + \cdots + a_1 + a_0)$. ∎

Theorem 4.2 extends to other bases the tests for divisibility of integers expressed in decimal notation by 3 and by 9.

Theorem 4.3. If $d \mid (b + 1)$, then $n = (a_k...a_1a_0)_b$ is divisible by d if and only if the alternating sum of digits $(-1)^k a_k + \cdots - a_1 + a_0$ is divisible by d.

Proof. Since $d \mid (b + 1)$, we have $b \equiv -1 \pmod{d}$. Hence, $b^j \equiv (-1)^j \pmod{d}$, and consequently, $n = (a_k \cdots a_1a_0)_b \equiv (-1)^k a_k + \cdots - a_1 + a_0 \pmod{d}$. Hence, $d \mid n$ if and only if $d \mid ((-1)^k a_k + \cdots - a_1 + a_0)$. ∎

Theorem 4.3 extends to other bases the test for divisibility by 11 of integers expressed in decimal notation.

Example 4.6. Let $n = (7F28A6)_{16}$ (in hex notation). Then, since $2 \mid 16$, from Theorem 4.1 we know that $2 \mid n$, since $2 \mid 6$. Likewise, since $4 \mid 16$, we see that $4 \nmid n$, since $4 \nmid 6$. By Theorem 4.2, since $3 \mid (16 - 1)$, $5 \mid (16 - 1)$, and $15 \mid (16 - 1)$, and $7 + F + 2 + 8 + A + 6 = (30)_{16}$, we know that $3 \mid n$, since $3 \mid (30)_{16}$, while $5 \nmid n$ and $15 \nmid n$, since $5 \nmid (30)_{16}$ and $15 \nmid (30)_{16}$. Furthermore, by Theorem 4.3, since $17 \mid (16 + 1)$ and $n \equiv 6 - A + 8 - 2 + F - 7 = (A)_{16} \pmod{17}$, we conclude that $17 \nmid n$, since $17 \nmid (A)_{16}$. □

Example 4.7. Let $n = (1001001111)_2$. Then, using Theorem 4.3 we see that $3 \mid n$, since $n \equiv 1 - 1 + 1 - 1 + 0 - 0 + 1 - 0 + 0 - 1 \equiv 0 \pmod{3}$ and $3 \mid (2 + 1)$. □

4.1 Exercises

1. Determine the highest power of 2 dividing each of the following positive integers.

   a) 201984   c) 89375744
   b) 1423408   d) 41578912246

2. Determine the highest power of 5 dividing each of the following positive integers.

   a) 112250   c) 235555790
   b) 4860625   d) 48126953125

3. Which of the following integers are divisible by 3? Of those that are, which are divisible by 9?

   a) 18381   c) 987654321
   b) 65412351   d) 78918239735

4. Which of the following integers are divisible by 11?

    a)   10763732   c)   674310976375
    b)   1086320015   d)   8924310064537

5. Find the highest power of two dividing each of the following integers.

    a) $(101111110)_2$ c) $(111000000)_2$
    b) $(1010000011)_2$ d) $(1011011101)_2$

6. Determine which of the integers in Exercise 5 are divisible by 3.

7. Which of the following integers are divisible by 2?

    a) $(1210122)_3$ c) $(1112201112)_3$
    b) $(211102101)_3$ d) $(10122222011101)_3$

8. Which of the integers in Exercise 7 are divisible by 4?

9. Which of the following integers are divisible by 3 and which are divisible by 5?

    a) $(3EA235)_{16}$ c) $(F117921173)_{16}$
    b) $(ABCDEF)_{16}$ d) $(10AB987301F)_{16}$

10. Which of the integers in Exercise 9 are divisible by 17?

A *repunit* is an integer with decimal expansion containing all 1's.

11. Determine which repunits are divisible by 3; and which are divisible by 9.

12. Determine which repunits are divisible by 11.

13. Determine which repunits are divisible by 1001. Which are divisible by 7? by 13?

14. Determine which repunits with fewer than 10 digits are prime.

A *base b repunit* is an integer with base b expansion containing all 1's.

15. Determine which base b repunits are divisible by factors of $b - 1$.

16. Determine which base b repunits are divisible by factors of $b + 1$.

A *base b palindromic integer* is an integer whose base b representation reads the same forward and backward.

17. Show that every decimal palindromic integer with an even number of digits is divisible by 11.

18. Show that every base 7 palindromic integer with an even number of digits is divisible by 8.

19. Develop a test for divisibility by 37, based on the fact that $10^3 \equiv 1 \pmod{37}$. Use this to check 443692 and 11092785 for divisibility by 37.

20. Devise a divisibility test for integers represented in base b notation for divisibility by n where n is a divisor of $b^2 + 1$. (*Hint:* Split the digits of the base b representation of the integer into blocks of two, starting on the right).

21. Use the test you developed in Exercise 20 to decide whether

   a) $(101110110)_2$ is divisible by 5.
   b) $(12100122)_3$ is divisible by 2, and whether it is divisible by 5.
   c) $(364701244)_8$ is divisible by 5, and whether it is divisible by 13.
   d) $(5837041320219)_{10}$ is divisible by 101.

22. An old receipt has faded. It reads 88 chickens at a total of $x4.2y$ where x and y are unreadable digits. How much did each chicken cost?

23. Use a congruence modulo 9 to find the missing digit, indicated by a question mark: $89878 \cdot 58965 = 5299 ? 56270$.

We can check a multiplication $c = ab$ by determining whether the congruence $c \equiv ab \pmod{m}$ is valid, where m is any modulus. If we find that c is not congruent to ab modulo m, then we know that an error has been made. When we take $m = 9$ and use the fact that an integer in decimal notation is congruent modulo 9 to the sum of its digits, this check is called *casting out nines*.

24. Check each of the following multiplications by casting out nines.

   a) $875961 \cdot 2753 = 2410520633$
   b) $14789 \cdot 23567 = 348532367$
   c) $24789 \cdot 43717 = 1092700713$

25. Is a check of a multiplication by casting out nines foolproof?

26. What combinations of digits of a decimal expansion of an integer are congruent to this integer modulo 99? Use your answer to devise a check for multiplication based on *casting out ninety nines*. Then use the test to check the multiplications in Exercise 24.

4.1 Computer Projects

Programming Projects

Write programs to do the following:

1. Given a positive integer n, determine the highest powers of 2 and of 5 that divide n.

2. Given a positive integer n, test n for divisibility by 3, 7, 9, 11, and 13. (Use congruences modulo 1001 for divisibility by 7 and 13.)

3. Given a positive integer n, determine the highest power of each factor of b that divides an integer from the base b expansion of n.

4. Test a positive integer n from its base b expansion, for divisibility by factors of $b - 1$ and of $b + 1$.

Computations and Explorations

Using the programs you have written or a computation program, carry out the following computations and explorations:

1. Determine whether the repunit with n digits is prime where n is a positive integer not exceeding 30. Can you go further?

4.2 The Perpetual Calendar

In this section we derive a formula that gives us the day of the week of any day of any year. Since the days of the week form a cycle of length seven, we use a congruence modulo 7. We denote each day of the week by a number in the set 0, 1, 2, 3, 4, 5, 6, setting

- *Sunday* = 0,

- *Monday* = 1,

- *Tuesday* = 2,

- *Wednesday* = 3,

- *Thursday* = 4,

- *Friday* = 5,

- *Saturday* = 6.

Julius Caesar changed the Egyptian calendar, which was based on a year of exactly 365 days, to a new calendar, called the *Julian calendar*, with a year of average length 365¼ days, with leap years every fourth year, to better reflect the true length of the year. However, more recent calculations have shown that the true length of the year is approximately 365.2422 days. As the centuries passed, the discrepancies of 0.0078 days per year added up, so that by the year 1582 approximately 10 extra days had been added unnecessarily in leap years. To remedy this, in 1582 Pope Gregory set up a new calendar. First, 10 days were added to the date, so that October 5, 1582, became October 15, 1582 (and the 6th through the 14th of October were skipped). It was decided that leap years would be precisely the years divisible by 4, except that those exactly divisible by 100, *i.e.*, the years that mark centuries, would be leap years only when divisible by 400. As an example, the years 1700, 1800, 1900, and 2100 are not leap years but 1600 and 2000 are. With this arrangement, the average length of a calendar year is 365.2425 days, rather close to the true year of 365.2422 days. An error of 0.0003 days per year remains, which is 3 days per 10000 years. In the future, this discrepancy will have to be accounted for, and various possibilities have been suggested to correct for this error.

In dealing with calendar dates for various parts of the world, we must also take into account the fact that the Gregorian calendar was not adopted everywhere in

1582. In Britain and what is now the United States, the Gregorian calendar was adopted only in 1752, and by then, it was necessary to add 11 days. In these places September 3, 1752 in the Julian calendar became September 14, 1752 in the Gregorian calendar. Japan changed over in 1873, Russia and nearby countries in 1917, while Greece held out until 1923.

We now set up our procedure for finding the day of the week in the Gregorian calendar for a given date. We first must make some adjustments, because the extra day in a leap year comes at the end of February. We take care of this by renumbering the months, starting each year in March, and considering the months of January and February part of the preceding year. For instance, February 1984, is considered the 12th month of 1983, and May 1984, is considered the 3rd month of 1984. With this convention, for the day of interest, let

- k = day of the month,

- m = month,

  with

  — *January* = 11

  — *February* = 12

  — *March* = 1

  — *April* = 2

  — *May* = 3

  — *June* = 4

  — *July* = 5

  — *August* = 6

  — *September* = 7

  — *October* = 8

  — *November* = 9

  — *December* = 10

- N = year,

where N is the current year unless the month is January or February in which case N is the previous year with $N = 100C + Y$, where

- C = century,

- Y = particular year of the century.

Example 4.8. With April 3, 1951 we have $k = 3$, $m = 2$, $N = 1951$, $C = 19$, and $Y = 51$. But note that for February 28, 1951 we have $k = 28$, $m = 12$, $N = 1950$, $C = 19$, and $Y = 50$, since, for our calculations, we consider February to be the 12th month of the previous year.   □

We use March 1, of each year as our basis. Let d_N represent the day of the week of March 1, in year N. We start with the year 1600 and compute the day of the week March 1, falls on in any given year. Note that between March 1 of year $N - 1$ and March 1 of year N, if year N is not a leap year, 365 days have passed, and since $365 \equiv 1 \pmod 7$, we see that $d_N \equiv d_{N-1} + 1$ (mod 7), while if year N is a leap year, since there is an extra day between the consecutive firsts of March, we see that

$$d_N \equiv d_{N-1} + 2 \pmod 7.$$

Hence, to find d_N from d_{1600}, we must first find out how many leap years have occurred between the year 1600 and the year N (not including 1600, but including N); let us call this number x. To compute x, first note that by the division algorithm there are $[(N - 1600)/4]$ years divisible by 4 between 1600 and N, there are $[(N-1600)/100]$ years divisible by 100 between 1600 and N, and there are $[(N - 1600)/400]$ years divisible by 400 between 1600 and N. Hence,

$$\begin{aligned}
x &= [(N - 1600)/4] - [(N - 1600)/100] + [(N - 1600)/400] \\
&= [N/4] - 400 - [N/100] + 16 + [N/400] - 4 \\
&= [N/4] - [N/100] + [N/400] - 388.
\end{aligned}$$

(We have used Exercise 12 of Section 1.1 to simplify this expression.) Now putting this in terms of C and Y, we see that

$$\begin{aligned}
x &= [25C + (Y/4)] - [C + (Y/100)] + [(C/4) + (Y/400)] - 388 \\
&= 25C + [Y/4] - C + [C/4] - 388 \\
&\equiv 3C + [C/4] + [Y/4] - 3 \pmod 7.
\end{aligned}$$

Here we have again used Exercise 12 of Section 1.1, the inequality $Y/100 < 1$, and the equation $[(C/4) + (Y/400)] = [C/4]$ (which follows from Exercise 19 of Section 1.5, since $Y/400 < 1/4$).

We can now compute d_N from d_{1600} by shifting d_{1600} by one day for every year that has passed, plus an extra day for each leap year between 1600 and N. This gives the following formula:

$$\begin{aligned}
d_N &\equiv d_{1600} + N - 1600 + x \\
&= d_{1600} + 100C + Y - 1600 + 3C + [C/4] + [Y/4] - 3 \pmod 7.
\end{aligned}$$

Simplifying, we have

$$d_N \equiv d_{1600} - 2C + Y + [C/4] + [Y/4] \pmod 7.$$

Now that we have a formula relating the day of the week for March 1 of any year, with the day of the week of March 1, 1600, we can use the fact that March 1, 1982, is a Monday to find the day of the week of March 1, 1600. For 1982, since

$N = 1982$, we have $C = 19$, and $Y = 82$, and since $d_{1982} = 1$, it follows that

$$1 \equiv d_{1600} - 38 + 82 + [19/4] + [82/4] \equiv d_{1600} - 2 \pmod{7}.$$

Hence, $d_{1600} = 3$, so that March 1, 1600, was a Wednesday. When we insert the value of d_{1600}, the formula for d_N becomes

$$d_N \equiv 3 - 2C + Y + [C/4] + [Y/4] \pmod{7}.$$

We now use this formula to compute the day of the week of the first day of each month of year N. To do this, we have to use the number of days of the week that the first of the month of a particular month is shifted from the first of the month of the preceding month. The months with 30 days shift the first of the following month up 2 days, because $30 \equiv 2 \pmod{7}$, and those with 31 days shift the first of the following month up 3 days, because $31 \equiv 3 \pmod{7}$. Therefore, we must add the following amounts:

from March 1 to April 1:	3 days
from April 1 to May 1:	2 days
from May 1 to June 1:	3 days
from June 1 to July 1:	2 days
from July 1 to August 1:	3 days
from August 1 to September 1:	3 days
from September 1 to October 1:	2 days
from October 1 to November 1:	3 days
from November 1 to December 1:	2 days
from December 1 to January 1:	3 days
from January 1 to February 1:	3 days.

We need a formula that gives us the same increments. Notice that we have 11 increments totaling 29 days, so that each increment averages 2.6 days. By inspection, we find that the function $[2.6m - 0.2] - 2$ has exactly the same increments as m goes from 2 to 12, and is zero when $m = 1$. (This formula was originally found by a Reverend Zeller; he apparently found it by trial and error.) Hence, the day of the week of the first day of month m of year N is given by the least nonnegative residue of $d_N + [2.6m - 0.2] - 2$ modulo 7.

To find W, the day of the week of day k of month m of year N, we simply add $k - 1$ to the formula we have devised for the day of the week of the first day of the same month. We obtain the formula:

$$W \equiv k + [2.6m - 0.2] - 2C + Y + [Y/4] + [C/4] \pmod{7}.$$

We can use this formula to find the day of the week of any date of any year in the Gregorian calendar.

Example 4.9. To find the day of the week of January 1, 1900, we have

$C = 18$, $Y = 99$, $m = 11$, and $k = 1$ (since we consider January as the eleventh month of the preceding year). Hence, we have $W \equiv 1 + 28 - 36 + 99 + 24 + 4 \equiv 1 \pmod 7$, so that January 1, 1990 was a Monday. ☐

4.2 Exercises

1. Find the day of the week of the day you were born, and of your birthday this year.

2. Find the day of the week of the following important dates in U. S. history (use the Julian calendar before September 3, 1752, and the Gregorian calendar from September 14, 1752 to the present)

   * a) October 12, 1492   (Columbus sights land in the Caribbean)
   * b) May 6, 1692   (Peter Minuit buys Manhattan from the natives)
   * c) June 15, 1752   (Benjamin Franklin invents the lightning rod)
     d) July 4, 1776   (U. S. Declaration of Independence)
     e) March 30, 1867   (U. S. buys Alaska from Russia)
     f) March 17, 1888   (Great blizzard in the Eastern U. S.)
     g) February 15, 1898 (U. S. Battleship Maine blown up in Havana Harbor)
     h) July 2, 1925   (Scopes convicted of teaching evolution)
     i) July 16, 1945   (First atomic bomb exploded)
     j) July 20, 1969   (First man on the moon)
     k) August 9, 1974   (Nixon resigns)
     l) March 28, 1979   (Three Mile Island nuclear mishap)
     m) January 1, 1984   (Ma Bell break up)
     n) December 25, 1991 (Death of the U.S.S.R.)
     o) June 5, 2013   (First man on Mars)

3. How many times will the 13th of the month fall on a Friday in the year 1999?

4. How many leap years will there be from the year 1 until the year 10,000, inclusive?

5. To correct the small discrepancy between the number of days in a year of the Gregorian calendar and an actual year, it has been suggested that the years exactly divisible by 4000 should not be leap years. Adjust the formula for the day of the week of a given date to take this correction into account.

6. Show that days with the same calendar date in two different years of the same century, 28, 56, or 84 years apart, fall on the identical day of the week.

7. Which of your birthdays, until your one hundredth, fall on the same day of the week as the day you were born?

* 8. A new calendar called the *International Fixed Calendar* has been proposed. In this calendar, there are 13 months, including all our present months, plus a new month, called *Sol*, which is placed between June and July. Each month has 28

days, except for the June of leap years which has an extra day (leap years are determined the same way as in the Gregorian calendar). There is an extra day, *Year End Day*, which is not in any month, which we may consider as December 29. Devise a perpetual calendar for the International Fixed Calendar to give the day of the week for any calendar date.

4.2 Computer Projects

Programming Projects

Write programs to do the following:

1. To give the day of the week of any date.

2. To print out a calendar of any year.

3. To print out a calendar for the International Fixed Calendar (See Exercise 8).

Computations and Explorations

Using the programs you have written or a computation program, carry out the following computations and explorations:

1. Find the number of times the 13th of a month falls on a Friday for all years between 1800 and 2300. Can you make and prove a conjecture based on your evidence?

4.3 Round-Robin Tournaments

Congruences can be used to schedule round-robin tournaments. In this section, we show how to schedule a tournament for N different teams, so that each team plays every other team exactly once. The method we describe was developed by Freund [107].

First note that if N is odd, not all teams can be scheduled in each round, since when teams are paired, the total number of teams playing is even. So, if N is odd, we add a dummy team, and if a team is paired with the dummy team during a particular round, it draws a bye in that round and does not play. Hence, we can assume that we always have an even number of teams, with the addition of a dummy team if necessary.

Now label the N teams with the integers 1, 2, 3, ..., $N-1$, N. We construct a schedule, pairing teams in the following way. We have team i, with $i \neq N$, play team j, with $j \neq N$ and $j \neq i$, in the kth round if $i + j \equiv k \pmod{N-1}$. This schedules games for all teams in round k, except for team N and the one team i for which $2i \equiv k \pmod{N-1}$. There is one such team because Theorem 3.10 tells us that the congruence $2x \equiv k \pmod{N-1}$ has exactly one solution with $1 \leq x \leq N-1$, since $(2, N-1) = 1$. We match this team i with team N in the kth round.

We must now show that each team plays every other team exactly once. We consider the first $N-1$ teams. Note that team i, where $1 \leq i \leq N-1$, plays team N in round k where $2i \equiv k \pmod{N-1}$, and this happens exactly once. In the other rounds, team i does not play the same team twice, for if team i played team j in both rounds k and k', then $i + j \equiv k \pmod{N-1}$, and $i + j \equiv k' \pmod{N-1}$ which is an obvious contradiction because $k \not\equiv k' \pmod{N-1}$. Hence, since each of the first $N-1$ teams plays $N-1$ games, and does not play any team more than once, it plays every team exactly once. Also, team N plays $N-1$ games, and since every other team plays team N exactly once, team N plays every other team exactly once.

Example 4.10. To schedule a round-robin tournament with 5 teams, labeled 1, 2, 3, 4, and 5, we include a dummy team labeled 6. In round one, team 1 plays team j where $1 + j \equiv 1 \pmod 5$. This is the team $j = 5$ so that team 1 plays team 5. Team 2 is scheduled in round one with team 4, since the solution of $2 + j \equiv 1 \pmod 5$ is $j = 4$. Since $i = 3$ is the solution of the congruence $2i \equiv 1 \pmod 5$, team 3 is paired with the dummy team 6, and hence, draws a bye in the first round. If we continue this procedure and finish scheduling the other rounds, we end up with the pairings shown in Figure 4.1, where the opponent of team i in round k is given in the kth row and ith column.   □

Team Round	1	2	3	4	5
1	5	4	bye	2	1
2	bye	5	4	3	2
3	2	1	5	bye	3
4	3	bye	1	5	4
5	4	3	2	1	bye

Figure 4.1. Round-Robin Schedule for Five Teams.

4.3 Exercises

1. Set up a round-robin tournament schedule for

   a) 7 teams.   c) 9 teams.
   b) 8 teams.   d) 10 teams.

2. In round-robin tournament scheduling, we wish to assign a *home team* and an *away team* for each game so that each of n teams, where n is odd, plays an equal number of home games and away games. Show that if when $i + j$ is odd, we assign the smaller of i and j as the home team, while if $i + j$ is even, we assign the larger of i and j as the home team, then each team plays an equal number of home and away games.

3. In a round-robin tournament scheduling, use Exercise 2 to determine the home team for each game when there are

   a) 5 teams.   b) 7 teams.   c) 9 teams.

4.3 Computer Projects

Programming Projects

Write programs to do the following:

1. Schedule round-robin tournaments for n teams where n is a positive integer.

2. Using Exercise 2, schedule round-robin tournaments for n teams, where n is an odd positive integer, specifying the home team for each game.

Computations and Explorations

Using the programs you have written or a computation program, carry out the following computations and explorations:

1. Construct a round-robin schedule for a tournament with 13 teams, specifying a home team for each game.

4.4 Computer File Storage and Hashing Functions

A university wishes to store a file for each of its students in its computer. The identifying number or *key* for each file is the social security number of the student enrolled. The social security number is a nine-digit integer, so it is extremely unfeasible to reserve a memory location for each possible social security number. Instead, a systematic way to arrange the files in memory, using a reasonable number of memory locations, should be used so that each file can be easily accessed. Systematic methods of arranging files have been developed based on *hashing functions*. A hashing function assigns to the key of each file a particular memory location. Various types of hashing functions have been suggested, but the type most commonly used involves modular arithmetic. We discuss this type of hashing function here. For a general discussion of hashing functions see Knuth [81] or Kronsjö [82].

Let k be the key of the file to be stored; in our example, k is the social security number of a student. Let m be a positive integer. We define the hashing function $h(k)$ by

$$h(k) \equiv k \pmod{m},$$

where $0 \le h(k) < m$, so that $h(k)$ is the least positive residue of k modulo m. We wish to pick m intelligently, so that the files are distributed in a reasonable way throughout the m different memory locations 0, 1, 2, ..., $m-1$.

The first thing to keep in mind is that m should not be a power of the base b which is used to represent the keys. For instance, when using social security numbers as keys, m should not be a power of 10, such as 10^3, because the value of the hashing function would simply be the last several digits of the key; this may not distribute the keys uniformly throughout the memory locations. For instance, the last three digits of early issued social security numbers may often be between 000 and 099, but seldom between 900 and 999. Likewise, it is unwise to use a number dividing $b^k \pm a$ where k and a are small integers for the modulus m. In such a case, $h(k)$ would depend too strongly on the particular digits of the key, and different keys with similar, but rearranged, digits may be sent to the same memory location. For instance, if $m = 111$, then, since $111 \mid (10^3 - 1) = 999$, we have $10^3 \equiv 1 \pmod{111}$, so that the social security numbers 064 212 848 and 064 848 212 are sent to the same memory location, since

$h(064\ 212\ 848) \equiv 064\ 212\ 848 \equiv 064 + 212 + 848 \equiv 1124 \equiv 14 \pmod{111}$

and

$h(064\ 848\ 212) \equiv 064\ 848\ 212 \equiv 064 + 848 + 212 \equiv 1124 \equiv 14 \pmod{111}.$

To avoid such difficulties, m should be a prime approximating the number of available memory locations devoted to file storage. For instance, if there are 5000 memory locations available for storage of 2000 student files we could pick m to be equal to the prime 4969.

We have avoided mentioning the problem that arises when the hashing function assigns the same memory location to two different files. When this occurs, we say that there is a *collision*. We need a method to resolve collisions, so that files are assigned to different memory locations. There are two kinds of collision resolution policies. In the first kind, when a collision occurs, extra memory locations are linked together to the first memory location. When one wishes to access a file where this collision resolution policy has been used, it is necessary to first evaluate the hashing function for the particular key involved. Then the list linked to this memory location is searched.

The second kind of collision resolution policy is to look for an open memory location when an occupied location is assigned to a file. Various suggestions, such as the following techniques have been made for accomplishing this.

Starting with our original hashing function $h_0(k) = h(k)$, we define a sequence of memory locations $h_1(k), h_2(k), \ldots$. We first attempt to place the file with key k at location $h_0(k)$. If this location is occupied, we move to location $h_1(k)$. If this is occupied, we move to location $h_2(k)$, and so on.

We can choose the sequence of functions $h_j(k)$ in various ways. The simplest way is to let

$$h_j(k) \equiv h(k) + j \ (\text{mod } m), \ 0 \le h_j(k) < m.$$

This places the file with key k as near as possible past location $h(k)$. Note that with this choice of $h_j(k)$, all memory locations are checked, so if there is an open location, it will be found. Unfortunately, this simple choice of $h_j(k)$ leads to difficulties; files tend to *cluster*. We see that if $k_1 \ne k_2$ and $h_i(k_1) = h_j(k_2)$ for nonnegative integers i and j, then $h_{i+k}(k_1) = h_{j+k}(k_2)$ for $k = 1,2,3,\ldots$, so that exactly the same sequence of locations is traced out once there is a collision. This lowers the efficiency of the search for files in the table. We would like to avoid this problem of clustering, so we choose the function $h_j(k)$ in a different way.

To avoid clustering, we use a technique called *double hashing*. We choose, as before,

$$h(k) \equiv k \ (\text{mod } m),$$

with $0 \le h(k) < m$, where m is prime, as the hashing function. We take a second hashing function

$$g(k) \equiv k + 1 \ (\text{mod } m-2),$$

where $0 < g(k) \le m - 1$, so that $(g(k), m) = 1$. We take as a *probing sequence*

$$h_j(k) \equiv h(k) + j \ g(k) \ (\text{mod } m),$$

where $0 \le h_j(k) < m$. Since $(g(k), m) = 1$, as j runs through the integers $0,1,2,\ldots, m - 1$, all memory locations are traced out. The ideal situation would be for $m - 2$ also to be prime, so that the values $g(k)$ are distributed in a reasonable way. Hence, we would like $m - 2$ and m to be twin primes.

Example 4.11. In our example using social security numbers, both $m = 4969$, and $m - 2 = 4967$ are prime. Our probing sequence is

$$h_j(k) \equiv h(k) + j \cdot g(k) \ (\text{mod } 4969),$$

where $0 \le h_j(k) < 4969$, $h(k) \equiv k \ (\text{mod } 4969)$, and $g(k) \equiv k + 1 \ (\text{mod } 4967)$.

Suppose we wish to assign memory locations to files for students with social security numbers:

$$k_1 = 344\ 401\ 659 \quad k_6 = 372\ 500\ 191$$
$$k_2 = 325\ 510\ 778 \quad k_7 = 034\ 367\ 980$$
$$k_3 = 212\ 228\ 844 \quad k_8 = 546\ 332\ 190$$
$$k_4 = 329\ 938\ 157 \quad k_9 = 509\ 496\ 993$$
$$k_5 = 047\ 900\ 151 \quad k_{10} = 132\ 489\ 973.$$

Since $k_1 \equiv 269$, $k_2 \equiv 1526$, and $k_3 \equiv 2854$ (mod 4969), we assign the first three files to locations 269, 1526, and 2854, respectively. Since $k_4 \equiv 1526$ (mod 4969), but memory location 1526 is taken, we compute $h_1(k_4) \equiv h(k_4) + g(k_4) = 1526 + 216 = 1742$ (mod 4969); this follows since $g(k_4) \equiv 1 + k_4 \equiv 216$ (mod 4967). Since location 1742 is free, we assign the fourth file to this location. The fifth, six, seventh, and eighth files go into the available locations 3960, 4075, 2376, and 578, respectively, because $k_5 \equiv 3960$, $k_6 \equiv 4075$, $k_7 \equiv 2376$, and $k_8 \equiv 578$ (mod 4969). We find that $k_9 \equiv 578$ (mod 4969); because location 578 is occupied, we compute $h_1(k_9) \equiv h(k_9) + g(k_9) = 578 + 2002 = 2580$ (mod 4969), where $g(k_9) \equiv 1 + k_9 \equiv 2002$ (mod 4967). Hence, we assign the ninth file to the free location 2580. Finally, we find that $k_{10} \equiv 1526$ (mod 4967), but location 1526 is taken. We compute $h_1(k_{10}) \equiv h(k_{10}) + g(k_{10}) = 1526 + 216 = 1742$ (mod 4969), because $g(k_{10}) \equiv 1 + k_{10} \equiv 216$ (mod 4967), but location 1742 is taken. Hence, we continue by finding $h_2(k_{10}) \equiv h(k_{10}) + 2g(k_{10}) \equiv 1958$ (mod 4969) and in this available location we place the tenth file.

Table 4.1 lists the assignments for the files of students by their social security numbers. In the table, the file locations are shown in boldface. □

Social Security Number	$h(k)$	$h_1(k)$	$h_2(k)$
344 401 659	**269**		
325 510 778	**1526**		
212 228 844	**2854**		
329 938 157	1526	**1742**	
047 900 151	**3960**		
372 500 191	**4075**		
034 367 980	**2376**		
546 332 190	**578**		
509 496 993	578	**2580**	
132 489 973	1526	1742	**1958**

Table 4.1. Hashing Function for Student Files.

We wish to find conditions where double hashing leads to clustering. Hence, we find conditions when

(4.1) $$h_i(k_1) = h_j(k_2)$$

and

(4.2) $$h_{i+1}(k_1) = h_{j+1}(k_2),$$

so that the two consecutive terms of two probe sequences agree. If both (4.1) and (4.2) occur, then

$$h(k_1) + ig(k_1) \equiv h(k_2) + jg(k_2) \pmod{m}$$

and

$$h(k_1) + (i + 1)g(k_1) \equiv h(k_2) + (j + 1)g(k_2) \pmod{m}.$$

Subtracting the first of these two congruences from the second, we obtain

$$g(k_1) \equiv g(k_2) \pmod{m},$$

so that

$$k_1 \equiv k_2 \pmod{m-2}.$$

Since $g(k_1) = g(k_2)$, we can substitute this into the first congruence to obtain

$$h(k_1) \equiv h(k_2) \pmod{m},$$

which shows that

$$k_1 \equiv k_2 \pmod{m}.$$

Consequently, since $(m-2, m) = 1$, Theorem 3.8 tells us that

$$k_1 \equiv k_2 \pmod{m(m-2)}.$$

Therefore, the only way that two probing sequences can agree for two consecutive terms is if the two keys involved, k_1 and k_2, are congruent modulo $m(m-2)$. Hence, clustering is extremely rare. Indeed, if $m(m-2) > k$ for all keys k, clustering will never occur.

4.4 Exercises

1. A parking lot has 101 parking places. A total of 500 parking stickers are sold and only 50-75 vehicles are expected to be parked at a time. Set up a hashing function and collision resolution policy for assigning parking places based on license plates displaying six-digit numbers.

2. Assign memory locations for students in your class, using as keys the day of the month of birthdays of students with hashing function $h(K) \equiv K \pmod{19}$,

   a) with probing sequence $h_j(K) \equiv h(K) + j \pmod{19}$.

   b) with probing sequence $h_j(K) \equiv h(K) + j \cdot g(K)$, $0 \leq j \leq 16$, where $g(K) \equiv 1 + K \pmod{17}$.

* 3. Let the hashing function be $h(K) \equiv K(\mod m)$, with $0 \le h(K) < m$, and let the probing sequence for collision resolution be $h_j(K) \equiv h(K) + jq \pmod{m}$, $0 \le h_j(K) < m$, for $j = 1,2, \cdots ,m-1$. Show that all memory locations are probed

   a) if m is prime and $1 \le q \le m - 1$.

   b) if $m = 2^r$ and q is odd.

* 4. A probing sequence for resolving collisions where the hashing function is $h(K) \equiv K \pmod{m}$, $0 \le h(K) < m$, is given by $h_j(K) \equiv h(K) + j(2h(K) + 1) \pmod{m}$, $0 \le h_j(K) < m$.

   a) Show that if m is prime, then all memory sequences are probed.

   b) Determine conditions for clustering to occur, i.e., when $h_j(K_1) = h_j(K_2)$ and $h_{j+r}(K_1) = h_{j+r}(K_2)$ for $r = 1,2,...$.

5. Using the hashing function and probing sequence of the example in the text, find open memory locations for the files of students with social security numbers: $k_{11} = 137612044$, $k_{12} = 505576452$, $k_{13} = 157170996$, $k_{14} = 131220418$. (Add these to the ten files already stored.)

4.4 Computer Projects

Programming Projects

Write programs to assign memory locations to student files, using the hashing function $h(k) \equiv k \pmod{1021}$, $0 \le h(k) < 1021$, where the keys are the social security numbers of students,

1. linking files together when collisions occur.

2. using $h_j(k) \equiv h(k) + j \pmod{1021}$, $j = 0,1,2, \cdots$ as the probing sequence.

3. using $h_j(k) \equiv h(k) + j \cdot g(k)$, $j = 0,1,2,...$, where $g(k) \equiv 1 + k \pmod{1019}$, as the probing sequence.

Computations and Explorations

Using the programs you have written or a computation program, carry out the following computations and explorations:

1. Assign memory locations to the files of all the students in your class using the hashing function and probing function from Example 4.11. After doing this, assign memory locations to other files with social security numbers that you make up.

4.5 Check Digits

Congruences are used to check for errors in strings of digits. In this section we will discuss error detection for bit strings, which are used to represent computer data. Then we will describe how congruences are used to detect errors in strings of decimal digits, which are used to identify passports, checks, books, and other types of objects.

Manipulating or transmitting bit strings can introduce errors. A simple error detection method is to append the bit string $x_1 x_2 \ldots x_n$ with a *parity check bit* x_{n+1} defined by

$$x_{n+1} \equiv x_1 + x_2 + \cdots + x_n \ (\mathrm{mod}\ 2),$$

so that $x_{n+1} = 0$ if an even number of the first n bits in the string are 1, while $x_{n+1} = 1$ if an odd number of these bits are 1. The appended string $x_1 x_2 \ldots x_n x_{n+1}$ satisfies the congruence

(4.3) $$x_1 + x_2 + \cdots + x_n + x_{n+1} \equiv 0 \ (\mathrm{mod}\ 2).$$

We use this congruence to look for errors.

Suppose we send $x_1 x_2 \ldots x_n x_{n+1}$, and the string $y_1 y_2 \ldots y_n y_{n+1}$ is received. These two strings are equal when there are no errors. But if an error was made, they differ in one or more positions. We check whether

(4.4) $$y_1 + y_2 + \cdots + y_n + y_{n+1} \equiv 0 \ (\mathrm{mod}\ 2)$$

holds. If this congruence fails, at least one error is present, but if it holds, errors may still be present. However, when errors are rare and random, the most common type of error is a single error, which is always detected. In general, we can detect an odd number of errors, but not an even number of errors (see Exercise 4).

Example 4.12. Suppose we receive 1101111 and 11001000, where the last bit in each string is a parity check bit.   For the first string note that $1 + 1 + 0 + 1 + 1 + 1 + 1 \equiv 0 \ (\mathrm{mod}\ 2)$, so that either the received string is what was transmitted or it contains an even number of errors. For the second string note that $1 + 1 + 0 + 0 + 1 + 0 + 0 + 0 \equiv 1 \ (\mathrm{mod}\ 2)$, so that the received string was not the string sent; we ask for retransmission.

Strings of decimal digits are used for identification numbers in many different contexts. Check digits, computed using a variety of schemes, are used to find errors in these strings. For instance, check digits are used to detect errors in passport numbers. In a scheme used by several European countries, if $x_1 x_2 x_3 x_4 x_5 x_6$ is the identification number of a passport, the check digit x_7 is chosen so that

$$x_7 \equiv 7x_1 + 3x_2 + x_3 + 7x_4 + 3x_5 + x_6 \ (\mathrm{mod}\ 10).$$

Example 4.13. Suppose that the identification number of a passport is 211894. To find the check digit x_7 we compute

$$x_7 \equiv 7 \cdot 2 + 3 \cdot 1 + 1 \cdot 1 + 7 \cdot 8 + 3 \cdot 9 + 1 \cdot 4 \equiv 5 \ (\mathrm{mod}\ 10),$$

so that the check digit is 5, and the seven-digit number 2118945 is printed on the passport.

We can always detect a single error in a passport identification number appended with a check digit computed in this way. To see this, suppose that we make an error of a in a digit, i.e. $y_j = x_j + a$ (mod 10) where x_j is the correct jth digit and y_j is the incorrect digit that replaces it. From the definition of the check digit, it follows that we change x_7 by either $7a$, $3a$, or a (mod 10) each of which changes x_7. However, errors caused by transposing two digits will be detected if and only if the difference between these two digits is not 5 or -5, that is, if they are not digits x_i and x_j with $|x_i - x_j| = 5$ (see Exercise 7). This scheme also detects a large number of possible errors involving the scrambling of three digits.

We now turn our attention to the use of check digits in publishing. Almost all recent books are identified by their *International Standard Book Number (ISBN)* which is a ten-digit code assigned by the publisher. For instance, the ISBN for the first edition of this text is 0-201-06561-4. Here the first block of digits, 0, represents the language of the book (English), the second block of digits, 201, represents the publishing company (Addison-Wesley), the third block of digits, 06561, is the number assigned to the book by the publishing company to this book, and the final digit, in this case 4, is the check digit. (The size of the blocks differs for different languages and publishers). The check digit in an ISBN can be used to detect the errors most commonly made when ISBNs are copied, namely single errors and errors made when two digits are transposed.

We are going to describe how this check digit is found and then we will show that it can be used to detect the commonly occurring types of errors. Suppose that the ISBN of a book is $x_1 x_2 ... x_{10}$, where x_{10} is the check digit. (We ignore the hyphens in the ISBN, since the grouping of digits does not effect how the check digit is computed.) The first nine digits are decimal digits, that is, belong to the set $\{0,1,2,3,4,5,6,7,8,9\}$, whereas the check digit x_{10} is a base 11 digit, belonging to the set $\{0,1,2,3,4,5,6,7,8,9,0,X\}$ where X is the base 11 digit representing the integer 10 (in decimal notation). The check digit is selected so that the congruence

$$\sum_{i=1}^{10} i\, x_i \equiv 0 \ (\text{mod } 11)$$

holds. As is easily seen (see Exercise 10), the check digit x_{10} can be computed from the congruence $x_{10} \equiv \sum_{i=1}^{9} i\, x_i$ (mod 11); i.e. the check digit is the remainder upon division by 11 of a weighted sum of the first nine digits.

Example 4.14. We find the check digit for the first edition of this text, which begins with 0-201-06561, by computing

$$x_{10} \equiv 1 \cdot 0 + 2 \cdot 2 + 3 \cdot 0 + 4 \cdot 1 + 5 \cdot 0 + 6 \cdot 6 + 7 \cdot 5 + 8 \cdot 6 + 9 \cdot 1 \equiv 4 \ (\text{mod } 11).$$

Hence the ISBN is 0-201-06561-4, as previously stated. Similarly, if the ISBN number of a book begins with 3-540-19102, we find the check digit using the congruence

$$x_{10} \equiv 1\cdot 3 + 2\cdot 5 + 3\cdot 4 + 4\cdot 0 + 5\cdot 1 + 6\cdot 9 + 7\cdot 1 + 8\cdot 0 + 9\cdot 2 \equiv 10 \pmod{11}.$$

This means that the check digit is X, the base 11 digit for the decimal number 10. Hence the ISBN number is 3-540-19102-X.

We will show that a single error, or a transposition of two digits, can be detected using the check digit of an ISBN. First, suppose that $x_1 x_2 ... x_{10}$ is a valid ISBN, but that this number has been printed as $y_1 y_2 ... y_{10}$. We know that $\sum_{i=1}^{10} ix_i \equiv 0 \pmod{11}$ since $x_1 x_2 ... x_{10}$ is a valid ISBN.

Suppose that exactly one error has been made in printing the ISBN. Then for some integer j we have $y_i = x_i$ for $i \neq j$ and $y_j = x_j + a$ where $-10 \leq a \leq 10$ and $a \neq 0$. Here $a = y_j - x_j$ is the error in the jth place. Note that

$$\sum_{i=1}^{10} iy_i = \sum_{i=1}^{10} ix_i + ja \equiv ja \not\equiv 0 \pmod{11}$$

since $\sum_{i=1}^{10} ix_i \equiv 0 \pmod{11}$ and by Lemma 2.4 it follows that $11 \nmid ja$ since $11 \nmid j$ and $11 \nmid a$. We conclude that $y_1 y_2 ... y_{10}$ is not a valid ISBN so that we can detect the error.

Now suppose that two unequal digits have been transposed. Then there are distinct integers j and k such that $y_j = x_k$ and $y_k = x_j$, and $y_i = x_i$ if $i \neq j$ and $i \neq k$. It follows that

$$\sum_{i=1}^{10} iy_i = \sum_{i=1}^{10} ix_i + (jx_k - jx_j) + (kx_j - kx_k) \equiv (j-k)(x_k - x_j) \not\equiv 0 \pmod{11}$$

since $\sum_{i=1}^{10} ix_i \equiv 0 \pmod{11}$ and $11 \nmid (j-k)$ and $11 \nmid (x_k - x_j)$. We see that $y_1 y_2 ... y_{10}$ is not a valid ISBN, so that we can detect the interchange of two unequal digits.

We have discussed how a single check digit can be used to detect errors in strings of digits. However, using a single check digit we cannot detect an error and then correct it, that is, replace the digit in error with the valid one. It is possible to detect and correct an error using additional digits satisfying certain congruences (see Exercises 14 and 16, for example). The reader is referred to any text on coding theory for more information on error detection and correction. Coding theory uses many results from different parts of mathematics, including number theory, abstract algebra, combinatorics, and even geometry.

4.5 Exercises

1. What is the parity check bit that should be added to each of the following bit strings?

   a) 111111
   b) 000000

   c) 101010
   d) 100000

   e) 11111111
   f) 11001011

2. Suppose you receive the following bit strings, where the last bit is a parity check bit. Which strings do you know are incorrect?

   a) 111111111

   b) 0101010101010

   c) 1111010101010101

3. Assume that each of the following strings, ending with a parity check bit, was received correctly except for a missing bit indicated with a question mark. What is the missing bit?

   a) 1?11111

   b) 000?10101

   c) ?0101010100

4. Show that a parity check bit can detect an odd number of errors, but not an even number of errors.

5. Using the check digit scheme described in the text find the check digit that should be added to the following passport identification numbers.

   a) 132999

   b) 805237

   c) 645153

6. Are the following passport identification numbers valid, where the seventh digit is the check digit computed as described in the text?

   a) 3300118

   b) 4501824

   c) 1873336

7. Show that the passport check digit scheme described in the text detects transposition of the digits x_i and x_j if and only if $|x_i - x_j| \neq 5$.

8. The identification number for a bank printed on a check consists of eight digits $x_1 x_2 \ldots x_8$, followed by a ninth check digit x_9, with $x_9 \equiv 7x_1 + 3x_2 + 9x_3 + 7x_4 + 3x_5 + 9x_6 + 7x_7 + 3x_8 \pmod{10}$.

   a) What is the check digit following the eight-digit identification number 00185403 for a bank?
   b) Which single errors in bank identification numbers does a check digit computed in this way detect?
   c) Which transpositions of two digits does this scheme detect?

9. What should the check digit be to complete ISBN starting with

   a) 2-113-54001
   b) 0-19-081082

   c) 1-2123-9940
   d) 0-07-038133

10. Show that the check digit x_{10} in an ISBN $x_1 x_2 \dots x_{10}$ can be computed from the congruence $x_{10} \equiv \sum_{i=1}^{9} i \, x_i \pmod{11}$.

11. Determine whether each of the following ISBNs are valid.

    a) 0-394-38049-5   c) 0-8218-0123-6   e) 90-6191-705-2

    b) 1-09-231221-3   d) 0-404-50874-X

12. Suppose that one digit, indicated with a question mark, in each of the following ISBNs has been smudged and cannot be read. What should this missing digit be?

    a) 0-19-8?3804-9   b) 91-554-212?-6   c) ?-261-05073-X

13. Suppose we specify that the valid ten-digit decimal codewords $x_1 x_2 \dots x_{10}$ are those satisfying the congruence $\sum_{i=1}^{10} x_i \equiv 0 \pmod{11}$.

    a) Can we detect all single errors in a codeword?

    b) Can we detect transposition of two digits in a codeword?

* 14. Suppose that the only valid ten-digit codewords $x_1 x_2 \dots x_{10}$ are those satisfying the congruences $\sum_{i=1}^{10} x_i \equiv \sum_{i=1}^{10} i x_i \equiv 0 \pmod{11}$.

    a) Show that the valid codewords, where the first digits are decimal digits, i.e. in the set $\{0,1,2,3,4,5,6,7,8,9\}$, are those where the last two digits satisfy the congruences $x_9 \equiv \sum_{j=1}^{8} (i+1) x_i \pmod{11}$ and $x_{10} \equiv \sum_{j=1}^{8} (9 - i) x_i \pmod{11}$.

    b) Find the number of valid decimal codewords.

    c) Show that any single error in a codeword can be detected and corrected, since the location and value of the error can be determined.

    d) Show that we can detect any error caused by transposing two digits in a codeword.

15. The government of Norway assigns an 11-digit decimal registration number $x_1 x_2 \dots x_{11}$ to each of its citizens using a scheme designed by the Norwegian number theorist E. Selmer. The digits $x_1 x_2 \dots x_6$ represent the date of birth, the digits $x_7 x_8 x_9$ identify the particular person born that day, and x_{10} and x_{11} are check digits that are computed using the congruences $x_{10} \equiv 8x_1 + 4x_2 + 5x_3 + 10x_4 + 3x_5 + 2x_6 + 7x_7 + 6x_8 + 9x_9 \pmod{11}$, and $x_{11} \equiv 6x_1 + 7x_2 + 8x_3 + 9x_4 + 4x_5 + 5x_6 + 6x_7 + 7x_8 + 8x_9 + 9x_{10} \pmod{11}$.

    a) Determine the check digits that follow the first nine digits 110491238.

    b) Show that this scheme detects all single errors in a registration number.

    * c) Which double errors are detected?

** **16.** Suppose that we specify that the valid ten-digit code words $x_1 x_2 ... x_{10}$, where each digit is a decimal digit, are those satisfying the congruences $\sum_{i=1}^{10} x_i \equiv$

$$\sum_{i=1}^{10} i x_i \equiv \sum_{i=1}^{10} i^2 x_i \equiv \sum_{i=1}^{10} i^3 x_i \equiv 0 \pmod{11}.$$

a) How many valid ten-digit code words are there?
b) Show how any two errors in a codeword can be corrected.
c) Suppose a codeword has been received as 0204906710. If two errors have been made, what is the correct codeword?

4.5 Computer Projects

Programming Projects

Write computer programs to do the following:

1. Determine whether a bit string, ending with a parity check bit, has either an odd or even number of errors.

2. Determine the check digit for an ISBN given the first nine digits.

3. Determine whether a ten-digit string, where the first nine digits are decimal digits and the last is a decimal digit or an X, is a valid ISBN.

Computations and Explorations

Using the programs you have written or a computation program, carry out the following computations and explorations:

1. Check the ISBN numbers of a selection of books to see whether the check digit was computed correctly.

5

Some Special Congruences

5.1 Wilson´s Theorem and Fermat´s Little Theorem

In this section we discuss two important congruences that are often useful in number theory. We first discuss a congruence for factorials called Wilson´s theorem.

Theorem 5.1. Wilson´s Theorem. If p is prime, then $(p-1)! \equiv -1 \pmod{p}$.

The first proof of Wilson´s Theorem was given by the French mathematician Joseph Lagrange in 1770*. The mathematician after whom the theorem is named, John Wilson, conjectured, but did not prove it. Before proving Wilson´s theorem, we use an example to illustrate the idea behind the proof.

---

* **JOSEPH LOUIS LAGRANGE** (1736-1813) was born in Italy and studied physics and mathematics at the University of Turin. Although he originally planned to pursue a career in physics, Lagrange's growing interest in mathematics led him to change course. At the age of 19 he was appointed as a mathematics professor at the Royal Artillery School in Turin. In 1766 he filled the post Euler vacated at the Royal Academy of Berlin when Frederick the Great sought him out. Lagrange directed the mathematics section of the Royal Academy for 20 years. In 1787, when his patron Frederick the Great died, Lagrange moved to France at the invitation of Louis XVI, to join the French Academy. In France he had a distinguished career in teaching and writing. He was a favorite of Marie Antoinette, but managed to win the favor of the new regime that came into power after the French Revolution. Lagrange's contributions to mathematics include unifying the mathematical theory of mechanics. He made fundamental discoveries in group theory and helped put calculus on a rigorous foundation. His contributions to number theory include the first proof of Wilson's theorem and the result that every positive integer can be written as the sum of four squares.

Example 5.1. Let $p=7$. We have $(7-1)! = 6! = 1 \cdot 2 \cdot 3 \cdot 4 \cdot 5 \cdot 6$. We will rearrange the factors in the product, grouping together pairs of inverses modulo 7. We note that $2 \cdot 4 \equiv 1 \pmod 7$ and $3 \cdot 5 \equiv 1 \pmod 7$. Hence, $6! \equiv 1 \cdot (2 \cdot 4) \cdot (3 \cdot 5) \cdot 6 \equiv 1 \cdot 6 \equiv -1 \pmod 7$. Thus, we have verified a special case of Wilson's theorem.
□

We now use the technique illustrated in the example to prove Wilson's theorem.

Proof. When $p=2$ we have $(p-1)! \equiv 1 \equiv -1 \pmod 2$. Hence, the theorem is true for $p=2$. Now let p be a prime greater than 2. Using Theorem 3.10, for each integer a with $1 \le a \le p-1$, there is an inverse $\bar{a}$, $1 \le \bar{a} \le p-1$, with $a\bar{a} \equiv 1 \pmod p$. By Theorem 3.11 the only positive integers less than p that are their own inverses are 1 and $p-1$. Therefore, we can group the integers from 2 to $p-2$ into $(p-3)/2$ pairs of integers, with the product of each pair congruent to 1 modulo p. Hence, we have

$$2 \cdot 3 \cdots (p-3) \cdot (p-2) \equiv 1 \pmod p.$$

We multiply both sides of the this congruence by 1 and $p-1$ to obtain

$$(p-1)! = 1 \cdot 2 \cdot 3 \cdots (p-3)(p-2)(p-1) \equiv 1 \cdot (p-1) \equiv -1 \pmod p.$$

This completes the proof. ∎

An interesting observation is that the converse of Wilson's theorem is also true, as the following theorem shows.

Theorem 5.2. If n is a positive integer such that $(n-1)! \equiv -1 \pmod n$, then n is prime.

Proof. Assume that n is a composite integer and that $(n-1)! \equiv -1 \pmod n$. Since n is composite, we have $n=ab$, where $1 < a < n$ and $1 < b < n$. Since $a < n$, we know that $a \mid (n-1)!$, because a is one of the $n-1$ numbers multiplied together to form $(n-1)!$. Since $(n-1)! \equiv -1 \pmod n$ it follows that $n \mid ((n-1)! + 1)$. This means, by Theorem 1.7, that a also divides $(n-1)! + 1$. By Theorem 1.8 since $a \mid (n-1)!$ and $a \mid ((n-1)! + 1)$, we conclude that $a \mid ((n-1)! + 1) - (n-1)! = 1$. This is an contradiction, since $a > 1$. ∎

We illustrate the use of this result with an example.

Example 5.2. Since $(6-1)! = 5! = 120 \equiv 0 \pmod 6$, Theorem 5.2 verifies the obvious fact that 6 is not prime.
□

As we can see, Wilson's theorem and its converse give us a primality test. To decide whether an integer n is prime, we determine whether $(n-1)! \equiv -1 \pmod n$. Unfortunately, this is an impractical test because $n-1$ multiplications modulo n are needed to find $(n-1)!$ requiring $O(n(\log_2 n)^2)$ bit operations.

When working with congruences involving exponents, the following theorem* is of great importance.

Theorem 5.3. **Fermat's Little Theorem.** If p is prime and a is a positive integer with $p \nmid a$, then $a^{p-1} \equiv 1 \pmod{p}$.

Proof. Consider the $p-1$ integers a, $2a$, ..., $(p-1)a$. None of these integers are divisible by p, for if $p \mid ja$, then by Lemma 2.3, $p \mid j$, since $p \nmid a$. This is impossible because $1 \le j \le p-1$. Furthermore, no two of the integers $a, 2a, ..., (p-1)a$ are congruent modulo p. To see this, assume that $ja \equiv ka \pmod{p}$ where $1 \le j < k \le p-1$. Then by Corollary 3.4.1, since $(a,p) = 1$, we have $j \equiv k \pmod{p}$. This is impossible, since j and k are positive integers less than $p - 1$.

Since the integers a, $2a$, ..., $(p-1)a$ are a set of $p-1$ integers all incongruent to zero, and no two congruent modulo p, we know that the least positive residues of a, $2a$, ..., $(p-1)a$, taken in some order, must be the integers 1, 2, ..., $p-1$. As a consequence, the product of the integers a, $2a$, ..., $(p-1)a$ is congruent modulo p to the product of the first $p-1$ positive integers. Hence,

$$a \cdot 2a \ \cdots \ (p-1)a \equiv 1 \cdot 2 \ \cdots \ (p-1) \pmod{p}.$$

Therefore,

$$a^{p-1}(p-1)! \equiv (p-1)! \pmod{p}.$$

Since $((p-1)!, p) = 1$, using Corollary 3.4.1, we cancel $(p-1)!$ to obtain

$$a^{p-1} \equiv 1 \pmod{p}. \qquad \square$$

We illustrate the ideas of the proof with an example.

Example 5.3. Let $p = 7$ and $a = 3$. Then, $1 \cdot 3 \equiv 3 \pmod 7$, $2 \cdot 3 \equiv 6 \pmod 7$, $3 \cdot 3 \equiv 2 \pmod 7$, $4 \cdot 3 \equiv 5 \pmod 7$, $5 \cdot 3 \equiv 1 \pmod 7$, and $6 \cdot 3 \equiv 4 \pmod 7$. Consequently,

$$(1 \cdot 3) \cdot (2 \cdot 3) \cdot (3 \cdot 3) \cdot (4 \cdot 3) \cdot (5 \cdot 3) \cdot (6 \cdot 3) \equiv 3 \cdot 6 \cdot 2 \cdot 5 \cdot 1 \cdot 4 \pmod 7,$$

so that $3^6 \cdot 1 \cdot 2 \cdot 3 \cdot 4 \cdot 5 \cdot 6 \equiv 3 \cdot 6 \cdot 2 \cdot 5 \cdot 1 \cdot 4 \pmod 7$. Hence, $3^6 \cdot 6! \equiv 6! \pmod 7$, and therefore $3^6 \equiv 1 \pmod 7$. $\qquad \square$

On occasion, we would like to have a congruence like Fermat's little theorem that holds for all integers a, given the prime p. This is supplied by the following result.

---

*   This is called Fermat's Little Theorem to distinguish it from the famous conjecture known as Fermat's Last Theorem, which we discuss in Chapter 11.

Theorem 5.4. If p is prime and a is a positive integer, then $a^p \equiv a \pmod{p}$.

Proof. If $p \nmid a$, by Fermat's little theorem we know that $a^{p-1} \equiv 1 \pmod{p}$. Multiplying both sides of this congruence by a, we find that $a^p \equiv a \pmod{p}$. If $p \mid a$, then $p \mid a^p$ as well, so that $a^p \equiv a \equiv 0 \pmod{p}$. This finishes the proof, since $a^p \equiv a \pmod{p}$ if $p \nmid a$ and if $p \mid a$. ∎

Fermat's little theorem is useful in finding the least positive residues of powers.

Example 5.4. We can find the least positive residue of 3^{201} modulo 11 with the help of Fermat's little theorem. We know that $3^{10} \equiv 1 \pmod{11}$. Hence, $3^{201} = (3^{10})^{20} \cdot 3 \equiv 3 \pmod{11}$. ∎

A useful application of Fermat's little theorem is provided by the following result.

Theorem 5.5. If p is prime and a is an integer with $p \nmid a$, then a^{p-2} is an inverse of a modulo p.

Proof. If $p \nmid a$, by Fermat's little theorem we have $a \cdot a^{p-2} = a^{p-1} \equiv 1 \pmod{p}$. Hence, a^{p-2} is an inverse of a modulo p. ∎

Example 5.5. By Theorem 5.5 we know that $2^9 = 512 \equiv 6 \pmod{11}$ is an inverse of 2 modulo 11. □

Theorem 5.5 gives us another way to solve linear congruences with respect to prime moduli.

Corollary 5.5.1. If a and b are positive integers and p is prime with $p \nmid a$, then the solutions of the linear congruence $ax \equiv b \pmod{p}$ are the integers x such that $x \equiv a^{p-2}b \pmod{p}$.

Proof. Suppose that $ax \equiv b \pmod{p}$. Since $p \nmid a$, we know from Theorem 5.5 that a^{p-2} is an inverse of $a \pmod{p}$. Multiplying both sides of the original congruence by a^{p-2}, we have

$$a^{p-2}ax \equiv a^{p-2}b \pmod{p}.$$

Hence,

$$x \equiv a^{p-2}b \pmod{p}.$$ ∎

Fermat's little theorem is the basis of a factorization method invented by J.M. Pollard in 1974. This method, known as the *Pollard $p - 1$ method*, can find a nontrivial factor of an integer n when n has a prime factor p such that the primes dividing $p - 1$ are relatively small.

To see how this method works, suppose that we want to find a factor of the positive integer n. Furthermore, suppose that n has a prime factor p such that $(p - 1)$ divides $k!$, where k is a positive integer. We want $p - 1$ to have only small prime factors so that there is such an integer k that is not too large. For example, if $p = 2269$ then $p-1 = 2268 = 2^2 3^4 7$, so that $p-1$ divides $9!$, but no smaller value of the factorial function.

The reasons we want $p-1$ to divide $k!$ is so that we can apply Fermat's little theorem. By Fermat's little theorem we know that $2^{p-1} \equiv 1 \pmod{p}$. Now since $p-1$ divides $k!$, $k! = (p-1)q$ for some integer q. Hence

$$2^{k!} = 2^{(p-1)q} = (2^{p-1})^q \equiv 1^q = 1 \pmod{p},$$

which implies that p divides $2^{k!} - 1$. Now let M be the least positive residue of $2^{k!} - 1$ modulo n, so that $M = (2^{k!} - 1) - nt$ for some integer t. We see that p divides M since it p divides both $2^{k!} - 1$ and n.

Now to find a divisor of n, we need only compute the greatest common diviisor of M and n, $d = (M,n)$. This can be done rapidly using the Euclidean algorithm. For this divisor d to be a nontrivial divisor, it is necessary that M not be 0. This is the case when n does not itself divide $2^{k!} - 1$, which is likely when n has large prime divisors.

To use this method, we must compute $2^{k!}$ where k is a positive integer. This can be done efficiently since modular exponentiation can be done efficiently. To find the least positive remainder of $2^{k!}$ modulo n we set $r_1 = 2$ and use the following sequence of computations: $r_2 \equiv r_1^2 \pmod{n}$, $r_3 \equiv r_2^3 \pmod{n}$, ..., $r_k \equiv r_{k-1}^k \pmod{n}$. We illustrate this procedure in the following example.

Example 5.6. To find $2^{9!} \pmod{5157437}$, we perform the following sequence of computations:

$$
\begin{aligned}
r_2 &\equiv r_1^2 = 2^2 \equiv 4 \pmod{5157437} \\
r_3 &\equiv r_2^3 = 4^3 \equiv 64 \pmod{5157437} \\
r_4 &\equiv r_3^4 = 64^4 \equiv 1304905 \pmod{5157437} \\
r_5 &\equiv r_4^5 = 1304905^5 \equiv 404913 \pmod{5157437} \\
r_6 &\equiv r_5^6 = 404913^6 \equiv 2157880 \pmod{5157437} \\
r_7 &\equiv r_6^7 = 2157880^7 \equiv 4879227 \pmod{5157437} \\
r_8 &\equiv r_7^8 = 4879227^8 \equiv 4379778 \pmod{5157437} \\
r_9 &\equiv r_8^9 = 4379778^9 \equiv 4381440 \pmod{5157437}.
\end{aligned}
$$

It follows that $2^{9!} \equiv 4381440 \pmod{5157437}$. $\qquad\square$

The following example illustrates the use of the Pollard $p-1$ method to find a factor the integer 5157437.

Example 5.7. To factor 5157437 using Pollard $p-1$ method, we successively find r_k, the least positive residue of $2^{k!}$ modulo 5157437, for $k = 1,2,3,...$, as was done in Example 5.4. We compute $(r_k - 1, 5157437)$ at each step. To find a factor of 5157437 requires nine steps, because $(r_k - 1, 5157437) = 1$ for

$k = 1,2,3,4,5,6,7,8$, (as the reader can verify) but $(r_9 - 1,5157437) = (4381439,5157437) = 2269$. It follows that 2269 is a divisor of 5157437.   □

The Pollard $p-1$ method does not always work. However, since nothing in the method depends on the choice of 2 as the base, we can extend the method and find a factor of more integers by using integers other than 2 as the base. In pratice, the Pollard $p-1$ method is used after trial divisions by small primes, but before the heavy artillery of such methods as the quadratic sieve and the elliptic curve method are used.

5.1 Exercises

1. Using Wilson's theorem, find the least positive residue of $8 \cdot 9 \cdot 10 \cdot 11 \cdot 12 \cdot 13$ modulo 7.

2. Using Fermat's little theorem, find the least positive residue of $2^{1000000}$ modulo 17.

3. Show that $3^{10} \equiv 1 \pmod{11^2}$.

4. Using Fermat's little theorem, find the last digit of the base 7 expansion of 3^{100}.

5. Using Fermat's little theorem, find the solutions of the following linear congruences.

   a) $7x \equiv 12 \pmod{17}$ b) $4x \equiv 11 \pmod{19}$

6. Show that if n is a composite integer with $n \neq 4$, then $(n-1)! \equiv 0 \pmod n$.

7. Show that if p is an odd prime, then $2(p-3)! \equiv -1 \pmod p$.

8. Show that if n is odd and $3 \nmid n$, then $n^2 \equiv 1 \pmod{24}$.

9. Show that $42 \mid (n^7 - n)$ for all positive integers n.

10. Use the Pollard $p-1$ method to find a divisor of 689.

11. Use the Pollard $p-1$ method to find a divisor of 7331117. (For this exercise, you will need to use either a calculator or computational software.)

12. Show that if p and q are distinct primes, then $p^{q-1} + q^{p-1} \equiv 1 \pmod{pq}$.

13. Show that p is prime and a and b are integers such that $a^p \equiv b^p \pmod p$, then $a^p \equiv b^p \pmod{p^2}$.

14. Show that if p is an odd prime, then $1^2 3^2 \cdots (p-4)^2 (p-2)^2 \equiv (-1)^{(p+1)/2} \pmod p$.

15. Show that if p is prime and $p \equiv 3 \pmod 4$, then $((p-1)/2)! \equiv \pm 1 \pmod p$.

16. a) Let p be prime and suppose that r is a positive integer less then p such that $(-1)^r r! \equiv -1 \pmod p$. Show that $(p-r+1)! \equiv -1 \pmod p$.

    b) Using part (a), show that $61! \equiv 63! \equiv -1 \pmod{71}$.

17. Using Wilson's theorem, show that if p is a prime and $p \equiv 1 \pmod 4$, then the congruence $x^2 \equiv -1 \pmod p$ has two incongruent solutions given by $x \equiv \pm ((p-1)/2)! \pmod p$.

18. Show that if p is a prime and $0 < k < p$, then $(p-k)!(k-1)! \equiv (-1)^k \pmod p$.

19. Show that if p is prime and a is an integer, then $p \mid (a^p + (p-1)! \, a)$.

* 20. For which positive integers n is $n^4 + 4^n$ prime?

21. Show that the pair of positive integers n and $n + 2$ are twin primes if and only if $4((n-1)! + 1) + n \equiv 0 \pmod{n(n+2)}$, where $n \neq 1$.

22. Show that if the positive integers n and $n + k$, where $n > k$ and k is an even positive integer, are both prime then $(k!)^2((n-1)! + 1) + n(k! - 1)(k-1)! \equiv 0 \pmod{n(n+k)}$.

23. Show that if p is prime, then $\begin{pmatrix} 2p \\ p \end{pmatrix} \equiv 2 \pmod p$.

24. Exercise 24 of Section 1.9 shows that if p is prime and k is a positive integer less than p, then the binomial coefficient $\begin{pmatrix} p \\ k \end{pmatrix}$ is divisible by p. Use this fact and the binomial theorem to show that if a and b are integers, then $(a + b)^p \equiv a^p + b^p \pmod p$.

25. Prove Fermat's little theorem by mathematical induction. (*Hint:* In the induction step use Exercise 24 to obtain a congruence for $(a + 1)^p$.)

* 26. Using Exercise 30 of Section 3.3, prove *Gauss' generalization of Wilson's theorem*, namely that the product of all the positive integers less than m that are relatively prime to m is congruent to $1 \pmod m$, unless $m = 4$, p^t, or $2p^t$ where p is an odd prime and t is a positive integer, in which case it is congruent to $-1 \pmod m$.

27. A deck of cards is shuffled by cutting the deck into two piles of 26 cards. Then, the new deck is formed by alternating cards from the two piles, starting with the bottom pile.

    a) Show that if a card begins in the cth position in the deck, it will be in the bth position in the new deck where $b \equiv 2c \pmod{53}$ and $1 \leq b \leq 52$.

    b) Determine the number of shuffles of the type described above that are needed to return the deck of cards to its original order.

28. Let p be prime and let a be a positive integer not divisible by p. We define the *Fermat quotient* $q_p(a)$ by $q_p(a) = (a^{p-1} - 1)/p$. Show that if a and b are positive integers not divisible by the prime p, then $q_p(ab) \equiv q_p(a) + q_p(b) \pmod p$.

29. Let p be prime and let $a_1, a_2, ..., a_p$ and $b_1, b_2, ..., b_p$ be complete systems of residues modulo p. Show that $a_1 b_1, a_2 b_2, ..., a_p b_p$ is not a complete system of residues modulo p.

* 30. Show that if n is a positive integer with $n \geq 2$ then n does not divide $2^n - 1$.

* 31. Let p be an odd prime. Show that $(p-1)!^{p^{n-1}} \equiv -1 \pmod{p^n}$.

5.1 Computer Projects

Programming Projects

Write programs to do the following:

1. Find all Wilson primes less than a given positive integer n. A *Wilson prime* is a prime p for which $(p-1)! \equiv -1 \pmod{p^2}$.

2. Find the primes p less than a given positive integer n for which $2^{p-1} \equiv 1 \pmod{p^2}$.

3. Solve linear congruences with prime moduli via Fermat's little theorem.

4. Factor a given positive integer n using the Pollard $p-1$ method.

Computations and Explorations

Using the programs you have written or a computation program, carry out the following computations and explorations:

1. Find all Wilson primes less than 10000.

2. Find all primes p less than 10000 for which $2^{p-1} \equiv 1 \pmod{p^2}$.

3. Find a factor of each of several different odd integers of your choice using the Pollard $p-1$ method.

5.2 Pseudoprimes

Fermat's little theorem tells us that if n is prime and b is any integer, then $b^n \equiv b \pmod{n}$. Consequently, if we can find an integer b such that $b^n \not\equiv b \pmod{n}$, then we know that n is composite.

Example 5.8. We can show that 63 is not prime by observing that

$$2^{63} = 2^{60} \cdot 2^3 = (2^6)^{10} \cdot 2^3 = 64^{10} 2^3 \equiv 2^3 \equiv 8 \not\equiv 2 \pmod{63}. \qquad \square$$

Using Fermat's little theorem, we can show that an integer is composite. It would be even more useful if it also provided a way to show that an integer is prime. The ancient Chinese believed that if $2^n \equiv 2 \pmod{n}$, then n must be prime. Unfortunately, the converse of Fermat's little theorem is not true, as the following example shows.

Example 5.9. Let $n = 341 = 11 \cdot 31$. By Fermat's little theorem we see that $2^{10} \equiv 1 \pmod{11}$, so that $2^{340} = (2^{10})^{34} \equiv 1 \pmod{11}$. Also $2^{340} = (2^5)^{68} \equiv (32)^{68} \equiv 1$

(mod 31). Hence, by Corollary 3.8.1, we have $2^{340} \equiv 1$ (mod 341). By multiplying both sides of this congruence by 2, we have $2^{341} \equiv 2$ (mod 341), even though 341 is not prime. ☐

Examples such as this lead to the following definition.

Definition. Let b be a positive integer. If n is a composite positive integer and $b^n \equiv b$ (mod n), then n is called a *pseudoprime to the base b*.

Note that if $(b,n) = 1$, then the congruence $b^n \equiv b$ (mod n) is equivalent to the congruence $b^{n-1} \equiv 1$ (mod n). To see this, note that by Corollary 3.4.1 we can divide both sides of the first congruence by b, since $(b,n) = 1$, to obtain the second congruence. By part (iii) of Theorem 3.3 we can multiply both sides of the second congruence by b to obtain the first. We will often use this equivalent condition.

Example 5.10. The integers $341 = 11 \cdot 31$, $561 = 3 \cdot 11 \cdot 17$ and $645 = 3 \cdot 5 \cdot 43$ are pseudoprimes to the base 2, since it is easily verified that $2^{340} \equiv 1$ (mod 341), $2^{560} \equiv 1$ (mod 561), and $2^{644} \equiv 1$ (mod 645). ☐

If there are relatively few pseudoprimes to the base b, then checking to see whether the congruence $b^n \equiv b$ (mod n) holds is a useful test; only a small fraction of composite numbers pass this test. In fact, there are far fewer pseudoprimes to the base b not exceeding a specified bound than prime numbers not exceeding this bound. In particular, there are 455052512 primes, but only 14884 pseudoprimes to the base 2, less than 10^{10}. Although pseudoprimes to any given base are rare, there are, nevertheless, infinitely many pseudoprimes to any given base. We will prove this for the base 2. The following lemma is useful in the proof.

Lemma 5.1. If d and n are positive integers such that d divides n, then $2^d - 1$ divides $2^n - 1$.

Proof. Since $d \mid n$, there is a positive integer t with $dt = n$. By setting $x = 2^d$ in the identity $x^t - 1 = (x - 1)(x^{t-1} + x^{t-2} + \cdots + 1)$, we find that $2^n - 1 = (2^d - 1)$ $(2^{d(t-1)} + 2^{d(t-2)} + \cdots + 2^d + 1)$. Consequently, we have $(2^d - 1) \mid (2^n - 1)$. ☐

We can now prove that there are infinitely many pseudoprimes to the base 2.

Theorem 5.6. There are infinitely many pseudoprimes to the base 2.

Proof. We will show that if n is an odd pseudoprime to the base 2, then $m = 2^n - 1$ is also an odd pseudoprime to the base 2. Since we have at least one odd pseudoprime to the base 2, namely $n_0 = 341$, we will be able to construct infinitely many odd pseudoprimes to the base 2 by taking $n_0 = 341$ and $n_{k+1} = 2^{n_k} - 1$ for $k = 0,1,2,3,....$ These integers are all different, since $n_0 < n_1 < n_2 < \cdots < n_k < n_{k+1} < \cdots$.

To continue the proof, let n be an odd pseudoprime, so that n is composite and $2^{n-1} \equiv 1$ (mod n). Since n is composite, we have $n = dt$ with $1 < d < n$ and $1 < t < n$. We will show that $m = 2^n - 1$ is also pseudoprime by first showing that it is composite, and then by showing that $2^{m-1} \equiv 1 \pmod{m}$.

To see that m is composite we use Lemma 5.1 to note that $(2^d - 1) \mid (2^n - 1) = m$. To show that $2^{m-1} \equiv 1 \pmod{m}$, note that since $2^n \equiv 2 \pmod{n}$, there is an integer k with $2^n - 2 = kn$. Hence, $2^{m-1} = 2^{2^n-2} = 2^{kn}$. By Lemma 5.1 it follows that $m = (2^n - 1) \mid (2^{kn} - 1) = 2^{m-1} - 1$. Hence, $2^{m-1} - 1 \equiv 0 \pmod{m}$, so that $2^{m-1} \equiv 1 \pmod{m}$. We conclude that m is also a pseudoprime to the base 2.   □

If we want to know whether an integer n is prime, and we find that $2^{n-1} \equiv 1 \pmod{n}$, we know that n either is prime or n is a pseudoprime to the base 2. One follow-up approach is to test n with other bases. That is, we check to see whether $b^{n-1} \equiv 1 \pmod{n}$ for various positive integers b. If we find any values of b with $(b,n) = 1$ and $b^{n-1} \not\equiv 1 \pmod{n}$, then we know that n is composite.

Example 5.11. We have seen that 341 is a pseudoprime to the base 2. Since

$$7^3 = 343 \equiv 2 \pmod{341}$$

and

$$2^{10} = 1024 \equiv 1 \pmod{341},$$

we have

$$7^{340} = (7^3)^{113}7 \equiv 2^{113}7 = (2^{10})^{11}\cdot 2^3 \cdot 7$$
$$\equiv 8\cdot 7 \equiv 56 \not\equiv 1 \pmod{341}.$$

Hence, by the contrapositive of Fermat's little theorem, we see that 341 is composite, since $7^{340} \not\equiv 1 \pmod{341}$.   □

Unfortunately, there are composite integers n that cannot be shown to be composite using the above approach, because there are integers that are pseudoprimes to every base, that is, there are composite integers n such that $b^{n-1} \equiv 1 \pmod{n}$, for all b with $(b,n) = 1$. This leads to the following definition.

Definition. A composite integer that satisfies $b^{n-1} \equiv 1 \pmod{n}$ for all positive integers b with $(b,n) = 1$ is called a *Carmichael number* after R. Carmichael* who studied them in the early part of the 20th century.

---

* **ROBERT DANIEL CARMICHAEL (1879-1967)** was born in Goodwater, Alabama. He received his B.A. from Lineville College in 1898 and his Ph.D. in 1911 from Princeton University. Carmichael taught at Indiana University from 1911 to 1915 and at the University of Illinois from 1915 until 1947. His thesis, written under the direction of G.D. Birkhoff, was considered the first significant American contribution to differential equations. Carmichael worked in a wide range of areas, including real analysis, differential equations, mathematical physics, group theory, and number theory.

Example 5.12. The integer $561 = 3 \cdot 11 \cdot 17$ is a Carmichael number. To see this, note that if $(b, 561) = 1$, then $(b,3) = (b,11) = (b,17) = 1$. Hence, from Fermat's little theorem, we have $b^2 \equiv 1 \pmod 3$, $b^{10} \equiv 1 \pmod{11}$, and $b^{16} \equiv 1 \pmod{17}$. Consequently, $b^{560} = (b^2)^{280} \equiv 1 \pmod 3$, $b^{560} = (b^{10})^{56} \equiv 1 \pmod{11}$, and $b^{560} = (b^{16})^{35} \equiv 1 \pmod{17}$. Therefore, by Corollary 3.8.1 $b^{560} \equiv 1 \pmod{561}$ for all b with $(b,n) = 1$. $\square$

In 1912 Carmichael conjectured that there are infinitely many Carmichael numbers. It took eighty years to resolve this conjecture. In mid-1992 Alford, Granville, and Pomerance showed that Carmichael was correct. Because of the complicated, non-elementary nature of their proof we will not describe it here. However, we will prove one of the key ingredients, a theorem which can be used to find Carmichael numbers.

Theorem 5.7. If $n = q_1 q_2 \cdots q_k$, where the q_j's are distinct primes that satisfy $(q_j - 1) \mid (n - 1)$ for all j, then n is a Carmichael number.

Proof. Let b be a positive integer with $(b,n) = 1$. Then $(b,q_j) = 1$ for $j = 1,2,\ldots,k$, and hence, by Fermat's little theorem, $b^{q_j - 1} \equiv 1 \pmod{q_j}$ for $j = 1, 2, \ldots, k$. Since $(q_j - 1) \mid (n - 1)$ for each integer $j = 1, 2, \ldots, k$, there are integers t_j with $t_j(q_j - 1) = n - 1$. Hence, for each j, we know that $b^{n-1} = b^{(q_j-1)t_j} \equiv 1 \pmod{q_j}$. Therefore, by Corollary 3.8.1 we see that $b^{n-1} \equiv 1 \pmod n$, and we conclude that n is a Carmichael number. $\square$

Example 5.13. Theorem 5.7 shows that $6601 = 7 \cdot 23 \cdot 41$ is a Carmichael number, because 7, 23, and 41 are all prime, $6 = (7 - 1) \mid 6600$, $22 = (23 - 1) \mid 6600$, and $40 = (41 - 1) \mid 6600$.

The converse of Theorem 5.7 is also true, that is, all Carmichael numbers are of the form $q_1 q_2 \cdots q_k$ where the q_j's are distinct primes and $(q_j - 1) \mid (n - 1)$ for all j. We will prove this fact in Chapter 8.

Once the congruence $b^{n-1} \equiv 1 \pmod n$, where n is an odd integer, has been verified, another possible approach is to consider the least positive residue of $b^{(n-1)/2}$ modulo n. We note that if $x = b^{(n-1)/2}$, then $x^2 = b^{n-1} \equiv 1 \pmod n$. If n is prime, by Theorem 3.11 we know that either $x \equiv 1$ or $x \equiv -1 \pmod n$. Consequently, once we have found that $b^{n-1} \equiv 1 \pmod n$, we can check to see whether $b^{(n-1)/2} \equiv \pm 1 \pmod n$. If this congruence does not hold, then n is composite.

Example 5.14. Let $b = 5$ and let $n = 561$, the smallest Carmichael number. We find that $5^{(561-1)/2} = 5^{280} \equiv 67 \pmod{561}$. Hence, 561 is composite. $\square$

We continue developing primality tests with the following definitions.

Definition. Let n be a positive integer with $n - 1 = 2^s t$, where s is a nonnegative integer and t is an odd positive integer. We say that n passes *Miller's test for the base b.* if either $b^t \equiv 1 \pmod n$ or $b^{2^j t} \equiv -1 \pmod n$ for some j with $0 \le j \le s - 1$.

The following example shows that 2047 passes Miller's test for the base 2.

Example 5.15. Let $n = 2047 = 23 \cdot 89$. Then $2^{2046} = (2^{11})^{186} = (2048)^{186} \equiv 1$ (mod 2047), so that 2047 is a pseudoprime to the base 2. Since $2^{2046/2} = 2^{1023} = (2^{11})^{93} = (2048)^{93} \equiv 1$ (mod 2047), 2047 passes Miller's test for the base 2. $\square$

We now show that if n is prime, then n passes Miller's test for all bases b with $n \nmid b$.

Theorem 5.8. If n is prime and b is a positive integer with $n \nmid b$, then n passes Miller's test for the base b.

Proof. Let $n - 1 = 2^s t$, where s is a nonnegative integer and t is an odd positive integer. Let $x_k = b^{(n-1)/2^k} = b^{2^{s-k} t}$, for $k = 0, 1, 2, ..., s$. Since n is prime, Fermat's little theorem tells us that $x_0 = b^{n-1} \equiv 1$ (mod n). By Theorem 3.11, since $x_1^2 = (b^{(n-1)/2})^2 = x_0 \equiv 1$ (mod n), either $x_1 \equiv -1$ (mod n) or $x_1 \equiv 1$ (mod n). If $x_1 \equiv 1$ (mod n), since $x_2^2 = x_1 \equiv 1$ (mod n), either $x_2 \equiv -1$ (mod n) or $x_2 \equiv 1$ (mod n). In general, if we have found that $x_0 \equiv x_1 \equiv x_2 \equiv \cdots \equiv x_k \equiv 1$ (mod n), with $k < s$, then, since $x_{k+1}^2 = x_k \equiv 1$ (mod n), we know that either $x_{k+1} \equiv -1$ (mod n) or $x_{k+1} \equiv 1$ (mod n).

Continuing this procedure for $k = 1, 2, ..., s$, we find that either $x_k \equiv 1$ (mod n), for $k = 0, 1, ..., s$, or $x_k \equiv -1$ (mod n) for some integer k. Hence, n passes Miller's test for the base b. $\square$

If the positive integer n passes Miller's test for the base b, then either $b^t \equiv 1$ (mod n) or $b^{2^j t} \equiv -1 \pmod{n}$ for some j with $0 \le j \le s - 1$, where $n - 1 = 2^s t$ and t is odd.

In either case, we have $b^{n-1} \equiv 1$ (mod n), since $b^{n-1} = (b^{2^j t})^{2^{s-j}}$ for $j = 0, 1, 2, ..., s$, so that an integer n that passes Miller's test for the base b is automatically a pseudoprime to the base b. With this observation, we are led to the following definition.

Definition. If n is composite and passes Miller's test for the base b, then we say n is a *strong pseudoprime to the base b*.

Example 5.16. By Example 5.15 we see that 2047 is a strong pseudoprime to the base 2. $\square$

Although strong pseudoprimes are exceedingly rare, there are still infinitely many of them. We demonstrate this for the base 2 with the following theorem.

Theorem 5.9. There are infinitely many strong pseudoprimes to the base 2.

Proof. We shall show that if n is a pseudoprime to the base 2, then $N = 2^n - 1$ is a *strong* pseudoprime to the base 2.

Let n be an odd integer which is a pseudoprime to the base 2. Hence, n is composite, and $2^{n-1} \equiv 1 \pmod{n}$. From this congruence, we see that $2^{n-1} - 1 = nk$ for some integer k; furthermore, k must be odd. We have

$$N - 1 = 2^n - 2 = 2(2^{n-1} - 1) = 2^1 nk;$$

this is the factorization of $N - 1$ into an odd integer and a power of 2.

We now note that

$$2^{(N-1)/2} = 2^{nk} = (2^n)^k \equiv 1 \pmod{N}$$

because $2^n = (2^n - 1) + 1 = N + 1 \equiv 1 \pmod{N}$. This demonstrates that N passes Miller's test.

In the proof of Lemma 5.1 we showed that if n is composite, then $N = 2^n - 1$ also is composite. Hence, N passes Miller's Test and is composite, so that N is a strong pseudoprime to the base 2. Since every pseudoprime n to the base 2 yields a strong pseudoprime $2^n - 1$ to the base 2 and since there are infinitely many pseudoprimes to the base 2, we conclude that there are infinitely many strong pseudoprimes to the base 2. □

The following observations are useful in combination with Miller's test for checking the primality of relatively small integers. The smallest odd strong pseudoprime to the base 2 is 2047, so that if $n < 2047$, n is odd, and n passes Miller's test to the base 2, then n is prime. Likewise, 1373653 is the smallest odd strong pseudoprime to both the bases 2 and 3, giving us a primality test for integers less than 1373653. The smallest odd strong pseudoprime to the bases 2, 3, and 5 is 25326001, and the smallest odd strong pseudoprime to all the bases 2, 3, 5, and 7 is 3215031751. Also, less than $25 \cdot 10^9$ the only odd integer which is a strong pseudoprime to all the bases 2, 3, 5, and 7 is 3251031751. (The reader should verify these statements.) This leads us to a primality test for integers less than $25 \cdot 10^9$. An odd integer n is prime if $n < 25 \cdot 10^9$, n passes Miller's test for the bases 2, 3, 5, and 7, and $n \neq 3215031751$.

There is no analogy of a Carmichael number for strong pseudoprimes. This is a consequence of the following theorem.

Theorem 5.10. If n is an odd composite positive integer, then n passes Miller's test for at most $(n-1)/4$ bases b with $1 \leq b \leq n - 1$.

We prove Theorem 5.10 in Chapter 8. Note that Theorem 5.10 tells us that if n passes Miller's tests for more than $(n-1)/4$ bases less than n, then n must be prime. However, this is a rather lengthy way, worse than performing trial divisions, to show that a positive integer n is prime. Miller's test does give an interesting and quick way of showing an integer n is "probably prime". To see this, take at random an integer b with $1 \leq b \leq n - 1$ (we will see how to make this "random" choice in Chapter 8). From Theorem 5.10, we see that if n is composite the probability that n passes Miller's test for the base b is less than $1/4$. If we pick k different bases less than n and perform Miller's tests for each of these bases we are led to the following result.

Theorem 5.11. Rabin's Probabilistic Primality Test. Let n be a positive integer. Pick k different positive integers less than n and perform Miller's test on n for each of these bases. If n is composite the probability that n passes all k tests is less than $(1/4)^k$.

Let n be a composite positive integer. Using Rabin's probabilistic primality test, if we pick 100 different integers at random between 1 and n and perform Miller's test for each of these 100 bases, then the probability than n passes all the tests is less than 10^{-60}, an extremely small number. In fact, it may be more likely that a computer error was made than that a composite integer passes all the 100 tests. Using Rabin's primality test does not definitely prove that an integer n that passes all 100 tests is prime, but does give extremely strong, indeed almost overwhelming, evidence that the integer is prime.

There is a famous conjecture in analytic number theory called the *generalized Riemann hypothesis*, which is a statement about the famous Riemann zeta function, named after the German mathematician Bernhard Riemann*. A consequence of this hypothesis is the following conjecture.

Conjecture 5.1. For every composite positive integer n, there is a base b with $b < 2\,(\log_2 n)^2$, such that n fails Miller's test for the base b.

If this conjecture is true, as many number theorists believe, the following result provides a rapid primality test.

Theorem 5.12. If the generalized Riemann hypothesis is valid, then there is an algorithm to determine whether a positive integer n is prime using $O((\log_2 n)^5)$ bit operations.

---

* **BERNHARD RIEMANN (1826-1866)**, the son of a minister, was born in Breselenz, Germany. His elementary education came from his father. After completing his secondary education, he entered Göttingen University to study theology. However, he also attended lectures on mathematics. After receiving the approval of his father to concentrate on mathematics, Riemann transfered to Berlin University where he studied under several prominent mathematicians, including Dirichlet and Jacobi. He subsequently returned to Göttingen where he obtained his Ph.D.

Riemann was one of the most imaginative and creative mathematicians of all time. He made fundamental contributions to geometry, mathematical physics, and analysis. He only wrote one paper on number theory, which was eight pages long, but this paper has had tremendous impact. Riemann died of tuberculosis at the early age of 39.

Proof. Let b be a positive integer less than n. To perform Miller's test for the base b on n takes $O((\log_2 n)^3)$ bit operations, because this test requires that we perform no more than $\log_2 n$ modular exponentiations, each using $O((\log_2 b)^2)$ bit operations. Assume that the generalized Riemann hypothesis is true. If n is composite, then by Conjective 5.1, there is a base b with $1 < b < 2 (\log_2 n)^2$ such that n fails Miller's test for b. To discover this b requires less than $O((\log_2 n)^3) \cdot O((\log_2 n)^2) = O((\log_2 n)^5)$ bit operations. Hence, using $O((\log_2 n)^5)$ bit operations we can determine whether n is composite or prime. □

The important point about Rabin's probabilistic primality test and Theorem 5.12 is that both results indicate that it is possible to check an integer n for primality using only $O((\log_2 n)^k)$ bit operations, where k is a positive integer. This contrasts strongly with the problem of factoring. The best algorithm known for factoring an integer requires a number of bit operations exponential in the square root of the logarithm of the number of bits in the integer being factored, while primality testing seems to require only a number of bit operations less than a polynomial in the number bits of the integer tested. We capitalize on this difference by presenting a recently invented cipher system in Chapter 7.

5.2 Exercises

1. Show that 91 is a pseudoprime to the base 3.

2. Show that 45 is a pseudoprime to the bases 17 and 19.

3. Show that the even integer $n = 161038 = 2 \cdot 73 \cdot 1103$ satisfies the congruence $2^n \equiv 2 \pmod{n}$. The integer 161038 is the smallest even pseudoprime to the base 2.

4. Show that every odd composite integer is a pseudoprime to both the base 1 and the base -1.

5. Show that if n is an odd composite integer and n is a pseudoprime to the base a, then n is a pseudoprime to the base $n - a$.

* 6. Show that if $n = (a^{2p} - 1)/(a^2 - 1)$, where a is an integer, $a > 1$, and p is an odd prime not dividing $a(a^2 - 1)$, then n is a pseudoprime to the base a. Conclude that there are infinitely many pseudoprimes to any base a. (*Hint:* To establish that $a^{n-1} \equiv 1 \pmod{n}$, show that $2p \mid (n - 1)$, and demonstrate that $a^{2p} \equiv 1 \pmod{n}$.)

7. Show that every composite Fermat number $F_m = 2^{2^m} + 1$ is a pseudoprime to the base 2.

8. Show that if p is prime and $2^p - 1$ is composite, then $2^p - 1$ is a pseudoprime to the base 2.

9. Show that if n is a pseudoprime to the bases a and b, then n is also a pseudoprime to the base ab.

10. Suppose that a and n are relatively prime positive integers. Show that if n is a pseudoprime to the base a, then n is a pseudoprime to the base $\bar{a}$, where $\bar{a}$ is an inverse of a modulo n.

11. a) Show that if n is a pseudoprime to the base a, but not a pseudoprime to the base b, where $(a,n) = (b,n) = 1$, then n is not a pseudoprime to the base ab.

    b) Show that if there is an integer b with $(b,n) = 1$ such that n is not a pseudoprime to the base b, then n is a pseudoprime to less than or equal to $\phi(n)/2$ different bases a with $(a,n) = 1$ and $1 \leq a < n$, where $\phi(n)$ is the number of positive integers not exceeding n that are relatively prime to n. (*Hint:* Show that the sets $a_1, a_2, ..., a_r$ and $ba_1, ba_2, ..., ba_r$ have no common elements, where $a_1, a_2, ..., a_r$ are the bases less than n and relatively prime to n to which n is a pseudoprime.)

12. Show that 25 is a strong pseudoprime to the base 7.

13. Show that 1387 is a pseudoprime, but not a strong pseudoprime to the base 2.

14. Show that 1373653 is a strong pseudoprime to both bases 2 and 3.

15. Show that 25326001 is a strong pseudoprime to bases 2, 3, and 5.

16. Show that the following integers are Carmichael numbers.

    a) $2821 = 7 \cdot 13 \cdot 31$
    b) $10585 = 5 \cdot 29 \cdot 73$
    c) $29341 = 13 \cdot 37 \cdot 61$
    d) $314821 = 13 \cdot 61 \cdot 397$
    e) $278545 = 5 \cdot 17 \cdot 29 \cdot 113$
    f) $172081 = 7 \cdot 13 \cdot 31 \cdot 61$
    g) $564651361 = 43 \cdot 3361 \cdot 3907$

17. Find a Carmichael number of the form $7 \cdot 23 \cdot q$ where q is an odd prime other than $q = 41$ or show that there are no others.

18. a) Show that every integer of the form $(6m+1)(12m+1)(18m+1)$, where m is a positive integer such that $6m+1$, $12m+1$, and $18m+1$ are all primes, is a Carmichael number.

    b) Conclude from part (a) that $1729 = 7 \cdot 13 \cdot 19$, $294409 = 37 \cdot 73 \cdot 109$, $56052361 = 211 \cdot 421 \cdot 631$, $118901521 = 271 \cdot 541 \cdot 811$, and $172947529 = 307 \cdot 613 \cdot 919$ are Carmichael numbers.

19. Show that if n is a positive integer with $n \equiv 3 \pmod 4$, then Miller's test takes $O((\log_2 n)^3)$ bit operations.

5.2 Computer Projects

Programming Projects

Write programs to do the following:

1. Given a positive integer n, determine whether n satisfies the congruence $b^{n-1} \equiv 1 \pmod{n}$ where b is a positive integer less than n; if it does, then n is either a prime or a pseudoprime to the base b.

2. Given a positive integer integer n, determine whether n passes Miller´s test to the base b; if it does then n is either prime or a strong pseudoprime to the base b.

3. Perform a primality test for integers less than $25 \cdot 10^9$ based on Miller´s tests for the bases 2, 3, 5, and 7. (Use the remarks that follow Theorem 5.9.)

4. Given an odd positive integer n, determine whether n passes Rabin´s probabilistic primality test.

5. Given a positive integer n, find all Carmichael numbers less than a given integer n.

Computations and Explorations

Using the programs you have written or a computation program, carry out the following computations and explorations:

1. Determine for which positive integers n, $n \leq 100$, the integer $n \cdot 2^n - 1$ is prime.

2. Find as many Carmichael numbers of the form $(6m+1)(12m+1)(18m+1)$, where $6m+1$, $12m+1$, and $18m+1$ are all prime, as you can.

3. Find as many even pseudoprimes to the base 2 which are the product of three primes as you can. Do you think there are infinitely many?

4. The integers of the form $n \cdot 2^n + 1$, where n is a positive integer greater than 1, are called *Cullen numbers*. Can you find a prime Cullen number?

5.3 Euler´s Theorem

Fermat´s little theorem tells us how to work with certain congruences involving exponents when the modulus is a prime. How do we work with the corresponding congruences modulo a composite integer? For this purpose, we first define a special counting function.

Definition. Let n be a positive integer. The *Euler phi-function* $\phi(n)$ is defined to be the number of positive integers not exceeding n that are relatively prime to n.

This function is named after the great Swiss mathematician Leonhard Euler* who introduced this function and proved the results of this section.

In Table 5.1 we display the values of $\phi(n)$ for $1 \leq n \leq 12$. The values of $\phi(n)$ for $1 \leq n \leq 100$ are given in Table 2 of the Appendix.

n	1	2	3	4	5	6	7	8	9	10	11	12
$\phi(n)$	1	1	2	2	4	2	6	4	6	4	10	4

Table 5.1. The Values of Euler's Phi-function for $1 \leq n \leq 12$.

In Chapter 6, we study the Euler phi-function further. In this section, we use the phi-function to give an analogue of Fermat's little theorem for composite moduli. To do this, we need to lay some groundwork.

Definition. A *reduced residue system modulo n* is a set of $\phi(n)$ integers such that each element of the set is relatively prime to n, and no two different elements of the set are congruent modulo n.

Example 5.17. The set 1, 3, 5, 7 is a reduced residue system modulo 8. The set $-3, -1, 1, 3$ is also such a set.   □

We will need the following theorem about reduced residue systems.

Theorem 5.13. If $r_1, r_2, ..., r_{\phi(n)}$ is a reduced residue system modulo n, and if a is a positive integer with $(a,n) = 1$, then the set $ar_1, ar_2, ..., ar_{\phi(n)}$ is also a reduced residue system modulo n.

Proof. To show that each integer ar_j is relatively prime to n, we assume that

---

* **LEONHARD EULER** (1707-1783) was the son of a minister from the vicinity of Basel, Switzerland, who, besides theology, had also studied mathematics. At 13 Euler entered the University of Basel with the aim of pursuing a career in theology, as his father wished. At the university, Euler was tutored in mathematics by Johann Bernoulli, of the famous Bernoulli family of mathematicians, and became friends with Johann's sons Nicklaus and Daniel. His interest in mathematics led him to abandon his plans to follow in his father's footsteps. Euler obtained his master's degree in philosophy at the age of 16. In 1727, Peter the Great invited Euler to join the Imperial Academy in St. Petersburg (now Leningrad), at the insistence of the Nicklaus and Daniel Bernoulli who entered the academy in 1725 when it was founded. Euler spent the years 1727-1741 and 1766-1783 at the Imperial Academy. He spent the interval 1741-1766 at the Royal Academy of Berlin. Euler was incredibly prolific. He wrote over 700 books and papers, and he left so much unpublished work that the Imperial Academy did not finish publication of Euler's work for 47 years after his death. During his life, his papers accumulated so rapidly that he kept a pile of papers to be published for the academy. They published the top papers in the pile first, so that later results were published before results they superseded or depended on. Euler was blind for the last 17 years of his life, but had a fantastic memory, so that his blindness did not deter his mathematical output. He also had 13 children, and was able to continue his research while a child or two bounced on his knees. The publication of the collected works of Euler by the Swiss Society of Natural Science will require more than 75 large volumes.

$(ar_j, n) > 1$. Then, there is a prime divisor p of (ar_j, n). Hence, either $p \mid a$ or $p \mid r_j$. Thus, we have either $p \mid a$ and $p \mid n$, or $p \mid r_j$ and $p \mid n$. However, we cannot have both $p \mid r_j$ and $p \mid n$, since r_j is a member of a reduced residue system modulo n, and both $p \mid a$ and $p \mid n$ cannot hold since $(a, n) = 1$. Hence, we can conclude that ar_j and n are relatively prime for $j = 1, 2, ..., \phi(n)$.

To demonstrate that no two ar_j's are congruent modulo n, we assume that $ar_j \equiv ar_k \pmod{n}$, where j and k are distinct positive integers with $1 \leq j \leq \phi(n)$ and $1 \leq k \leq \phi(n)$. Since $(a, n) = 1$, by Corollary 3.4.1 we see that $r_j \equiv r_k \pmod{n}$. This is a contradiction, since r_j and r_k come from the original set of reduced residues modulo n, so that $r_j \not\equiv r_k \pmod{n}$. □

We illustrate the use of Theorem 5.13 by the following example.

Example 5.18. The set $1, 3, 5, 7$ is a reduced residue system modulo 8. Since $(3, 8) = 1$, from Theorem 5.13, the set $3 \cdot 1 = 3, 3 \cdot 3 = 9, 3 \cdot 5 = 15, 3 \cdot 7 = 21$ is also a reduced residue system modulo 8. □

We now state Euler's theorem.

Theorem 5.14. Euler's Theorem. If m is a positive integer and a is an integer with $(a, m) = 1$, then $a^{\phi(m)} \equiv 1 \pmod{m}$. □

Before we prove Euler's theorem, we illustrate the idea behind the proof with an example.

Example 5.19. We know that both the sets $1, 3, 5, 7$ and $3 \cdot 1, 3 \cdot 3, 3 \cdot 5, 3 \cdot 7$ are reduced residue systems modulo 8. Hence, they have the same least positive residues modulo 8. Therefore,

$$(3 \cdot 1) \cdot (3 \cdot 3) \cdot (3 \cdot 5) \cdot (3 \cdot 7) \equiv 1 \cdot 3 \cdot 5 \cdot 7 \pmod{8},$$

and

$$3^4 \cdot 1 \cdot 3 \cdot 5 \cdot 7 \equiv 1 \cdot 3 \cdot 5 \cdot 7 \pmod{8}.$$

Since $(1 \cdot 3 \cdot 5 \cdot 7, 8) = 1$, we conclude that

$$3^4 = 3^{\phi(8)} \equiv 1 \pmod{8}.$$ □

We now use the ideas illustrated by this example to prove Euler's theorem.

Proof. Let $r_1, r_2, ..., r_{\phi(m)}$ denote the reduced residue system made up of the positive integers not exceeding m that are relatively prime to m. By Theorem 5.13, since $(a, m) = 1$, the set $ar_1, ar_2, ..., ar_{\phi(m)}$ is also a reduced residue system modulo m. Hence, the least positive residues of $ar_1, ar_2, ..., ar_{\phi(m)}$ must be the integers $r_1, r_2, ..., r_{\phi(m)}$ in some order. Consequently, if we multiply together all terms in each of these reduced residue systems, we obtain

$$ar_1 ar_2 \cdots ar_{\phi(m)} \equiv r_1 r_2 \cdots r_{\phi(m)} \pmod{m} .$$

Thus,

$$a^{\phi(m)} r_1 r_2 \cdots r_{\phi(m)} \equiv r_1 r_2 \cdots r_{\phi(m)} \pmod{m}.$$

Since $(r_1 r_2 \cdots r_{\phi(m)}, m) = 1$, from Corollary 3.4.1, we can conclude that $a^{\phi(m)} \equiv 1 \pmod{m}$. $\qquad\square$

We can use Euler's Theorem to find inverses modulo m. If a and m are relatively prime, we know that

$$a \cdot a^{\phi(m)-1} = a^{\phi(m)} \equiv 1 \pmod{m}.$$

Hence, $a^{\phi(m)-1}$ is an inverse of a modulo m.

Example 5.20. We know that $2^{\phi(9)-1} = 2^{6-1} = 2^5 = 32 \equiv 5 \pmod{9}$ is an inverse of 2 modulo 9. $\qquad\square$

We can solve linear congruences using this observation. To solve $ax \equiv b \pmod{m}$, where $(a,m) = 1$, we multiply both sides of this congruence by $a^{\phi(m)-1}$ to obtain

$$a^{\phi(m)-1} ax \equiv a^{\phi(m)-1} b \pmod{m}.$$

Therefore, the solutions are those integers x such that $x \equiv a^{\phi(m)-1} b \pmod{m}$.

Example 5.21. The solutions of $3x \equiv 7 \pmod{10}$ are given by $x \equiv 3^{\phi(10)-1} \cdot 7 \equiv 3^3 \cdot 7 \equiv 9 \pmod{10}$, since $\phi(10) = 4$. $\qquad\square$

5.3 Exercises

1. Find a reduced residue system modulo each of the following integers.

   a) 6   d) 14
   b) 9   e) 16
   c) 10   f) 17

2. Find a reduced residue system modulo 2^m, where m is a positive integer.

3. Show that if $c_1, c_2, \ldots, c_{\phi(m)}$ is a reduced residue system modulo m, where m is a positive integer with $m \neq 2$, then $c_1 + c_2 + \cdots + c_{\phi(m)} \equiv 0 \pmod{m}$.

4. Show that if a and m are positive integers with $(a,m) = (a-1,m) = 1$, then $1 + a + a^2 + \cdots + a^{\phi(m)-1} \equiv 0 \pmod{m}$.

5. Use Euler's theorem to find the least positive residue of 3^{100000} modulo 35.

6. Show that if a is an integer such that a is not divisble by 3 or such that a is divisible by 9, then $a^7 \equiv a \pmod{63}$.

7. Show that if a is an integer relatively prime to 32760, then $a^{12} \equiv 1 \pmod{32760}$.

8. Show that $a^{\phi(b)} + b^{\phi(a)} \equiv 1 \pmod{ab}$, if a and b are relatively prime positive integers.

9. Solve each of the following linear congruences using Euler's theorem.

   a) $5x \equiv 3 \pmod{14}$
   b) $4x \equiv 7 \pmod{15}$
   c) $3x \equiv 5 \pmod{16}$

10. Show that the solutions to the simultaneous system of congruences

$$x \equiv a_1 \pmod{m_1}$$
$$x \equiv a_2 \pmod{m_2}$$
$$\cdot$$
$$\cdot$$
$$\cdot$$
$$x \equiv a_r \pmod{m_r},$$

where the m_j are pairwise relatively prime, are given by

$$x \equiv a_1 M_1^{\phi(m_1)} + a_2 M_2^{\phi(m_2)} + \cdots + a_r M_r^{\phi(m_r)} \pmod{M},$$

where $M = m_1 m_2 \cdots m_r$ and $M_j = M/m_j$ for $j = 1, 2, \ldots, r$.

11. Use Exercise 10 to solve each of the systems of congruences in Exercise 4 of Section 3.3.

12. Use Exercise 10 to solve the system of congruences in Exercise 5 of Section 3.3.

13. Use Euler's theorem to find the last digit in the decimal expansion of 7^{1000}

14. Use Euler's theorem to find the last digit in the hexadecimal expansion of $5^{1000000}$.

15. Find $\phi(n)$ for the integers n with $13 \leq n \leq 20$.

16. Show that every positive integer relatively prime to 10 divides infinitely many repunits (see the preamble to Exercise 5 of Section 4.1). (*Hint*: Note that the n-digit repunit $111 \cdots 11 = (10^n - 1)/9$.)

17. Show that every positive integer relatively prime to b divides infinitely many base b repunits (see the preamble to Exercise 9 of Section 4.1).

* 18. Show that if m is a positive integer, $m > 1$, then $a^m \equiv a^{m-\phi(m)} \pmod{m}$ for all positive integers a.

5.3 Computer Projects

Programming Projects

Write programs to do the following:

    1. Construct a reduced residue set modulo n for a given positive integer n.

    2. Solve linear congruences using Euler's theorem.

    3. Find the solutions of a system of linear congruences using Euler's theorem and the Chinese remainder theorem (see Exercise 10).

Computations and Explorations

Using the programs you have written or a computation program, carry out the following computations and explorations:

    1. Find $\phi(n)$ for all integers n less than 10000. What conjectures can you make about the values of $\phi(n)$?

6

Multiplicative Functions

6.1 The Euler Phi-function

The Euler phi-function has the property that its value at an integer n is the product of the values of the Euler phi-function at the prime powers that occur in the factorization of n. Functions with this property are called multiplicative; such functions arise throughout number theory. In this chapter we will show that the Euler phi-function is multiplicative. From this fact we will derive a formula for its values based on prime factorizations. Later in this chapter we will study other multiplicative functions, including the number of divisors function and the sum of divisors function.

We first present some definitions.

Definition. An *arithmetic function* is a function that is defined for all positive integers.

Throughout this chapter, we are interested in arithmetic functions that have a special property.

Definition. An arithmetic function f is called *multiplicative* if $f(mn) = f(m)f(n)$ whenever m and n are relatively prime positive integers. It is called *completely multiplicative* if $f(mn) = f(m)f(n)$ for all positive integers m and n.

Example 6.1. The function $f(n) = 1$ for all n is completely multiplicative, and hence also multiplicative, because $f(mn) = 1$, $f(m) = 1$, and $f(n) = 1$, so that $f(mn) = f(m)f(n)$. Similarly, the function $g(n) = n$ is completely multiplicative, and hence multiplicative, since $g(mn) = mn = g(m)g(n)$. $\qquad \square$

If f is a multiplicative function, then we can find a simple formula for $f(n)$ given the prime-power factorization of n.

Theorem 6.1. If f is a multiplicative function and if $n = p_1^{a_1} p_2^{a_2} \cdots p_s^{a_s}$ is the prime-power factorization of the positive integer n, then $f(n) = f(p_1^{a_1}) f(p_2^{a_2}) \cdots f(p_s^{a_s})$.

Proof. We will prove the theorem using mathematical induction on the number of different primes in the prime factorization of the integer n. If n has one prime in its prime-power factorization, then $n = p_1^{a_1}$ for some prime p_1. It follows that the result is trivially true.

Suppose that the theorem is true for all integers with k different primes in their prime-power factorization. Now suppose that n has $k+1$ different primes in its prime-power factorization, say $n = p_1^{a_1} p_2^{a_2} \cdots p_k^{a_k} p_{k+1}^{a_{k+1}}$. Since f is multiplicative and $(p_1^{a_1} p_2^{a_2} \cdots p_k^{a_k}, p_{k+1}^{a_{k+1}}) = 1$, we see that $f(n) = f(p_1^{a_1} p_2^{a_2} \cdots p_k^{a_k}) f(p_{k+1}^{a_{k+1}})$. By the inductive hypothesis, we know that $f(p_1^{a_1} p_2^{a_2} p_3^{a_3} \cdots p_k^{a_k}) = f(p_1^{a_1}) f(p_2^{a_2}) f(p_3^{a_3}) \cdots f(p_k^{a_k})$. It follows that $f(n) = f(p_1^{a_1}) f(p_2^{a_2}) \cdots f(p_k^{a_k}) f(p_{k+1}^{a_{k+1}})$. This completes the inductive proof. ∎

We now return to the Euler phi-function. We first consider its values at primes and then at prime powers.

Theorem 6.2. If p is prime, then $\phi(p) = p - 1$. Conversely, if p is a positive integer with $\phi(p) = p - 1$, then p is prime.

Proof. If p is prime then every positive integer less than p is relatively prime to p. Since there are $p - 1$ such integers, we have $\phi(p) = p - 1$.

Conversely, if p is composite, then p has a divisor d with $1 < d < p$, and, of course, p and d are not relatively prime. Since we know that at least one of the $p - 1$ integers $1, 2, \ldots, p - 1$, namely d, is not relatively prime to p, $\phi(p) \leq p - 2$. Hence, if $\phi(p) = p - 1$, then p must be prime. ∎

We now find the values of the phi-function at prime powers.

Theorem 6.3. Let p be a prime and a a positive integer. Then $\phi(p^a) = p^a - p^{a-1}$.

Proof. The positive integers less than p^a that are not relatively prime to p are those integers not exceeding p^a that are divisible by p. These are the integers kp where $1 \leq k \leq p^{a-1}$. Since there are exactly p^{a-1} such integers, there are $p^a - p^{a-1}$ integers less than p^a that are relatively prime to p^a. Hence, $\phi(p^a) = p^a - p^{a-1}$. ∎

Example 6.2. Using Theorem 6.3, we find that $\phi(5^3) = 5^3 - 5^2 = 100$, $\phi(2^{10}) = 2^{10} - 2^9 = 512$, and $\phi(11^2) = 11^2 - 11 = 110$. □

To find a formula for $\phi(n)$, given the prime factorization of n, it suffices to show that ϕ is multiplicative. We illustrate the idea behind the proof with the following example.

Example 6.3. Let $m = 4$ and $n = 9$, so that $mn = 36$. We list the integers from 1 to 36 in a rectangular chart, as shown in Figure 6.1.

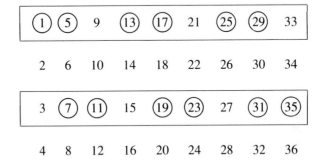

Figure 6.1. Demonstrating that $\phi(36) = \phi(4)\phi(9)$.

Neither the second nor fourth row contains integers relatively prime to 36, since each element in these rows is not relatively prime to 4, and hence not relatively prime to 36. We enclose the other two rows; each element of these rows is relatively prime to 4. Within each of these rows, there are 6 integers relatively prime to 9. We circle these; they are the 12 integers in the list relatively prime to 36. Hence $\phi(36) = 2 \cdot 6 = \phi(4)\phi(9)$. □

We now state and prove the theorem that shows that ϕ is multiplicative.

Theorem 6.4. Let m and n be relatively prime positive integers. Then $\phi(mn) = \phi(m)\phi(n)$.

Proof. We display the positive integers not exceeding mn in the following way.

1	$m+1$	$2m+1$	...	$(n-1)m+1$
2	$m+2$	$2m+2$	...	$(n-1)m+2$
3	$m+3$	$2m+3$	...	$(n-1)m+3$
.	.	.		.
.	.	.		.
.	.	.		.
r	$m+r$	$2m+r$	...	$(n-1)m+r$
.	.	.		.
.	.	.		.
.	.	.		.
m	$2m$	$3m$	...	mn

Now suppose that r is a positive integer not exceeding m. Suppose $(m,r) = d > 1$. Then no number in the rth row is relatively prime to mn, since any element of this row is of the form $km + r$, where k is an integer with $1 \leq k \leq n - 1$, and $d \mid (km+r)$, since $d \mid m$ and $d \mid r$.

Consequently, to find those integers in the display that are relatively prime to mn, we need to look at the rth row only if $(m,r) = 1$. If $(m,r) = 1$ and $1 \leq r \leq m$, we must determine how many integers in this row are relatively prime to mn. The elements in this row are $r, m + r, 2m + r, ..., (n-1)m + r$. Since $(r,m) = 1$, each of these integers is relatively prime to m. By Theorem 3.6 the n integers in the rth row form a complete system of residues modulo n. Hence, exactly $\phi(n)$ of these integers are relatively prime to n. Since these $\phi(n)$ integers are also relatively prime to m, they are relatively prime to mn.

Since there are $\phi(m)$ rows, each containing $\phi(n)$ integers relatively prime to mn, we can conclude that $\phi(mn) = \phi(m)\phi(n)$. ∎

Combining Theorems 6.3 and 6.4, we derive the following formula for $\phi(n)$.

Theorem 6.5. Let $n = p_1^{a_1} p_2^{a_2} \cdots p_k^{a_k}$ be the prime-power factorization of the positive integer n. Then

$$\phi(n) = n(1 - \frac{1}{p_1})(1 - \frac{1}{p_2}) \cdots (1 - \frac{1}{p_k}).$$

Proof. Since ϕ is multiplicative, Theorem 6.1 tells us that

$$\phi(n) = \phi(p_1^{a_1})\phi(p_2^{a_2}) \cdots \phi(p_k^{a_k}).$$

In addition, by Theorem 6.3 we know that

$$\phi(p_j^{a_j}) = p_j^{a_j} - p_j^{a_j-1} = p_j^{a_j}(1 - \frac{1}{p_j})$$

for $j = 1, 2, \dots, k$. Hence,

$$\phi(n) = p_1^{a_1}(1 - \frac{1}{p_1})p_2^{a_2}(1 - \frac{1}{p_2}) \cdots p_k^{a_k}(1 - \frac{1}{p_k})$$

$$= p_1^{a_1}p_2^{a_2} \cdots p_k^{a_k}(1 - \frac{1}{p_1})(1 - \frac{1}{p_2}) \cdots (1 - \frac{1}{p_k})$$

$$= n(1 - \frac{1}{p_1})(1 - \frac{1}{p_2}) \cdots (1 - \frac{1}{p_k}).$$

This is the desired formula for $\phi(n)$. ∎

We illustrate the use of Theorem 6.5 with the following example.

Example 6.4. Using Theorem 6.5 we note that

$$\phi(100) = \phi(2^2 5^2) = 100(1 - \frac{1}{2})(1 - \frac{1}{5}) = 40$$

and

$$\phi(720) = \phi(2^4 3^2 5) = 720(1 - \frac{1}{2})(1 - \frac{1}{3})(1 - \frac{1}{5}) = 192. \quad \square$$

Note that $\phi(n)$ is even except when $n = 2$ as the following theorem shows.

Theorem 6.6. Let n be a postive integer greater than 2. Then $\phi(n)$ is even.

Proof. Suppose that $n = p_1^{a_1}p_2^{a_2} \cdots p_s^{a_s}$ is the prime-power factorization of n. Since ϕ is multiplicative, it follows that $\phi(n) = \prod_{j=1}^{s} \phi(p_j^{a_j})$. By Theorem 6.3 we know that $\phi(p_j^{a_j}) = p_j^{a_j-1}(p_j - 1)$. We can see that $\phi(p_j^{a_j})$ is even if p_j is an odd prime, since then $p_j - 1$ is even, or if $p_j = 2$ and $a_j \geq 1$, since then $p_j^{a_j-1}$ is even. Since $n > 2$, at least one of these two conditions holds, so that $\phi(p_j^{a_j})$ is even for at least one integer j, $1 \leq j \leq s$. We conclude that $\phi(n)$ is even. ∎

Let f be an arithmetic function. Then

$$\sum_{d|n} f(d)$$

represents the sum of the values of f at all the positive divisors of n.

Example 6.5. If f is an arithmetic function, then

$$\sum_{d|12} f(d) = f(1) + f(2) + f(3) + f(4) + f(6) + f(12).$$

For instance,

$$\sum_{d \mid 12} d^2 = 1^2 + 2^2 + 3^2 + 4^2 + 6^2 + 12^2$$

$$= 1 + 4 + 9 + 16 + 36 + 144 = 210. \qquad \square$$

The following result, which states that n is the sum of the values of the phi-function at all the positive divisors of n, will also be useful in the sequel.

Theorem 6.7. Let n be a positive integer. Then

$$\sum_{d \mid n} \phi(d) = n.$$

Proof. We split the set of integers from 1 to n into classes. Put the integer m into the class C_d if the greatest common divisor of m and n is d. We see that m is in C_d, i.e. $(m,n) = d$, if and only if $(m/d, n/d) = 1$. Hence, the number of integers in C_d is the number of positive integers not exceeding n/d that are relatively prime to the integer n/d. From this observation, we see that there are $\phi(n/d)$ integers in C_d. Since we divided the integers 1 to n into disjoint classes and each integer is in exactly one class, n is the sum of the numbers of elements in the different classes. Consequently, we see that

$$n = \sum_{d \mid n} \phi(n/d).$$

As d runs through the positive integers that divide n, n/d also runs through these divisors, so that

$$n = \sum_{d \mid n} \phi(n/d) = \sum_{d \mid n} \phi(d).$$

This proves the theorem.   ■

Example 6.6. We illustrate the proof of Theorem 6.7 when $n = 18$. The integers from 1 to 18 can be split into classes C_d where $d \mid 18$ such that the class C_d contains those integers m with $(m, 18) = d$. We have

$$\begin{aligned}
C_1 &= \{1, 5, 7, 11, 13, 17\} \quad & C_6 &= \{6, 12\} \\
C_2 &= \{2, 4, 8, 10, 14, 16\} \quad & C_9 &= \{9\} \\
C_3 &= \{3, 15\} \quad & C_{18} &= \{18\}.
\end{aligned}$$

We see that the class C_d contains $\phi(18/d)$ integers, as the six classes contain $\phi(18) = 6$, $\phi(9) = 6$, $\phi(6) = 2$, $\phi(3) = 2$, $\phi(2) = 1$, and $\phi(1) = 1$ integers, respectively. We note that $18 = \phi(18) + \phi(9) + \phi(6) + \phi(3) + \phi(2) + \phi(1) = \sum_{d \mid 18} \phi(d)$. $\square$

6.1 Exercises

1. Determine whether each of the following arithmetic functions is completely multiplicative. Prove your answers.

   a) $f(n) = 0$ d) $f(n) = \log n$ g) $f(n) = n + 1$
   b) $f(n) = 2$ e) $f(n) = n^2$ h) $f(n) = n^n$
   c) $f(n) = n/2$ f) $f(n) = n!$ i) $f(n) = \sqrt{n}$

2. Find the value of the Euler phi-function at each of the following integers.

   a) 100   d) $2 \cdot 3 \cdot 5 \cdot 7 \cdot 11 \cdot 13$
   b) 256   e) 10!
   c) 1001   f) 20!

3. Show that $\phi(5186) = \phi(5187) = \phi(5188)$.

* 4. Find all positive integers n such that $\phi(n)$ has the given value.

   a) 1   d) 6
   b) 2   e) 14
   c) 3   f) 24

5. For which positive integers n does $\phi(3n) = 3\phi(n)$?

6. For which positive integers n is $\phi(n)$ divisible by 4?

7. For which positive integers n is $\phi(n)$ equal to $n/2$?

8. For which positive integers n does $\phi(n)|n$?

9. Show that if n is a positive integer, then

$$\phi(2n) = \begin{cases} \phi(n) & \text{if } n \text{ is odd} \\ 2\phi(n) & \text{if } n \text{ is even.} \end{cases}$$

10. Show that if n is a positive integer having k distinct odd prime divisors, then $\phi(n)$ is divisible by 2^k.

11. For which positive integers n is $\phi(n)$ a power of 2?

12. Show that if m and k are positive integers, then $\phi(m^k) = m^{k-1}\phi(m)$.

13. Show that if a and b are positive integers, then

$$\phi(ab) = (a,b)\phi(a)\phi(b)/\phi((a,b)).$$

14. Find the least positive integer n such that

   a) $\phi(n) \geq 100$ c) $\phi(n) \geq 10000$
   b) $\phi(n) \geq 1000$ d) $\phi(n) \geq 100000$

* 15. Show that if n is a composite positive integer and $\phi(n) \mid (n-1)$ then n is square-free and is the product of at least three distinct primes.

16. Show that if m and n are positive integers with $m \mid n$, then $\phi(m) \mid \phi(n)$.

* 17. Prove Theorem 6.5, using the principle of inclusion-exclusion (see Exercise 18 of Section 1.4).

18. Show that a positive integer n is composite if and only if $\phi(n) \le n - \sqrt{n}$.

19. Let n be a positive integer. Define the sequence of positive integers $n_1, n_2, n_3, \ldots$ recursively by $n_1 = \phi(n)$ and $n_{k+1} = \phi(n_k)$ for $k = 1, 2, 3, \ldots$. Show that there is a positive integer r such that $n_r = 1$.

A multiplicative function is called *strongly multiplicative* if and only if $f(p^k) = f(p)$ for every prime p and every positive integer k.

20. Show that $f(n) = \phi(n)/n$ is a strongly multiplicative function.

Two arithmetic functions f and g may be multiplied using the *Dirichlet product* which is defined by

$$(f*g)(n) = \sum_{d\mid n} f(d)g(n/d).$$

21. Show that $f*g = g*f$.

22. Show that $(f*g)*h = f*(g*h)$.

We define the ι *function* by

$$\iota(n) = \begin{cases} 1 & \text{if } n = 1 \\ 0 & \text{if } n > 1, \end{cases}$$

23. a) Show that ι is a multiplicative function.

b) Show that $\iota*f = f*\iota = f$ for all arithmetic functions f.

24. The arithmetic function g is said to be the *inverse* of the arithmetic function f if $f*g = g*f = \iota$. Show that the arithmetic function f has an *inverse* if and only if $f(1) \ne 0$. Show that if f has an inverse it is unique. (*Hint*: When $f(1) \ne 0$, find the inverse f^{-1} of f by calculating $f(n)$ recursively, using the fact that

$$\iota(n) = \sum_{d\mid n} f(d)f^{-1}(n/d).)$$

25. Show that if f and g are multiplicative functions, then the Dirichlet product $f*g$ is also multiplicative.

The *Möbius function*, named after A.F. Möbius*, is defined by

$$\mu(n) = \begin{cases} 1 & \text{if } n = 1 \\ (-1)^s & \text{if } n \text{ is square}-\text{free with prime factorization} \\ & \qquad n = p_1 p_2 \cdots p_s \\ 0 & \text{if } n \text{ has square factor larger than } 1. \end{cases}$$

26. Find the value of the Möbius function at each of the following integers.

a) 100
b) 105
c) 1001

d) $2 \cdot 3 \cdot 5 \cdot 7 \cdot 11 \cdot 13$
e) 740
f) 10!

27. Find $\sum\limits_{d \mid n} \mu(d)$ where n takes each of the values in Exercise 25.

28. Show that the Möbius function is multiplicative.

29. Show that if n is a positive integer greater than one, then $\sum\limits_{d \mid n} \mu(d) = 0$.

* **30.** Let f be an arithmetic function. Show that if F is the arithmetic function defined by

$$F(n) = \sum_{d \mid n} f(d),$$

then

$$f(n) = \sum_{d \mid n} \mu(d) F(n/d).$$

This result is called the *Möbius inversion formula*.

* **31.** Use the Möbius inversion formula to show that if f is an arithmetic function and F is the arithmetic function defined by

---

* **AUGUST FERDINAND MÖBIUS (1790-1868)** was born in the town of Schulpforta, near Naumburg, Germany. His father was a dancing teacher and his mother was a descendant of Luther. Möbius was taught at home until he was 13, displaying an interest and talent in mathematics at a young age. He received formal training in mathematics from 1803 until 1809, when he entered Leipzig University. He intended to study law, but instead, decided to concentrate on subjects more to his interest, mathematics, physics, and astronomy. After pursuing further studies at Göttingen, where he studied astronomy with Gauss, and at Halle, where he studied mathematics with Pfaff, he became professor of astronomy at Leipzig, remaining there until his death. Möbius made contributions to a wide range of subjects, including astronomy, mechanics, projective geometry, optics, statics, and number theory. Today, he is best known for his discovery of a surface with one side, called the *Möbius strip*, which can be formed by taking a strip of paper and connecting two opposite ends after twisting it.

$$F(n) = \sum_{d|n} f(d),$$

then if F is multiplicative, so is f.

32. Using the Möbius inversion formula and the fact that $n = \sum_{d|n} \phi(n/d)$, prove that

    a) $\phi(p^t) = p^t - p^{t-1}$, where p is a prime and t is a positive integer.

    b) $\phi(n)$ is multiplicative.

33. Show that the function $f(n) = n^k$ is completely multiplicative for every real number k.

We define *Liouville's function* $\lambda(n)$ by $\lambda(1) = 1$ and for $n > 1$ by $\lambda(n) = (-1)^{a_1 + a_2 + \cdots + a_m}$, if the prime-power factorization of n is $n = p_1^{a_1} p_2^{a_2} \cdots p_m^{a_m}$.

34. Show that $\lambda(n)$ is completely multiplicative.

35. Show that if n is a positive integer then $\sum_{d|n} \lambda(d)$ equals 0 if n is not a perfect square, and equals 1 if n is a perfect square.

36. Show that if f and g are multiplicative functions then fg is also multiplicative, where $(fg)(n) = f(n)g(n)$ for every positive integer n.

37. Show that if f and g are completely multiplicative functions then fg is also completely multiplicative.

38. Show that if f is completely multiplicative, then $f(n) = f(p_1)^{a_1} f(p_2)^{a_2} \cdots f(p_m)^{a_m}$ when the prime-power factorization of n is $n = p_1^{a_1} p_2^{a_2} \cdots p_m^{a_m}$.

A function f that satisfies the equation $f(mn) = f(m) + f(n)$ for all relatively prime positive integers m and n is called *additive*, and if the above equation holds for all positive integers m and n, f is called *completely additive*.

39. Show that the function $f(n) = \log n$ is completely additive.

40. Show that if $\omega(n)$ is the function that denotes the number of distinct prime factors of n, then ω is additive, but not completely additive.

41. Show that if f is an additive function and if $g(n) = 2^{f(n)}$, then g is multiplicative.

6.1 Computer Projects

Programming Projects

Write programs to do the following:

  1. Given a positive integer n, find the value of $\phi(n)$.

  2. Given a positive integer n, find the number of iterations of the phi function, starting with n, required to reach 1. (This is the the integer r in Exercise 19.)

Computations and Explorations

Using the programs you have written or a computation program, carry out the following computations and explorations:

  1. Find the largest integer n such that $\phi(n) \leq 10000$.

  2. Find as many positive integers n such that $\phi(n) = \phi(n+1)$. Can you formulate any conjectures based on the evidence you found?

  3. Can you find a positive integer n other than 5186 such that $\phi(n) = \phi(n+1) = \phi(n+2)$? Can you find four consecutive positive integers n, $n+1$, $n+2$, $n+3$, $n+4$ such that $\phi(n+1) = \phi(n+2) = \phi(n+3) = \phi(n+4)$?

  4. Lehmer conjectured that n is prime if $\phi(n)$ divides $n-1$. Gather evidence for this conjecture or find a counterexample if you can.

  5. Carmichael conjectured that for every positive integer n there is a positive integer m distinct from n such that $\phi(m) = \phi(n)$. Gather evidence for this conjecture or find a counterexample if you can.

6.2 The Sum and Number of Divisors

As we mentioned in Section 6.1, the number of divisors function and the sum of divisors function are both multiplicative functions. We will show these functions are multiplicative and derive formulae for their values at a positive integer n from the prime factorization of n.

Definition. The *sum of divisors function*, denoted by σ, is defined by setting $\sigma(n)$ equal to the sum of all the positive divisors of n.

In Table 6.1 we give $\sigma(n)$ for $1 \leq n \leq 12$. The values of $\sigma(n)$ for $1 \leq n \leq 100$ are given in Table 2 of the Appendix.

n	1	2	3	4	5	6	7	8	9	10	11	12
$\sigma(n)$	1	3	4	7	6	12	8	15	13	18	12	28

Table 6.1. The Sum of the Divisors for $1 \le n \le 12$.

The other function which we will study is the number of divisors function.

Definition. The *number of divisors function*, denoted by τ, is defined by setting $\tau(n)$ equal to the number of positive divisors of n.

In Table 6.2 we give $\tau(n)$ for $1 \le n \le 12$. The values of $\tau(n)$ for $1 \le n \le 100$ are given in Table 2 of the Appendix.

n	1	2	3	4	5	6	7	8	9	10	11	12
$\tau(n)$	1	2	2	3	2	4	2	4	3	4	2	6

Table 6.2. The Number of Divisors for $1 \le n \le 12$.

Note that we can express $\sigma(n)$ and $\tau(n)$ in terms of summation notation. It is simple to see that

$$\sigma(n) = \sum_{d \mid n} d$$

and

$$\tau(n) = \sum_{d \mid n} 1.$$

To prove that σ and τ are multiplicative, we use the following theorem.

Theorem 6.8. If f is a multiplicative function, then the arithmetic function $F(n) = \sum_{d \mid n} f(d)$ is also multiplicative.

Before we prove the theorem, we illustrate the idea behind its proof with the following example. Let f be a multiplicative function, and let $F(n) = \sum_{d \mid n} f(d)$. We will show that $F(60) = F(4)F(15)$. Each of the divisors of 60 may be written as the product of a divisor of 4 and a divisor of 15 in the following way: $1 = 1 \cdot 1$, $2 = 2 \cdot 1$, $3 = 1 \cdot 3$, $4 = 4 \cdot 1$, $5 = 1 \cdot 5$, $6 = 2 \cdot 3$, $10 = 2 \cdot 5$, $12 = 4 \cdot 3$, $15 = 1 \cdot 15$, $20 = 4 \cdot 5$, $30 = 2 \cdot 15$, $60 = 4 \cdot 15$ (in each product, the first factor is the divisor of 4 , and the second is the divisor of 15). Hence,

$$F(60) = f(1) + f(2) + f(3) + f(4) + f(5) + f(6) + f(10) + f(12)$$
$$+ f(15) + f(20) + f(30) + f(60)$$
$$= f(1 \cdot 1) + f(2 \cdot 1) + f(1 \cdot 3) + f(4 \cdot 1) + f(1 \cdot 5) + f(2 \cdot 3)$$
$$+ f(2 \cdot 5) + f(4 \cdot 3) + f(1 \cdot 15) + f(4 \cdot 5) + f(2 \cdot 15) + f(4 \cdot 15)$$
$$= f(1)f(1) + f(2)f(1) + f(1)f(3) + f(4)f(1) + f(1)f(5)$$
$$+ f(2)f(3) + f(2)f(5) + f(4)f(3) + f(1)f(15) + f(4)f(5)$$
$$+ f(2)f(15) + f(4)f(15)$$
$$= (f(1) + f(2) + f(4))(f(1) + f(3) + f(5) + f(15))$$
$$= F(4)F(15).$$

We now prove Theorem 6.8 using the idea illustrated by the example.

Proof. To show that F is a multiplicative function, we must show that if m and n are relatively prime positive integers, then $F(mn) = F(m)F(n)$. So let us assume that $(m,n) = 1$. We have

$$F(mn) = \sum_{d \mid mn} f(d).$$

By Lemma 2.5, since $(m,n) = 1$, each divisor of mn can be written uniquely as the product of relatively prime divisors d_1 of m and d_2 of n, and each pair of divisors d_1 of m and d_2 of n corresponds to a divisor $d = d_1 d_2$ of mn. Hence, we can write

$$F(mn) = \sum_{\substack{d_1 \mid m \\ d_2 \mid n}} f(d_1 d_2).$$

Since f is multiplicative and since $(d_1, d_2) = 1$, we see that

$$F(mn) = \sum_{\substack{d_1 \mid m \\ d_2 \mid n}} f(d_1)f(d_2)$$

$$= \sum_{d_1 \mid m} f(d_1) \sum_{d_2 \mid n} f(d_2)$$

$$= F(m)F(n). \qquad \blacksquare$$

We can now use Theorem 6.8 to show that σ and τ are multiplicative.

Corollary 6.7.1. The sum of divisor function σ and the number of divisor function τ are multiplicative functions.

Proof. Let $f(n) = n$ and $g(n) = 1$. Both f and g are multiplicative. By Theorem 6.8 we see that $\sigma(n) = \sum_{d \mid n} f(d)$ and $\tau(n) = \sum_{d \mid n} g(d)$ are multiplicative. $\qquad \blacksquare$

Now that we know that σ and τ are multiplicative, we can derive formulae for their values based on prime factorizations. First, we find formulae for $\sigma(n)$ and $\tau(n)$ when n is the power of a prime.

Lemma 6.1. Let p be prime and a a positive integer. Then

$$\sigma(p^a) = (1+p+p^2+ \cdots +p^a) = \frac{p^{a+1}-1}{p-1}$$

and

$$\tau(p^a) = a + 1.$$

Proof. The divisors of p^a are 1, p, p^2 ,...., p^{a-1}, p^a. Consequently, p^a has exactly $a + 1$ divisors, so that $\tau(p^a) = a + 1$. Also, we note that $\sigma(p^a) = 1 + p + p^2 + \cdots + p^{a-1} + p^a = \frac{p^{a+1}-1}{p-1}$, using the formula in Example 1.10 for the sum of terms of a geometric progression.   ∎

Example 6.7. When we apply Lemma 6.1 with $p = 5$ and $a = 3$ we find that $\sigma(5^3)$ $= 1 + 5 + 5^2 + 5^3 = \dfrac{5^4-1}{5-1} = 156$ and $\tau(5^3) = 1 + 3 = 4$.   □

Lemma 6.1 and Corollary 6.7.1 lead to the following formulae.

Theorem 6.9. Let the positive integer n have prime factorization $n = p_1^{a_1} p_2^{a_2} \cdots p_s^{a_s}$. Then

$$\sigma(n) = \frac{p_1^{a_1+1}-1}{p_1-1} \cdot \frac{p_2^{a_2+1}-1}{p_2-1} \cdot \ldots \cdot \frac{p_s^{a_s+1}-1}{p_s-1} = \prod_{j=1}^{s} \frac{p_j^{a_j+1}-1}{p_j-1}$$

and

$$\tau(n) = (a_1+1)(a_2+1) \cdots (a_s+1) = \prod_{j=1}^{s} (a_j+1).$$

Proof. Since both σ and τ are multiplicative, we see that $\sigma(n) = \sigma(p_1^{a_1} p_2^{a_2} \cdots p_s^{a_s}) = \sigma(p_1^{a_1})\sigma(p_2^{a_2}) \cdots \sigma(p_s^{a_s})$ and $\tau(n) = \tau(p_1^{a_1} p_2^{a_2} \cdots p_s^{a_s}) = \tau(p_1^{a_1}) \tau(p_2^{a_2}) \cdots \tau(p_s^{a_s})$. Inserting the values for $\sigma(p_i^{a_i})$ and $\tau(p_i^{a_i})$ found in Lemma 6.1, we obtain the desired formulae.   ∎

We illustrate how to use Theorem 6.8 with the following example.

Example 6.8. Using Theorem 6.8 we find

$$\sigma(200) = \sigma(2^3 5^2) = \frac{2^4-1}{2-1} \cdot \frac{5^3-1}{5-1} = 15 \cdot 31 = 465,$$

$$\tau(200) = \tau(2^3 5^2) = (3+1)(2+1) = 12.$$

Similarly, we have

$$\sigma(720) = \sigma(2^4 \cdot 3^2 \cdot 5) = \frac{2^5-1}{2-1} \cdot \frac{3^3-1}{3-1} \cdot \frac{5^2-1}{5-1} = 31 \cdot 13 \cdot 6 = 2418,$$

$$\tau(2^4 \cdot 3^2 \cdot 5) = (4+1)(2+1)(1+1) = 30.$$ □

6.2 Exercises

1. Find the sum of the positive integer divisors of each of the following integers.

   a) 35
   b) 196
   c) 1000
   d) 2^{100}

   e) $2 \cdot 3 \cdot 5 \cdot 7 \cdot 11$
   f) $2^5 3^4 5^3 7^2 11$
   g) 10!
   h) 20!

2. Find the number of positive integer divisors of each of the following integers.

   a) 36
   b) 99
   c) 144

   d) $2 \cdot 3 \cdot 5 \cdot 7 \cdot 11 \cdot 13 \cdot 17 \cdot 19$
   e) $2 \cdot 3^2 \cdot 5^3 \cdot 7^4 \cdot 11^5 \cdot 13^4 \cdot 17^5 \cdot 19^5$
   f) 20!

3. Which positive integers have an odd number of positive divisors?

4. For which positive integers n is the sum of divisors of n odd?

* 5. Find all positive integers n with $\sigma(n)$ equal to each of the following integers.

   a) 12
   b) 18
   c) 24

   d) 48
   e) 52
   f) 84

* 6. Find the smallest positive integer n with $\tau(n)$ equal to each of the following integers.

   a) 1
   b) 2
   c) 3

   d) 6
   e) 14
   f) 100

7. Show that if $k > 1$ is an integer, then the equation $\tau(n) = k$ has infinitely many solutions.

8. Which positive integers have exactly two positive divisors?

9. Which positive integers have exactly three positive divisors?

10. Which positive integers have exactly four positive divisors?

11. What is the product of the positive divisors of a positive integer n?

Let $\sigma_k(n)$ denote the sum of the kth powers of the divisors of n, so that $\sigma_k(n) = \sum_{d|n} d^k$. Note that $\sigma_1(n) = \sigma(n)$.

12. Find $\sigma_3(4)$, $\sigma_3(6)$ and $\sigma_3(12)$.

13. Give a formula for $\sigma_k(p)$, where p is prime.

14. Give a formula for $\sigma_k(p^a)$, where p is prime, and a is a positive integer.

15. Show that the function σ_k is multiplicative.

16. Using Exercises 14 and 15, find a formula for $\sigma_k(n)$, where n has prime-power factorization $n = p_1^{a_1} p_2^{a_2} \cdots p_m^{a_m}$.

* 17. Find all positive integers n such that $\phi(n) + \sigma(n) = 2n$.

* **18.** Show that no two positive integers have the same product of divisors.

19. Show that the number of ordered pairs of positive integers with least common multiple equal to the positive integer n is $\tau(n^2)$.

20. Let n be a positive integer, $n \geq 2$. Define the sequence of integers $n_1, n_2, n_3, \ldots$ by $n_1 = \tau(n)$ and $n_{k+1} = \tau(n_k)$ for $k = 1, 2, 3, \ldots$. Show that there is a positive integer r such that $2 = n_r = n_{r+1} = n_{r+2} = \cdots$.

21. Show that a positive integer n is composite if and only if $\sigma(n) > n + \sqrt{n}$.

22. Let n be a positive integer. Show that $\tau(2^n - 1) \geq \tau(n)$.

* **23.** Show that $\displaystyle\sum_{j=1}^{n} \tau(j) = 2 \sum_{j=1}^{[\sqrt{n}]} [n/j] - [\sqrt{n}]^2$ whenever n is a positive integer. Then use this formula to find $\displaystyle\sum_{j=1}^{100} \tau(j)$.

* **24.** Let a and b be positive integers. Show that $\sigma(a)/a \leq \sigma(ab)/(ab) \leq \sigma(a)\sigma(b)/(ab)$.

* **25.** Show that if a and b are positive integers then $\sigma(a)\sigma(b) = \displaystyle\sum_{d|(a,b)} d\sigma(ab/d^2)$.

* **26.** Show that if n is a positive integer then $\displaystyle\left[\sum_{d|n} \tau(d)\right]^2 = \sum_{d|n} \tau(d)^3$.

* **27.** Find the determinant of the $n \times n$ matrix with (i,j)th entry equal to (i,j).

* **28.** Let n be a positive integer such that $24 \mid (n+1)$. Show that $\sigma(n)$ is divisible by 24.

29. Show that there are infinitely many pairs of positive integers m, n with $\phi(m) = \sigma(n)$ if there were infinitely many pairs of twin primes or if there were infinitely many Mersenne primes.

6.2 Computer Projects

Programming Projects

Write programs to do the following:

1. Given a positive integer n, find $\tau(n)$, the number of positive divisors of n.

2. Given a positive integer n find $\sigma(n)$, the sum of the positive divisors of n.

3. Given a positive integer n and a positive integer k find $\sigma_k(n)$, the sum of the kth powers of the positive divisors of n.

4. Given a positive integer n, find the integer r defined in Exercise 20.

Computations and Explorations

Using the programs you have written or a computation program, carry out the following computations and explorations:

1. Find as many pairs, triples, and quadruples as you can of consecutive integers each with the same number of positive divisors.

2. Determine the number of iterations required for the sequence $n_1 = \tau(n)$, $n_2 = \tau(n_1),..., n_{k+1} = \tau(n_k),...$ to reach the integer 2 for all positive integers n not exceeding 1000. Formulate some conjectures based on your evidence.

6.3 Perfect Numbers and Mersenne Primes

Because of certain mystical beliefs, the ancient Greeks were interested in those integers that are equal to the sum of all their proper positive divisors. These integers are called *perfect numbers*.

Definition. If n is a positive integer and $\sigma(n) = 2n$, then n is called a *perfect number*.

Example 6.9. Since $\sigma(6) = 1 + 2 + 3 + 6 = 12$, we see that 6 is perfect. We also note that $\sigma(28) = 1 + 2 + 4 + 7 + 14 + 28 = 56$, so that 28 is another perfect number. ☐

The ancient Greeks knew how to find all even perfect numbers. The following theorem tells us which even positive integers are perfect.

Theorem 6.10. The positive integer n is an even perfect number if and only if

$$n = 2^{m-1}(2^m - 1)$$

where m is an integer such that $m \geq 2$ and $2^m - 1$ is prime.

Proof. First, we show that if $n = 2^{m-1}(2^m - 1)$ where $2^m - 1$ is prime, then n is perfect. We note that since $2^m - 1$ is odd, we have $(2^{m-1}, 2^m - 1) = 1$. Since σ is a multiplicative function, we see that

$$\sigma(n) = \sigma(2^{m-1})\sigma(2^m - 1).$$

Lemma 6.1 tells us that $\sigma(2^{m-1}) = 2^m - 1$ and $\sigma(2^m - 1) = 2^m$, since we are assuming that $2^m - 1$ is prime. Consequently,

$$\sigma(n) = (2^m - 1)2^m = 2n,$$

demonstrating that n is a perfect number.

To show that the converse is true, let n be an even perfect number. Write $n = 2^s t$ where s and t are positive integers and t is odd. Since $(2^s, t) = 1$, we see from Lemma 6.1 that

(6.1) $\qquad \sigma(n) = \sigma(2^s t) = \sigma(2^s)\sigma(t) = (2^{s+1} - 1)\sigma(t).$

Since n is perfect, we have

(6.2) $\qquad\qquad\qquad \sigma(n) = 2n = 2^{s+1} t.$

Combining (6.1) and (6.2) shows that

(6.3) $\qquad\qquad\qquad (2^{s+1} - 1)\sigma(t) = 2^{s+1} t.$

Since $(2^{s+1}, 2^{s+1} - 1) = 1$, from Lemma 2.3 we see that $2^{s+1} \mid \sigma(t)$. Therefore, there is an integer q such that $\sigma(t) = 2^{s+1} q$. Inserting this expression for $\sigma(t)$ into (6.3) tells us that

$$(2^{s+1} - 1)2^{s+1} q = 2^{s+1} t,$$

and, therefore,

(6.4) $\qquad\qquad\qquad (2^{s+1} - 1)q = t.$

Hence, $q \mid t$ and $q \neq t$.

When we replace t by the expression on the left-hand side of (6.4), we find that

(6.5) $\qquad t + q = (2^{s+1} - 1)q + q = 2^{s+1} q = \sigma(t).$

We will show that $q = 1$. Note that if $q \neq 1$, then there are at least three distinct positive divisors of t, namely 1, q, and t. This implies that $\sigma(t) \geq t + q + 1$, which contradicts (6.5). Hence, $q = 1$ and, from (6.4), we conclude that $t = 2^{s+1} - 1$. Also, from (6.5), we see that $\sigma(t) = t + 1$, so that t must be prime, since its only positive divisors are 1 and t. Therefore, $n = 2^s(2^{s+1} - 1)$, where $2^{s+1} - 1$ is prime. ∎

By Theorem 6.10 we see that to find even perfect numbers, we must find primes of the form $2^m - 1$. In our search for primes of this form, we first show that the exponent m must be prime.

Theorem 6.11. If m is a positive integer and $2^m - 1$ is prime, then m must be prime.

Proof. Assume that m is not prime, so that $m = ab$ where $1 < a < m$ and $1 < b < m$. Then

$$2^m - 1 = 2^{ab} - 1 = (2^a - 1)(2^{a(b-1)} + 2^{a(b-2)} + \cdots + 2^a + 1).$$

Since both factors on the right side of the equation are greater than 1, we see that $2^m - 1$ is composite if m is not prime. Therefore, if $2^m - 1$ is prime, then m must also be prime. ∎

By Theorem 6.11 we see that to search for primes of the form $2^m - 1$, we need to consider only integers m that are prime. Integers of the form $2^m - 1$ have been studied in great depth; these integers are named after a French monk of the seventeenth century, Mersenne*, who studied them.

Definition. If m is a positive integer, then $M_m = 2^m - 1$ is called the mth *Mersenne number*, and, if p is prime and $M_p = 2^p - 1$ is also prime, then M_p is called a *Mersenne prime*.

Example 6.10. The Mersenne number $M_7 = 2^7 - 1$ is prime, whereas the Mersenne number $M_{11} = 2^{11} - 1 = 2047 = 23 \cdot 89$ is composite. □

It is possible to prove various theorems that help decide whether Mersenne numbers are prime. One such theorem will now be given. Related results are found in the Exercises 37-39 in Chapter 9.

Theorem 6.12. If p is an odd prime, then any divisor of the Mersenne number $M_p = 2^p - 1$ is of the form $2kp + 1$ where k is a positive integer.

Proof. Let q be a prime dividing $M_p = 2^p - 1$. By Fermat's little theorem, we know that $q \mid (2^{q-1} - 1)$. Also, from Lemma 3.2 we know that

$$(6.6) \qquad (2^p - 1, 2^{q-1} - 1) = 2^{(p, q-1)} - 1.$$

Since q is a common divisor of $2^p - 1$ and $2^{q-1} - 1$, we know that $(2^p - 1, 2^{q-1} - 1) > 1$. Hence, $(p, q-1) = p$, since the only other possibility, namely $(p, q-1) = 1$, would imply from (6.6) that $(2^p - 1, 2^{q-1} - 1) = 1$. Hence $p \mid (q-1)$, and, therefore, there is a positive integer m with $q - 1 = mp$. Since q is odd we see that m must be even, so that $m = 2k$, where k is a positive integer. Hence, $q = mp + 1 = 2kp + 1$. ∎

We can use Theorem 6.12 to help decide whether Mersenne numbers are prime. We illustrate this with the following examples.

Example 6.11. To decide whether $M_{13} = 2^{13} - 1 = 8191$ is prime, we only need

---

* **MARIN MERSENNE** (1588-1648) was a Franciscan friar who lived most of his life in Parisian cloisters. He was the author of *Cognitata Physico-Mathematica* which stated without proof that M_p is prime for $p = 2,3,5,7,13,17,19,31,67,127,257$ and for no other primes p with $p < 257$. It took over 300 years to totally settle this claim made by Mersenne. Finally, in 1947 it was shown that Mersenne made five errors in his work (namely that M_{61} is prime, M_{67} is composite, M_{89} is prime, M_{107} is prime, and M_{257} is composite). Besides his famous statement about primes of the form M_p, Mersenne contributed to the development of number theory through his extensive correspondence with many mathematicians, including Fermat. Mersenne effectively served as a clearing house and a disseminator of new mathematical ideas in the 17th century.

look for a prime factor not exceeding $\sqrt{8191} = 90.504...$. Furthermore, by Theorem 6.12, any such prime divisor must be of the form $26k + 1$. The only candidates for primes dividing M_{13} less than or equal to $\sqrt{M_{13}}$ are 53 and 79. Trial division easily rules out these cases, so that M_{13} is prime.   □

Example 6.12. To decide whether $M_{23} = 2^{23} - 1 = 8388607$ is prime, we only need to determine whether M_{23} is divisible by a prime less than or equal to $\sqrt{M_{23}} = 2896.309...$ of the form $46k + 1$. The first prime of this form is 47. A trial division shows that $8388607 = 47 \cdot 178481$, so that M_{23} is composite.   □

Because there are special primality tests for Mersenne numbers, it has been possible to determine whether extremely large Mersenne numbers are prime.

Following is one such primality test, known as the Lucas-Lehmer test after Edouard Lucas* who developed the theory the test in based on in the 1870s and Derrick H. Lehmer** who developed a simplified version of the test in 1930. This test has been used to find the largest known Mersenne primes. For most of recent history the largest known Mersenne prime was the largest known prime as is currently the case. However, from late 1990 until early 1992, the largest known prime was $391581 \cdot 2^{216193} - 1$. Because this number is of the form $k \cdot 2^n - 1$ it was possible to use special tests to show that it is prime.

Theorem 6.13. The Lucas-Lehmer Test. Let p be a prime and let $M_p = 2^p - 1$ denote the pth Mersenne number. Define a sequence of integers recursively by setting $r_1 = 4$, and for $k \geq 2$,

---

*   **FRANÇOIS-EDOUARD-ANATOLE LUCAS (1842-1891)** was born in Amiens, France and was educated at the École Normale. After finishing his studies, he worked as an assistant at the Paris Observatory and during the Franco-Prussian war he served as an artillery officer. After the war he became a teacher at a secondary school. He was considered to be an excellent, and entertaining, teacher. Lucas was extremely fond of calculating and devised plans for a computer, which unfortunately were never realized. Besides his contributions to number theory, Lucas is also remembered for his work in recreational mathematics. The most famous of his contributions in this area is the well known tower of Hanoi problem. A freak accident led to Lucas' death. He was gashed in the check by a piece of a plate which was accidentally dropped at a banquet. An infection in the resulting wound killed him several days later.

**   **DERRICK H. LEHMER (1905-1991)** was born in Berkeley, California. He received his undergraduate degree in 1927 from the University of California and his masters and doctorate degrees from Brown University, in 1929 and 1930, respectively. He served on the staffs of the California Institute of Technology, the Institute for Advanced Study, Lehigh University, and Cambridge University before joining the mathematics department at the University of California, Berkeley in 1940. Lehmer has made many contributions to number theory. He invented many special purpose devices for number theoretic computations, some with his father who was also a mathematician. Lehmer was the thesis advisor of Harold Stark, who in turn was the thesis advisor of the author.

$$r_k \equiv r_{k-1}^2 - 2 \ (\text{mod } M_p), \ \ 0 \leq r_k < M_p.$$

Then M_p is prime if and only if $r_{p-1} \equiv 0 \ (\text{mod } M_p)$.

The proof of the Lucas-Lehmer test may be found in Lenstra [115] and Sierpiński [42]. We use an example to illustrate how the Lucas-Lehmer test is used.

Example 6.13. Consider the Mersenne number $M_5 = 2^5 - 1 = 31$. Then $r_1 = 4$, $r_2 \equiv 4^2 - 2 = 14 \ (\text{mod } 31)$, $r_3 \equiv 14^2 - 2 \equiv 8 \ (\text{mod } 31)$, and $r_4 \equiv 8^2 - 2 \equiv 0 \ (\text{mod } 31)$. Since $r_4 \equiv 0 \ (\text{mod } 31)$, we conclude that $M_5 = 31$ is prime. $\qquad \square$

The Lucas-Lehmer test can be performed quite rapidly as the following corollary states.

Corollary 6.12.1. Let p be prime and let $M_p = 2^p - 1$ denote the pth Mersenne number. It is possible to determine whether M_p is prime using $O(p^3)$ bit operations.

Proof. To determine whether M_p is prime using the Lucas-Lehmer test requires $p - 1$ squarings modulo M_p, each requiring $O((\log M_p)^2) = O(p^2)$ bit operations. Hence, the Lucas-Lehmer test requires $O(p^3)$ bit operations. $\qquad \blacksquare$

Much activity has been directed toward the discovery of Mersenne primes, especially since each new Mersenne prime often becomes the largest prime known, and for each new Mersenne prime, there is a new perfect number. At the present time, a total of 32 Mersenne primes are known.

It has been conjectured but has not been proved, that there are infinitely many Mersenne primes.

Computers were used to find the 20 largest Mersenne primes known. The discovery by high school students of the 25th and 26th Mersenne primes received much publicity, including coverage on the nightly news of a major television network. An interesting account of the search for the 27th Mersenne prime and related historical and computational information may be found in [125]. A report of the discovery of the 28th Mersenne prime is given in [106]. For more information on the current status of Mersenne numbers the reader is referred to the article by Colquitt and Welsh [104] which describes the discovery of the Mersenne prime M_{110503}.

The largest Mersenne prime found is M_{756839}. It was discovered in March, 1992 by a Cray-2 supercomputer at Harwell Laboratory in Oxfordshire, England, using 19 hours of computer time to run a program written by Slowinski. The next largest known Mersenne known is M_{216091} which was shown to be prime in 1985 by Slowinski using three hours of time on a supercomputer. These tests were carried out using trial divisions followed by the Lucas-Lehmer test. In particular, potential

Mersenne primes of the form $2^p - 1$ were first tested for divisibility by integers of the form $2kp + 1$ which are also congruent to ± 1 modulo 8, where k is a positive integer. (All such divisors must be of this form by Theorem 6.12 and an exercise in Section 9.1).

The known Mersenne primes, the number of decimal digits in these primes, and their dates of discovery are displayed in Table 6.3. Note that there may be undiscovered Mersenne primes greater than M_{110503} but less than M_{216091} or M_{756839}. Such primes may exist, but yet not be known, becasue as of mid-1992, researchers have only completely checked M_p for primality when $p < 170000$, $216091 < p < 360000$, or $430000 < p < 520000$. The ranges $170000 < p < 216091$ and $360000 < p < 430000$ have only been partially checked and the range $520000 < p < 750000$ has been checked only sparsely. The discovery that M_{756839} is prime was an extremely lucky coincidence, since only a few values of M_p were tested for primality with p of this order of magnitude.

We have reduced the study of even perfect numbers to the study of Mersenne primes. We may ask whether there are odd perfect numbers. The answer is still unknown. It is possible to demonstrate that if they exist, odd perfect numbers must have certain properties (see Exercises 28-32 for example). Furthermore, it is known that there are no odd perfect numbers less than 10^{300}, and it has been shown that any odd perfect number must have at least eight different prime factors. A discussion of odd perfect numbers may be found in Guy [20], and information concerning recent results about odd perfect numbers is given by Hagis [110].

p	Number of decimal digits in M_p	Date of Discovery
2	1	ancient times
3	1	ancient times
5	2	ancient times
7	3	ancient times
13	4	15th century
17	6	1603
19	6	1603
31	10	1772
61	19	1883
89	27	1911
107	33	1914
127	39	1876
521	157	1952
607	183	1952
1279	386	1952
2203	664	1952
2281	687	1952
3217	969	1957
4253	1281	1961
4423	1332	1961
9689	2917	1963
9941	2993	1963
11213	3376	1963
19937	6002	1971
21701	6533	1978
23209	6987	1979
44497	13395	1979
86243	25962	1983
110503	33265	1988
132049	39751	1983
216091	65050	1985
756839	227832	1992

Table 6.3. The Known Mersenne Primes.

6.3 Exercises

1. Find the six smallest even perfect numbers.

2. Find the seventh and eighth even perfect numbers.

3. Find a factor of each of the following integers.

a) $2^{15} - 1$
b) $2^{91} - 1$
c) $2^{1001} - 1$

4. Find a factor of each of the following integers.

a) $2^{111} - 1$
b) $2^{289} - 1$
c) $2^{46189} - 1$

If n is a positive integer, then we say that n is *deficient* if $\sigma(n) < 2n$, and we say that n is *abundant* if $\sigma(n) > 2n$. Every integer is either deficient, perfect, or abundant.

5. Find the six smallest abundant positive integers.

* 6. Find the smallest odd abundant positive integer.

7. Show that every prime power is deficient.

8. Show that any divisor of a deficient or perfect number is deficient.

9. Show that any multiple of an abundant or perfect number, other than the perfect number itself, is abundant.

10. Show that if $n = 2^{m-1}(2^m - 1)$, where m is a positive integer such that $2^m - 1$ is composite, then n is abundant.

Two positive integers m and n are called an *amicable pair* if $\sigma(m) = \sigma(n) = m + n$.

11. Show that each of the following pairs of integers are amicable pairs.

a) 220, 284
b) 1184, 1210
c) 79750, 88730

12. a) Show that if n is a positive integer with $n \geq 2$, such that $3 \cdot 2^{n-1} - 1$, $3 \cdot 2^n - 1$, and $3^2 \cdot 2^{2n-1} - 1$ are all prime, then $2^n(3 \cdot 2^{n-1} - 1)(3 \cdot 2^n - 1)$ and $2^n(3^2 \cdot 2^{2n-1} - 1)$ form an amicable pair.

b) Find three amicable pairs using part (a).

An integer n is called *k-perfect* if $\sigma(n) = kn$. Note that a perfect number is 2-perfect.

13. Show that $120 = 2^3 \cdot 3 \cdot 5$ is 3-perfect.

14. Show that $30240 = 2^5 \cdot 3^3 \cdot 5 \cdot 7$ is 4-perfect.

15. Show that $14182439040 = 2^7 \cdot 3^4 \cdot 5 \cdot 7 \cdot 11^2 \cdot 17 \cdot 19$ is 5-perfect.

16. Find all 3-perfect numbers of the form $n = 2^k \cdot 3 \cdot p$, where p is an odd prime.

17. Show that if n is 3-perfect and $3 \nmid n$, then $3n$ is 4-perfect.

An integer n is *k-abundant* if $\sigma(n) > (k+1)n$.

18. Find a 3-abundant integer.

19. Find a 4-abundant integer.

** 20. Show that for each positive integer k there are an infinite number of k-abundant integers.

A positive integer n is called *superperfect* if $\sigma(\sigma(n)) = 2n$.

21. Show that 16 is superperfect.

22. Show that if $n = 2^q$ where $2^{q+1} - 1$ is prime, then n is superperfect.

23. Show that every even superperfect number is of the form $n = 2^q$ where $2^{q+1} - 1$ is prime.

* 24. Show that if $n = p^2$ where p is an odd prime, then n is not superperfect.

25. Use Theorem 6.12 to determine whether each of the following Mersenne numbers is prime.

a) M_7 c) M_{17}

b) M_{11} d) M_{29}

26. Use the Lucas-Lehmer test, Theorem 6.13, to determine whether each of the following Mersenne numbers is prime.

a) M_3 c) M_{11}

b) M_7 d) M_{13}

* 27. a) Show that if n is a positive integer and $2n + 1$ is prime, then either $(2n+1) \mid M_n$ or $(2n+1) \mid (M_n + 2)$. (*Hint:* Use Fermat's little theorem to show that $M_n(M_n + 2) \equiv 0 \pmod{2n+1}$.)

b) Use part (a) to show that M_{11} and M_{23} are composite.

* 28. a) Show that if n is an odd perfect number, then $n = p^a m^2$ where p is an odd prime, $p \equiv a \equiv 1 \pmod 4$, and m is an integer.

b) Use part (a) to show that if n is an odd perfect number, then $n \equiv 1 \pmod 4$.

* 29. Show that if $n = p^a m^2$ is an odd perfect number where p is prime, then $n \equiv p \pmod 8$.

* 30. Show that if n is an odd perfect number, then 3, 5, and 7 are not all divisors of n.

* **31.** Show that if n is an odd perfect number then n has at least three different prime divisors.

** **32.** Show that if n is an odd perfect number then n has at least four different prime divisors.

33. Find all positive integers n such that the product of all divisors of n other than n is exactly n^2. (These integers are multiplicative analogues of perfect numbers.)

34. Let n be a positive integer. Define the *aliquot sequence* $n_1, n_2, n_3, \ldots,$ recursively by $n_1 = \sigma(n) - n$ and $n_{k+1} = \sigma(n_k) - n_k$ for $k = 1, 2, 3, \ldots$. (The word *aliquot* is an adjective that means "contained an exact number of times in something else." Archaically, the *aliquot parts* of an integer are the divisors of this integer.

a) Show that if n is perfect, then $n = n_1 = n_2 = n_3 = \cdots$.

b) Show that if n and m are an amicable pair, then $n_1 = m$, $n_2 = n$, $n_3 = m$, $n_4 = n, \ldots$ and so on, *i.e.*, the sequence $n_1, n_2, n_3, \ldots$ is periodic with period 2.

c) Find the aliquot sequence of integers generated if $n = 12496 = 2^4 \cdot 11 \cdot 71$.

Before computers were used to examine the behavior of aliquot sequences, it was conjectured that for all integers n the aliquot sequence of integers $n_1, n_2, n_3, \ldots$ is bounded. However, evidence obtained from calculations with large integers suggest that some of these sequences are unbounded.

* **35.** Show that if n is a positive integer greater than 1, then the Mersenne number M_n cannot be the power of a positive integer.

6.3 Computer Projects

Programming Projects

Write programs to do the following:

1. Classify positive integers according to whether they are deficient, perfect, or abundant (see the preamble to Exercise 3).

2. Use Theorem 6.12 to look for factors of Mersenne numbers.

3. Use the Lucas-Lehmer test determine whether the Mersenne number $2^p - 1$ is prime, where p is a prime.

4. Given a positive integer n, determine if the aliquot sequence defined in Exercise 32 is periodic

5. Given a positive integer n, find all amicable pairs of integers a, b where $a \le n$ and $b \le n$.

Computations and Explorations

Using the programs you have written or a computation program, carry out the following computations and explorations:

1. Find factors of as many Mersenne numbers of the form M_p, where p is prime, as you can using Theorem 6.12.

2. Verify the primality of as many Mersenne primes as you can using the Lucas-Lehmer test.

3. Find all amicable pairs where both integers in the pair are less than 10000.

4. Show that the aliquot sequence, defined in Exercise 32, obtained by taking $n = 14316$ is periodic with period 28.

5. Find as many aliquot sequences as you can that are periodic with period 4.

6. Find the number of terms in the aliquot sequence obtained by taking $n = 138$ before this sequence reaches the integer 1. What is the largest term of the sequence? Can you answer the same question for $n = 276$?

7

Cryptology

7.1 Character Ciphers

From ancient times to the present, secret messages have been sent. Classically, the need for secret communication has occurred in diplomacy and in military affairs. Now, with electronic communication coming into widespread use, secrecy has become an important issue. Just recently, with the advent of electronic banking, secrecy has become necessary even for financial transactions. Hence, there is a great deal of interest in the techniques of making messages unintelligible to everyone except the intended receiver.

Before discussing specific secrecy systems, we present some terminology. The discipline devoted to secrecy systems is called *cryptology*. *Cryptography* is the part of cryptology that deals with the design and implementation of secrecy systems, while *cryptanalysis* is aimed at breaking these systems. A message that is to be altered into a secret form is called *plaintext*. A *cipher* is a method for altering a plaintext message into *ciphertext* by changing the letters of the plaintext using a transformation. The *key* determines a particular transformation from a set of possible transformations. The process of changing plaintext into ciphertext is called *encryption* or *enciphering*, while the reverse process of changing the ciphertext back to the plaintext by the intended receiver, possessing knowledge of the method for doing this, is called *decryption* or *deciphering*. This, of course, is different from the process someone other than the intended receiver uses to make the message intelligible through cryptanalysis.

In this chapter, we present secrecy systems based on modular arithmetic. The first of these had its origin with Julius Caesar. The newest secrecy system we will discuss was invented in the late 1970's. In all these systems we start by translating letters into numbers. We take as our standard alphabet the letters of English and translate them into the integers from 0 to 25, as shown in Table 7.1.

letter	A	B	C	D	E	F	G	H	I	J	K	L	M	N	O	P	Q	R	S	T	U	V	W	X	Y	Z
numerical equivalent	0	1	2	3	4	5	6	7	8	9	10	11	12	13	14	15	16	17	18	19	20	21	22	23	24	25

Table 7.1. The Numerical Equivalents of Letters.

Of course, if we were sending messages in Russian, Greek, Hebrew, or any other language we would use the appropriate alphabet range of integers. Also, we may want to include punctuation marks, a symbol to indicate blanks, and perhaps the digits for representing numbers as part of the message. However, for the sake of simplicity, we restrict ourselves to the letters of the English alphabet.

First, we discuss secrecy systems based on transforming each letter of the plaintext message into a different letter to produce the ciphertext. Such ciphers are called *character* or *monographic ciphers*, since each letter is changed individually to another letter by a *substitution*. Altogether there are 26! possible ways to produce a monographic transformation. We will discuss a set that is based on modular arithmetic.

A cipher that was used by Julius Caesar is based on the substitution in which each letter is replaced by the letter three further down the alphabet, with the last three letters shifted to the first three letters of the alphabet. To describe this cipher using modular arithmetic, let P be the numerical equivalent of a letter in the plaintext and C the numerical equivalent of the corresponding ciphertext letter. Then

$$C \equiv P + 3 \pmod{26}, \quad 0 \le C \le 25.$$

The correspondence between plaintext and ciphertext is given in Table 7.2.

plaintext	A	B	C	D	E	F	G	H	I	J	K	L	M	N	O	P	Q	R	S	T	U	V	W	X	Y	Z
	0	1	2	3	4	5	6	7	8	9	10	11	12	13	14	15	16	17	18	19	20	21	22	23	24	25
	3	4	5	6	7	8	9	10	11	12	13	14	15	16	17	18	19	20	21	22	23	24	25	0	1	2
ciphertext	D	E	F	G	H	I	J	K	L	M	N	O	P	Q	R	S	T	U	V	W	X	Y	Z	A	B	C

Table 7.2. The Correspondence of Letters for the Caesar Cipher.

To encipher a message using this transformation, we first change it to its numerical equivalent, grouping letters in blocks of five. Then we transform each number. The grouping of letters into blocks helps to prevent successful cryptanalysis based on recognizing particular words. We illustrate this procedure by enciphering the message

THIS MESSAGE IS TOP SECRET.

Broken into groups of five letters, the message is

THISM ESSAG EISTO PSECR ET.

Converting the letters into their numerical equivalents, we obtain

19   7   8   18   12   4   18   18   0   6   4   8   18   19   14
15   18   4   2   17   4   19 .

Using the Caesar transformation $C \equiv P + 3 \pmod{26}$, this becomes

22   10   11   21   15   7   21   21   3   9   7   11   21   22   17
18   21   7   5   20   7   22 .

Translating back to letters, we have

WKLVP HVVDJ HLVWR SVHFU HW.

This is the message we send.

The receiver deciphers it in the following manner. First, the letters are converted to numbers. Then, the relationship $P \equiv C - 3 \pmod{26}$, $0 \le P \le 25$, is used to change the ciphertext back to the numerical version of the plaintext, and finally the message is converted to letters.

We illustrate the deciphering procedure with the following message enciphered by the Caesar cipher:

WKLVL VKRZZ HGHFL SKHU.

First, we change these letters into their numerical equivalents, to obtain

22 10 11 21 11   21 10 17 25 25   7 6 7 5 11   18 10 7 20.

Next, we perform the transformation $P \equiv C - 3 \pmod{26}$ to change this to plaintext, and we obtain

$$19\ 7\ 8\ 18\ 8 \quad 18\ 7\ 14\ 22\ 22 \quad 4\ 3\ 4\ 2\ 8 \quad 15\ 7\ 4\ 17.$$

We translate this back to letters and recover the plaintext message

THISI SHOWW EDECI PHER.

By combining the appropriate letters into words, we find that the message reads

THIS IS HOW WE DECIPHER.

The Caesar cipher is one of a family of similar ciphers described by a *shift transformation.*

$$C \equiv P + k \pmod{26}, \quad 0 \le C \le 25,$$

where k is the key representing the size of the shift of letters in the alphabet. There are 26 different transformations of this type, including the case of $k \equiv 0 \pmod{26}$, where letters are not altered, since in this case $C \equiv P \pmod{26}$.

More generally, we will consider transformations of the type

(7.1) $$C \equiv aP + b \pmod{26}, \quad 0 \le C \le 25,$$

where a and b are integers with $(a, 26) = 1$. These are called *affine transformations.* Shift transformations are affine transformations with $a = 1$. We require that $(a, 26) = 1$, so that as P runs through a complete system of residues modulo 26, C also does. There are $\phi(26) = 12$ choices for a, and 26 choices for b, giving a total of $12 \cdot 26 = 312$ transformations of this type (one of these is $C \equiv P \pmod{26}$ obtained when $a = 1$ and $b = 0$). If the relationship between plaintext and ciphertext is described by (7.1), then the inverse relationship is given by

$$P \equiv \bar{a}(C - b) \pmod{26}, \quad 0 \le P \le 25,$$

where $\bar{a}$ is an inverse of $a \pmod{26}$.

As an example of such a cipher, let $a = 7$ and $b = 10$, so that $C \equiv 7P + 10 \pmod{26}$. Hence, $P \equiv 15(C - 10) \equiv 15C + 6 \pmod{26}$, since 15 is an inverse of 7 modulo 26. The correspondence between letters is given in Table 7.3.

	A	B	C	D	E	F	G	H	I	J	K	L	M	N	O	P	Q	R	S	T	U	V	W	X	Y	Z
plaintext	0	1	2	3	4	5	6	7	8	9	10	11	12	13	14	15	16	17	18	19	20	21	22	23	24	25
	10	17	24	5	12	19	0	7	14	21	2	9	16	23	4	11	18	25	6	13	20	1	8	15	22	3
ciphertext	K	R	Y	F	M	T	A	H	O	V	C	J	Q	X	E	L	S	Z	G	N	U	B	I	P	W	D

Table 7.3. The Correspondence of Letters for the Cipher with $C \equiv 7P + 10$ (mod 26).

To illustrate how we obtained this correspondence, note that the plaintext letter L with numerical equivalent 11 corresponds to the ciphertext letter J, since $7 \cdot 11 + 10 = 87 \equiv 9$ (mod 26) and 9 is the numerical equivalent of J.

To illustrate how to encipher, note that

PLEASE SEND MONEY

is transformed to

LJMKG MGMFQ EXMW.

Also note that the ciphertext

FEXEN ZMBMK JNHMG MYZMN

corresponds to the plaintext

DONOT REVEA LTHES ECRET,

or combining the appropriate letters

DO NOT REVEAL THE SECRET.

We now discuss some of the techniques directed at the cryptanalysis of ciphers based on affine transformations. In attempting to break a monographic cipher, the frequency of letters in the ciphertext is compared with the frequency of letters in ordinary text. This gives information concerning the correspondence between letters. In various frequency counts of English text, one finds the percentages listed in Table 7.4 for the occurrence of the 26 letters of the alphabet. Counts of letter frequencies in other languages may be found in Friedman [74] and Kullback [79].

letter	A	B	C	D	E	F	G	H	I	J	K	L	M	N	O	P	Q	R	S	T	U	V	W	X	Y	Z
frequency (in %)	7	1	3	4	13	3	2	3	8	<1	<1	4	3	8	7	3	<1	8	6	9	3	1	1	<1	2	<1

Table 7.4. The Frequencies of Occurrence of the Letters of the Alphabet.

From this information, we see that the most frequently occurring letters in typical English text are E, T, N, R, I, O, and A, with E occurring substantially more than the other letters, 13% of the time, with T, N, R, I, O, and A each occurring between 7% and 9% of the time. We can use this information to determine which cipher based on an affine transformation has been used to encipher a message.

First, suppose that we know in advance that a shift cipher has been employed to encipher a message; each letter of the message has been transformed by a correspondence $C \equiv P + k \pmod{26}$, $0 \le C \le 25$. To cryptanalyze the ciphertext

$$\begin{array}{lllll} YFXMP & CESPZ & CJTDF & DPQFW & QZCPY \\ NTASP & CTYRX & PDDLR & PD \end{array}$$

we first count the number of occurrences of each letter in the ciphertext. This is displayed in Table 7.5.

letter	A	B	C	D	E	F	G	H	I	J	K	L	M	N	O	P	Q	R	S	T	U	V	W	X	Y	Z
number of occurrences	1	0	4	5	1	3	0	0	0	1	0	1	1	1	0	7	2	2	2	3	0	0	1	2	3	2

Table 7.5. The Number of Occurrences of Letters in a Ciphertext.

We notice that the most frequently occurring letter in the ciphertext is P with the letters C,D,F,T, and Y occurring with relatively high frequency. Our initial guess would be that P represents E, since E is the most frequently occurring letter in English text. If this is so, then $15 \equiv 4+k \pmod{26}$, so that $k \equiv 11 \pmod{26}$. Consequently, we would have $C \equiv P+11 \pmod{26}$ and $P \equiv C-11 \pmod{26}$. This correspondence is given in Table 7.6.

ciphertext	A	B	C	D	E	F	G	H	I	J	K	L	M	N	O	P	Q	R	S	T	U	V	W	X	Y	Z
	0	1	2	3	4	5	6	7	8	9	10	11	12	13	14	15	16	17	18	19	20	21	22	23	24	25
plaintext	15	16	17	18	19	20	21	22	23	24	25	0	1	2	3	4	5	6	7	8	9	10	11	12	13	14
	P	Q	R	S	T	U	V	W	Z	Y	Z	A	B	C	D	E	F	G	H	I	J	K	L	M	N	O

Table 7.6. Correspondence of Letters for the Sample Ciphertext.

Using this correspondence, we attempt to decipher the message. We obtain

NUMBE RTHEO RYISU SEFUL FOREN
CIPHE RINGM ESSAG ES

This can easily be read as

NUMBER THEORY IS USEFUL FOR
ENCIPHERING MESSAGES.

Consequently, we made the correct guess. If we had tried this transformation, and instead of the plaintext, it produced garbled text, we would have tried another likely transformation based on the frequency count of letters in the ciphertext.

Now suppose we know that an affine transformation of the form $C \equiv aP + b \pmod{26}$, $0 \le C \le 25$, has been used for enciphering. For instance, suppose we wish to cryptanalyze the enciphered message

USLEL JUTCC YRTPS URKLT YGGFV
ELYUS LRYXD JURTU ULVCU URJRK
QLLQL YXSRV LBRYZ CYREK LVEXB
RYZDG HRGUS LJLLM LYPDJ LJTJU
FALGU PTGVT JULYU SLDAL TJRWU
SLJFE OLPU

The first thing to do is to count the occurrences of each letter; this count is displayed in Table 7.7.

letter	A	B	C	D	E	F	G	H	I	J	K	L	M	N	O	P	Q	R	S	T	U	V	W	X	Y	Z
number of occurrences	2	2	4	4	5	3	6	1	0	10	3	22	1	0	1	4	2	12	7	8	16	5	1	3	10	2

Table 7.7. The Number of Occurrences of Letters in a Ciphertext.

With this information, we guess that the letter L, which is the most frequently occurring letter in the ciphertext, corresponds to E, while the letter U, which occurs with the second highest frequency, corresponds to T. This implies, if the transformation is of the form $C \equiv aP + b \pmod{26}$, the pair of congruences

$$4a + b \equiv 11 \pmod{26}$$
$$19a + b \equiv 20 \pmod{26}.$$

By Theorem 3.14 we see that the solution of this system is $a \equiv 11 \pmod{26}$ and $b \equiv 19 \pmod{26}$.

If this is the correct enciphering transformation, then using the fact that 19 is an inverse of 11 modulo 26, the deciphering transformation is

$$P \equiv 19 (C - 19) \equiv 19C - 361 \equiv 19C + 3 \pmod{26}, \ 0 \leq P \leq 25.$$

This gives the correspondence found in Table 7.8.

	A	B	C	D	E	F	G	H	I	J	K	L	M	N	O	P	Q	R	S	T	U	V	W	X	Y	Z
ciphertext	0	1	2	3	4	5	6	7	8	9	10	11	12	13	14	15	16	17	18	19	20	21	22	23	24	25
	3	22	15	8	1	20	13	6	25	18	11	4	23	16	9	2	21	14	7	0	19	12	5	24	17	10
plaintext	D	W	P	I	B	U	N	G	Z	S	L	E	X	Q	J	C	V	O	H	A	T	M	P	Y	R	K

Table 7.8. The Correspondence of Letters for the Sample Ciphertext.

With this correspondence, we try to read the ciphertext. The ciphertext becomes

```
THEBE  STAPP  ROACH  TOLEA  RNNUM
BERTH  EORYI  STOAT  TEMPT  TOSOL
VEEVE  RYHOM  EWORK  PROBL  EMBYW
ORKIN  GONTH  ESEEX  ERCIS  ESAST
UDENT  CANMA  STERT  HEIDE  ASOFT
HESUB  JECT
```

We leave it to the reader to combine the appropriate letters into words to see that the message is intelligible.

The methods described in this section can be extended to construct cipher systems more difficult to break than character ciphers. For example, plaintext letters can be shifted by different shifts, as is done in Vigenère ciphers, described in the preamble to Exercise 17. There are many additional methods that are based on enciphering blocks of letters, rather than individual characters. Such methods will be described in Section 7.2 and in subsequent sections of this chapter.

7.1 Exercises

1. Using the Caesar cipher, encipher the message ATTACK AT DAWN.

2. Decipher the ciphertext message LFDPH LVDZL FRQTX HUHG that has been enciphered using the Caesar cipher.

3. Encipher the message SURRENDER IMMEDIATELY using the affine transformation $C \equiv 11P + 18 \pmod{26}$.

4. Encipher the message THE RIGHT CHOICE using the affine transformation $C \equiv 15P + 14 \pmod{26}$.

5. Decipher the message YLFQX PCRIT, which was enciphered using the affine transformation $C \equiv 21P + 5 \pmod{26}$.

6. Decipher the message RTOLK TOIK, which was enciphered using the affine transformation $C \equiv 3P + 24 \pmod{26}$.

7. If the most common letter in a long ciphertext, enciphered by a shift transformation $C \equiv P + k \pmod{26}$ is Q, then what is the most likely value of k?

8. The message KYVMR CLVFW KYVBV PZJJV MVEKV VE was enciphered using a shift transformation $C \equiv P + k \pmod{26}$. Use frequencies of letters to determine the value of k. What is the plaintext message?

9. The message IVQLM IQATQ SMIKP QTLVW VMQAJ MBBMZ BPIVG WCZWE VNZWU KPQVM AMNWZ BCVMK WWSQM was enciphered using a shift transformation $C \equiv P + k \pmod{26}$. Use frequencies of letter to determine the value of k. What is the plaintext message?

10. If the two most common letters in a long ciphertext, enciphered by an affine transformation $C \equiv aP + b \pmod{26}$ are X and Q, respectively, then what are the most likely values for a and b?

11. If the two most common letters in a long ciphertext, enciphered by an affine transformation $C \equiv aP + b \pmod{26}$ are W and B, respectively, then what are the most likely values for a and b?

12. The message MJMZK CXUNM GWIRY VCPUW MPRRW GMIOP MSNYS RYRAZ PXMCD WPRYE YXD was enciphered using an affine transformation $C \equiv aP + b \pmod{26}$. Use frequencies of letters to determine the values of a and b. What is the plaintext message?

13. The message WEZBF TBBNJ THNBT ADZQE TGTYR BZAJN ANOOZ ATWGN ABOVG FNWZV A was enciphered using an affine transformation $C \equiv aP + b \pmod{26}$. The most common letters in the plaintext are A, E, N and S. What is the plaintext message?

14. The message PJXFJ SWJNX JMRTJ FVSUJ OOJWF OVAJR WHEOF JRWJO DJFFZ BJF was enciphered using an affine transformation $C \equiv aP + b \pmod{26}$. Use frequencies of letters to determine the values of a and b. What is the plaintext message?

Given two ciphers, plaintext may be enciphered by using one of the ciphers, and by then using the other cipher. This procedure produces a *product cipher*.

15. Find the product cipher obtained by using the transformation $C \equiv 5P + 13 \pmod{26}$ followed by the transformation $C \equiv 17P + 3 \pmod{26}$.

16. Find the product cipher obtained by using the transformation $C \equiv aP + b \pmod{26}$ followed by the transformation $C \equiv cP + d \pmod{26}$, where $(a, 26) = (c, 26) = 1$.

Instead of enciphering each letter of a plaintext message in the same way, we can vary how we encipher letters. For example, a *Vigenère* cipher operates in the following way. A sequence of letters $l_1, l_2, \ldots, l_n$, with numerical equivalents $k_1, k_2, \ldots, k_n$, serves as the key. Plaintext messages are split into blocks of length n. To encipher a plaintext block of letters with numerical equivalents $p_1, p_2, \ldots, p_n$ to obtain a ciphertext block of letters with numerical equivalents $c_1, c_2, \ldots, c_n$, we use a sequence of shift ciphers with

$$c_i \equiv p_i + k_i \pmod{26}, \ 0 \le c_i \le 25,$$

for $i = 1, 2, \ldots, n$. In Exercises 17 and 18 use the word SECRET as the key for a Vigenère cipher.

17. Using this Vigenère cipher, encipher the message

DO NOT OPEN THIS ENVELOPE.

18. Decipher the following message which was enciphered using this Vigenère cipher:

WBRCS LAZGJ MGKMF V.

* 19. Describe how cryptanalysis of ciphertext, which was enciphered using a Vigenère cipher, can be carried out. (*Hint*: Begin by determining the length of the key. To do this look for repeated patterns of letters in the ciphertext that might correspond to the same blocks of plaintext.)

* 20. Using the procedure you described in Exercise 18 decipher the following ciphertext which was enciphered using a Vigenère cipher:

UCYFC OOCQU CYFHE BHFTH EFERF
GQJCK XVBUV BSHFT BLCZB SWKUV
BNWKE HLTIC GSOUV BTZFO UPBBA
BFOPK PPTLV HOBUB PIPGC OUIKF

7.1 Computer Projects

Programming Projects

Write programs to do the following:

1. Encipher messages using the Caesar cipher.

2. Encipher messages using the transformation $C \equiv P + k$ (mod 26), where k is a given integer.

3. Encipher messages using the transformation $C \equiv aP + b$ (mod 26), where a and b are integers with $(a, 26) = 1$.

4. Decipher messages that have been enciphered using the Caesar cipher.

5. Decipher messages that have been enciphered using the transformation $C \equiv P + k$ (mod 26), where k is a given integer.

6. Decipher messages that have been enciphered using the transformation $C \equiv aP + b$ (mod 26), where a and b are integers with $(a, 26) = 1$.

* 7. Cryptanalyze, using frequency counts, ciphertext that was enciphered using a transformation of the form $C \equiv P + k$ (mod 26) where k is an unknown integer.

* 8. Cryptanalyze, using frequency counts, ciphertext that was enciphered using a transformation of the form $C \equiv aP + b$ (mod 26) where a and b are unknown integers with $(a, 26) = 1$.

9. Encipher messages using Vigenère ciphers (see the preamble to Exercise 17).

* 10. Decipher messages that have been enciphered using Vigenère ciphers.

Computations and Explorations

Using the programs you have written or a computation program, carry out the following computations and explorations:

1. Find the frequency of the letters of the English alphabet in different types of English text, such as in this book, in computer programs, and in a novel.

2. Encipher some messages using affine transformations as messages for your classmates to decipher.

3. Decipher messages that were enciphered by your classmates using affine transformations using letter frequency analysis.

4. Encipher some messages using Vigenère ciphers for your classmates to decipher.

* 5. Decipher messages enciphered by your classmates using Vigenère ciphers.

7.2 Block Ciphers

We have seen that monographic ciphers based on substitution are vulnerable to cryptanalysis based on the frequency of occurrence of letters in the ciphertext. To avoid this weakness, cipher systems were developed that substitute for each block of plaintext letters of a specified length, a block of ciphertext letters of the same length. Ciphers of this sort are called *block* or *polygraphic ciphers*. In this section, we will discuss some polygraphic ciphers based on modular arithmetic; these were developed by Hill [144] around 1930.

First, we consider *digraphic ciphers*; in these ciphers each block of two letters of plaintext is replaced by a block of two letters of ciphertext. We illustrate this process with an example.

The first step is to split the message into blocks of two letters (adding a dummy letter, say X, at the end of the message, if necessary, so that the final block has two letters). For instance, the message

THE GOLD IS BURIED IN ORONO

is split up as

TH EG OL DI SB UR IE DI NO RO NO.

Next, these letters are translated into their numerical equivalents (as previously done) to obtain

$$19 \ 7 \quad 4 \ 6 \quad 14 \ 11 \quad 3 \ 8 \quad 18 \ 1 \quad 20 \ 17 \quad 8 \ 4 \quad 3 \ 8$$
$$13 \ 14 \quad 17 \ 14 \quad 13 \ 14.$$

Each block of two plaintext numbers $P_1 P_2$ is converted into a block of two ciphertext numbers $C_1 C_2$ by defining C_1 to be the least nonnegative residue modulo 26 of a linear combination of P_1 and P_2 and by defining C_2 to be the least nonnegative residue modulo 26 of a different linear combination of P_1 and P_2. For example, we can let

$$C_1 \equiv 5P_1 + 17P_2 \pmod{26}, \ 0 \le C_1 < 26$$
$$C_2 \equiv 4P_1 + 15P_2 \pmod{26}, \ 0 \le C_2 < 26.$$

For instance, the first block 19 7 is converted to 6 25, because

$$C_1 \equiv 5 \cdot 19 + 17 \cdot 7 \equiv 6 \pmod{26}$$
$$C_2 \equiv 4 \cdot 19 + 15 \cdot 7 \equiv 25 \pmod{26}.$$

After performing this operation on the entire message, the following ciphertext is obtained:

6 25 18 2 23 13 21 2 3 9 25 23 4 14 21 2 17 2 11 18 17 2.

When these blocks are translated into letters, we have the ciphertext message

GZ SC XN VC DJ ZX EO VC RC LS RC.

The deciphering procedure for this cipher system is obtained by using Theorem 3.14. To find the plaintext block $P_1 P_2$ corresponding to the ciphertext block $C_1 C_2$, we use the relationship

$$P_1 \equiv 17C_1 + 5C_2 \pmod{26}$$
$$P_2 \equiv 18C_1 + 23C_2 \pmod{26}.$$

(The reader should verify that this relationship is implied by Theorem 3.14.)

The digraphic cipher system we have presented here is conveniently described using matrices. For this cipher system, we have

$$\begin{bmatrix} C_1 \\ C_2 \end{bmatrix} \equiv \begin{bmatrix} 5 & 17 \\ 4 & 15 \end{bmatrix} \begin{bmatrix} P_1 \\ P_2 \end{bmatrix} \pmod{26}.$$

By Theorem 3.16, we see that the matrix $\begin{bmatrix} 17 & 5 \\ 18 & 23 \end{bmatrix}$ is an inverse of $\begin{bmatrix} 5 & 17 \\ 4 & 15 \end{bmatrix}$ modulo 26. Hence, Theorem 3.15 tells us that deciphering can be done using the relationship

$$\begin{bmatrix} P_1 \\ P_2 \end{bmatrix} \equiv \begin{bmatrix} 17 & 5 \\ 18 & 23 \end{bmatrix} \begin{bmatrix} C_1 \\ C_2 \end{bmatrix} \pmod{26}.$$

In general, a Hill cipher system may be obtained by splitting plaintext into blocks of n letters, translating the letters into their numerical equivalents, and forming ciphertext using the relationship

$$\mathbf{C} \equiv \mathbf{AP} \pmod{26},$$

where $\mathbf{A}$ is an $n \times n$ matrix, $(\det \mathbf{A}, 26) = 1$, $\mathbf{C} = \begin{bmatrix} C_1 \\ C_2 \\ \cdot \\ \cdot \\ \cdot \\ C_n \end{bmatrix}$ and $\mathbf{P} = \begin{bmatrix} P_1 \\ P_2 \\ \cdot \\ \cdot \\ \cdot \\ P_n \end{bmatrix}$, where

$C_1 C_2 ... C_n$ is the ciphertext block that corresponds to the plaintext block $P_1 P_2 ... P_n$. Finally, the ciphertext numbers are translated back to letters. For deciphering, we use the matrix $\overline{\mathbf{A}}$, an inverse of $\mathbf{A}$ modulo 26, which may be obtained using Theorem 3.18. Since $\overline{\mathbf{A}}\mathbf{A} \equiv \mathbf{I} \pmod{26}$, we have

$$\overline{\mathbf{A}}\mathbf{C} \equiv \overline{\mathbf{A}}(\mathbf{AP}) \equiv (\overline{\mathbf{A}}\mathbf{A})\mathbf{P} \equiv \mathbf{P} \pmod{26}.$$

Hence, to obtain plaintext from ciphertext, we use the relationship

$$\mathbf{P} \equiv \overline{\mathbf{A}}\mathbf{C} \pmod{26}.$$

We illustrate this procedure using $n = 3$ and the enciphering matrix

$$\mathbf{A} = \begin{bmatrix} 11 & 2 & 19 \\ 5 & 23 & 25 \\ 20 & 7 & 1 \end{bmatrix}.$$

Since $\det \mathbf{A} \equiv 5 \pmod{26}$, we have $(\det \mathbf{A}, 26) = 1$. To encipher a plaintext block of length three, we use the relationship

$$\begin{bmatrix} C_1 \\ C_2 \\ C_3 \end{bmatrix} \equiv \mathbf{A} \begin{bmatrix} P_1 \\ P_2 \\ P_3 \end{bmatrix} \pmod{26}.$$

To encipher the message STOP PAYMENT, we first split the message into blocks of three letters, adding a final dummy letter X to fill out the last block. We have plaintext blocks

STO PPA YME NTX.

We translate these letters into their numerical equivalents

$$18\ 19\ 14 \qquad 15\ 15\ 0 \qquad 24\ 12\ 4 \qquad 13\ 19\ 23.$$

We obtain the first block of ciphertext in the following way:

$$\begin{bmatrix} C_1 \\ C_2 \\ C_3 \end{bmatrix} \equiv \begin{bmatrix} 11 & 2 & 19 \\ 5 & 23 & 25 \\ 20 & 7 & 1 \end{bmatrix} \begin{bmatrix} 18 \\ 19 \\ 14 \end{bmatrix} \equiv \begin{bmatrix} 8 \\ 19 \\ 13 \end{bmatrix} \pmod{26}.$$

Enciphering the entire plaintext message in the same manner, we obtain the ciphertext message

$$8\ 19\ 13 \qquad 13\ 4\ 15 \qquad 0\ 2\ 22 \qquad 20\ 11\ 0.$$

Translating this message into letters, we have our ciphertext message

$$\text{ITN}\ \ \text{NEP}\ \ \text{ACW}\ \ \text{ULA}.$$

The deciphering process for this polygraphic cipher system takes a ciphertext block and obtains a plaintext block using the transformation

$$\begin{bmatrix} P_1 \\ P_2 \\ P_3 \end{bmatrix} \equiv \overline{A} \begin{bmatrix} C_1 \\ C_2 \\ C_3 \end{bmatrix} \pmod{26}$$

where

$$\overline{A} = \begin{bmatrix} 6 & -5 & 11 \\ -5 & -1 & -10 \\ -7 & 3 & 7 \end{bmatrix}$$

is an inverse of A modulo 26, which may be obtained using Theorem 3.18.

Because polygraphic ciphers operate with blocks, rather than with individual letters, they are not vulnerable to cryptanalysis based on letter frequency. However, polygraphic ciphers operating with blocks of size n are vulnerable to cryptanalysis based on frequencies of blocks of size n. For instance, with a digraphic cipher system, there are $26^2 = 676$ digraphs, blocks of length two. Studies have been done to compile the relative frequencies of digraphs in typical English text. By comparing the frequencies of digraphs in the ciphertext with the average frequencies of digraphs, it is often possible to successfully attack digraphic ciphers. For example, according to some counts, the most common digraph in English is TH, followed closely by HE. If a Hill digraphic cipher system has been employed and the most common digraph is KX, followed by VZ, we may guess that the ciphertext digraphs KX and VZ correspond to TH and HE, respectively. This would mean that the blocks 19 7 and 7 4 are sent to 10 23 and 21 25, respectively. If A is the enciphering matrix, this implies that

$$\mathbf{A} \begin{pmatrix} 19 & 7 \\ 7 & 4 \end{pmatrix} \equiv \begin{pmatrix} 10 & 21 \\ 23 & 25 \end{pmatrix} \pmod{26}.$$

Since $\begin{pmatrix} 4 & 19 \\ 19 & 19 \end{pmatrix}$ is an inverse of $\begin{pmatrix} 19 & 7 \\ 7 & 4 \end{pmatrix} \pmod{26}$, we find that

$$\mathbf{A} \equiv \begin{pmatrix} 10 & 21 \\ 23 & 25 \end{pmatrix} \begin{pmatrix} 4 & 19 \\ 19 & 19 \end{pmatrix} \equiv \begin{pmatrix} 23 & 17 \\ 21 & 2 \end{pmatrix} \pmod{26},$$

which gives a possible key. After attempting to decipher the ciphertext using $\overline{\mathbf{A}} = \begin{pmatrix} 2 & 9 \\ 5 & 23 \end{pmatrix}$ to transform the ciphertext, we would know whether our guess was correct.

In general, if we know n correspondences between plaintext blocks of size n and ciphertext blocks of size n, for instance if we know that the ciphertext blocks $C_{1j}C_{2j}...C_{nj}$, $j = 1,2,...,n$, correspond to the plaintext blocks $P_{1j}P_{2j}...P_{nj}$, $j = 1,2,...,n$, respectively, then we have

$$\mathbf{A} \begin{pmatrix} P_{1j} \\ . \\ . \\ . \\ P_{nj} \end{pmatrix} \equiv \begin{pmatrix} C_{1j} \\ . \\ . \\ . \\ C_{nj} \end{pmatrix} \pmod{26},$$

for $j = 1,2,...,n$.

These n congruences can be succinctly expressed using the matrix congruence

$$\mathbf{AP} \equiv \mathbf{C} \pmod{26},$$

where $\mathbf{P}$ and $\mathbf{C}$ are $n \times n$ matrices with ijth entries P_{ij} and C_{ij}, respectively. If $(\det \mathbf{P}, 26) = 1$, then we can find the enciphering matrix $\mathbf{A}$ via

$$\mathbf{A} \equiv \mathbf{C}\overline{\mathbf{P}} \pmod{26},$$

where $\overline{\mathbf{P}}$ is an inverse of $\mathbf{P}$ modulo 26.

Cryptanalysis using frequencies of polygraphs is only worthwhile for small values of n, where n is the size of the polygraphs. When $n = 10$, for example, there are 26^{10}, which is approximately 1.4×10^{14}, polygraphs of this length. Any analysis of the relative frequencies of these polygraphs is extremely infeasible.

7.2 Exercises

1. Using the digraphic cipher that sends the plaintext block $P_1 P_2$ to the ciphertext block $C_1 C_2$ with

$$C_1 \equiv 3P_1 + 10P_2 \pmod{26}$$
$$C_2 \equiv 9P_1 + 7P_2 \pmod{26},$$

   encipher the message BEWARE OF THE MESSENGER.

2. Using the digraphic cipher that sends the plaintext block $P_1 P_2$ to the ciphertext block $C_1 C_2$ with

$$C_1 \equiv 8P_1 + 9P_2 \pmod{26}$$
$$C_2 \equiv 3P_1 + 11P_2 \pmod{26},$$

   encipher the message DO NOT SHOOT THE MESSENGER.

3. Decipher the ciphertext message RD SR QO VU QP CZ AN QW RD DS AK OB which was enciphered using the digraphic cipher which sends the plaintext block $P_1 P_2$ into the ciphertext block $C_1 C_2$ with

$$C_1 \equiv 13P_1 + 4P_2 \pmod{26}$$
$$C_2 \equiv 9P_1 + P_2 \pmod{26}.$$

4. Decipher the ciphertext message UW DM NK QB EK, which was enciphered using the digraphic cipher which sends the plaintext block $P_1 P_2$ into the ciphertext block $C_1 C_2$ with

$$C_1 \equiv 23P_1 + 3P_2 \pmod{26}$$
$$C_2 \equiv 10P_1 + 25P_2 \pmod{26}.$$

5. A cryptanalyst has determined that the two most common digraphs in a ciphertext message are RH and NI and guesses that these ciphertext digraphs correspond to the two most common diagraphs in English text, TH and HE. If the plaintext was enciphered using a Hill digraphic cipher described by

$$C_1 \equiv aP_1 + bP_2 \pmod{26}$$
$$C_2 \equiv cP_1 + dP_2 \pmod{26},$$

   what are $a, b, c,$ and d?

6. How many pairs of letters remain unchanged when encryption is performed using each of the following digraphic ciphers?

   a) $C_1 \equiv 4P_1 + 5P_2 \pmod{26}$
      $C_2 \equiv 3P_1 + P_2 \pmod{26}$

   b) $C_1 \equiv 7P_1 + 17P_2 \pmod{26}$
      $C_2 \equiv P_1 + 6P_2 \pmod{26}$

   c) $C_1 \equiv 3P_1 + 5P_2 \pmod{26}$
      $C_2 \equiv 6P_1 + 3P_2 \pmod{26}$

7. Show that if the enciphering matrix $\mathbf{A}$ in the Hill cipher system is involutory modulo 26, $i.e.$, $\mathbf{A}^2 \equiv \mathbf{I} \pmod{26}$, then $\mathbf{A}$ also serves as a deciphering matrix for this cipher system.

8. A cryptanalyst has determined that the three most common trigraphs (blocks of length three) in a ciphertext are, LME, WRI and ZYC and guesses that these ciphertext trigraphs correspond to the three most common trigraphs in English text, THE, AND, and THA. If the plaintext was enciphered using a Hill trigraphic cipher described by $C \equiv AP$ (mod 26), what are the entries of the 3×3 enciphering matrix A?

9. Find the product cipher obtained by using the digraphic Hill cipher with enciphering matrix $\begin{bmatrix} 2 & 3 \\ 1 & 17 \end{bmatrix}$ followed by using on the result the digraphic Hill cipher with enciphering matrix $\begin{bmatrix} 5 & 1 \\ 25 & 4 \end{bmatrix}$.

* 10. Show that the product cipher obtained from two digraphic Hill ciphers is again a digraphic Hill cipher.

* 11. Show that the product cipher obtained by enciphering first using a Hill cipher with blocks of size m and then using a Hill cipher with blocks of size n is again a Hill cipher using blocks of size $[m,n]$.

12. Find the 6×6 enciphering matrix corresponding to the product cipher obtained by first using the Hill cipher with enciphering matrix $\begin{bmatrix} 3 & 1 \\ 2 & 1 \end{bmatrix}$, followed by using the Hill cipher with enciphering matrix $\begin{bmatrix} 1 & 1 & 0 \\ 1 & 0 & 1 \\ 0 & 1 & 1 \end{bmatrix}$.

* 13. A *transposition cipher* is a cipher where blocks of a specified size are enciphered by permuting their characters in a specified manner. For instance, plaintext blocks of length five, $P_1 P_2 P_3 P_4 P_5$, may be sent to ciphertext blocks $C_1 C_2 C_3 C_4 C_5 = P_4 P_5 P_2 P_1 P_3$. Show that every such transposition cipher is a Hill cipher with an enciphering matrix that contains only 0's and 1's as entries with the property that each row and each column contains exactly one 1.

Hill ciphers are special cases of block ciphers based on *affine transformations*. To form such a transformation, let A be an $n \times n$ matrix with integer entries and (det A, 26) $= 1$ and let B be an $n \times 1$ matrix with integer entries. To encipher a message we split it into blocks of length n and put the numerical equivalents of the letters in each block into an $n \times 1$ matrix P (padding the last block with dummy letters, if necessary). We find the corresponding ciphertext block by computing $C \equiv (AP + B)$ (mod 26) and translating the entries in C back into letters.

14. Encipher the message HAVE A NICE DAY using the affine transformation $C \equiv \begin{bmatrix} 3 & 2 \\ 7 & 11 \end{bmatrix} P + \begin{bmatrix} 8 \\ 19 \end{bmatrix}$ (mod 26) to encipher blocks of two successive letters.

15. What is the deciphering transformation associated with the affine transformation in Exercise 14?

16. What is the deciphering transformation associated with the enciphering transformation $C \equiv (AP + B)$ (mod 26) where A is an $n \times n$ matrix with integer entries and (det A, 26) $= 1$ and B is an $n \times 1$ matrix with integer entries?

17. Decipher the message HG PM QR YN NM that was enciphered using the affine transformation $\mathbf{C} \equiv \begin{bmatrix} 5 & 2 \\ 11 & 15 \end{bmatrix} \mathbf{P} + \begin{bmatrix} 14 \\ 3 \end{bmatrix}$ (mod 26).

18. Explain how you would go about deciphering a message enciphered in blocks of length two using an affine transformation $\mathbf{C} \equiv \mathbf{AP} + \mathbf{B}$ (mod 26) where $\mathbf{A}$ is a 2×2 matrix with integer entries and $(\det \mathbf{A}, 26) = 1$ and $\mathbf{B}$ is a 2×1 matrix.

19. Explain how you would go about deciphering a message enciphered in blocks of length three using an affine transformation $\mathbf{C} \equiv \mathbf{AP} + \mathbf{B}$ (mod 26) where $\mathbf{A}$ is a 3×3 matrix with integer entries and $(\det \mathbf{A}, 26) = 1$ and $\mathbf{B}$ is a 3×1 matrix.

20. Is the product cipher of two digraphic block ciphers based on affine transformations also a digraphic block cipher based on an affine transformation?

* 21. Is the product cipher of two block ciphers based on affine transformations, enciphering blocks of length m and blocks of length n, respectively, also a block cipher based on an affine transformation?

7.2 Computer Projects

Programming Projects

Write programs to do the following:

1. Encipher messages using a Hill cipher.

2. Decipher messages that were enciphered using a Hill cipher.

* 3. Cryptanalyze messages that were enciphered using a digraphic Hill cipher, by analyzing the frequency of digraphs in the ciphertext.

4. Encipher messages using a cipher based on an affine transformation. (See the preamble to Exercise 14.)

5. Decipher messages that were enciphered using an affine transformation.

6. By analyzing the frequency of digraphs in ciphertext, cryptanalyze messages enciphered using a digraphic block cipher based on an affine transformation.

Computations and Explorations

Using the programs you have written or a computation program, carry out the following computations and explorations:

1. Find the frequencies of digraphs in various types of English texts, such as this text, computer programs, and a novel.

2. Find the frequencies of trigraphs in various types of English texts, such as this text, computer programs, and a novel.

3. Encipher some messages using Hill ciphers for your classmates to decipher.

4. Decipher messages encrypted by your classmates using Hill ciphers.

7.3 Exponentiation Ciphers

In this section, we discuss a cipher, based on modular exponentiation, that was invented in 1978 by Pohlig and Hellman [148]. We will see that ciphers produced by this system are resistant to cryptanalysis.

Let p be an odd prime and let e, the enciphering key, be a positive integer with $(e, p-1) = 1$. To encipher a message, we first translate the letters of the message into numerical equivalents (retaining initial zeros in the two-digit numerical equivalents of letters). We use the same relationship we have used before, as shown in Table 7.9.

letter	A	B	C	D	E	F	G	H	I	J	K	L	M	N	O	P	Q	R	S	T	U	V	W	X	Y	Z
numerical equivalent	00	01	02	03	04	05	06	07	08	09	10	11	12	13	14	15	16	17	18	19	20	21	22	23	24	25

Table 7.9. Two-digit Numerical Equivalents of Letters.

Next, we group the resulting numbers into blocks of $2m$ decimal digits, where $2m$ is the largest positive even integer such that all blocks of numerical equivalents corresponding to m letters (viewed as a single integer with $2m$ decimal digits) are less than p, e.g. if $2525 < p < 252525$, then $m = 2$.

For each plaintext block P, which is an integer with $2m$ decimal digits, we form a ciphertext block C using the relationship

$$C \equiv P^e \pmod{p}, \ 0 \le C < p.$$

The ciphertext message consists of these ciphertext blocks which are integers less than p. Notice that different values of e determine different ciphers and hence e is aptly called the enciphering key. We illustrate the enciphering technique with the following example.

Example 7.1. Let the prime to be used as the modulus in the enciphering procedure be $p = 2633$ and let the enciphering key to be used as the exponent in the modular exponentiation be $e = 29$, so that $(e, p-1) = (29, 2632) = 1$. To encipher the plaintext message,

THIS IS AN EXAMPLE OF AN EXPONENTIATION CIPHER,

we first convert the letters of the message into their numerical equivalents, and then form blocks of length four from these digits, to obtain

1907	0818	0818	0013	0423
0012	1511	0414	0500	1304
2315	1413	0413	1908	0019
0814	1302	0815	0704	1723.

Note that we have added the two digits 23, corresponding to the letter X, at the end of the message to fill out the final block of four digits.

We next translate each plaintext block P into a ciphertext block C using the relationship

$$C \equiv P^{29} \pmod{2633}, \ 0 < C < 2633.$$

For instance, to decipher the first plaintext block we compute

$$C \equiv 1907^{29} \equiv 2199 \pmod{2633}.$$

To efficiently carry out the modular exponentiation, we use the algorithm given in Section 3.1. When we encipher the blocks we obtain the ciphertext:

2199	1745	1745	1206	2437
2425	1729	1619	0935	0960
1072	1541	1701	1553	0735
2064	1351	1704	1841	1459. □

To decipher a ciphertext block C, we need to know a deciphering key, namely an integer d such that $de \equiv 1 \pmod{p-1}$, so that d is an inverse of $e \pmod{p-1}$, which exists since $(e, p-1) = 1$. If we raise the ciphertext block C to the dth power modulo p, we recover our plaintext block P, since

$$C^d \equiv (P^e)^d = P^{ed} \equiv P^{k(p-1)+1} \equiv (P^{p-1})^k P \equiv P \pmod{p},$$

where $de = k(p-1) + 1$, for some integer k, since $de \equiv 1 \pmod{p-1}$. (Note that we have used Fermat's little theorem to see that $P^{p-1} \equiv 1 \pmod{p}$.)

Example 7.2. To decipher the ciphertext blocks generated using the prime modulus $p = 2633$ and the enciphering key $e = 29$, we need an inverse of e modulo $p - 1 = 2632$. An easy computation, as done in Section 3.2, shows that $d = 2269$ is such an inverse. To decipher the ciphertext block C in order to find the corresponding plaintext block P, we use the relationship

$$P \equiv C^{2269} \pmod{2633}.$$

For instance, to decipher the ciphertext block 2199, we have

$$P \equiv 2199^{2269} \equiv 1907 \pmod{2633}.$$

Again, the modular exponentiation is carried out using the algorithm given in Section 3.2.   □

For each plaintext block P that we encipher by computing $P^e \pmod{p}$, we use only $O((\log_2 p)^3)$ bit operations, as Theorem 3.9 demonstrates. Before we decipher we need to find an inverse d of e modulo $p - 1$. This can be done using $O(\log^3 p)$ bit operations (see Exercise 15 of Section 3.2), and this needs to be done only once.

Then, to recover the plaintext block P from a ciphertext block C, we simply need to compute the least positive residue of C^d modulo p; we can do this using $O((\log_2 p)^3)$ bit operations. Consequently, the processes of enciphering and deciphering using modular exponentiation can be done rapidly.

On the other hand, cryptanalysis of messages enciphered using modular exponentiation generally cannot be done rapidly. To see this, suppose we know the prime p used as the modulus, and moreover, suppose we know the plaintext block P corresponding to a ciphertext block C, so that

$$(7.2) \qquad C \equiv P^e \pmod{p}.$$

For successful cryptanalysis, we need to find the enciphering key e. When the relationship (7.2) holds, we say that e is the *logarithm of C to the base P modulo p*. There are various algorithms for finding logarithms to a given base modulo a prime, a problem known as the *discrete logarithm problem*. The fastest such algorithm requires approximately $\exp(\sqrt{\log p \log \log p})$ bit operations (see [138] or the excellent survey article by McCurley in [84] on the discrete logarithm problem). To find logarithms modulo a prime with n decimal digits using the fastest known algorithm requires approximately the same number of bit operations as factoring integers with the same number of decimal digits, when the fastest known factoring algorithm is used. Consulting Table 2.1, we see that finding logarithms modulo a prime p requires an extremely long time. For instance, when p has 100 decimal digits, finding logarithms modulo p requires approximately 74 years, whereas when p has 200 decimal digits, approximately 3.8×10^9 years are required.

We should mention that for primes p where $p-1$ has only small prime factors, it is possible to use special techniques to find logarithms modulo p using $O(\log^2 p)$ bit operations. Clearly, this sort of prime should not be used as a modulus in this cipher system. Taking a prime $p = 2q + 1$, where q is also prime, obviates this difficulty.

Modular exponentiation is useful for establishing *common keys* to be used by two or more individuals. These common keys may, for instance, be used as keys in a cipher system for sessions of data communication, and should be constructed so that unauthorized individuals cannot discover them in a feasible amount of computer time.

Let p be a large prime and let a be an integer relatively prime to p. Each individual in the network picks a key k that is an integer relatively prime to $p-1$. When two individuals with keys k_1 and k_2 wish to exchange a key, the first individual sends the second the integer y_1, where

$$y_1 \equiv a^{k_1} \pmod{p}, \qquad 0 < y_1 < p,$$

and the second individual finds the common key K by computing

$$K \equiv y_1^{k_2} \equiv a^{k_1 k_2} \pmod{p}, \qquad 0 < K < p.$$

Similarly, the second individual sends the first the integer y_2 where

$$y_2 \equiv a^{k_2} \pmod{p}, \quad 0 < y_2 < p,$$

and the first individual finds the common key K by computing

$$K \equiv y_2^{k_1} \equiv a^{k_1 k_2} \pmod{p}, \quad 0 < K < p.$$

We note that other individuals in the network cannot find this common key K in a feasible amount of computer time, since they must compute logarithms modulo p to find K.

In a similar manner, a common key can be shared by any group of n individuals. If these individuals have keys $k_1, k_2, \ldots, k_n$, they can share the common key

$$K = a^{k_1 k_2 \cdots k_n} \pmod{p}.$$

We leave an explicit description of a method used to produce this common key K as a problem for the reader.

An amusing application of exponentiation ciphers has been described by Shamir, Rivest, and Adleman [153]. They show that by using exponentiation ciphers, a fair game of poker may be played by two players communicating via computers. Suppose Alex and Betty wish to play poker. First, they jointly choose a large prime p. Next, they individually choose secret keys e_1 and e_2, to be used as exponents in modular exponentiation. Let E_{e_1} and E_{e_2} represent the corresponding enciphering transformations, so that

$$E_{e_1}(M) \equiv M^{e_1} \pmod{p}$$
$$E_{e_2}(M) \equiv M^{e_2} \pmod{p},$$

where M is a plaintext message. Let d_1 and d_2 be the inverses of e_1 and e_2 modulo p respectively, and let D_{e_1} and D_{e_2} be the corresponding deciphering transformations, so that

$$D_{e_1}(C) \equiv C^{d_1} \pmod{p}$$
$$D_{e_2}(C) \equiv C^{d_2} \pmod{p},$$

where C is a ciphertext message.

Note that enciphering transformations commute, that is

$$E_{e_1}(E_{e_2}(M)) = E_{e_2}(E_{e_1}(M)),$$

since

$$(M^{e_2})^{e_1} \equiv (M^{e_1})^{e_2} \pmod{p}.$$

To play electronic poker, the deck of cards is represented by the 52 messages

$$M_1 = \text{"TWO OF CLUBS"}$$
$$M_2 = \text{"THREE OF CLUBS"}$$

.

.

.

$$M_{52} = \text{"ACE OF SPADES."}$$

When Alex and Betty wish to play poker electronically, they use the following sequence of steps. We suppose Betty is the dealer.

i. Betty uses her enciphering transformation to encipher the 52 messages for the cards. She obtains $E_{e_2}(M_1)$, $E_{e_2}(M_2)$,...,$E_{e_2}(M_{52})$. Betty shuffles the deck, by randomly reordering the enciphered messages. Then she sends the 52 shuffled enciphered messages to Alex.

ii. Alex selects, at random, five of the enciphered messages that Betty has sent him. He returns these five messages to Betty and she deciphers them to find her hand, using her deciphering transformation D_{e_2}, since $D_{e_2}(E_{e_2}(M)) = M$ for all messages M. Alex cannot determine which cards Betty has, since he cannot decipher the enciphered messages $E_{e_2}(M_j)$, $j = 1,2,...,52$.

iii. Alex selects five other enciphered messages at random. Let these messages be C_1, C_2, C_3, C_4, and C_5, where

$$C_j = E_{e_2}(M_{i_j}),$$

$j = 1,2,3,4,5$. Alex enciphers these five previously enciphered messages using his enciphering transformation. He obtains the five messages

$$C_j^* = E_{e_1}(C_j) = E_{e_1}(E_{e_2}(M_{i_j}))$$

$j = 1,2,3,4,5$. Alex sends these five messages that have been enciphered twice (first by Betty and afterwards by Alex) to Betty.

iv. Betty uses her deciphering transformation D_{e_2} to find

$$\begin{aligned}
D_{e_2}(C_j^*) &= D_{e_2}(E_{e_1}(E_{e_2}(M_{i_j}))) \\
&= D_{e_2}(E_{e_2}(E_{e_1}(M_{i_j}))) \\
&= E_{e_1}(M_{i_j}),
\end{aligned}$$

since $E_{e_1}(E_{e_2}(M)) = E_{e_2}(E_{e_1}(M))$ and $D_{e_2}(E_{e_2}(M)) = M$ for all messages M. Betty sends the fives message $E_{e_1}(M_{i_j})$ back to Alex.

v. Alex uses his deciphering transformation D_{e_1} to obtain his hand, since

$$D_{e_1}(E_{e_1}(M_{i_j})) = M_{i_j}.$$

When a game is played where it is necessary to deal additional cards, such as draw poker, the same steps are followed to deal additional cards from the remaining deck. Note that using the procedure we have described, neither player knows the cards in the hand of the other player, and all hands are equally likely for each player. To guarantee that no cheating has occurred, at the end of the game both players reveal their keys, so that each player can verify that the other player was actually dealt the cards claimed.

A description of a possible weakness in this scheme, and how it may be overcome, may be found in the exercise set of Section 9.1.

7.3 Exercises

1. Using the prime $p = 101$ and enciphering key $e = 3$, encipher the message GOOD MORNING using modular exponentiation.

2. Using the prime $p = 2621$ and enciphering key $e = 7$, encipher the message SWEET DREAMS using modular exponentiation.

3. What is the plaintext message that corresponds to the ciphertext 01 09 00 12 12 09 24 10 that is produced using modular exponentiation with modulus $p = 29$ and enciphering exponent $e = 5$?

4. What is the plaintext message that corresponds to the ciphertext 1213 0902 0539 1208 1234 1103 1374 that is produced using modular exponentiation with modulus $p = 2591$ and enciphering key $e = 13$?

5. Show that the enciphering and deciphering procedures are identical when enciphering is done using modular exponentiation with modulus $p = 31$ and enciphering key $e = 11$.

6. With modulus $p = 29$ and unknown enciphering key e, modular exponentiation produces the ciphertext 04 19 19 11 04 24 09 15 15. Cryptanalyze the above cipher, if it is also known that the ciphertext block 24 corresponds to the plaintext letter U (with numerical equivalent 20). (*Hint:* First find the logarithm of 24 to the base 20 modulo 29 using some guesswork.)

7. Using the method described in the text for exchanging common keys, what is the common key that can be used by individuals with keys $k_1 = 27$ and $k_2 = 31$ when the modulus is $p = 101$ and the base is $a = 5$?

8. What is the group key K that can be shared by four individuals with keys $k_1 = 11, k_2 = 12, k_3 = 17, k_4 = 19$ using the modulus $p = 1009$ and base $a = 3$?

* 9. Describe a procedure to allow n individuals to share the common key described in the text.

7.3 Computer Projects

Programming Projects

Write programs to do the following:

1. Encipher messages using modular exponentiation.

2. Decipher messages that have been enciphered using modular exponentiation.

3. Cryptanalyze ciphertext that has been enciphered using modular exponentiation when a correspondence between a plaintext block P and a ciphertext block C is known.

4. Produce common keys for individuals in a network.

* 5. Play electronic poker using encryption via modular exponentiation.

Computations and Explorations

Using the programs you have written or a computation program, carry out the following computations and explorations:

1. Solve examples of the discrete logarithm problem.

2. Encipher some messages for your classmates to decipher using exponentiation ciphers.

3. Decipher messages enciphered by your classmates using exponentiation ciphers, given the enciphering key and prime modulus.

7.4 Public-Key Cryptography

If one of the cipher systems previously described in this chapter is used to establish secure communications within a network, then each pair of communicants must employ an enciphering key that is kept secret from the other individuals in the network, since once the enciphering key in one of those cipher systems is known, the deciphering key can be found using a small amount of computer time. Consequently, to maintain secrecy the enciphering keys must themselves be transmitted over a channel of secure communications.

To avoid assigning a key to each pair of individuals that must be kept secret from the rest of the network, a new type of cipher system, called a *public-key cipher system*, has been recently introduced. In this type of cipher system, enciphering keys can be made public, since an unrealistically large amount of computer time is required to find a deciphering transformation from an enciphering transformation. To use a public-key cipher system to establish secret communications in a network of n individuals, each individual produces a key of the type specified by the cipher system, retaining certain private information that went into the construction of the enciphering transformation $E(k)$, obtained from the key k according to a specified rule. Then a directory of the n keys $k_1, k_2,...,k_n$ is published. When individual i wishes to send a message to individual j, the letters of the message are translated into their numerical equivalents and combined into blocks of specified size. Then, for each plaintext block P a corresponding ciphertext block $C = E_{k_j}(P)$ is computed using the enciphering transformation E_{k_j}. To decipher the message, individual j applies the deciphering transformation D_{k_j} to each ciphertext block C to find P, i.e.

$$D_{k_j}(C) = D_{k_j}(E_{k_j}(P)) = P.$$

Since the deciphering transformation D_{k_j} cannot be found in a realistic amount of time by anyone other than individual j, no unauthorized individuals can decipher the message, even though they know the key k_j. Furthermore, cryptanalysis of the ciphertext message, even with knowledge of k_j, is extremely infeasible due to the large amount of computer time needed.

The *RSA cipher system*, invented by Rivest, Shamir, and Adleman [150] in the 1970s, is a public-key cipher system based on modular exponentiation where the keys are pairs (e,n), consisting of an exponent e and a modulus n that is the product of two large primes, *i.e.* $n = pq$, where p and q are large primes, so that $(e,\phi(n)) = 1$. To encipher a message, we first translate the letters into their numerical equivalents and then form blocks of the largest possible size (with an even number of digits). To encipher a plaintext block P, we form a ciphertext block C by

$$E(P) = C \equiv P^e \pmod{n}, \quad 0 < C < n.$$

The deciphering procedure requires knowledge of an inverse d of e modulo $\phi(n)$, which exists since $(e,\phi(n)) = 1$. To decipher the ciphertext block C, we find

$$D(C) \equiv C^d = (P^e)^d = P^{ed} = P^{k\phi(n) + 1}$$
$$\equiv (P^{\phi(n)})^k P \equiv P \pmod{n},$$

where $ed = k\phi(n) + 1$ for some integer k, since $ed \equiv 1 \pmod{\phi(n)}$, and by Euler's theorem, we have $P^{\phi(n)} \equiv 1 \pmod{n}$, when $(P, n) = 1$ (the probability that P and n are not relatively prime is extremely small; see Exercise 4 at the end of this section). The pair (d,n) is a *deciphering key*.

To illustrate how the RSA cipher system works, we present an example where the enciphering modulus is the product of the two primes 43 and 59 (which are smaller than the large primes that would actually be used). We have $n = 43 \cdot 59 = 2537$ as the modulus and $e = 13$ as the exponent for the RSA cipher. Note that we have $(e, \phi(n)) = (13, 42 \cdot 58) = 1$. To encipher the message

PUBLIC KEY CRYPTOGRAPHY,

we first translate the letters into their numerical equivalents, and then group these numbers together into blocks of four. We obtain

1520	0111	0802	1004
2402	1724	1519	1406
1700	1507	2423,	

where we have added the dummy letter $X = 23$ at the end of the passage to fill out the final block.

We encipher each plaintext block into a ciphertext block, using the relationship

$$C \equiv P^{13} \pmod{2537}.$$

For instance, when we encipher the first plaintext block 1520, we obtain the ciphertext block

$$C \equiv (1520)^{13} \equiv 95 \pmod{2537}.$$

Enciphering all the plaintext blocks, we obtain the ciphertext message

0095	1648	1410	1299
0811	2333	2132	0370
1185	1457	1084.	

To decipher messages that were enciphered using the RSA cipher, we must find an inverse of $e = 13$ modulo $\phi(2537) = \phi(43 \cdot 59) = 42 \cdot 58 = 2436$. A short computation using the Euclidean algorithm, as done in Section 3.2, shows that $d = 937$ is an inverse of 13 modulo 2436. Consequently, to decipher the cipher text block C, we use the relationship

$$P \equiv C^{937} \pmod{2537}, \ 0 \le P \le 2537,$$

which is valid because

$$C^{937} \equiv (P^{13})^{937} \equiv (P^{2436})^5 P \equiv P \pmod{2537};$$

note that we have used Euler's theorem to see that

$$P^{\phi(2537)} = P^{2436} \equiv 1 \pmod{2537},$$

when $(P, 2537) = 1$ (which is true for all of the plaintext blocks in our example).

To understand how the RSA cipher system fulfills the requirements of a public-key cipher system, first note that each individual can find two large primes p and q, with 100 decimal digits, in just a few minutes of computer time. These primes can be found by picking odd integers with 100 digits at random; by the prime number theorem, the probability that such an integer is prime is approximately $2/\log 10^{100}$. Hence, we expect to find a prime after examining an average of $1/(2/\log 10^{100})$, or approximately 115, such integers. To test these randomly chosen odd integers for primality, we use Rabin's probabilistic primality test discussed in Section 5.2. For each of these 100-digit odd integers we perform Miller's test for 100 bases less than the integer; the probability that a composite integer passes all these tests is less than 10^{-60}. The procedure we have just outlined requires only a few minutes of computer time to find a 100-digit prime, and each individual need do it only twice.

Once the primes p and q have been found, an enciphering exponent e should be chosen with $(e, \phi(pq)) = 1$. One suggestion for choosing e is to take any prime greater than both p and q. No matter how e is found, it should be true that $2^e > n = pq$, so that it is impossible to recover the plaintext block P, $P \ne 0$ or 1, just by taking the eth root of the integer C with $C \equiv P^e \pmod{n}$, $0 < C < n$. As long as $2^e > n$, every message other than $P = 0$ and 1, is enciphered by exponentiation followed by a reduction modulo n.

We note that the modular exponentiation needed for enciphering messages using the RSA cipher system can be done using only a few seconds of computer time when the modulus, exponent, and base in the modular exponentiation have as many as 200 decimal digits. Also, using the Euclidean algorithm, we can rapidly find an inverse d of the enciphering exponent e modulo $\phi(n)$ when the primes p and q are known, so that $\phi(n) = \phi(pq) = (p-1)(q-1)$ is known.

To see why knowledge of the enciphering key (e, n) does not easily lead to the deciphering key (d, n), note that to find d, an inverse of e modulo $\phi(n)$, requires that we first find $\phi(n) = \phi(pq) = (p-1)(q-1)$. Note that finding $\phi(n)$ is not easier than factoring the integer n. To see why, note that $p + q = n - \phi(n) + 1$ and $p - q = \sqrt{(p+q)^2 - 4pq} = \sqrt{(p+q)^2 - 4n}$. It follows that $p = \frac{1}{2}[(p+q) + (p-q)]$ and $q = \frac{1}{2}[(p+q) + (p-q)]$. Consequently p and q can easily be found when $n = pq$ and $\phi(n) = (p-1)(q-1)$ are known. Note that when p and q both have around 100 decimal digits, $n = pq$ has around 200 decimal digits. Using the fastest factorization algorithm known, millions of years of computer time are required to factor an integer of this size. Also, if the integer d is known, but $\phi(n)$ is not, then n may also be factored easily, since $ed - 1$ is a multiple of $\phi(n)$ and there are special algorithms for factoring an integer n using any multiple of $\phi(n)$ (see Miller [118]).

It has not been proven that it is impossible to decipher messages enciphered using the RSA cipher system without factoring n, but so far no such method has been discovered. As yet, all deciphering methods suggested that work in general are equivalent to factoring n, and as we have remarked, factoring large integers seems to be an intractable problem, requiring tremendous amounts of computer time. If no method of deciphering RSA messages is found without factoring the modulus n, the security of the RSA system can be maintained as factoring methods and computational power improve by increasing the size of the modulus. Unfortunately, messages enciphering using the RSA will become vulnerable to attack when factoring the modulus n becomes feasible. This means that extra care should be taken, say by using primes p and q each with several hundred digits, to protect the secrecy of messages that need to be kept secret for tens, or hundreds, of years.

Note that a few extra precautions should be taken in choosing the primes p and q to be used in the RSA cipher system to prevent the use of special rapid techniques to factor $n = pq$. For example, both $p - 1$ and $q - 1$ should have large prime factors, $(p - 1, q - 1)$ should be small, and p and q should have decimal expansions differing in length by a few digits.

As we have remarked, the security of the RSA cipher system depends on the difficulty of factoring large integers. In particular, for the RSA cipher system, once the modulus n has been factored it is easy to find the deciphering transformation from the enciphering transformation. However, it may be possible to somehow find the deciphering transformation from the enciphering transformation without factoring n, although this seems unlikely.

Rabin [149] has discovered a variant of the RSA cipher system for which factorization of the modulus n has almost the same computational complexity as obtaining the deciphering transformation from the enciphering transformation. To describe Rabin's cipher system, let $n = pq$, where p and q are odd primes, and let b be an integer with $0 \le b < n$. To encipher the plaintext message P, we form

$$C \equiv P(P+b) \pmod{n}.$$

We will not discuss the deciphering procedure for Rabin ciphers here, because it relies on some concepts we have not yet developed (see Exercise 49 in Section 9.1). However, we remark that there are four possible values of P for each ciphertext C such that $C \equiv P(P+b) \pmod{n}$, an ambiguity which complicates the deciphering process. When p and q are known, the deciphering procedure for a Rabin cipher can be carried out rapidly since $O(\log n)$ bit operations are needed.

Rabin has shown that if there is an algorithm for deciphering in this cipher system, without knowledge of the primes p and q, that requires $f(n)$ bit operations, then there is an algorithm for the factorization of n requiring only $2(f(n) + \log n)$ bit operations. Hence the process of deciphering messages enciphered with a Rabin cipher without knowledge of p and q is a problem of computational complexity similar to that of factorization.

Public-key cipher systems can also be used to send signed messages. When signatures are used, the recipient of a message is sure that the message came from the sender, and can convince an impartial judge that only the sender could be the source of the message. This authentication is needed for electronic mail, electronic banking, and electronic stock market transactions. To see how the RSA cipher system can be used to send signed messages, suppose that individual i wishes to send a signed message to individual j. The first thing that individual i does to a plaintext block P is to compute

$$S = D_{k_i}(P) \equiv P^{d_i} \pmod{n_i},$$

where (d_i, n_i) is the deciphering key for individual i, which only individual i knows. Then, if $n_j > n_i$, where (e_j, n_j) is the enciphering key for individual j, individual i enciphers S by forming

$$C = E_{k_j}(S) \equiv S^{e_j} \pmod{n_j}, \qquad 0 < C < n_j.$$

When $n_j < n_i$ individual i splits S into blocks of size less than n_j and enciphers each block using the enciphering transformation E_{k_j}.

For deciphering, individual j first uses the private deciphering transformation D_{k_j} to recover S, since

$$D_{k_j}(C) = D_{k_j}(E_{k_j}(S)) = S.$$

To find the plaintext message P, supposedly sent by individual i, individual j next uses the public enciphering transformation E_{k_i}, since

$$E_{k_i}(S) = E_{k_i}(D_{k_i}(P)) = P.$$

Here, we have used the identity $E_{k_i}(D_{k_i}(P)) = P$, which follows from the fact that

$$E_{k_i}(D_{k_i}(P)) \equiv (P^{d_i})^{e_i} \equiv P^{d_i e_i} \equiv P \pmod{n_i},$$

since

$$d_i e_i \equiv 1 \pmod{\phi(n_i)}.$$

The combination of the plaintext block P and the signed version S convinces individual j that the message actually came from individual i. Also, individual i cannot deny sending the message, since no one other than individual i could have produced the signed message S from the original message P.

The RSA cipher system relies on the difference in the computer time needed to find primes and the computer time needed to factor. In Chapter 9, we will use this same difference to develop a technique to "flip coins" electronically.

7.4 Exercises

1. Find the primes p and q if $n = pq = 14647$ and $\phi(n) = 14400$.

2. Find the primes p and q if $n = pq = 4386607$ and $\phi(n) = 4382136$.

3. Suppose a cryptanalyst discovers a message P that is not relatively prime to the enciphering modulus $n = pq$ used in a RSA cipher. Show that the cryptanalyst can factor n.

4. Show that it is extremely unlikely that a message such as that described in Exercise 3 can be discovered. Do this by demonstrating that the probability that a message P is not relatively prime to n is $\dfrac{1}{p} + \dfrac{1}{q} - \dfrac{1}{pq}$, and if p and q are both larger than 10^{100}, this probability is less than 10^{-99}.

5. What is the ciphertext that is produced when the RSA cipher with key $(e,n) = (3,2669)$ is used to encipher the message BEST WISHES?

6. What is the ciphertext that is produced when the RSA cipher with key $(e,n) = (7,2627)$ is used to encipher the message LIFE IS A DREAM?

7. If the ciphertext message produced by the RSA cipher with key $(e,n) = (13,2747)$ is 2206 0755 0436 1165 1737, what is the plaintext message?

8. If the ciphertext message produced by the RSA cipher with key $(e,n) = (5,2881)$ is 0504 1874 0347 0515 2088 2356 0736 0468, what is the plaintext message?

9. Harold and Audrey have as their RSA keys $(3,23 \cdot 47)$ and $(7,31 \cdot 59)$, respectively.

a) Using the method in the text, what is the signed ciphertext sent by Harold to Audrey, when the plaintext message is CHEERS HAROLD?

b) Using the method in the text, what is the signed ciphertext sent by Audrey to Harold when the plaintext message is SINCERELY AUDREY?

In Exercises 10 and 11 we present two methods for sending signed messages using the RSA cipher system, avoiding possible changes in block sizes.

* **10.** Let H be a fixed integer. Let each individual have two pairs of enciphering keys: $k = (e,n)$ and $k* = (e,n*)$ with $n < H < n*$, where n and $n*$ are both the product of two primes. Using the RSA cipher system, individual i can send a signed message P to individual j by sending $E_{k_j^*}(D_{k_i}(P))$.

a) Show that it is not necessary to change block sizes when the transformation $E_{k_j^*}$ is applied after D_{k_i} has been applied.

b) Explain how individual j can recover the plaintext message P, and why no one other than individual i could have sent the message.

c) Let individual i have enciphering keys $(3,11 \cdot 71)$ and $(3,29 \cdot 41)$ so that $781 = 11 \cdot 71 < 1000 < 1189 = 29 \cdot 41$, and let individual j have enciphering keys $(7,19 \cdot 47)$ and $(7,31 \cdot 37)$, so that $893 = 19 \cdot 47 < 1000 < 1147 = 31 \cdot 37$. What ciphertext message does individual i send to individual j using the method given in this problem when the signed plaintext message is HELLO ADAM? What ciphertext message does individual j send to individual i when the signed plaintext message is GOODBYE ALICE?

* **11.** a) Show that if individuals i and j have enciphering keys $k_i = (e_i,n_i)$ and $k_j = (e_j,n_j)$, respectively, where both n_i and n_j are products of two distinct primes, then individual i can send a signed message P to individual j without needing to change the size of blocks by sending

$$E_{k_j}(D_{k_i}(P)) \text{ if } n_i < n_j$$
$$D_{k_i}(E_{k_j}(P)) \text{ if } n_i > n_j \ .$$

b) How can individual j recover P?

c) How can individual j guarantee that a message came from individual i?

d) Let $k_i = (11,47 \cdot 61)$ and $k_j = (13,43 \cdot 59)$. Using the method described in part (a), what does individual i send to individual j if the message is REGARDS FRED, and what does individual j send to individual i if the message is REGARDS ZELDA?

12. Encipher the message SELL NOW using the Rabin cipher $C \equiv P(P + 5) \pmod{2573}$.

13. Encipher the message LEAVE TOWN using the Rabin cipher $C \equiv P(P + 11) \pmod{3901}$.

7.4 Computer Projects

Programming Projects

Write programs to do the following:

1. Encipher messages with an RSA cipher.

2. Decipher messages that were enciphered using an RSA cipher.

3. Send signed messages using an RSA cipher and the method described in the text.

4. Send signed messages using an RSA cipher and the method in Exercise 10.

5. Send signal messages using an RSA cipher and the method in Exercise 11.

6. Encipher messages using a Rabin cipher.

Computations and Explorations

Using the programs you have written or a computation program, carry out the following computations and explorations:

1. Construct a key for the RSA cipher for inclusion in a directory of enciphering keys for the members of your class.

2. For each member of your class encipher a message using the RSA cipher using the public keys published in the directory.

3. Decipher the messages sent to you by your classmates that were enciphered using your RSA enciphering key.

7.5 Knapsack Ciphers

In this section, we discuss cipher systems based on the knapsack problem. Given a set of positive integers $a_1, a_2, ..., a_n$ and an integer S, the *knapsack problem* asks which of these integers, if any, add together to give S. Another way to phrase the knapsack problem is to ask for values of $x_1, x_2, ..., x_n$, each either 0 or 1, such that

$$(7.3) \qquad S = a_1x_1 + a_2x_2 + \cdots + a_nx_n.$$

We use an example to illustrate the knapsack problem.

Example 7.3. Let $(a_1, a_2, a_3, a_4, a_5) = (2, 7, 8, 11, 12)$. By inspection, we see that there are two subsets of these five integers that add together to give 21, namely $21 = 2 + 8 + 11 = 2 + 7 + 12$. Equivalently, there are exactly two solutions to the equation $2x_1 + 7x_2 + 8x_3 + 11x_4 + 12x_5 = 21$, with $x_i = 0$ or 1 for $i = 1, 2, 3, 4, 5$. These solutions are $x_1 = x_3 = x_4 = 1$, $x_2 = x_5 = 0$, and $x_1 = x_2 = x_5 = 1, x_3 = x_4 = 0$. $\qquad \square$

To verify that equation (7.3) holds, where each x_i is either 0 or 1, requires that we perform at most n additions. On the other hand, to search by trial and error for solutions of (7.3) may require that we check all 2^n possibilities for $(x_1, x_2, ..., x_n)$. The best method known for finding a solution of the knapsack problem requires

$O(2^{n/2})$ bit operations, which makes a computer solution of a general knapsack problem extremely infeasible even when $n = 100$.

Certain values of the integers $a_1, a_2, ..., a_n$ make the solution of the knapsack problem much easier than the solution in the general case. For instance, if $a_j = 2^{j-1}$, to solve $S = a_1 x_1 + a_2 x_2 + \cdots + a_n x_n$, where $x_i = 0$ or 1 for $i = 1, 2, ..., n$, simply requires that we find the binary expansion of S. We can also produce easy knapsack problems by choosing the integers $a_1, a_2, ..., a_n$ so that the sum of the first $j-1$ of these integers is always less than the jth integer, *i.e.*, so that

$$\sum_{i=1}^{j-1} a_i < a_j, \qquad j = 2, 3, ..., n.$$

If a sequence of integers $a_1, a_2, ..., a_n$ satisfies this inequality, we call the sequence *super-increasing*.

Example 7.4. The sequence 2, 3, 7, 14, 27 is super-increasing because $3 > 2$, $7 > 3+2$, $14 > 7+3+2$, and $27 > 14+7+3+2$. □

To see that knapsack problems involving super-increasing sequences are easy to solve, we first consider an example.

Example 7.5. Let us find the integers from the set 2, 3, 7, 14, 27 that have 37 as their sum. First, we note that since $2 + 3 + 7 + 14 < 27$, a sum of integers from this set can only be greater than 27 if the sum contains the integer 27. Hence, if $2x_1 + 3x_2 + 7x_3 + 14x_4 + 27x_5 = 37$ with each $x_i = 0$ or 1, we must have $x_5 = 1$ and $2x_1 + 3x_2 + 7x_3 + 14x_4 = 10$. Since $14 > 10$, x_4 must be 0 and we have $2x_1 + 3x_2 + 7x_3 = 10$. Since $2 + 3 < 7$, we must have $x_3 = 1$ and therefore $2x_1 + 3x_2 = 3$. Obviously, we have $x_2 = 1$ and $x_1 = 0$. The solution is $37 = 3 + 7 + 27$. □

In general, to solve knapsack problems for a super-increasing sequence a_1, $a_2, ..., a_n$, *i.e.*, to find the values of $x_1, x_2, ..., x_n$ with $S = a_1 x_1 + a_2 x_2 + \cdots + a_n x_n$ and $x_i = 0$ or 1 for $i = 1, 2, ..., n$ when S is given, we use the following algorithm. First, we find x_n by noting that

$$x_n = \begin{cases} 1 & \text{if } S \geq a_n \\ 0 & \text{if } S < a_n. \end{cases}$$

Then, we find $x_{n-1}, x_{n-2}, ..., x_1$, in succession, using the equations

$$x_j = \begin{cases} 1 \text{ if } S - \sum_{i=j+1}^{n} x_i a_i \geq a_j \\ \\ 0 \text{ if } S - \sum_{i=j+1}^{n} x_i a_i < a_j, \end{cases}$$

for $j = n-1, n-2, ..., 1$.

To see that this algorithm works, first note that if $x_n = 0$ when $S \geq a_n$, then $\sum_{i=1}^{n} a_i x_i \leq \sum_{i=1}^{n-1} a_i < a_n \leq S$, contradicting the condition $\sum_{j=1}^{n} a_j x_j = S$. Similarly, if $x_j = 0$ when $S - \sum_{i=j+1}^{n} x_i a_i \geq a_j$, then $\sum_{i=1}^{n} a_i x_i \leq \sum_{i=1}^{j-1} x_i + \sum_{i=j+1}^{n} x_i a_i < a_j + \sum_{i=j+1}^{n} x_i a_i \leq S$, which is again a contradiction.

Using this algorithm, knapsack problems based on super-increasing sequences can be solved extremely quickly. We now discuss a cipher system based on this observation. This cipher system was invented by Merkle and Hellman [147], and was considered a good choice for a public-key cipher system until recently. We will comment more about this later.

The ciphers that we describe here are based on transformed super-increasing sequences. To be specific, let $a_1, a_2, ..., a_n$ be super-increasing and let m be a positive integer with $m > 2a_n$. Let w be an integer relatively prime to m with inverse $\overline{w}$ modulo m. We form the sequence $b_1, b_2, ..., b_n$ where $b_j \equiv w a_j \pmod{m}$ and $0 < b_j < m$. We cannot use this special technique to solve a knapsack problem of the type $S = \sum_{i=1}^{n} b_i x_i$ where S is a positive integer, since the sequence $b_1, b_2, ..., b_n$ is not super-increasing. However, when $\overline{w}$ is known, we can find

(7.4) $$\overline{w}S = \sum_{i=1}^{n} \overline{w} b_i x_i \equiv \sum_{i=1}^{n} a_i x_i \pmod{m},$$

since $\overline{w} b_j \equiv a_j \pmod{m}$. From (7.4) we see that

$$S_0 = \sum_{i=1}^{n} a_i x_i$$

where S_0 is the least positive residue of $\overline{w}S$ modulo m. We can easily solve the equation

$$S_0 = \sum_{i=1}^{n} a_i x_i,$$

since $a_1, a_2, ..., a_n$ is super-increasing. This solves the knapsack problem

$$S = \sum_{i=1}^{n} b_i x_i,$$

since $b_j \equiv wa_j \pmod{m}$ and $0 \le b_j < m$. We illustrate this procedure with an example.

Example 7.6. The super-increasing sequence $(a_1, a_2, a_3, a_4, a_5) = (3,5,9,20,44)$ can be transformed into the sequence $(b_1, b_2, b_3, b_4, b_5) = (23,68,69,5,11)$ by taking $b_j \equiv 67a_j \pmod{89}$, for $j = 1,2,3,4,5$. To solve the knapsack problem $23x_1 + 68x_2 + 69x_3 + 5x_4 + 11x_5 = 84$, we can multiply both sides of this equation by 4, an inverse of 67 modulo 89, and then reduce modulo 89, to obtain the congruence $3x_1 + 5x_2 + 9x_3 + 20x_4 + 44x_5 \equiv 336 \equiv 69 \pmod{89}$. Since $89 > 3 + 5 + 9 + 20 + 44$, we can conclude that $3x_1 + 5x_2 + 9x_3 + 20x_4 + 44x_5 = 69$. The solution of this easy knapsack problem is $x_5 = x_4 = x_2 = 1$ and $x_3 = x_1 = 0$. Hence, the original knapsack problem has as its solution $68 + 5 + 11 = 84$.

The cipher system based on the knapsack problem works as follows. Each individual chooses a super-increasing sequence of positive integers of a specified length, say N, e.g. $a_1, a_2,..., a_N$, as well as a modulus m with $m > 2a_N$ and a multiplier w with $(m,w) = 1$. The transformed sequence $b_1, b_2,..., b_N$, where $b_j \equiv wa_j \pmod{m}$, $0 < b_j < m$, for $j = 1,2,...,N$, is made public. When someone wishes to send a message P to this individual, the message is first translated into a string of zeros and ones using the binary equivalents of letters, as shown in Table 7.10. This string of zeros and ones is next split into segments of length N (for simplicity we suppose that the length of the string is divisible by N; if not, we can simply fill out the last block with all ones). For each block, a sum is computed using the sequence $b_1, b_2,..., b_N$; for instance, the block $x_1 x_2 ... x_N$ gives $S = b_1 x_1 + b_2 x_2 + \cdots + b_N x_N$. Finally, the sums generated by each block form the ciphertext message.

We note that to decipher ciphertext generated by the knapsack cipher, without knowledge of m and w, requires that a group of hard knapsack problems of the form

(7.5) $$S = b_1 x_1 + b_2 x_2 + \cdots + b_N x_N$$

be solved. On the other hand, when m and w are known, the knapsack problem (7.5) can be transformed into an easy knapsack problem, since

$$\overline{w}S = \overline{w}b_1 x_1 + \overline{w}b_2 x_2 + \cdots + \overline{w}b_N x_N$$
$$\equiv a_1 x_1 + a_2 x_2 + \cdots + a_N x_N \pmod{m},$$

where $\overline{w}b_j \equiv a_j \pmod{m}$, where $\overline{w}$ is an inverse of w modulo m, so that

(7.6) $$S_0 = a_1 x_1 + a_2 x_2 + \cdots + a_N x_N,$$

where S_0 is the least positive residue of $\overline{w}S$ modulo m. We have equality in (7.6), since both sides of the equation are positive integers less than m which are congruent modulo m.

letter	binary equivalent	letter	binary equivalent
A	00000	N	01101
B	00001	O	01110
C	00010	P	01111
D	00011	Q	10000
E	00100	R	10001
F	00101	S	10010
G	00110	T	10011
H	00111	U	10100
I	01000	V	10101
J	01001	W	10110
K	01010	X	10111
L	01011	Y	11000
M	01100	Z	11001

Table 7.10. The Binary Equivalents of Letters.

We illustrate the enciphering and deciphering procedures of the knapsack cipher with an example. We start with the super-increasing sequence $(a_1,a_2,a_3,a_4,a_5,a_6,a_7,a_8,a_9,a_{10}) = (2,11,14,29,58,119,241,480,959,1917)$. We take $m = 3837$ as the enciphering modulus, so that $m > 2a_{10}$, and $w = 1001$ as the multiplier, so that $(m,w) = 1$, to transform the super-increasing sequence into the sequence $(2002,3337,2503,2170,503,172,3347,855,709,417)$.

To encipher the message

REPLY IMMEDIATELY,

we first translate the letters of the message into their five digit binary equivalents, as shown in Table 7.10, and then group these digits into blocks of ten, to obtain

1000100100   0111101011   1100001000
0110001100   0010000011   0100000000
1001100100   0101111000.

For each block of ten binary digits, we form a sum by adding together the appropriate terms of the sequence (2002, 3337, 2503, 2170, 503, 172, 3347, 855, 709, 417) in the slots corresponding to positions of the block containing a digit equal to 1. This gives us

3360   12986   8686   10042   3629   3337   5530   9529.

For instance, we compute the first sum, 3360, by adding 2002, 503, and 855.

To decipher, we find the least positive residue modulo 3837 of 23 times each sum, since 23 is an inverse of 1001 modulo 3837, and then we solve the corresponding easy knapsack problem with respect to the original super-increasing sequence $(2,11,14,29,58,119,241,480,959,1917)$. For example, to decipher the first block, we find that $3360 \cdot 23 \equiv 540 \pmod{3837}$, and then note that $540 = 480 + 58 + 2$. This tells us that the first block of plaintext binary digits is 1000100100.

Knapsack ciphers originally seemed to be excellent candidates for use in public key cryptosystems. However, Shamir [151] has shown that they are not satisfactory for public-key cryptography. The reason is that there is an efficient algorithm for solving knapsack problems involving sequences $b_1, b_2,..., b_n$ with $b_j \equiv wa_j \pmod{m}$, where w and m are relatively prime positive integers and $a_1, a_2,..., a_n$ is a super-increasing sequence. The algorithm found by Shamir can solve these knapsack problems using only $O(P(n))$ bit operations, where P is a polynomial, instead of requiring exponential time, as is required for known algorithms for general knapsack problems, involving sequences of a general nature. Although we will not go into details of the algorithm found by Shamir here, the reader can find these details by consulting the paper by Odlyzko in [84].

There are several possibilities for altering this cipher system to avoid the weakness found by Shamir. One such possibility is to choose a sequence of pairs of relatively prime integers (w_1,m_1), (w_2,m_2) ,..., (w_r,m_r), and then form the series of sequences

$$b_j^{(1)} \equiv w_1 a_j \pmod{m_1}$$
$$b_j^{(2)} \equiv w_2 b_j^{(1)} \pmod{m_2}$$
$$\cdot$$
$$\cdot$$
$$\cdot$$
$$b_j^{(r)} \equiv w_r b_j^{(r-1)} \pmod{m_r},$$

for $j = 1,2,...,n$. We then use the final sequence $b_1^{(r)}, b_2^{(r)},..., b_n^{(r)}$ as the enciphering sequence. Unfortunately, efficient algorithms have been found for solving knapsack problems involving sequences obtained by iterating modular multiplications with different moduli.

A comprehensive discussion of knapsack ciphers can be found in the article "The Rise and Fall of Knapsack Cryptosystems" by Odlyzko in [84]. This article describes knapsack ciphers and their generalizations and goes on to explain the attacks that have been found for breaking them. Moreover, it includes a description of a variant of the knapsack cipher, the Chor-Rivest knapsack, that has not yet been fully broken, even though various modes of attack are known. It will be interesting to see whether this variant yields to attack in the next few years.

7.5 Exercises

1. Decide whether each of the following sequences is super-increasing.

    a) (3,5,9,19,40)   c) (3,7,17,30,59)
    b) (2,6,10,15,36)   d) (11,21,41,81,151)

2. Show that if $a_1, a_2, ..., a_n$ is a super-increasing sequence, then $a_j \geq 2^{j-1}$ for $j = 1, 2, ..., n$.

3. Show that the sequence $a_1, a_2, ..., a_n$ is super-increasing if $a_{j+1} > 2a_j$ for $j = 1, 2, ..., n - 1$.

4. Find all subsets of the integers $2, 3, 4, 7, 11, 13, 16$ that have 18 as their sum.

5. Find the sequence obtained from the super-increasing sequence $(1, 3, 5, 10, 20, 41, 80)$ when modular multiplication is applied with multiplier $w = 17$ and modulus $m = 162$.

6. Encipher the message BUY NOW using the knapsack cipher based on the sequence obtained from the super-increasing sequence $(17, 19, 37, 81, 160)$, by performing modular multiplication with multiplier $w = 29$ and modulus $m = 331$.

7. Decipher the ciphertext 402 105 150 325 that was enciphered by the knapsack cipher based on the sequence $(306, 374, 233, 19, 259)$. This sequence is obtained by using modular multiplication with multiplier $w = 17$ and modulus $m = 464$, to transform the super-increasing sequence $(18, 22, 41, 83, 179)$.

8. Find the sequence obtained by applying successively the modular multiplications with multipliers and moduli $(7, 92)$, $(11, 95)$, and $(6, 101)$, respectively, on the super-increasing sequence $(3, 4, 8, 17, 33, 67)$.

9. What process can be employed to decipher messages that have been enciphered using knapsack ciphers that involve sequences arising from iterating modular multiplications with different moduli?

A *multiplicative knapsack problem* is a problem of the following type: Given positive integers $a_1, a_2, ..., a_n$ and a positive integer P, find the subset, or subsets, of these integers with product P, or equivalently, find all solutions of

$$P = a_1^{x_1} a_2^{x_2} \cdots a_n^{x_n}$$

where $x_j = 0$ or 1 for $j = 1, 2, ..., n$.

10. Find all products of subsets of the integers $2, 3, 5, 6, 10$ equal to 60.

11. Find all products of subsets of the integers $8, 13, 17, 21, 95, 121$ equal to 15960.

12. Show that if the integers $a_1, a_2, ..., a_n$ are pairwise relatively prime, then the multiplicative knapsack problem $P = a_1^{x_1} a_2^{x_2} \cdots a_n^{x_n}$, $x_j = 0$ or 1 for $j = 1, 2, ..., n$, is easily solved from the prime factorizations of the integers $P, a_1, a_2, ..., a_n$, and show that if there is a solution, then it is unique.

13. Show that by taking logarithms to the base b modulo m, where $(b, m) = 1$ and $0 < b < m$, the multiplicative knapsack problem

$$P = a_1^{x_1} a_2^{x_2} \cdots a_n^{x_n}$$

is converted into an additive knapsack problem

$$S = \alpha_1 x_1 + \alpha_2 x_2 + \cdots + \alpha_n x_n$$

where S, α_1, α_2,..., α_n are the logarithms of P, a_1, a_2,..., a_n to the base b modulo m, respectively.

14. Explain how Exercises 12 and 13 can be used to produce ciphers where messages are easily deciphered when the mutually relatively prime integers a_1, a_2,..., a_n are known, but cannot be deciphered quickly when the integers α_1, α_2,..., α_n are known.

7.5 Computer Projects

Programming Projects

Write programs to do the following:

1. Solve knapsack problems by trial and error.

2. Solve knapsack problems involving super-increasing sequences.

3. Encipher messages using knapsack ciphers.

4. Decipher messages that were enciphered using knapsack ciphers.

5. Encipher and decipher messages using knapsack ciphers involving sequences arising from iterating modular multiplications with different moduli.

6. Solve multiplicative knapsack problems involving sequences of mutually relatively prime integers (see Exercise 14).

Computations and Explorations

Using the programs you have written or a computation program, carry out the following computation and explorations:

1. Starting with a super-increasing sequence that you have contructed, perform modular multiplication with modulus m and multiplier w to find a sequence to serve as your public key for the knapsack cipher.

2. For each of your classmates encipher a message using their public key for the knapsack ciphers.

3. Decipher the messages that were sent to you by classmates.

** 4. Using algorithms described in the article by Odlyzko [84] solve knapsack problems based on a sequence obtained by modular multiplication of a super-increasing sequence.

7.6 Some Applications to Computer Science

In this section we describe two applications of cryptography to computer science. The Chinese remainder theorem is used in both applications.

The first application involves the enciphering of a database. A *database* is a collection of computer files or records. Here we will show how to encipher an entire database so that individual files may be deciphered without jeopardizing the security of other files in the database.

Suppose that a database B contains the n files F_1, F_2, ..., F_n. Since each file is a string of 0's and 1's, we can consider each file to be a binary integer. We first choose n distinct primes m_1, m_2, ..., m_n with $m_j > F_j$ for $j = 1, 2, ..., n$. As the ciphertext we use an integer C that is congruent to F_j modulo m_j for $j = 1, 2, ..., n$; the existence of such an integer is guaranteed by the Chinese remainder theorem. We let $M = m_1 m_2 \cdots m_n$ and $M_j = M/m_j$ for $j = 1, 2, ..., n$. Furthermore, let $e_j = M_j \cdot y_j$ where y_j is an inverse of M_j modulo m_j. For the ciphertext, we take the integer C with

$$C \equiv \sum_{j=1}^{n} e_j F_j \pmod{M}, \quad 0 \le C < M.$$

The integers e_1, e_2, ..., e_n serve as the *write subkeys* of the cipher.

To retrieve the jth file F_j from the ciphertext C, we simply note that

$$F_j \equiv C \pmod{m_j}, \quad 0 \le F_j < m_j.$$

We call the moduli m_1, m_2, ..., m_n the *read subkeys* of the cipher. Note that knowledge of m_j permits access only to file j; for access to the other files, it is necessary to know the moduli other than m_j.

We illustrate the enciphering and deciphering procedures for databases with the following example.

Example 7.7. Suppose our database contains four files F_1, F_2, F_3, and F_4, represented by the binary integers $(0111)_2$, $(1001)_2$, $(1100)_2$, and $(1111)_2$, or in decimal notation $F_1 = 7$, $F_2 = 9$, $F_3 = 12$ and $F_4 = 15$. We pick four primes, $m_1 = 11$, $m_2 = 13$, $m_3 = 17$, and $m_4 = 19$, greater than the corresponding integers representing the files. To encipher this database, we use the Chinese remainder theorem to find the ciphertext C which is the positive integer with $C \equiv 7 \pmod{11}$, $C \equiv 9 \pmod{13}$, $C \equiv 12 \pmod{17}$, and $C \equiv 15 \pmod{19}$, less than $M = 11 \cdot 13 \cdot 17 \cdot 19 = 46189$. To compute C we first find $M_1 = 13 \cdot 17 \cdot 19 = 4199$, $M_2 = 11 \cdot 17 \cdot 19 = 3553$, $M_3 = 11 \cdot 13 \cdot 19 = 2717$, and $M_4 = 11 \cdot 13 \cdot 17 = 2431$. We easily find that $y_1 = 7$, $y_2 = 10$, $y_3 = 11$ and $y_4 = 18$ are inverses of M_j modulo m_j for $j = 1, 2, 3, 4$. Hence, the write subkeys are $e_1 = 4199 \cdot 7 = 29393$, $e_2 = 3553 \cdot 10 = 35530$, $e_3 = 2717 \cdot 11 = 29887$, and $e_4 = 2431 \cdot 18 = 43758$. To construct the ciphertext, we note that

$$C \equiv e_1 F_1 + e_2 F_2 + e_3 F_3 + e_4 F_4$$
$$\equiv 29393 \cdot 7 + 35530 \cdot 9 + 29887 \cdot 12 + 43758 \cdot 15$$
$$\equiv 1540535$$
$$\equiv 16298 \pmod{46189},$$

so that $C = 16298$. The read subkeys are the integers m_j, $j = 1,2,3,4$. To recover the file F_j from C, we simply find the least positive residue of C modulo m_j. For instance, we find F_1 by noting that

$$F_1 \equiv 16298 \equiv 7 \pmod{11}. \qquad \square$$

We now discuss another application of cryptography, namely a method for sharing secrets. Suppose that in a communications network, there is some vital, but extremely sensitive information. If this information is distributed to several individuals, it becomes much more vulnerable to exposure; on the other hand, if this information is lost, there are serious consequences. An example of such information is the *master key K* used for access to the password file in a computer system.

In order to protect this master key K from both loss and exposure, we construct *shadows* k_1, k_2, ..., k_r which are given to r different individuals. We will show that the key K can be produced easily from any s of these shadows, where s is a positive integer less than r, whereas the knowledge of less than s of these shadows does not permit the key K to be found. Because at least s different individuals are needed to find K, the key is not vulnerable to exposure. In addition, the key K is not vulnerable to loss, since any s individuals from the r individuals with shadows can produce K. Schemes with the properties we have just described are called *(s,r) threshold schemes.*

To develop a system that can be used to generate shadows with these properties, we use the Chinese remainder theorem. We choose a prime p greater than the key K and a sequence of pairwise relatively prime integers m_1, m_2, ..., m_r that are not divisible by p, such that

$$m_1 < m_2 < \cdots < m_r,$$

and

(7.7) $$m_1 m_2 \cdots m_s > p m_r m_{r-1} \cdots m_{r-s+2}.$$

Note that the inequality (7.7) states that the product of the s smallest of the integers m_j is greater than the product of p and the $s-1$ largest of the integers m_j. From (7.7), we see that if $M = m_1 m_2 \cdots m_s$, then M/p is greater than the product of any set of $s-1$ of the integers m_j.

Now let t be a nonnegative integer less than M/p that is chosen at random. Let

$$K_0 = K + tp,$$

so that $0 \le K_0 \le M-1$ (since $0 \le K_0 = K + tp < p + tp = (t+1)p \le (M/p)p = M$).

To produce the shadows k_1, k_2, ..., k_r, we let k_j be the integer with

$$k_j \equiv K_0 \pmod{m_j}, \ 0 \leq k_j < m_j,$$

for $j = 1,2,...,r$. To see that the master key K can be found by any s individuals possessing shadows, from the total of r individuals with shadows, suppose that the s shadows $k_{j_1}, k_{j_2}, ..., k_{j_s}$ are available. Using the Chinese remainder theorem, we can easily find the least positive residue of K_0 modulo M_i where $M_j = m_{j_1} m_{j_2} \cdots m_{j_s}$. Since we know that $0 \leq K_0 < M \leq M_j$, we can determine K_0, and then find $K = K_0 - tp$.

On the other hand, suppose that we know only the $s-1$ shadows $k_{i_1}, k_{i_2}, ..., k_{i_{s-1}}$. By the Chinese remainder theorem, we can determine the least positive residue a of K_0 modulo M_i where $M_i = m_{i_1} m_{i_2} \cdots m_{i_{s-1}}$. With these shadows, the only information we have about K_0 is that a is the least positive residue of K_0 modulo M_j and $0 \leq K_0 < M$. Consequently, we only know that

$$K_0 = a + xM_i,$$

where $0 \leq x < M/M_i$. From (7.7), we can conclude that $M/M_i > p$, so that as x ranges through the positive integers less than M/M_i, x takes every value in a full set of residues modulo p. Since $(m_j,p) = 1$ for $j = 1, 2, ..., s$, we know that $(M_i,p) = 1$, and consequently, $a + xM_i$ runs through a full set of residues modulo p as x does. Hence, we see that the knowledge of $s-1$ shadows is insufficient to determine K_0, as K_0 could be in any of the p congruence classes modulo p.

We use an example to illustrate this threshold scheme.

Example 7.8. Let $K = 4$ be the master key. We will use a $(2,3)$ threshold scheme of the kind just described with $p = 7$, $m_1 = 11$, $m_2 = 12$, and $m_3 = 17$, so that $M = m_1 m_2 = 132 > pm_3 = 119$. We pick $t = 14$ randomly from among the positive integers less than $M/p = 132/7$. This gives us

$$K_0 = K + tp = 4 + 14 \cdot 7 = 102.$$

The three shadows k_1, k_2, and k_3 are the least positive residues of K_0 modulo m_1, m_2, and m_3, i.e.

$$k_1 \equiv 102 \equiv 3 \pmod{11}$$
$$k_2 \equiv 102 \equiv 6 \pmod{12}$$
$$k_3 \equiv 102 \equiv 0 \pmod{17},$$

so that the three shadows are $k_1 = 3$, $k_2 = 6$, and $k_3 = 0$.

We can recover the master key K from any two of the three shadows. Suppose we know that $k_1 = 3$ and $k_3 = 0$. Using the Chinese remainder theorem, we can determine K_0 modulo $m_1 m_3 = 11 \cdot 17 = 187$, i.e. since $K_0 \equiv 3 \pmod{11}$ and $K_0 \equiv 0 \pmod{17}$ we have $k_0 \equiv 102 \pmod{187}$. Since $0 \leq K_0 < M = 132 < 187$, we know that $K_0 = 102$, and consequently the master key is $K = K_0 - tp = 102 - 14 \cdot 7 = 4$. □

We will develop another threshold scheme in Exercise 12 of Section 8.2. The interested reader should also consult Denning [73] for related topics in cryptography.

7.6 Exercises

1. Suppose that the database B contains four files, $F_1 = 4$, $F_2 = 6$, $F_3 = 10$, and $F_4 = 13$. Let $m_1 = 5$, $m_2 = 7$, $m_3 = 11$, and $m_4 = 16$ be the read subkeys of the cipher used to encipher the database.

   a) What are the write subkeys of the cipher?

   b) What is the ciphertext C corresponding to the database?

2. When the database B with three files F_1, F_2, and F_3 is enciphered using the method described in the text, with read subkeys $m_1 = 14$, $m_2 = 15$, and $m_3 = 19$, the corresponding ciphertext is $C = 619$. If file F_3 is changed from $F_3 = 11$ to $F_3 = 12$, what is the updated value of the ciphertext C?

3. Decompose the master key $K = 3$ into three shadows using a $(2,3)$ threshold scheme of the type described in the text with $p = 5$, $m_1 = 8$, $m_2 = 9$, $m_3 = 11$ and with $t = 13$.

4. Show how to recover the master key K from each of the three pairs of shadows found in Exerise 3.

7.6 Computer Projects

Programming Projects

Write programs to do the following:

1. Using the system described in the text, encipher databases and recover files from the ciphertext version of databases.

2. Update files in the ciphertext version of databases (see Exercise 2).

3. Find the shadows in a threshold scheme of the type described in the text.

4. Recover the master key from a set of shadows.

Computations and Explorations

Using the programs you have written or a computation program, carry out the following computations and explorations:

1. Construct a $(6,4)$-threshold scheme that decomposes a master key into six shadows. Distribute these shadows to six members of your class and then select three different groups of four of these six people, reconstructing the key from the four shadows of the people in each group.

8

Primitive Roots

8.1 The Order of an Integer and Primitive Roots

In this section we begin our study of the least positive residues modulo n of powers of an integer a relatively prime to n, where n is a positive integer greater than 1. We will start by studying the *order* of a modulo n, the exponent of the least power of a congruent to one modulo n. Then we will study integers a such that the least positive residues of these powers run through all positive integers less than m that are relatively prime to m. Such integers, when they exist, are called *primitive roots* of m. One of our major goals in this chapter will be to determine which positive integers have primitive roots.

The Order of an Integer

By Euler's theorem if m is a positive integer and if a is an integer relatively prime to m, then $a^{\phi(m)} \equiv 1 \pmod{m}$. Therefore, at least one positive integer x satisfies the congruence $a^x \equiv 1 \pmod{m}$. Consequently, by the well-ordering property, there is a least positive integer x satisfying this congruence.

Definition. Let a and m be relatively prime positive integers. Then, the least positive integer x such that $a^x \equiv 1 \pmod{m}$ is called the *order of a modulo m*.

We denote the order of a modulo m by $\operatorname{ord}_m a$.

Example 8.1. To find the order of 2 modulo 7, we compute the least positive residues modulo 7 of powers of 2. We find that

$$2^1 \equiv 2 \pmod{7}, \ 2^2 \equiv 4 \pmod{7}, \ 2^3 \equiv 1 \pmod{7}.$$

Therefore, $\operatorname{ord}_7 2 = 3$.

Similarly, to find the order of 3 modulo 7 we compute

$$3^1 \equiv 3 \ (\text{mod } 7), \ 3^2 \equiv 2 \ (\text{mod } 7), \ 3^3 \equiv 6 \ (\text{mod } 7)$$
$$3^4 \equiv 4 \ (\text{mod } 7), \ 3^5 \equiv 5 \ (\text{mod } 7), \ 3^6 \equiv 1 \ (\text{mod } 7).$$

$\square$

We see that $\text{ord}_7 3 = 6$.

In order to find all solutions of the congruence $a^x \equiv 1 \ (\text{mod } m)$, we need the following theorem.

Theorem 8.1. If a and n are relatively prime integers with $n > 0$, then the positive integer x is a solution of the congruence $a^x \equiv 1 \ (\text{mod } n)$ if and only if $\text{ord}_n a \mid x$.

Proof. If $\text{ord}_n a \mid x$, then $x = k \cdot \text{ord}_n a$ where k is a positive integer. Hence,

$$a^x = a^{k \cdot \text{ord}_n a} = (a^{\text{ord}_n a})^k \equiv 1 \ (\text{mod } n).$$

Conversely, if $a^x \equiv 1 \ (\text{mod } n)$, we first use the division algorithm to write

$$x = q \cdot \text{ord}_n a + r, \quad 0 \le r < \text{ord}_n a.$$

From this equation, we see that

$$a^x = a^{q \cdot \text{ord}_n a + r} = (a^{\text{ord}_n a})^q a^r \equiv a^r \ (\text{mod } n).$$

Since $a^x \equiv 1 \ (\text{mod } n)$, we know that $a^r \equiv 1 \ (\text{mod } n)$. From the inequality $0 \le r < \text{ord}_n a$, we conclude that $r = 0$, since, by definition, $y = \text{ord}_n a$ is the least positive integer such that $a^y \equiv 1 \ (\text{mod } n)$. Because $r = 0$, we have $x = q \cdot \text{ord}_n a$. Therefore, $\text{ord}_n a \mid x$. ■

Example 8.2. We can use Theorem 8.1 and Example 8.1 to determine whether $x = 10$ and $x = 15$ are solutions of $2^x \equiv 1 \ (\text{mod } 7)$. By Example 8.1 we know that $\text{ord}_7 2 = 3$. Since 3 does not divide 10, but 3 divides 15, by Theorem 8.1 we see that $x = 10$ is not a solution of $2^x \equiv 1 \ (\text{mod } 7)$, but $x = 15$ is a solution of this congruence. $\square$

Theorem 8.1 leads to the following corollary.

Corollary 8.1.1. If a and n are relatively prime integers with $n > 0$, then $\text{ord}_n a \mid \phi(n)$.

Proof. Since $(a,n) = 1$, Euler's theorem tells us that

$$a^{\phi(n)} \equiv 1 \ (\text{mod } n).$$

Using Theorem 8.1 we conclude that $\text{ord}_n a \mid \phi(n)$. ■

We can use Corollary 8.1.1 as a shortcut when we compute orders. The following example illustrates the procedure.

Example 8.3. To find the order of 5 modulo 17, we first note that $\phi(17) = 16$. Since the only positive divisors of 16 are 1, 2, 4, 8, and 16, by Corollary 8.1.1 these are the only possible values of $\text{ord}_{17} 5$. Since

$$5^1 \equiv 5 \pmod{17}, \ 5^2 \equiv 8 \pmod{17}, \ 5^4 \equiv 13 \pmod{17},$$
$$5^8 \equiv 16 \pmod{17}, \ 5^{16} \equiv 1 \pmod{17},$$

we conclude that $\operatorname{ord}_{17} 5 = 16$. □

The following theorem will be useful in our subsequent discussions.

Theorem 8.2. If a and n are relatively prime integers with $n > 0$, then $a^i \equiv a^j$ (mod n), where i and j are nonnegative integers, if and only if $i \equiv j$ (mod $\operatorname{ord}_n a$).

Proof. Suppose that $i \equiv j$ (mod $\operatorname{ord}_n a$) and $0 \le j \le i$. Then we have $i = j + k \cdot \operatorname{ord}_n a$, where k is a positive integer. Hence

$$a^i = a^{j+k \cdot \operatorname{ord}_n a} = a^j (a^{\operatorname{ord}_n a})^k \equiv a^j \pmod{n},$$

since $a^{\operatorname{ord}_n a} \equiv 1 \pmod{n}$.

Conversely, assume that $a^i \equiv a^j$ (mod n) with $i \ge j$. Since $(a,n) = 1$, we know that $(a^j, n) = 1$. Hence, using Corollary 3.4.1, the congruence

$$a^i \equiv a^j a^{i-j} \equiv a^j \pmod{n}$$

implies, by cancellation of a^j, that

$$a^{i-j} \equiv 1 \pmod{n}.$$

By Theorem 8.1 it follows that $\operatorname{ord}_n a$ divides $i - j$, or equivalently, $i \equiv j$ (mod $\operatorname{ord}_n a$). ∎

The next example illustrates the use of Theorem 8.2.

Example 8.4. Let $a = 3$ and $n = 14$. By Theorem 8.2 we see that $3^5 \equiv 3^{11}$ (mod 14), but $3^9 \not\equiv 3^{20}$ (mod 14), since $\phi(14) = 6$ and $5 \equiv 11$ (mod 6) but $9 \not\equiv 20$ (mod 6). □

Primitive Roots

Given an integer n, we are interested in integers a with order modulo n equal to $\phi(n)$, the largest possible order modulo n. As we will show later, when such an integer exists, the least positive residues of its powers run through all positive integers relatively prime to n and less than n.

Definition. If r and n are relatively prime integers with $n > 0$ and if $\operatorname{ord}_n r = \phi(n)$, then r is called a *primitive root modulo n*.

Example 8.5. We have previously shown that $\operatorname{ord}_7 3 = 6 = \phi(7)$. Consequently, 3 is a primitive root modulo 7. Likewise, since $\operatorname{ord}_7 5 = 6$, as can easily be verified, 5 is also a primitive root modulo 7. □

Not all integers have primitive roots. For instance, there are no primitive roots modulo 8. To see this, note that the only integers less than 8 and relatively prime to 8 are 1, 3, 5, and 7, and $\text{ord}_8 1 = 1$, while $\text{ord}_8 3 = \text{ord}_8 5 = \text{ord}_8 7 = 2$. Since $\phi(8) = 4$, there are no primitive roots modulo 8. In our subsequent discussions, we will find all integers possessing primitive roots.

To indicate one way in which primitive roots are useful, we give the following theorem.

Theorem 8.3. If r and n are relatively prime positive integers with $n > 0$ and if r is a primitive root modulo n, then the integers

$$r^1, r^2, ..., r^{\phi(n)}$$

form a reduced residue set modulo n.

Proof. To demonstrate that the first $\phi(n)$ powers of the primitive root r form a reduced residue set modulo n, we only need to show that they are all relatively prime to n, and that no two are congruent modulo n.

Since $(r,n) = 1$, it follows from Exercise 14 of Section 2.1 that $(r^k,n) = 1$ for any positive integer k. Hence, these powers are all relatively prime to n. To show that no two of these powers are congruent modulo n, assume that

$$r^i \equiv r^j \pmod{n}.$$

By Theorem 8.2 we see that $i \equiv j \pmod{\phi(n)}$. However, for $1 \le i \le \phi(n)$ and $1 \le j \le \phi(n)$, the congruence $i \equiv j \pmod{\phi(n)}$ implies that $i = j$. Hence, no two of these powers are congruent modulo n. This shows that we do have a reduced residue system modulo n. ■

Example 8.6. We see that 2 is a primitive root modulo 9, since $2^2 \equiv 4$, $2^3 \equiv 8$, and $2^6 \equiv 1 \pmod 9$. By Theorem 8.3 the first $\phi(9) = 6$ powers of 2 form a reduced residue system modulo 9. These are $2^1 \equiv 2 \pmod 9$, $2^2 \equiv 4 \pmod 9$, $2^3 \equiv 8 \pmod 9$, $2^4 \equiv 7 \pmod 9$, $2^5 \equiv 5 \pmod 9$, and $2^6 \equiv 1 \pmod 9$. □

When an integer possesses a primitive root, it usually has many primitive roots. To demonstrate this, we first prove the following theorem.

Theorem 8.4. If $\text{ord}_m a = t$ and if u is a positive integer, then

$$\text{ord}_m(a^u) = t/(t,u).$$

Proof. Let $s = \text{ord}_m(a^u)$, $v = (t,u)$, $t = t_1 v$, and $u = u_1 v$. By Theorem 2.1 we know that $(t_1,u_1) = 1$.

Note that

$$(a^u)^{t_1} = (a^{u_1 v})^{(t/v)} = (a^t)^{u_1} \equiv 1 \pmod{m},$$

since $\text{ord}_m a = t$. Hence Theorem 8.1 tells us that $s \mid t_1$.

On the other hand, since

$$(a^u)^s = a^{us} \equiv 1 \pmod{m},$$

we know that $t \mid us$. Hence, $t_1 v \mid u_1 vs$, and consequently, $t_1 \mid u_1 s$. Since $(t_1, u_1) = 1$, using Lemma 2.3 we see that $t_1 \mid s$.

Now, since $s \mid t_1$ and $t_1 \mid s$, we conclude that $s = t_1 = t/v = t/(t,u)$. This proves the result. ∎

Example 8.7. By Theorem 8.4 we see that $\text{ord}_7 3^4 = 6/(6,4) = 6/2 = 3$ since we showed in Example 8.1 that $\text{ord}_7 3 = 6$. □

The following corollary of Theorem 8.4 tells us which powers of a primitive root are also primitive roots.

Corollary 8.4.1. Let r be a primitive root modulo m where m is an integer, $m > 1$. Then r^u is a primitive root modulo m if and only if $(u, \phi(m)) = 1$.

Proof. By Theorem 8.4 we know that

$$\text{ord}_m r^u = \text{ord}_m r/(u, \text{ord}_m r)$$
$$= \phi(m)/(u, \phi(m)).$$

Consequently, $\text{ord}_m r^u = \phi(m)$, and r^u is a primitive root modulo m, if and only if $(u, \phi(m)) = 1$. ∎

This leads immediately to the following theorem.

Theorem 8.5. If the positive integer m has a primitive root, then it has a total of $\phi(\phi(m))$ incongruent primitive roots.

Proof. Let r be a primitive root modulo m. Then Theorem 8.3 tells us that the integers $r, r^2, ..., r^{\phi(m)}$ form a reduced residue system modulo m. By Corollary 8.4.1 we know that r^u is a primitive root modulo m if and only if $(u, \phi(m)) = 1$. Since there are exactly $\phi(\phi(m))$ such integers u, there are exactly $\phi(\phi(m))$ primitive roots modulo m. ∎

Example 8.8. Let $m = 11$. Note that 2 is a primitive root modulo 11 (see Exercise 3 at the end of this section). Since 11 has a primitive root, by Theorem 8.5 we know that 11 has $\phi(\phi(11)) = 4$ incongruent primitive roots. Since $\phi(11) = 10$, by the proof of Theorem 8.5 we see that we can find these primitive roots by taking the least nonnegative residues of 2^1, 2^3, 2^7, and 2^9, which are 2, 8, 7, and 6, respectively. In other words, the integers $2, 6, 7, 8$ form a complete set of incongruent primitive roots modulo 11. □

8.1 Exercises

1. Determine the following orders.

   a) $\text{ord}_5 2$ c) $\text{ord}_{13} 10$
   b) $\text{ord}_{10} 3$ d) $\text{ord}_{10} 7$

2. Determine the following orders.

   a) $\text{ord}_{11} 3$ c) $\text{ord}_{21} 10$
   b) $\text{ord}_{17} 2$ d) $\text{ord}_{25} 9$

3. Show that

   a)   5 is a primitive root of 6.

   b)   2 is a primitive root of 11.

4. Find a primitive root modulo each of the following integers.

   a) 4   d) 13
   b) 5   e) 14
   c) 10   f) 18

5. Show that the integer 12 has no primitive roots.

6. Show that the integer 20 has no primitive roots.

7. How many incongruent primitive roots does 14 have ? Find a set of this many incongruent primitive roots modulo 14.

8. How many incongruent primitive roots does 13 have? Find a set of this many incongruent primitive roots modulo 13.

9. Show that if $\bar{a}$ is an inverse of a modulo n, then $\text{ord}_n a = \text{ord}_n \bar{a}$.

10. Show that if n is a positive integer and a and b are integers relatively prime to n such that $(\text{ord}_n a, \text{ord}_n b) = 1$, then $\text{ord}_n (ab) = \text{ord}_n a \cdot \text{ord}_n b$.

11. What can be said about $\text{ord}_n (ab)$ if a and b are integers relatively prime to n when $\text{ord}_n a$ and $\text{ord}_n b$ are not necessarily relatively prime?

12. Decide whether it is true that if n is a positive integer and d is a divisor of $\phi(n)$, then there is an integer a with $\text{ord}_n a = d$. Give reasons for your answers.

13. Show that if a is an integer relatively prime to the positive integer m and $\text{ord}_m a = st$, then $\text{ord}_m a^t = s$.

14. Show that if m is a positive integer and a is an integer relatively prime to m such that $\text{ord}_m a = m - 1$, then m is prime.

15. Show that r is a primitive root modulo the odd prime p if and only if

$$r^{(p-1)/q} \not\equiv 1 \pmod{p}$$

for all prime divisors q of $p-1$.

16. Show that if r is a primitive root modulo the positive integer m, then $\bar{r}$ is also a primitive root modulo m, if $\bar{r}$ is an inverse of r modulo m.

17. Show that $\operatorname{ord}_{F_n} 2 \le 2^{n+1}$, where $F_n = 2^{2^n} + 1$ is the nth Fermat number.

* 18. Let p be a prime divisor of the Fermat number $F_n = 2^{2^n} + 1$.

 a) Show that $\operatorname{ord}_p 2 = 2^{n+1}$.

 b) From part (a), conclude that $2^{n+1} \mid (p-1)$, so that p must be of the form $2^{n+1}k + 1$.

19. Let $m = a^n - 1$, where a and n are positive integers. Show that $\operatorname{ord}_m a = n$ and conclude that $n \mid \phi(m)$.

* 20. a) Show that if p and q are distinct odd primes, then pq is a pseudoprime to the base 2 if and only if $\operatorname{ord}_q 2 \mid (p-1)$ and $\operatorname{ord}_p 2 \mid (q-1)$.

 b) Use part (a) to decide which of the following integers are pseudoprimes to the base 2: $13 \cdot 67$, $19 \cdot 73$, $23 \cdot 89$, $29 \cdot 97$.

* 21. Show that if p and q are distinct odd primes, then pq is a pseudoprime to the base 2 if and only if $M_p M_q = (2^p - 1)(2^q - 1)$ is a pseudoprime to the base 2.

There is a method for deciphering messages that were enciphered by an RSA cipher, without knowledge of the deciphering key. This method is based on iteration. Suppose that the public key (e,n) used for enciphering is known, but the deciphering key (d,n) is not. To decipher a ciphertext block C, we form a sequence $C_1, C_2, C_3,\dots$ setting $C_1 \equiv C^e \pmod{n}$, $0 < C_1 < n$, and $C_{j+1} \equiv C_j^e \pmod{n}$, $0 < C_{j+1} < n$ for $j = 1,2,3,\dots$.

22. Show that $C_j \equiv C^{e^j} \pmod{n}$, $0 < C_j < n$.

* 23. Show that there is an index j such that $C_j = C$ and $C_{j-1} = P$, where P is the original plaintext message. Show that this index j is a divisor of $\operatorname{ord}_{\phi(n)} e$.

24. Let $n = 47 \cdot 59$ and $e = 17$. Using iteration, find the plaintext corresponding to the ciphertext 1504.

(*Note:* This iterative method for attacking RSA ciphers is seldom successful in a reasonable amount of time. Moreover, the primes p and q may be chosen so that this attack is almost always futile. See Exercise 19 of Section 8.2.)

8.1 Computer Projects

Programming Projects

Write projects to do the following:

1. Find the order of a modulo m, when a and m are relatively prime positive integers.

   2. Find primitive roots when they exist.

   3. Attempt to decipher RSA ciphers by iteration (see the preamble to Exercise 22).

Computations and Explorations

   Using the programs you have written or a computation program, carry out the following computations and explorations:

   1. Find as many integers you can for which 2 is a primitive root. Do you think that there are infinitely many such integers?

8.2 Primitive Roots for Primes

   In this section and in the one following, our objective is to determine which integers have primitive roots. In this section, we show that every prime has a primitive root. To do this, we first need to study polynomial congruences.

   Let $f(x)$ be a polynomial with integer coefficients. We say that an integer c is a *root of* $f(x)$ *modulo m* if $f(c) \equiv 0 \pmod{m}$. It is easy to see that if c is a root of $f(x)$ modulo m, then every integer congruent to c modulo m is also a root.

Example 8.9. The polynomial $f(x) = x^2 + x + 1$ has exactly two incongruent roots modulo 7, namely $x \equiv 2 \pmod{7}$ and $x \equiv 4 \pmod{7}$.   □

Example 8.10. The polynomial $g(x) = x^2 + 2$ has no roots modulo 5.   □

Example 8.11. Fermat's little theorem tells us that if p is prime, then the polynomial $h(x) = x^{p-1} - 1$ has exactly $p-1$ incongruent roots modulo p, namely $x \equiv 1, 2, 3, ..., p-1 \pmod{p}$.   □

   We will need the following important theorem concerning roots of polynomials modulo p where p is a prime.

Theorem 8.6. Lagrange's Theorem. Let $f(x) = a_n x^n + a_{n-1} x^{n-1} + \cdots + a_1 x + a_0$ be a polynomial of degree n, $n \geq 1$, with integer coefficients and with leading coefficient a_n not divisible by p. Then $f(x)$ has at most n incongruent roots modulo p.

Proof. We use mathematical induction to prove the theorem. When $n = 1$, we have $f(x) = a_1 x + a_0$ with $p \nmid a_1$. A root of $f(x)$ modulo p is a solution of the linear congruence $a_1 x \equiv -a_0 \pmod{p}$. By Theorem 3.10 since $(a_1, p) = 1$, this linear congruence has exactly one solution, so that there is exactly one root modulo p of $f(x)$. Clearly, the theorem is true for $n = 1$.

   Now suppose that the theorem is true for polynomials of degree $n - 1$, and let $f(x)$ be a polynomial of degree n with leading coefficient not divisible by p.

Assume that the polynomial $f(x)$ has $n + 1$ incongruent roots modulo p, say $c_0, c_1, ..., c_n$, so that $f(c_k) \equiv 0 \pmod{p}$ for $k = 0, 1, ..., n$. We have

$$
\begin{aligned}
f(x) - f(c_0) &= a_n(x^n - c_0^n) + a_{n-1}(x^{n-1} - c_0^{n-1}) + \cdots + a_1(x - c_0) \\
&= a_n(x - c_0)(x^{n-1} + x^{n-2}c_0 + \cdots + xc_0^{n-2} + c_0^{n-1}) \\
&\quad + a_{n-1}(x - c_0)(x^{n-2} + x^{n-3}c_0 + \cdots + xc_0^{n-3} + c_0^{n-2}) \\
&\quad + \cdots + a_1(x - c_0) \\
&= (x - c_0)g(x),
\end{aligned}
$$

where $g(x)$ is a polynomial of degree $n - 1$ with leading coefficient a_n. We now show that $c_1, c_2, ..., c_n$ are all roots of $g(x)$ modulo p. Let k be an integer, $1 \le k \le n$. Since $f(c_k) \equiv f(c_0) \equiv 0 \pmod{p}$, we have

$$
f(c_k) - f(c_0) = (c_k - c_0)g(c_k) \equiv 0 \pmod{p}.
$$

It follows that $g(c_k) \equiv 0 \pmod{p}$, since $c_k - c_0 \not\equiv 0 \pmod{p}$. Hence, c_k is a root of $g(x)$ modulo p. This shows that the polynomial $g(x)$, which is of degree $n - 1$ and has a leading coefficient not divisible by p, has n incongruent roots modulo p. This contradicts the induction hypothesis. Hence, $f(x)$ must have no more than n incongruent roots modulo p. The induction argument is complete. ∎

We use Lagrange's theorem to prove the following result.

Theorem 8.7. Let p be prime and let d be a divisor of $p - 1$. Then the polynomial $x^d - 1$ has exactly d incongruent roots modulo p.

Proof. Let $p - 1 = de$. Then

$$
\begin{aligned}
x^{p-1} - 1 &= (x^d - 1)(x^{d(e-1)} + x^{d(e-2)} + \cdots + x^d + 1) \\
&= (x^d - 1)g(x).
\end{aligned}
$$

From Fermat's little theorem, we see that $x^{p-1} - 1$ has $p - 1$ incongruent roots modulo p. Furthermore, any root of $x^{p-1} - 1$ modulo p is either a root of $x^d - 1$ modulo p or a root of $g(x)$ modulo p.

Lagrange's theorem tells us that $g(x)$ has at most $d(e - 1) = p - d - 1$ roots modulo p. Since every root of $x^{p-1} - 1$ modulo p that is not a root of $g(x)$ modulo p must be a root of $x^d - 1$ modulo p, we know that the polynomial $x^d - 1$ has at least $(p - 1) - (p - d - 1) = d$ incongruent roots modulo p. On the other hand, Lagrange's theorem tells us that it has at most d incongruent roots modulo p. Consequently, $x^d - 1$ has precisely d incongruent roots modulo p. ∎

Theorem 8.7 can be used to prove the following result which tells us how many incongruent integers have a given order modulo p.

Theorem 8.8. Let p be a prime and let d be a positive divisor of $p - 1$. Then the number of incongruent integers of order d modulo p is equal to $\phi(d)$.

Proof. For each positive integer d dividing $p - 1$, let $F(d)$ denote the number of positive integers of order d modulo p that are less than p. Since the order modulo p of an integer not divisible by p divides $p - 1$, it follows that

$$p - 1 = \sum_{d \mid p-1} F(d).$$

By Theorem 6.6 we know that

$$p - 1 = \sum_{d \mid p-1} \phi(d).$$

We will show that $F(d) \leq \phi(d)$ when $d \mid (p-1)$. This inequality, together with the equality

$$\sum_{d \mid p-1} F(d) = \sum_{d \mid p-1} \phi(d),$$

implies that $F(d) = \phi(d)$ for each positive divisor d of $p - 1$.

Let $d \mid (p-1)$. If $F(d) = 0$, it is clear that $F(d) \leq \phi(d)$. Otherwise, there is an integer a of order d modulo p. Since $\mathrm{ord}_p a = d$, the integers

$$a, a^2, \ldots, a^d$$

are incongruent modulo p. Furthermore, each of these powers of a is a root of $x^d - 1$ modulo p, since $(a^k)^d \equiv (a^d)^k \equiv 1 \pmod{p}$ for all positive integers k. By Theorem 8.7 we know that $x^d - 1$ has exactly d incongruent roots modulo p, so every root modulo p is congruent to one of these powers of a. However, by Theorem 8.4 we know that the powers of a with order d are those of the form a^k with $(k,d) = 1$. There are exactly $\phi(d)$ such integers k with $1 \leq k \leq d$, and consequently, if there is one element of order d modulo p, there must be exactly $\phi(d)$ such positive integers less than d. Hence, $F(d) \leq \phi(d)$.

Therefore, we can conclude that $F(d) = \phi(d)$, which tells us that there are precisely $\phi(d)$ incongruent integers of order d modulo p. ∎

The following corollary is derived immediately from Theorem 8.8.

Corollary 8.8.1. Every prime has a primitive root.

Proof. Let p be a prime. By Theorem 8.7 we know that there are $\phi(p-1)$ incongruent integers of order $p - 1$ modulo p. Since each of these is, by definition, a primitive root, p has $\phi(p-1)$ primitive roots. ∎

The smallest positive primitive root of each prime less than 1000 is given in Table 3 of the Appendix.

8.2 Exercises

1. Find the number of incongruent roots modulo 11 of each of the following polynomials.

   a)　$x^2 + 2$　　　　　　　c)　$x^3 + x^2 + 2x + 2$
   b)　$x^2 + 10$　　　　　　d)　$x^4 + x^2 + 1$

2. Find the number of incongruent roots modulo 13 of each of the following polynomials.

   a) $x^2 + 1$

   b) $x^2 + 3x + 2$

   c) $x^3 + 12$

   d) $x^4 + x^2 + x + 1$

3. Find the number of primitive roots of each of the following primes.

   a)   7

   b)   13

   c)   17

   d)   19

   e)   29

   f)   47

4. Find a complete set of incongruent primitive roots of 7.

5. Find a complete set of incongruent primitive roots of 13.

6. Find a complete set of incongruent primitive roots of 17.

7. Find a complete set of incongruent primitive roots of 19.

8. Let r be a primitive root of the prime p with $p \equiv 1 \pmod 4$. Show that $-r$ is also a primitive root.

9. Show that if p is a prime and $p \equiv 1 \pmod 4$, there is an integer x such that $x^2 \equiv -1 \pmod p$. (*Hint*: Use Theorem 8.8 to show that there is an integer x of order 4 modulo p.)

10. a)   Find the number of incongruent roots modulo 6 of the polynomial $x^2 - x$.

    b)   Explain why the answer to part (a) does not contradict Lagrange's theorem.

11. a)   Use Lagrange's theorem to show that if p is a prime and $f(x)$ is a polynomial of degree n with integer coefficients and more than n roots modulo p, then p divides every coefficient of $f(x)$.

    b)   Let p be prime. Using part (a), show that every coefficient of the polynomial $f(x) = (x-1)(x-2) \cdots (x-p+1) - x^{p-1} + 1$ is divisible by p.

    c)   Using part (b), give a proof of Wilson's theorem. (*Hint*: Consider the constant term of $f(x)$.)

12. Find the least positive residue of the product of a set of $\phi(p-1)$ incongruent primitive roots modulo a prime p.

* 13. A systematic method for constructing a primitive root modulo a prime p is outlined in this problem. Let the prime factorization of $\phi(p) = p-1$ be $p - 1 = q_1^{t_1} q_2^{t_2} \cdots q_r^{t_r}$ where $q_1, q_2, \ldots, q_r$ are prime.

    a)   Use Theorem 8.8 to show that there are integers $a_1, a_2, \ldots, a_r$ such that $\mathrm{ord}_p a_1 = q_1^{t_1}, \mathrm{ord}_p a_2 = q_2^{t_2}, \ldots, \mathrm{ord}_p a_r = q_r^{t_r}$.

    b)   Use Exercise 10 of Section 8.1 to show that $a = a_1 a_2 \cdots a_r$ is a primitive root modulo p.

    c)   Follow the procedure outlined in parts (a) and (b) to find a primitive root modulo 29.

* **14.** Suppose that the composite positive integer n has prime-power factorization $n = p_1^{a_1} p_2^{a_2} \cdots p_r^{a_r}$. Show that the number of incongruent bases modulo n for which n is a pseudoprime to that base is $\prod\limits_{j=1}^{r} (n-1, p_j - 1)$.

15. Use Exercise 14 to show that every odd composite integer that is not a power of 3 is a pseudoprime to at least two bases other than ± 1.

16. Show that if p is prime and $p = 2q + 1$, where q is prime and a is a positive integer with $1 < a < p-1$, then $p - a^2$ is a primitive root modulo p.

* **17.** a) Suppose that $f(x)$ is a polynomial with integer coefficients of degree $n-1$. Let $x_1, x_2, \ldots, x_n$ be n incongruent integers modulo p. Show that for all integers x, the congruence

$$f(x) \equiv \sum_{j=1}^{n} f(x_j) \prod_{\substack{i=1 \\ i \neq j}}^{n} (x - x_i) \overline{(x_j - x_i)} \pmod{p}$$

holds, where $\overline{x_j - x_i}$ is an inverse of $x_j - x_i$ modulo p. This technique for finding $f(x)$ modulo p is called *Lagrange interpolation*.

b) Find the least positive residue of $f(5)$ modulo 11 if $f(x)$ is a polynomial of degree 3 with $f(1) \equiv 8$, $f(2) \equiv 2$, and $f(3) \equiv 4 \pmod{11}$.

18. In this exercise we develop a threshold scheme for protection of master keys in a computer system, different from the scheme discussed in Section 7.6. Let $f(x)$ be a randomly chosen polynomial of degree $r-1$, with the condition that K, the master key, is the constant term of the polynomial. Let p be a prime, such that $p > K$ and $p > s$. The s shadows $k_1, k_2, \ldots, k_s$ are computed by finding the least positive residue of $f(x_j)$ modulo p for $j = 1, 2, \ldots, s$ where $x_1, x_2, \ldots, x_s$ are randomly chosen integers incongruent modulo p, i.e.,

$$k_j \equiv f(x_j) \pmod{p}, \quad 0 \le k_j < p,$$

for $j = 1, 2, \ldots, s$.

a) Use Lagrange interpolation, described in Exercise 17, to show that the master key K can be determined from any r shadows.

b) Show that the master key K cannot be determined from fewer than r shadows.

c) Let $K = 33$, $p = 47$, $r = 4$, and $s = 7$. Let $f(x) = 4x^3 + x^2 + 31x + 33$. Find the seven shadows corresponding to the values of $f(x)$ at $1, 2, 3, 4, 5, 6, 7$.

d) Show how to find the master key from the four shadows $f(1), f(2), f(3)$, and $f(4)$.

19. Show that an RSA cipher with enciphering modulus $n = pq$ is resistant to attack by iteration (see the preamble to Exercise 22 of Section 8.1) if $p = 2p' + 1$ and $q = 2q' + 1$, where p' and q' are primes.

8.2 Computer Projects

Programming Projects

Write programs to do the following:

1. Given a prime p, use Exercise 13 to find a primitive root of p.

2. Implement the threshold scheme given in Exercise 18.

Computations and Explorations

Using the programs you have written or a computation program, carry out the following computations and explorations:

1. Erdös has asked whether for each sufficiently large prime p there is a prime q for which q is a primitive root of p. What evidence can you find for this conjecture? For which small primes p is the statement in the conjecture false?

8.3 The Existence of Primitive Roots

In the previous section, we showed that every prime has a primitive root. In this section, we will find all positive integers having primitive roots. First, we will show that every power of an odd prime possesses a primitive root. We begin by considering squares of primes.

Theorem 8.9. If p is an odd prime with primitive root r, then either r or $r + p$ is a primitive root modulo p^2.

Proof. Since r is a primitive root modulo p, we know that

$$\operatorname{ord}_p r = \phi\,(p) = p - 1.$$

Let $n = \operatorname{ord}_{p^2} r$, so that

$$r^n \equiv 1 \pmod{p^2}.$$

Since a congruence modulo p^2 obviously holds modulo p, we have

$$r^n \equiv 1 \pmod{p}.$$

By Theorem 8.1 it follows that

$$p - 1 = \operatorname{ord}_p r \mid n.$$

On the other hand, Corollary 8.1.1 tells us that

$$n \mid \phi(p^2) = p(p - 1).$$

Since $n \mid p(p-1)$ and $p - 1 \mid n$, either $n = p - 1$ or $n = p(p - 1)$. If $n = p(p - 1)$, then r is a primitive root modulo p^2, since $\operatorname{ord}_{p^2} r = \phi(p^2)$. Otherwise, we have $n = p - 1$, so that

(8.1)
$$r^{p-1} \equiv 1 \pmod{p^2}.$$

Let $s = r+p$. Then, since $s \equiv r \pmod{p}$, s is also a primitive root modulo p. Hence, $\text{ord}_{p^2} s$ equals either $p-1$ or $p(p-1)$. We will show that $\text{ord}_{p^2} s \neq p-1$. The binomial theorem tells us that

$$s^{p-1} = (r+p)^{p-1} = r^{p-1} + (p-1)r^{p-2}p + \binom{p-1}{2}r^{p-3}p^2 + \cdots + p^{p-1}$$

$$\equiv r^{p-1} + (p-1)p \cdot r^{p-2} \pmod{p^2}.$$

Hence, using (8.1), we see that

$$s^{p-1} \equiv 1 + (p-1)p \cdot r^{p-2} \equiv 1 - pr^{p-2} \pmod{p^2}.$$

From this last congruence, we can conclude that

$$s^{p-1} \not\equiv 1 \pmod{p^2}.$$

To see this, note that if $s^{p-1} \equiv 1 \pmod{p^2}$, then $pr^{p-2} \equiv 0 \pmod{p^2}$. This last congruence implies that $r^{p-2} \equiv 0 \pmod{p}$, which is impossible, since $p \nmid r$ (remember r is a primitive root of p). Hence, $\text{ord}_{p^2} s = p(p-1) = \phi(p^2)$. Consequently, $s = r+p$ is a primitive root of p^2. ∎

Example 8.12. The prime $p = 7$ has $r = 3$ as a primitive root. Using observations made it the proof of Theorem 8.9 either $\text{ord}_{49} 3 = 6$ or $\text{ord}_{49} 3 = 42$. However

$$r^{p-1} = 3^6 \not\equiv 1 \pmod{49}.$$

It folllows that $\text{ord}_{49} 3 = 42$. Hence 3 is also a primitive root of $p^2 = 49$. □

We note that it is extremely rare for the congruence

$$r^{p-1} \equiv 1 \pmod{p^2}$$

to hold when r is a primitive root modulo the prime p. Consequently, it is very seldom that a primitive root r modulo the prime p is not also a primitive root modulo p^2. When this occurs, Theorem 8.9 tell us that $r + p$ is a primitive root modulo p^2. The following example illustrates this.

Example 8.13. Let $p = 487$. For the primitive root 10 modulo 487, we have

$$10^{486} \equiv 1 \pmod{487^2}.$$

Hence, 10 is not a primitive root modulo 487^2, but by Theorem 8.9, we know that $497 = 10 + 487$ is a primitive root modulo 487^2. □

We now turn our attention to arbitrary powers of primes.

Theorem 8.10. Let p be an odd prime. Then p^k has a primitive root for all positive integers k. Moreover, if r is a primitive root modulo p^2, then r is a primitive root modulo p^k, for all positive integers k.

Proof. By Theorem 8.9 we know that p has a primitive root r that is also a primitive root modulo p^2, so that

(8.2) $r^{p-1} \not\equiv 1 \pmod{p^2}$.

Using mathematical induction, we will prove that for this primitive root r,

(8.3) $r^{p^{k-2}(p-1)} \not\equiv 1 \pmod{p^k}$

for all positive integers k, $k \geq 2$. Once we have established this congruence, we can show that r is also a primitive root modulo p^k by the following reasoning. Let

$$n = \operatorname{ord}_{p^k} r.$$

By Theorem 6.8 we know that $n \mid \phi(p^k) = p^{k-1}(p-1)$. On the other hand, since

$$r^n \equiv 1 \pmod{p^k},$$

we also know that

$$r^n \equiv 1 \pmod{p}.$$

By Theorem 8.1 we see that $p-1 = \phi(p) \mid n$. Because $(p-1) \mid n$, and $n \mid p^{k-1}(p-1)$, we know that $n = p^t(p-1)$, where t is an integer such that $0 \leq t \leq k-1$. If $n = p^t(p-1)$ with $t \leq k-2$, then

$$r^{p^{k-2}(p-1)} = (r^{p^t(p-1)})^{p^{k-2-t}} \equiv 1 \pmod{p^k},$$

which would contradict (8.3). Hence, $\operatorname{ord}_{p^k} r = p^{k-1}(p-1) = \phi(p^k)$. Consequently, r is also a primitive root modulo p^k.

All that remains is to prove (8.3) using mathematical induction. The case of $k=2$ follows from (8.2). Let us assume that the assertion is true for the positive integer $k \geq 2$. Then

$$r^{p^{k-2}(p-1)} \not\equiv 1 \pmod{p^k}.$$

Since $(r,p) = 1$, we know that $(r,p^{k-1}) = 1$. Consequently, from Euler's theorem, we know that

$$r^{p^{k-2}(p-1)} = r^{\phi(p^{k-1})} \equiv 1 \pmod{p_{k-1}}.$$

Therefore, there is an integer d such that

$$r^{p^{k-2}(p-1)} = 1 + dp^{k-1},$$

where $p \nmid d$, since by hypothesis $r^{p^{k-2}(p-1)} \not\equiv 1 \pmod{p^k}$. We take the pth power of both sides of the above equation, to obtain, via the binomial theorem,

$$r^{p^{k-1}(p-1)} = (1 + dp^{k-1})^p$$
$$= 1 + p(dp^{k-1}) + \binom{p}{2}(dp^{k-1})^2 + \cdots + (dp^{k-1})^p$$
$$\equiv 1 + dp^k \pmod{p^{k+1}}.$$

Since $p \nmid d$, we can conclude that

$$r^{p^{k-1}(p-1)} \not\equiv 1 \pmod{p^{k+1}}.$$

This completes the proof by induction. ∎

Example 8.14. By Example 8.12 we know that $r = 3$ is a primitive root modulo 7 and 7^2. Hence Theorem 8.10 tells us that $r = 3$ is also a primitive root modulo 7^k for all positive integers k. □

It is now time to discuss whether there are primitive roots modulo powers of 2. We first note that both 2 and $2^2 = 4$ have primitive roots, namely 1 and 3, respectively. For higher powers of 2, the situation is different, as the following theorem shows; there are no primitive roots modulo these powers of 2.

Theorem 8.11. If a is an odd integer, and if k is an integer, $k \geq 3$, then
$$a^{\phi(2^k)/2} = a^{2^{k-2}} \equiv 1 \pmod{2^k}.$$

Proof. We prove this result using mathematical induction. If a is an odd integer, then $a = 2b + 1$, where b is an integer. Hence,
$$a^2 = (2b + 1)^2 = 4b^2 + 4b + 1 = 4b(b + 1) + 1.$$

Since either b or $b + 1$ is even, we see that $8 \mid 4b\,(b + 1)$. By Exercise 5 of Section 3.1 it follows that
$$a^2 \equiv 1 \pmod 8.$$

This is the congruence of interest when $k = 3$.

Now to complete the induction argument, let us assume that
$$a^{2^{k-2}} \equiv 1 \pmod{2^k}.$$

Then there is an integer d such that
$$a^{2^{k-2}} = 1 + d \cdot 2^k.$$

Squaring both sides of the above equality, we obtain
$$a^{2^{k-1}} = 1 + d2^{k+1} + d^2 2^{2k}.$$

This yields
$$a^{2^{k-1}} \equiv 1 \pmod{2^{k+1}},$$

which completes the induction argument. ∎

Theorem 8.11 tells us that no power of 2, other than 2 and 4, has a primitive root, since when a is an odd integer, $\mathrm{ord}_{2^k} a \neq \phi(2^k)$, since $a^{\phi(2^k)/2} \equiv 1 \pmod{2^k}$.

Even though there are no primitive roots modulo 2^k for $k \geq 3$, there always is an element of largest possible order, namely $\phi(2^k)/2$, as the following theorem shows.

Theorem 8.12. Let $k \geq 3$ be an integer. Then

$$\operatorname{ord}_{2^k} 5 = \phi(2^k)/2 = 2^{k-2}.$$

Proof. Theorem 8.11 tells us that

$$5^{2^{k-2}} \equiv 1 \pmod{2^k},$$

for $k \geq 3$. By Theorem 8.1 we see that $\operatorname{ord}_{2^k} 5 \mid 2^{k-2}$. Therefore, if we show that $\operatorname{ord}_{2^k} 5 \nmid 2^{k-3}$, we can conclude that

$$\operatorname{ord}_{2^k} 5 = 2^{k-2}.$$

To show that $\operatorname{ord}_{2^k} 5 \nmid 2^{k-3}$, we will prove by mathematical induction that for $k \geq 3$,

$$5^{2^{k-3}} \equiv 1 + 2^{k-1} \not\equiv 1 \pmod{2^k}.$$

For $k = 3$, we have

$$5 \equiv 1 + 4 \pmod 8.$$

Now assume that

$$5^{2^{k-3}} \equiv 1 + 2^{k-1} \pmod{2^k}.$$

This means that there is a positive integer d such that

$$5^{2^{k-3}} = (1 + 2^{k-1}) + d2^k.$$

Squaring both sides, we find that

$$5^{2^{k-2}} = (1 + 2^{k-1})^2 + 2(1 + 2^{k-1})d2^k + (d2^k)^2,$$

so that

$$5^{2^{k-2}} \equiv (1 + 2^{k-1})^2 = 1 + 2^k + 2^{2k-2} \equiv 1 + 2^k \pmod{2^{k+1}}.$$

This completes the induction argument and shows that

$$\operatorname{ord}_{2^k} 5 = \phi(2^k)/2. \qquad \blacksquare$$

We have now demonstrated that all powers of odd primes possess primitive roots, while the only powers of 2 having primitive roots are 2 and 4. Next, we determine which integers not powers of primes, *i.e.*, those integers divisible by two or more primes, have primitive roots. We will demonstrate that the only positive integers not powers of primes possessing primitive roots are twice powers of odd primes.

We first narrow down the set of positive integers we need consider with the following result.

Theorem 8.13. If n is a positive integer that is not a prime power or twice a prime power, then n does not have a primitive root.

Proof. Let n be a positive integer with prime-power factorization

$$n = p_1^{t_1} p_2^{t_2} \cdots p_m^{t_m}.$$

Let us assume that the integer n has a primitive root r. This means that $(r,n) = 1$ and $\text{ord}_n r = \phi(n)$. Since $(r,n) = 1$, we know that $(r,p') = 1$, whenever p' is one of the prime powers occurring in the factorization of n. By Euler's theorem, we know that

$$r^{\phi(p')} \equiv 1 \pmod{p'}.$$

Now let U be the least common multiple of $\phi(p_1^{t_1})$, $\phi(p_2^{t_2}),...,\phi(p_m^{t_m})$, i.e.,

$$U = [\phi(p_1^{t_1}),\phi(p_2^{t_2}),...,\phi(p_m^{t_m})].$$

Since $\phi(p_i^{t_i}) \mid U$, we know that

$$r^U \equiv 1 \pmod{p_i^{t_i}}$$

for $i = 1,2,...,m$. Using the Chinese remainder theorem, it now follows that

$$r^U \equiv 1 \pmod{n}$$

This implies that

$$\text{ord}_n r = \phi(n) \le U.$$

By Theorem 6.4, since ϕ is multiplicative, we have

$$\phi(n) = \phi(p_1^{t_1} p_2^{t_2} \cdots p_m^{t_m}) = \phi(p_1^{t_1})\phi(p_2^{t_2}) \cdots \phi(p_m^{t_m}).$$

This formula for $\phi(n)$ and the inequality $\phi(n) \le U$ imply that

$$\phi(p_1^{t_1})\ \phi(p_2^{t_2}) \cdots \phi(p_m^{t_m}) \le [\phi(p_1^{t_1}),\phi(p_2^{t_2}),...,\ \phi(p_m^{t_m})].$$

Since the product of a set of integers is less than or equal to their least common multiple only if the integers are pairwise relatively prime (and then the less than or equal to relation is really just an equality), the integers $\phi(p_1^{t_1}),\phi(p_2^{t_2}),...,\ \phi(p_m^{t_m})$ must be pairwise relatively prime.

We note that $\phi(p') = p^{t-1}(p-1)$, so that $\phi(p')$ is even if p is odd, or if $p = 2$ and $t \ge 2$. Hence, the numbers $\phi(p_1^{t_1})$, $\phi(p_2^{t_2}),...,\ \phi(p_m^{t_m})$ are not pairwise relatively prime unless $m = 1$ and n is a prime power or $m = 2$ and n is $n = 2p'$, where p is an odd prime and t is a positive integer. ∎

We have now limited consideration to integers of the form $n = 2p'$, where p is an odd prime and t is a positive integer. We now show that all such integers have primitive roots.

Theorem 8.14. If p is an odd prime and t is a positive integer, then $2p'$ possesses a primitive root. In fact, if r is a primitive root modulo p', then if r is odd it is also a primitive root modulo $2p'$, while if r is even, $r + p'$ is a primitive root modulo $2p'$.

Proof. If r is a primitive root modulo p', then

$$r^{\phi(p')} \equiv 1 \pmod{p'},$$

and no positive exponent smaller than $\phi(p^t)$ has this property. By Theorem 6.4 we note that $\phi(2p^t) = \phi(2) \, \phi(p^t) = \phi(p^t)$, so that $r^{\phi(2p^t)} \equiv 1 \pmod{p^t}$.

If r is odd, then

$$r^{\phi(2p^t)} \equiv 1 \pmod{2}.$$

Thus, by Corollary 3.8.1 we see that $r^{\phi(2p^t)} \equiv 1 \pmod{2p^t}$. No smaller power of r is congruent to 1 modulo $2p^t$. Such power would also be congruent to 1 modulo p_t, contradicting that r is a primitive root of p_t. It follows that r is a primitive root modulo $2p^t$.

On the other hand, if r is even, then $r + p^t$ is odd. Hence,

$$(r + p^t)^{\phi(2p^t)} \equiv 1 \pmod{2}.$$

Since $r + p^t \equiv r \pmod{p^t}$, we see that

$$(r + p^t)^{\phi(2p^t)} \equiv 1 \pmod{p^t}.$$

Therefore, $(r + p^t)^{\phi(2p^t)} \equiv 1 \pmod{2p^t}$, and as no smaller power of $r + p^t$ is congruent to 1 modulo $2p^t$, we see that $r + p^t$ is a primitive root modulo $2p^t$. ∎

Example 8.15. Earlier in this section we showed that 3 is a primitive root modulo 7^t for all positive integers t. Hence, since 3 is odd, Theorem 8.14 tells us that 3 is also a primitive root modulo $2 \cdot 7^t$ for all positive integers t. For instance, 3 is a primitive root modulo 14.

Similarly, we know that 2 is a primitive root modulo 5^t for all positive integers t. Since $2 + 5^t$ is odd, Theorem 8.14 tells us that $2 + 5^t$ is a primitive root modulo $2 \cdot 5^t$ for all positive integers t. For example, 27 is a primitive root modulo 50. □

Combining Corollary 8.8.1 and Theorems 8.10, 8.13, and 8.14, we can now describe which positive integers have a primitive root.

Theorem 8.15. The positive integer n, $n > 1$, possesses a primitive root if and only if

$$n = 2, 4, p^t, \text{ or } 2p^t,$$

where p is an odd prime and t is a positive integer.

8.3 Exercises

1. Which of the integers $4, 10, 16, 22, 28$ have a primitive root?

2. Which of the integers $8, 9, 12, 26, 27, 31, 33$ have a primitive root?

3. Find a primitive root modulo each of the following moduli.

   a) 3^2 c) 23^2
   b) 5^2 d) 29^2

4. Find a primitive root modulo each of the following moduli.

    a) 11^2 c) 17^2
    b) 13^2 d) 19^2

5. Find a primitive root, for all positive integers k modulo each of the following moduli.

    a) 3^k c) 13^k
    b) 11^k d) 17^k

6. Find a primitive root, for all positive integers k modulo each of following moduli.

    a) 23^k c) 31^k
    b) 29^k d) 37^k

7. Find a primitive root modulo each of the following moduli.

    a)   10   c)   38
    b)   34   d)   50

8. Find a primitive root modulo each of the following moduli.

    a)   6   c)   26
    b)   18   d)   338

9. Find all the primitive roots modulo 22.

10. Find all the primitive roots modulo 25.

11. Find all the primitive roots modulo 38.

12. Show that there are the same number of primitive roots modulo $2p^t$ as there are of p^t, where p is an odd prime and t is a positive integer.

13. Show that the integer m has a primitive root if and only if the only solutions of the congruence $x^2 \equiv 1 \pmod{m}$ are $x \equiv \pm 1 \pmod{m}$.

* 14. Let n be a positive integer possessing a primitive root. Using this primitive root, prove that the product of all positive integers less than n and relatively prime to n is congruent to -1 modulo n. (When n is prime, this result is Wilson's Theorem.)

* 15. Show that although there are no primitive roots modulo 2^k where k is an integer, $k \geq 3$, every odd integer is congruent to exactly one of the integers $(-1)^\alpha 5^\beta$, where $\alpha = 0$ or 1 and β is an integer satisfying $0 \leq \beta \leq 2^{k-2} - 1$.

16. Find the smallest odd prime p that has a primitive root r that is not also a primitive root modulo p^2.

8.3 Computer Projects

Programming Projects

Write computer programs to do the following:

1. Find primitive roots modulo powers of odd primes.

2. Find primitive roots modulo twice powers of odd primes.

Computations and Explorations

Using the programs you have written or a computation program, carry out the following computations and explorations:

1. Find as many examples as you can where r is a primitive root of the prime p, but r is not a primitive root of p^2. Can you make any conjectures about how often this occurs?

8.4 Index Arithmetic

In this section we demonstrate how primitive roots may be used to do modular arithmetic. Let r be a primitive root modulo the positive integer m (so that m is of the form described in Theorem 8.15). By Theorem 8.3 we know that the integers

$$r, \ r^2, \ r^3, \ ..., \ r^{\phi(m)}$$

form a reduced system of residues modulo m. From this fact, we see that if a is an integer relatively prime to m, then there is a unique integer x with $1 \le x \le \phi(m)$ such that

$$r^x \equiv a \ (\text{mod } m).$$

This leads to the following definition.

Definition. Let m be a positive integer with primitive root r. If a is a positive integer with $(a,m) = 1$, then the unique integer x with $1 \le x \le \phi(m)$ and $r^x \equiv a \ (\text{mod } m)$ is called the *index of a to the base r modulo m*. With this definition, we have $a \equiv r^{\text{ind}_r a} \ (\text{mod } m)$.

If x is the index of a to the base r modulo m, then we write $x = \text{ind}_r a$, where we do not indicate the modulus m in the notation, since it is assumed to be fixed. From the definition, we know that if a and b are integers relatively prime to m and $a \equiv b \ (\text{mod } m)$, then $\text{ind}_r a = \text{ind}_r b$.

Example 8.16. Let $m = 7$. We have seen that 3 is a primitive root modulo 7 and that $3^1 \equiv 3 \ (\text{mod } 7)$, $3^2 \equiv 2 \ (\text{mod } 7)$, $3^3 \equiv 6 \ (\text{mod } 7)$, $3^4 \equiv 4 \ (\text{mod } 7)$, $3^5 \equiv 5 \ (\text{mod } 5)$, and $3^6 \equiv 1 \ (\text{mod } 7)$.

Hence, modulo 7 we have

$$\text{ind}_3 1 = 6, \ \text{ind}_3 2 = 2, \ \text{ind}_3 3 = 1,$$
$$\text{ind}_3 4 = 4, \ \text{ind}_3 5 = 5, \ \text{ind}_3 6 = 3.$$

With a different primitive root modulo 7, we obtain a different set of indices. For instance, calculations show that with respect to the primitive root 5,

$$\text{ind}_5 1 = 6, \ \text{ind}_5 2 = 4, \ \text{ind}_5 3 = 5,$$
$$\text{ind}_5 4 = 2, \ \text{ind}_5 5 = 1, \ \text{ind}_5 6 = 3. \qquad \square$$

We now develop some properties of indices. These properties are somewhat similar to those of logarithms, but instead of equalities, we have congruences modulo $\phi(m)$.

Theorem 8.16. Let m be a positive integer with primitive root r, and let a and b be integers relatively prime to m. Then

    (i) $\text{ind}_r 1 \equiv 0 \ (\text{mod} \ \phi(m))$.
    (ii) $\text{ind}_r(ab) \equiv \text{ind}_r a + \text{ind}_r b \ (\text{mod} \ \phi(m))$
    (iii) $\text{ind}_r a^k \equiv k \cdot \text{ind}_r a \ (\text{mod} \ \phi(m))$ if k is a positive integer.

Proof of (i). From Euler's theorem, we know that $r^{\phi(m)} \equiv 1 \ (\text{mod} \ m)$. Since r is a primitive root modulo m, no smaller positive power of r is congruent to 1 modulo m. Hence, $\text{ind}_r 1 = \phi(m) \equiv 0 \ (\text{mod} \ \phi(m))$.

Proof of (ii). To prove this congruence, note that from the definition of indices,

$$r^{\text{ind}_r(ab)} \equiv ab \ (\text{mod} \ m)$$

and

$$r^{\text{ind}_r a + \text{ind}_r b} \equiv r^{\text{ind}_r a} \cdot r^{\text{ind}_r b} \equiv ab \ (\text{mod} \ m).$$

Hence,

$$r^{\text{ind}_r(ab)} \equiv r^{\text{ind}_r a + \text{ind}_r b} \ (\text{mod} \ m).$$

Using Theorem 8.2, we conclude that

$$\text{ind}_r(ab) \equiv \text{ind}_r a + \text{ind}_r b \ (\text{mod} \ \phi(m)).$$

Proof of (iii). To prove the congruence of interest, first note that, by definition, we have

$$r^{\text{ind}_r a^k} \equiv a^k \ (\text{mod} \ m)$$

and

$$r^{k \cdot \text{ind}_r a} \equiv (r^{\text{ind}_r a})^k \equiv a^k \ (\text{mod} \ m).$$

Hence,

$$r^{\text{ind}_r a^k} \equiv r^{k \cdot \text{ind}_r a} \ (\text{mod} \ m).$$

Using Theorem 8.2, this leads us immediately to the congruence we want, namely

$$\text{ind}_r a^k \equiv k \cdot \text{ind}_r a \ (\text{mod} \ \phi(m)). \qquad \blacksquare$$

Example 8.17. From the previous examples, we see that modulo 7, $\text{ind}_5 2 = 4$ and $\text{ind}_5 3 = 5$. Since $\phi(7) = 6$, part (ii) of Theorem 8.16 tells us that

$$\text{ind}_5 6 = \text{ind}_5 2 \cdot 3 = \text{ind}_5 2 + \text{ind}_5 3 = 4 + 5 = 9 \equiv 3 \pmod{6}.$$

Note that this agrees with the value previously found for $\text{ind}_5 6$.

From part (iii) of Theorem 8.16, we see that

$$\text{ind}_5 3^4 \equiv 4 \cdot \text{ind}_5 3 \equiv 4 \cdot 5 = 20 \equiv 2 \pmod{6}.$$

Note that direct computation gives the same result, since

$$\text{ind}_5 3^4 = \text{ind}_5 81 = \text{ind}_5 4 = 2. \qquad \square$$

Indices are helpful in the solution of certain types of congruences. Consider the following examples.

Example 8.18. We will use indices to solve the congruence $6x^{12} \equiv 11 \pmod{17}$. We find that 3 is a primitive root of 17 (since $3^8 \equiv -1 \pmod{17}$). The indices of integers to the base 3 modulo 17 are given in Table 8.1.

a	1	2	3	4	5	6	7	8	9	10	11	12	13	14	15	16
$\text{ind}_3 a$	16	14	1	12	5	15	11	10	2	3	7	13	4	9	6	8

Table 8.1. Indices to the Base 3 Modulo 17.

Taking the index of each side of the congruence to the base 3 modulo 17, we obtain a congruence modulo $\phi(17) = 16$, namely

$$\text{ind}_3 (6x^{12}) \equiv \text{ind}_3 11 = 7 \pmod{16}.$$

Using (ii) and (iii) of Theorem 8.16, we obtain

$$\text{ind}_3 (6x^{12}) \equiv \text{ind}_3 6 + \text{ind}_3 (x^{12}) \equiv 15 + 12 \cdot \text{ind}_3 x \pmod{16}.$$

Hence,

$$15 + 12 \cdot \text{ind}_3 x \equiv 7 \pmod{16}$$

or

$$12 \cdot \text{ind}_3 x \equiv 8 \pmod{16}.$$

From this congruence it follows (as the reader should show) that

$$\text{ind}_3 x \equiv 2 \pmod{4}.$$

Hence,

$$\text{ind}_3 x \equiv 2, 6, 10, \text{ or } 14 \pmod{16}.$$

Consequently, from the definition of indices, we find that

$$x \equiv 3^2, 3^6, 3^{10} \text{ or } 3^{14} \pmod{17},$$

(note that this congruence holds modulo 17). Since $3^2 \equiv 9$, $3^6 \equiv 15$, $3^{10} \equiv 8$, and

$3^{14} \equiv 2 \pmod{17}$, we conclude that

$$x \equiv 9, \ 15, \ 8, \ \text{or} \ 2 \pmod{17}.$$

Since each step in the computations is reversible, there are four incongruent solutions of the original congruence modulo 17. □

Example 8.19. We wish to find all solutions of the congruence $7^x \equiv 6 \pmod{17}$. When we take indices to the base 3 modulo 17 of both sides of this congruence, we find that

$$\mathrm{ind}_3 (7^x) \equiv \mathrm{ind}_3 6 = 15 \pmod{16}.$$

By part (iii) of Theorem 8.16 we obtain

$$\mathrm{ind}_3 (7^x) \equiv x \cdot \mathrm{ind}_3 7 \equiv 11x \pmod{16}.$$

Hence,

$$11x \equiv 15 \pmod{16}.$$

Since 3 is an inverse of 11 modulo 16, we multiply both sides of the linear congruence above by 3, to find that

$$x \equiv 3 \cdot 15 = 45 \equiv 13 \pmod{16}.$$

All steps in this computation are reversible. Therefore, the solutions of

$$7^x \equiv 6 \pmod{17}$$

are given by

$$x \equiv 13 \pmod{16}.$$ □

Next, we discuss congruences of the form $x^k \equiv a \pmod{m}$, where m is a positive integer with a primitive root and $(a,m) = 1$. First, we present a definition.

Definition. If m and k are positive integers and a is an integer relatively prime to m, then we say that a is a *kth power residue of* m if the congruence $x^k \equiv a \pmod{m}$ has a solution.

When m is an integer possessing a primitive root, the following theorem gives a useful criterion for an integer a relatively prime to m to be a kth power residue of m.

Theorem 8.17. Let m be a positive integer with a primitive root. If k is a positive integer and a is an integer relatively prime to m, then the congruence $x^k \equiv a \pmod{m}$ has a solution if and only if

$$a^{\phi(m)/d} \equiv 1 \pmod{m}$$

where $d = (k,\phi(m))$. Furthermore, if there are solutions of $x^k \equiv a \pmod{m}$, then there are exactly d incongruent solutions modulo m.

Proof. Let r be a primitive root modulo the positive integer m. We note that the congruence

$$x^k \equiv a \pmod{m}$$

holds if and only if

(8.1) $k \cdot \operatorname{ind}_r x \equiv \operatorname{ind}_r a \pmod{\phi(m)}.$

Now let $d = (k, \phi(m))$ and $y = \operatorname{ind}_r x$, so that $x \equiv r^y \pmod{m}$. By Theorem 3.10, we note that if $d \nmid \operatorname{ind}_r a$, then the linear congruence

(8.2) $ky \equiv \operatorname{ind}_r a \pmod{\phi(m)}$

has no solutions, and hence, there are no integers x satisfying (8.1). If $d \mid \operatorname{ind}_r a$, then there are exactly d integers y incongruent modulo $\phi(m)$ such that (8.2) holds, and hence, exactly d integers x incongruent modulo m such that (8.1) holds. Since $d \mid \operatorname{ind}_r a$ if and only if

$$(\phi(m)/d) \operatorname{ind}_r a \equiv 0 \pmod{\phi(m)},$$

and this congruence holds if and only if

$$a^{\phi(m)/d} \equiv 1 \pmod{m},$$

the theorem is true.   ■

We note that Theorem 8.17 tells us that if p is a prime, k is a positive integer, and a is an integer relatively prime to p, then a is a kth power residue of p if and only if

$$a^{(p-1)/d} \equiv 1 \pmod{p},$$

where $d = (k, p-1)$. We illustrate this observation with an example.

Example 8.20. To determine whether 5 is a sixth power residue of 17, *i.e.* whether the congruence

$$x^6 \equiv 5 \pmod{17}$$

has a solution, we determine that

$$5^{16/(6,16)} = 5^8 \equiv -1 \pmod{17}. \qquad \square$$

Hence, 5 is not a sixth power residue of 17.

A table of indices with respect to the least primitive root modulo each prime less than 100 is given in Table 4 of the Appendix.

We now present the proof of Theorem 5.10. We state this theorem again for convenience.

Theorem 5.10. If n is an odd composite positive integer, then n passes Miller's test for at most $(n-1)/4$ bases b with $1 \le b < n-1$.

We need the following lemma in the proof.

Lemma 8.1. Let p be an odd prime and let e and q be positive integers. Then the number of incongruent solutions of the congruence $x^{q-1} \equiv 1 \pmod{p^e}$ is $(q, p^{e-1}(p-1))$.

Proof. Let r be a primitive root of p^e. By taking indices with respect to r, we see that $x^q \equiv 1 \pmod{p^e}$ if and only if $qy \equiv 0 \pmod{\phi(p^e)}$ where $y = \text{ind}_r x$. Using Theorem 3.10, we see that there are exactly $(q, \phi(p^e))$ incongruent solutions of $qy \equiv 0 \pmod{\phi(p^e)}$. Consequently, there are $(q, \phi(p^e)) = (q, p^{e-1}(p-1))$ incongruent solutions of $x^q \equiv 1 \pmod{p^e}$. ∎

We now proceed with a proof of Theorem 5.10.

Proof. Let $n - 1 = 2^s t$, where s is a positive integer and t is an odd positive integer. For n to be a strong pseudoprime to the base b, either

$$b^t \equiv 1 \pmod{n}$$

or

$$b^{2^j t} \equiv -1 \pmod{n}$$

for some integer j with $0 \leq j \leq s - 1$. In either case, we have

$$b^{n-1} \equiv 1 \pmod{n}.$$

Let the prime-power factorization of n be $n = p_1^{e_1} p_2^{e_2} \cdots p_r^{e_r}$. By Lemma 8.1 we know that there are $(n - 1, p_j^{e_j}(p_j - 1)) = (n - 1, p_j - 1)$ incongruent solutions of $x^{n-1} \equiv 1 \pmod{p_j^{e_j}}$, $j = 1, 2, \dots, r$. Consequently, the Chinese remainder theorem tells us that there are exactly $\prod_{j=1}^{r} (n - 1, p_j - 1)$ incongruent solutions of $x^{n-1} \equiv 1 \pmod{n}$.

To prove the theorem, we first consider the case where the prime-power factorization of n contains a prime power $p_k^{e_k}$ with exponent $e_k \geq 2$. Since

$$(p_k - 1)/p_k^{e_k} = \left[1/p_k^{e_k - 1} \right] - \left[1/p_k^{e_k} \right] \leq 2/9$$

(the largest possible value occurs when $p_j = 3$ and $e_j = 2$), we see that

$$\prod_{j=1}^{r} (n - 1, p_j - 1) \leq \prod_{j=1}^{r} (p_j - 1)$$

$$\leq \left[\prod_{\substack{j=1 \\ j \neq k}}^{r} p_j \right] \left[\frac{2}{9} p_k^{e_k} \right]$$

$$\leq \frac{2}{9} n.$$

Since $\frac{2}{9} n \leq \frac{1}{4}(n - 1)$ for $n \geq 9$, it follows that

$$\prod_{j=1}^{r} (n-1, p_j - 1) \le (n-1)/4.$$

Consequently, there are at most $(n-1)/4$ integers b, $1 \le b \le n$, for which n is a strong pseudoprime to the base b.

The other case to consider is when $n = p_1 p_2 \cdots p_r$, where $p_1, p_2, \ldots, p_r$ are distinct odd primes. Let

$$p_i - 1 = 2^{s_i} t_i, \quad i = 1, 2, \ldots, r,$$

where s_i is a positive integer and t_i is an odd positive integer. We reorder the primes $p_1, p_2, \ldots, p_r$ (if necessary) so that $s_1 \le s_2 \le \cdots \le s_r$. We note that

$$(n-1, p_i - 1) = 2^{\min(s, s_i)} (t, t_i).$$

The number of incongruent solutions of $x^t \equiv 1 \pmod{p_i}$ is $T = (t, t_i)$. From Exercise 22 at the end of this section, there are $2^j t_i$ incongruent solutions of $x^{2^{j}} \equiv -1 \pmod{p_i}$ when $0 \le j \le s_i - 1$, and no solutions otherwise. Hence, using the Chinese remainder theorem, there are $T_1 T_2 \cdots T_r$ incongruent solutions of $x^t \equiv 1 \pmod{n}$, and $2^{jr} T_1 T_2 \cdots T_r$ incongruent solutions of $x^{2^{j}t} \equiv -1 \pmod{n}$ when $0 \le j \le s_1 - 1$. Therefore, there are a total of

$$T_1 T_2 \cdots T_r \left[1 + \sum_{j=0}^{s_1 - 1} 2^{jr} \right] = T_1 T_2 \cdots T_r \left[1 + \frac{2^{rs_1} - 1}{2^r - 1} \right]$$

integers b with $1 \le b \le n-1$, for which n is a strong pseudoprime to the base b.

Now note that

$$\phi(n) = (p_1 - 1)(p_2 - 1) \cdots (p_r - 1) = t_1 t_2 \cdots t_r 2^{s_1 + s_2 + \cdots + s_r}.$$

We will show that

$$T_1 T_2 \cdots T_r \left[1 + \frac{2^{rs_1} - 1}{2^r - 1} \right] \le \phi(n)/4,$$

which proves the desired result. Because $T_1 T_2 \cdots T_r \le t_1 t_2 \cdots t_r$, we can achieve our goal by showing that

(8.3) $$\left[1 + \frac{2^{rs_1} - 1}{2^r - 1} \right] / 2^{s_1 + s_2 + \cdots + s_r} \le 1/4.$$

Since $s_1 \le s_2 \le \cdots \le s_r$, we see that

$$\left[1 + \frac{2^{rs_1} - 1}{2^r - 1}\right] / 2^{s_1 + s_2 + \cdots + s_r} \le \left[1 + \frac{2^{rs_1} - 1}{2^r - 1}\right] / 2^{rs_1}$$

$$= \frac{1}{2^{rs_1}} + \frac{2^{rs_1} - 1}{2^{rs_1}(2^r - 1)}$$

$$= \frac{1}{2^{rs_1}} + \frac{1}{2^{r-1}} - \frac{1}{2^{rs_1}(2^r - 1)}$$

$$= \frac{1}{2^r - 1} + \frac{2^r - 2}{2^{rs_1}(2^r - 1)}$$

$$\le \frac{1}{2^{r-1}}.$$

From this inequality, we conclude that (8.3) is valid when $r \le 3$.

When $r = 2$, we have $n = p_1 p_2$ with $p_1 - 1 = 2^{s_1} t_1$ and $p_2 - 1 = 2^{s_2} t_2$, with $s_1 \le s_2$. If $s_1 < s_2$, then (8.3) is again valid, since

$$\left[1 + \frac{2^{2s_1} - 1}{3}\right] / 2^{s_1 + s_2} = \left[1 + \frac{2^{2s_1} - 1}{3}\right] / \left(2^{2s_1} \cdot 2^{s_2 - s_1}\right)$$

$$= \left[\frac{1}{3} + \frac{1}{3 \cdot 2^{2s_1 - 1}}\right] / 2^{s_2 - s_1}$$

$$\le \frac{1}{4}.$$

When $s_1 = s_2$, we have $(n-1, p_1 - 1) = 2^s T_1$ and $(n-1, p_2 - 1) = 2^s T_2$. Let us assume that $p_1 > p_2$. Note that $T_1 \ne t_1$, for if $T_1 = t_1$, then $(p_1 - 1) \mid (n - 1)$, so that

$$n = p_1 p_2 \equiv p_2 \equiv 1 \pmod{p_1 - 1},$$

which implies that $p_2 > p_1$, a contradiction. Since $T_1 \ne t_1$, we know that $T_1 \le t_1/3$. Similarly, if $p_1 < p_2$ then $T_2 \ne t_2$, so that $T_2 \le t_2/3$. Hence, $T_1 T_2 \le t_1 t_2/3$, and since $\left[1 + \frac{2^{2s_1} - 1}{3}\right] / 2^{2s_1} \le \frac{1}{2}$, we have

$$T_1 T_2 \left[1 + \frac{2^{2s_1} - 1}{3}\right] \le t_1 t_2 \, 2^{2s_1}/6 = \phi(n)/6,$$

proving the theorem for this final case, since $\phi(n)/6 \le (n-1)/6 < (n-1)/4$. ∎

By analyzing the inequalities in the proof of Theorem 5.10, we can see that the probability that n is a strong pseudoprime to the randomly chosen base b, $1 \le b \le n - 1$, is close to $1/4$ only for integers n with prime factorizations of the form $n = p_1 p_2$ with $p_1 = 1 + 2q_1$ and $p_2 = 1 + 4q_2$, where q_1 and q_2 are odd primes, or $n = q_1 q_2 q_3$ with $p_1 = 1 + 2q_1$, $p_2 = 1 + 2q_2$, and $p_3 = 1 + 2q_3$, where q_1, q_2, and q_3 are distinct odd primes (see Exercise 23).

8.4 Exercises

1. Write out a table of indices modulo 23 with respect to the primitive root 5.

2. Find all the solutions of the congruences

   a) $3x^5 \equiv 1 \pmod{23}$ b) $3x^{14} \equiv 2 \pmod{23}$.

3. Find all the solutions of the congruences

   a) $3^x \equiv 2 \pmod{23}$ b) $13^x \equiv 5 \pmod{23}$.

4. For which positive integers a is the congruence $ax^4 \equiv 2 \pmod{13}$ solvable?

5. For which positive integers b is the congruence $8x^7 \equiv b \pmod{29}$ solvable?

6. Find the solutions of $2^x \equiv x \pmod{13}$, using indices to the base 2 modulo 13.

7. Find all the solutions of $x^x \equiv x \pmod{23}$.

8. Show that if p is an odd prime and r is a primitive root of p, then $\text{ind}_r(p-1) = (p-1)/2$.

9. Let p be an odd prime. Show that the congruence $x^4 \equiv -1 \pmod{p}$ has a solution if and only if p is of the form $8k + 1$.

10. Prove that there are infinitely many primes of the form $8k+1$. (*Hint*: Assume that $p_1, p_2, ..., p_n$ are the only primes of this form. Let $Q = (p_1 p_2 \cdots p_n)^4 + 1$. Show that Q must have an odd prime factor different than $p_1, p_2, ..., p_n$, and by Exercise 9, necessarily of the form $8k+1$.)

By Exercise 15 of Section 8.3, we know that if a is an odd positive integer, then there are unique integers α and β with $\alpha = 0$ or 1 and $0 \le \beta \le 2^{k-2}-1$ such that $a \equiv (-1)^\alpha 5^\beta \pmod{2^k}$. Define the *index system of a modulo 2^k* to be equal to the pair (α, β).

11. Find the index systems of 7 and 9 modulo 16.

12. Develop rules for the index systems modulo 2^k of products and powers analogous to the rules for indices.

13. Use the index system modulo 32 to find all solutions of $7x^9 \equiv 11 \pmod{32}$ and $3^x \equiv 17 \pmod{32}$.

Let $n = 2^{t_0} p_1^{t_1} p_2^{t_2} \cdots p_m^{t_m}$ be the prime-power factorization of n. Let a be an integer relatively prime to n. Let $r_1, r_2, ..., r_m$ be primitive roots of $p_1^{t_1}, p_2^{t_2}, ..., p_m^{t_m}$, respectively, and let $\gamma_1 = \text{ind}_{r_1} a \pmod{p_1^{t_1}}$, $\gamma_2 = \text{ind}_{r_2} a \pmod{p_2^{t_2}}, ...,$ $\gamma_m = \text{ind}_{r_m} a \pmod{p_m^{t_m}}$. If $t_0 \le 2$, let r_0 be a primitive root of 2^{t_0}, and let $\gamma_0 = \text{ind}_{r_0} a \pmod{2^t}$. If $t_0 \ge 3$, let (α, β) be the index system of a modulo 2^k, so that $a \equiv (-1)^\alpha 5^\beta \pmod{2^k}$. Define the *index system of a modulo n* to be $(\gamma_0, \gamma_1, \gamma_2, ..., \gamma_m)$ if $t_0 \le 2$ and $(\alpha, \beta, \gamma_1, \gamma_2, ..., \gamma_m)$ if $t_0 \ge 3$.

14. Show that if n is a positive integer, then every integer has a unique index system modulo n.

15. Find the index systems of 17 and 41 (mod 120) (in your computations, use 2 as a primitive root of the prime factor 5 of 120).

16. Develop rules for the index systems modulo n of products and powers analogous to those for indices.

17. Use an index system modulo 60 to find the solutions of $11x^7 \equiv 43$ (mod 60).

18. Let p be a prime, $p > 3$. Show that if $p \equiv 2$ (mod 3) then every integer not divisible by 3 is a third-power, or *cubic*, residue of p, while if $p \equiv 1$ (mod 3), an integer a is a cubic residue of p if and only if $a^{(p-1)/3} \equiv 1$ (mod p).

19. Let e be a positive integer with $e \geq 2$. Show that if k is an odd positive integer, then every odd integer a is a kth power residue of 2^e.

* 20. Let e be a positive integer with $e \geq 2$. Show that if k is even, then an integer a is a kth power residue of 2^e if and only if $a \equiv 1$ (mod $(4k, 2^e)$).

* 21. Let e be a positive integer with $e \geq 2$. Show that if k is a positive integer, then the number of incongruent kth power residues of 2^e is

$$\frac{2^{e-1}}{(n,2)(n,2^{e-2})}.$$

☞ 22. Let $N = 2^j u$ be a positive integer with j a nonnegative integer and u an odd positive integer and let $p - 1 = 2^s t$, where s and t are positive integers with t odd. Show that there are $2^j(t,u)$ incongruent solutions of $x^N \equiv -1$ (mod p) if $0 \leq j \leq s - 1$, and no solutions otherwise.

* 23. a) Show that the probability that n is a strong pseudoprime for a base b randomly chosen with $1 \leq b \leq n - 1$ is near $(n-1)/4$ only when n has a prime factorization of the form $n = p_1 p_2$ where $p_1 = 1 + 2q_1$ and $p_2 = 1 + 4q_2$ with q_1 and q_2 prime or $n = p_1 p_2 p_3$ where $p_1 = 1 + 2q_1, p_2 = 1 + 2q_2, p_3 = 1 + 2q_3$ with q_1, q_2, q_3 distinct odd primes.

    b) Find the probability that $n = 49939 \cdot 99877$ is a strong pseudoprime to the base b randomly chosen with $1 \leq b \leq n - 1$.

8.4 Computer Projects

Programming Projects

Write programs to do the following:

1. Construct a table of indices modulo a particular primitive root of an integer.

2. Using indices, solve congruences of the form $ax^b \equiv c$ (mod m) where a, b, c, and m are integers with $c > 0$, $m > 0$, and where m has a primitive root.

3. Find kth power residues of a positive integer m having a primitive root, where k is a positive integer.

4. Find index systems modulo powers of 2 (see preamble to Exercise 11).

5. Find index systems modulo arbitrary positive integers (see preamble to Exercise 14).

Computations and Explorations

Using the programs you have written or a computation program, carry out the following computations and explorations:

1. Find integers n for which the probability that n is a strong pseudoprime to the randomly chosen base b, $1 \le b \le n - 1$ is close to 1/4.

8.5 Primality Tests Using Primitive Roots

From the concepts of orders of integers and primitive roots, we can produce useful primality tests. The following theorem, which is essentially due to Lucas, presents such a test.

Theorem 8.18. If n is a positive integer and if an integer x exists such that

$$x^{n-1} \equiv 1 \pmod{n}$$

and

$$x^{(n-1)/q} \not\equiv 1 \pmod{n}$$

for all prime divisors q of $n - 1$, then n is prime.

Proof. Since $x^{n-1} \equiv 1 \pmod{n}$, Theorem 8.1 tells us that $\mathrm{ord}_n x \mid (n-1)$. We will show that $\mathrm{ord}_n x = n - 1$. Suppose that $\mathrm{ord}_n x \ne n - 1$. Since $\mathrm{ord}_n x \mid (n-1)$, there is an integer k with $n - 1 = k \cdot \mathrm{ord}_n x$ and since $\mathrm{ord}_n x \ne n - 1$, we know that $k > 1$. Let q be a prime divisor of k. Then

$$x^{(n-1)/q} = x^{k/(\mathrm{ord}_n x \cdot q)} = (x^{\mathrm{ord}_n x})^{(k/q)} \equiv 1 \pmod{n}.$$

However, this contradicts the hypotheses of the theorem, so we must have $\mathrm{ord}_n x = n - 1$. Now, since $\mathrm{ord}_n x \le \phi(n)$ and $\phi(n) \le n - 1$, it follows that $\phi(n) = n - 1$. By Theorem 6.2 we know that n must be prime. ∎

Note that Theorem 8.18 is equivalent to the fact that if there is an integer with order modulo n equal to $n - 1$, then n must be prime. We illustrate the use of Theorem 8.18 with an example.

Example 8.20. Let $n = 1009$. Then $11^{1008} \equiv 1 \pmod{1009}$. The prime divisors of 1008 are 2, 3, and 7. We see that $11^{1008/2} = 11^{504} \equiv -1 \pmod{1009}$, $11^{1008/3} = 11^{336} \equiv 374 \pmod{1009}$, and $11^{1008/7} = 11^{144} \equiv 935 \pmod{1009}$. Hence, by Theorem 8.18 we know that 1009 is prime. □

The following corollary of Theorem 8.18 gives a slightly more efficient primality test.

Corollary 8.18.1. If n is an odd positive integer and if x is a positive integer such that

$$x^{(n-1)/2} \equiv -1 \pmod{n}$$

and

$$x^{(n-1)/q} \not\equiv 1 \pmod{n}$$

for all odd prime divisors q of $n - 1$, then n is prime.

Proof. Since $x^{(n-1)/2} \equiv -1 \pmod{n}$, we see that

$$x^{n-1} = (x^{(n-1)/2})^2 \equiv (-1)^2 \equiv 1 \pmod{n}.$$

Since the hypotheses of Theorem 8.18 are met, we know that n is prime. ∎

Example 8.21. Let $n = 2003$. The odd prime divisors of $n-1 = 2002$ are 7, 11, and 13. Since $5^{2002/2} = 5^{1001} \equiv -1 \pmod{2003}$, $5^{2002/7} = 5^{286} \equiv 874 \pmod{2003}$, $5^{2002/11} = 5^{183} \equiv 886 \pmod{2003}$, and $5^{2002/13} = 5^{154} \equiv 633 \pmod{2003}$, we see from Corollary 8.18.1 that 2003 is prime. □

To determine whether an integer n is prime using either Theorem 8.18 or Corollary 8.18.1 it is necessary to know the prime factorization of $n - 1$. As we have remarked before, finding the prime factorization of an integer is a time-consuming process. Only when we have some *a priori* information about the factorization of $n - 1$ are the primality tests given by these results practical. Indeed, with such information these tests can be useful. Such a situation occurs with the Fermat numbers; in Chapter 9 we give a primality test for these numbers based on the ideas of this section.

How quickly a computer can verify primality or compositeness? We answer this question as follows.

Theorem 8.19. If n is composite, this can be proved with $O((\log_2 n)^2)$ bit operations.

Proof. If n is composite, there are integers a and b with $1 < a < n$, $1 < b < n$, and $n = ab$. Hence, given the two integers a and b, we multiply a and b and verify that $n = ab$, taking $O((\log_2 n)^2)$ bit operations to prove that n is composite. ∎

We can use Corollary 8.18.1 to estimate the number of bit operations needed to prove primality when the appropriate information is known.

Theorem 8.20. If n is prime, this can be proven using $O((\log_2 n)^4)$ bit operations.

Proof. We use the second principle of mathematical induction. The induction hypothesis is an estimate for $f(n)$, where $f(n)$ is the total number of multiplications and modular exponentiations needed to verify that the integer n is prime.

We demonstrate that

$$f(n) \leq 3 \ (\log n / \log 2) - 2.$$

First, we note that $f(2) = 1$. We assume that for all primes q, with $q < n$, the inequality

$$f(q) \leq 3 \ (\log q / \log 2) - 2$$

holds.

To prove that n is prime, we use Corollary 8.18.1. Once we have the numbers 2^a, $q_1, \ldots, q_t$, and x that supposedly satisfy

(i) $n - 1 = 2^a q_1 q_2 \cdots q_t$,

(ii) q_i is prime for $i = 1, 2, \ldots, t$,

(iii) $x^{(n-1)/2} \equiv -1 \pmod{n}$,

and

(iv) $x^{(n-1)/q_i} \equiv 1 \pmod{n}$, for $i = 1, 2, \ldots, t$,

we need to do t multiplications to check (i), $t + 1$ modular exponentiations to check (iii) and (iv), and $f(q_i)$ multiplications and modular exponentiations to check (ii), that q_i is prime for $i = 1, 2, \ldots, t$. Hence,

$$f(n) = t + (t+1) + \sum_{i=1}^{t} f(q_i)$$
$$\leq 2t + 1 + \sum_{i=1}^{t} ((3 \log q_i / \log 2) - 2)$$

Now each multiplication requires $O((\log_2 n)^2)$ bit operations and each modular exponentiation requires $O((\log_2 n)^3)$ bit operations. Since the total number of multiplications and modular exponentiations needed is $f(n) = O(\log_2 n)$, the total number of bit operations needed is $O((\log_2 n)(\log_2 n)^3) = O((\log_2 n)^4)$. ■

Theorem 8.20 was discovered by Pratt. He interpreted the result as showing that every prime has a "succinct certification of primality." It should be noted that Theorem 8.20 cannot be used to find this short proof of primality, for the factorization of $n - 1$ and the primitive root x of n are required. More information on this subject may be found in Lenstra [101].

Recently, an extremely efficient primality test has been developed by Adleman, Pomerance, and Rumely. We will not describe the test here because it relies on concepts not developed in this book. We note that to determine whether an integer is prime using this test requires fewer than $(\log_2 n)^{c \ \log_2 \log_2 \log_2 n}$ bit operations, where c is a constant. For instance, to determine whether a 100-digit integer is prime requires just 40 seconds and to determine whether a 200-digit integer is prime requires just 10 minutes. Even a 1000-digit integer may be checked for primality in a reasonable amount of time, one week. For more information about this test see [90] and [105].

8.5 Exercises

1. Show that 101 is prime using Theorem 8.18 with $x = 2$.

2. Show that 211 is prime using Theorem 8.18 with $x = 2$.

3. Show that 233 is prime using Corollary 8.18.1 with $x = 3$.

4. Show that 257 is prime using Corollary 8.18.1 with $x = 3$.

5. Show that if an integer x exists such that

$$x^{2^{2^n}} \equiv 1 \pmod{F_n}$$

and

$$x^{2^{(2^n-1)}} \not\equiv 1 \pmod{F_n},$$

then the Fermat number $F_n = 2^{2^n} + 1$ is prime.

* 6. Let n be a positive integer. Show that if the prime-power factorization of $n - 1$ is $n - 1 = p_1^{a_1} p_2^{a_2} \cdots p_t^{a_t}$ and for $j = 1, 2,\ldots, t$, there exists an integer x_j such that

$$x_j^{(n-1)/p_j} \not\equiv 1 \pmod{n}$$

and

$$x_j^{n-1} \equiv 1 \pmod{n},$$

then n is prime.

* 7. Let n be a positive integer such that

$$n - 1 = m \prod_{j=1}^{r} q_j^{a_j}$$

where m is a positive integer, $a_1, a_2,\ldots, a_r$ are positive integers, and $q_1, q_2,\ldots, q_r$ are relatively prime integers greater than one. Furthermore, let $b_1, b_2,\ldots, b_r$ be positive integers such that there exist integers $x_1, x_2,\ldots, x_r$ with

$$x_j^{n-1} \equiv 1 \pmod{n}$$

and

$$(x_j^{(n-1)/q_j} - 1, n) = 1$$

for $j = 1, 2,\ldots, r$, where every prime factor of q_j is greater than or equal to b_j for $j = 1, 2,\ldots, r$, and

$$n < (1 + \prod_{j=1}^{r} b_j^{a_j})^2.$$

Show that n is prime.

8.5 Computer Projects

Programming Projects

Write programs to show that a positive integer n is prime using

1. Theorem 8.18.

2. Corollary 8.18.1.

3. Exercise 4.

4. Exercise 5.

Computations and Explorations

Using the programs you have written or a computation program, carry out the following computations and explorations:

1. Give a succinct certification of primality of $F_4 = 2^{2^4} + 1 = 65537$.

8.6 Universal Exponents

Let n be a positive integer with prime-power factorization

$$n = p_1^{t_1} p_2^{t_2} \cdots p_m^{t_m}.$$

If a is an integer relatively prime to n, then Euler's theorem tells us that

$$a^{\phi(p^t)} \equiv 1 \pmod{p^t}$$

whenever p^t is one of the prime powers occurring in the factorization of n . As in the proof of Theorem 8.13, let

$$U = [\phi(p_1^{t_1}), \phi(p_2^{t_2}), \ldots, \phi(p_m^{t_m})],$$

the least common multiple of the integers $\phi(p_i^{t_i})$, $i = 1, 2, \ldots, m$. Since

$$\phi(p_i^{t_i}) \mid U$$

for $i = 1, 2, \ldots, m$, using Theorem 8.1 we see that

$$a^U \equiv 1 \pmod{p_i^{t_i}}$$

for $i = 1, 2, \ldots, m$. Hence, by Corollary 3.8.1 it follows that

$$a^U \equiv 1 \pmod{n}.$$

This leads to the following definition.

Definition. A *universal exponent* of the positive integer n is a positive integer U

such that

$$a^U \equiv 1 \ (\text{mod } n),$$

for all integers a relatively prime to n.

Example 8.22. Since the prime power factorization of 600 is $2^3 \cdot 3 \cdot 5^2$, it follows that $U = [\phi(2^3), \phi(3), \phi(5^2)] = [2, 2, 20] = 20$ is a universal exponent of 600. □

From Euler's theorem, we know that $\phi(n)$ is a universal exponent. As we have already demonstrated, the integer $U = [\phi(p_1^{t_1}), \phi(p_2^{t_2}), ..., \phi(p_m^{t_m})]$ is also a universal exponent of $n = p_1^{t_1} p_2^{t_2} \cdots p_m^{t_m}$. We are interested in finding the *smallest* positive universal exponent of n.

Definition. The least universal exponent of the positive integer n is called the *minimal universal exponent of n*, and is denoted by $\lambda(n)$.

We now find a formula for the minimal universal exponent $\lambda(n)$, based on the prime-power factorization of n.

First, note that if n has a primitive root, then $\lambda(n) = \phi(n)$. Since powers of odd primes possess primitive roots, we know that

$$\lambda(p^t) = \phi(p^t),$$

whenever p is an odd prime and t is a positive integer. Similarly, we have $\lambda(2) = \phi(2) = 1$ and $\lambda(4) = \phi(4) = 2$, since both 2 and 4 have primitive roots. On the other hand, if $t \geq 3$, then we know by Theorem 8.11 that

$$a^{2^{t-2}} \equiv 1 \ (\text{mod } 2^t)$$

and $\text{ord}_{2^t} a = 2^{t-2}$, so that we can conclude that $\lambda(2^t) = 2^{t-2}$ if $t \geq 3$.

We have found $\lambda(n)$ when n is a power of a prime. Next, we turn our attention to arbitrary positive integers n.

Theorem 8.21. Let n be a positive integer with prime-power factorization

$$n = 2^{t_0} p_1^{t_1} p_2^{t_2} \cdots p_m^{t_m}.$$

Then $\lambda(n)$, the minimal universal exponent of n, is given by

$$\lambda(n) = [\lambda(2^{t_0}), \phi(p_1^{t_1}), ..., \phi(p_m^{t_m})],$$

Moreover, there exists an integer a such that $\text{ord}_n a = \lambda(n)$, the largest possible order of an integer modulo n.

Proof. Let a be an integer with $(a, n) = 1$. For convenience, let

$$M = [\lambda(2^{t_0}), \phi(p_1^{t_1}), \phi(p_2^{t_2}), ..., \phi(p_m^{t_m})].$$

Since M is divisible by all of the integers $\lambda(2^{t_0})$, $\phi(p_1^{t_1}) = \lambda(p_1^{t_1})$, $\phi(p_2^{t_2}) = \lambda(p_2^{t_2}), ..., \phi(p_m^{t_m}) = \lambda(p_m^{t_m})$, and since $a^{\lambda(p^t)} \equiv 1 \ (\text{mod } p^t)$ for all prime-powers in the factorization of n, we see that

$$a^M \equiv 1 \pmod{p^t},$$

whenever p^t is a prime-power occurring in the factorization of n.

Consequently, by Corollary 3.8.1 we can conclude that

$$a^M \equiv 1 \pmod{n}.$$

The last congruence establishes the fact that M is a universal exponent. We must now show that M is the *least* universal exponent. To do this, we find an integer a such that no positive power smaller than the Mth power of a is congruent to 1 modulo n. With this in mind, let r_i be a primitive root of $p_i^{t_i}$.

We consider the system of simultaneous congruences

$$x \equiv 3 \pmod{2^{t_0}}$$
$$x \equiv r_1 \pmod{p_1^{t_1}}$$
$$x \equiv r_2 \pmod{p_2^{t_2}}$$

$$\cdot$$
$$\cdot$$
$$\cdot$$

$$x \equiv r_m \pmod{p_m^{t_m}}.$$

By the Chinese remainder theorem, there is a simultaneous solution a of this system which is unique modulo $n = 2^{t_0} p_1^{t_1} p_2^{t_2} \cdots p_m^{t_m}$; we will show that $\mathrm{ord}_n\, a = M$. To prove this claim, assume that N is a positive integer such that

$$a^N \equiv 1 \pmod{n}.$$

Then, if p^t is a prime-power divisor of n, we have

$$a^N \equiv 1 \pmod{p^t},$$

so that

$$\mathrm{ord}_{p^t}\, a \mid N.$$

But, since a satisfies each of the $m + 1$ congruences of the system, we have

$$\mathrm{ord}_{p^t}\, a = \lambda(p^t),$$

for each prime power in the factorization. Hence by Theorem 8.1 we have

$$\lambda(p^t) \mid N$$

for all prime powers p^t in the factorization of n. Therefore, by Corollary 3.8.1, we know that $M = [\lambda(2^{t_0}),\lambda(p_1^{t_1}),\lambda(p_2^{t_2}),...,\lambda(p_m^{t_m})] \mid N$

Since $a^M \equiv 1 \pmod{n}$ and $M \mid N$ whenever $a^N \equiv 1 \pmod{n}$, we can conclude that

$$\mathrm{ord}_n\, a = M.$$

This shows that $M = \lambda(n)$ and simultaneously produces a positive integer a with $\mathrm{ord}_n\, a = \lambda(n)$.   ∎

Example 8.23. Since the prime-power factorization of 180 is $2^2 \cdot 3^2 \cdot 5$, from Theorem 8.21 it follows that

$$\lambda(180) = [\phi(2^2), \phi(3^2), \phi(5)] = [2, 6, 4] = 12.$$

To find an integer a with $\text{ord}_{180}\, a = 12$, first we find primitive roots modulo 3^2 and 5. For instance, we take 2 and 3 as primitive roots modulo 3^2 and 5, respectively. Then, using the Chinese remainder theorem, we find a solution of the system of congruences

$$a \equiv 3 \pmod{4}$$
$$a \equiv 2 \pmod{9}$$
$$a \equiv 3 \pmod{5},$$

obtaining $a \equiv 83 \pmod{180}$. From the proof of Theorem 8.21 we see that $\text{ord}_{180}\, 83 = 12$. $\qquad\square$

Example 8.24. Let $n = 2^6 3^2 5 \cdot 7 \cdot 13 \cdot 17 \cdot 19 \cdot 37 \cdot 73$. Then, we have

$$\begin{aligned}
\lambda(n) &= [\lambda(2^6), \phi(3^2), \phi(5), \phi(17), \phi(19), \phi(37), \phi(73)] \\
&= [2^4, 2 \cdot 3, 2^2, 2^4, 2 \cdot 3^2, 2^2 3^2, 2^3 3^2] \\
&= 2^4 \cdot 3^2 \\
&= 144.
\end{aligned}$$

Hence, whenever a is a positive integer relatively prime to $2^6 \cdot 3^2 \cdot 5 \cdot 17 \cdot 17 \cdot 19 \cdot 37 \cdot 73$ we know that $a^{144} \equiv 1 \pmod{2^6 \cdot 3^2 \cdot 5 \cdot 17 \cdot 19 \cdot 37 \cdot 37 \cdot 73}$. $\qquad\square$

We now return to the Carmichael numbers that we discussed in Section 5.2. Recall that a Carmichael number is a composite integer that satisfies $b^{n-1} \equiv 1 \pmod{n}$ for all positive integers b with $(b, n) = 1$. We proved that if $n = q_1 q_2 \cdots q_k$, where $q_1, q_2, \ldots, q_k$ are distinct primes satisfying $(q_j - 1) \mid (n-1)$ for $j = 1, 2, \ldots, k$, then n is a Carmichael number. Here, we prove the converse of this result.

Theorem 8.22. If $n > 2$ is a Carmichael number, then $n = q_1 q_2 \cdots q_k$, where the q_j's are distinct primes such that $(q_j - 1) \mid (n-1)$ for $j = 1, 2, \ldots, k$.

Proof. If n is a Carmichael number, then

$$b^{n-1} \equiv 1 \pmod{n}$$

for all positive integers b with $(b, n) = 1$. Theorem 8.21 tells us that there is an integer a with $\text{ord}_n a = \lambda(n)$, where $\lambda(n)$ is the minimal universal exponent, and since $a^{n-1} \equiv 1 \pmod{n}$, Theorem 8.1 tells us that

$$\lambda(n) \mid (n-1).$$

Now n must be odd, for if n was even, then $n-1$ would be odd, but $\lambda(n)$ is even (since $n > 2$), contradicting the fact that $\lambda(n) \mid (n-1)$.

We now show that n must be the product of distinct primes. Suppose n has a prime-power factor p^t with $t \geq 2$. Then

$$\lambda(p^t) = \phi(p^t) = p^{t-1}(p-1) \mid \lambda(n) = n-1.$$

This implies that $p \mid (n-1)$, which is impossible since $p \mid n$. Consequently, n must be the product of distinct odd primes, say

$$n = q_1 q_2 \cdots q_k.$$

We conclude the proof by noting that

$$\lambda(q_i) = \phi(q_i) = (q_j - 1) \mid \lambda(n) = n-1. \qquad \blacksquare$$

We can easily prove more about the prime factorizations of Carmichael numbers.

Theorem 8.23. A Carmichael number must have at least three different odd prime factors.

Proof. Let n be a Carmichael number. Then n cannot have just one prime factor, since it is composite, and is the product of distinct primes. So assume that $n = pq$, where p and q are odd primes with $p > q$. Then

$$n-1 = pq-1 = (p-1)q + (q-1) \equiv q-1 \not\equiv 0 \pmod{p-1},$$

which shows that $(p-1) \nmid (n-1)$. Hence, n cannot be a Carmichael number if it has just two different prime factors. $\qquad \blacksquare$

8.6 Exercises

1. Find $\lambda(n)$, the minimal universal exponent of n, for the following values of n.

   a) 100
   b) 144
   c) 222
   d) 884

   e) $2^4 \cdot 3^3 \cdot 5^2 \cdot 7$
   f) $2^5 \cdot 3^2 \cdot 5^2 \cdot 7^3 \cdot 11^2 \cdot 13 \cdot 17 \cdot 19$
   g) 10!
   h) 20!

2. Find all positive integers n such that $\lambda(n)$ is equal to each of the following integers.

   a) 1
   b) 2
   c) 3

   d) 4
   e) 5
   f) 6

3. Find the largest integer n with $\lambda(n) = 12$.

4. Find an integer with the largest possible order modulo

   a) 12
   b) 15
   c) 20

   d) 36
   e) 40
   f) 63.

5. Show that if m is a positive integer, then $\lambda(m)$ divides $\phi(m)$.

6. Show that if m and n are relatively prime positive integers, then $\lambda(mn) = [\lambda(m),\lambda(n)]$.

7. Let n be the largest positive integer satisfying the equation $\lambda(n) = a$, where a is a fixed positive integer. Show that if m is another solution of $\lambda(m) = a$, then m divides n.

8. Suppose that n is a positive integer. How many incongruent integers are there with maximal order modulo n?

9. Show that if a and m are relatively prime positive integers, then the solutions of the congruence $ax \equiv b \pmod{m}$ are the integers x such that $x \equiv a^{\lambda(m)-1}b \pmod{m}$.

10. Show that if c is a positive integer greater than one, then the integers $1^c, 2^c,..., (m-1)^c$ form a complete system of residues modulo m if and only if m is square-free and $(c,\lambda(m)) = 1$.

* 11. a) Show that if c and m are positive integers then the congruence $x^c \equiv x \pmod{m}$ has exactly

$$\prod_{j=1}^{r} (1 + (c-1, \phi(p_j^{a_j})))$$

incongruent solutions, where m has prime-power factorization $m = p_1^{a_1}p_2^{a_2} \cdots p_r^{a_r}$.

b) Show that $x^c \equiv x \pmod{m}$ has exactly 3^r solutions if $(c-1, \phi(m)) = 2$.

12. Use Exercise 11 to show that there are always at least 9 plaintext messages that are not changed when enciphered using an RSA cipher.

* 13. Show that 561 is the only Carmichael number of the form $3pq$ where p and q are primes.

* 14. Find all Carmichael numbers of the form $5pq$ where p and q are primes.

* 15. Show that there are only a finite number of Carmichael numbers of the form $n = pqr$, where p is a fixed prime, and q and r are also primes.

16. Show that the deciphering exponent d for an RSA cipher with enciphering key (e,n) can be taken to be an inverse of e modulo $\lambda(n)$.

Let n be a positive integer. When $(a,n) = 1$ we define the *generalized Fermat quotient* $q_n(a)$ by $q_n(a) \equiv (a^{\lambda(n)} - 1)/n \pmod{n}$ and $0 \leq q_n(a) < n$.

17. Show that if $(a,n) = (b,n) = 1$ then $q_n(ab) \equiv q_n(a) + q_n(b) \pmod{n}$.

18. Show that if $(a,n) = 1$ then $q_n(a + nc) \equiv q_n(a) + \lambda(n)c\bar{a} \pmod{n}$, where $\bar{a}$ is the inverse of a modulo n.

8.6 Computer Projects

Programming Projects

Write programs to do the following:

1. Find the minimal universal exponent of a positive integer.

2. Find an integer with the minimal universal exponent of n as its order modulo n.

3. Given a positive integer M, find all positive integers n with minimal universal exponent equal to M.

4. Solve linear congruences using the method of Exercise 9.

Computations and Explorations

Using the programs you have written or a computation program, carry out the following computations and explorations:

1. Find the universal exponent of all integers less than 1000.

2. Find Carmichael numbers with at least four different prime factors.

8.7 Pseudo-Random Numbers

Numbers chosen randomly are often useful in computer simulation of complicated phenomena. To perform simulations, some method for generating random numbers is needed. There are various mechanical means for generating random numbers, but these are inefficient for computer use. Instead, a systematic method using computer arithmetic is preferable. One such method, called the *middle-square method*, introduced by John von Neumann*, works as follows. To generate four-digit random numbers, we start with an arbitrary four-digit number, say 6139. We square this number to obtain 37687321, and we take the middle four digits 6873 as the second random number. We iterate this procedure to obtain a sequence of random numbers, always squaring and removing the middle four-digits to obtain a new random number from the preceding one. (The square of a four-digit number has eight or fewer digits. Those with fewer than eight digits are considered eight-digit numbers by adding initial digits of 0.)

---

* **JOHN VON NEUMANN** (1903-1957) was born in Budapest, Hungary. In 1930, after holding several positions at universities in Germany, he came to the United States. In 1933 Von Neumann became, along with Albert Einstein, one of the first members of the famous Institute for Advanced Study in Princeton, New Jersey. Von Neumann was one of the most versatile mathematical talents of the twentieth century. He invented the mathematical discipline known as game theory. Using game theory, he made many important discoveries in mathematical ecomonics. Von Neumann made fundamental contributions to the development of the first computers, and participated in the early development of atomic weapons.

Sequences produced by the middle-square method are, in reality, not randomly chosen. When the initial four-digit number is known, the entire sequence is determined. However, the sequence of numbers produced appears to be random, and the numbers produced are useful for computer simulations. The integers in sequences that have been chosen in some methodical manner, but appear to be random, are called *pseudo-random numbers.*

It turns out that the middle-square method has some unfortunate weaknesses. The most undesirable feature of this method is that, for many choices of the initial integer, the method produces the same small set of numbers over and over. For instance, starting with the four-digit integer 4100 and using the middle-square method, we obtain the sequence 8100,6100,2100,4100,8100,6100,2100,... which only gives four different numbers before repeating.

The most commonly used method for generating pseudo-random numbers is called the *linear congruential method* which works as follows. Integers m, a, c, and x_0 are chosen so that $2 \leq a < m$, $0 \leq c < m$, and $0 \leq x_0 \leq m$. The sequence of pseudo-random numbers is defined recursively by

$$x_{n+1} \equiv ax_n + c \ (\text{mod } m), \quad 0 \leq x_{n+1} < m,$$

for $n = 0,1,2,3,....$ We call m the *modulus*, a the *multiplier*, c the *increment*, and x_0 the *seed* of the pseudo-random number generator. The following examples illustrate the linear congruential method.

Example 8.25. When we take $m = 12$, $a = 3$, $c = 4$, and $x_0 = 5$ in the linear congruential generator, we have $x_1 \equiv 3 \cdot 5 + 4 \equiv 7 \ (\text{mod } 12)$, so that $x_1 = 7$. Similarly, we find that $x_2 = 1$, since $x_2 \equiv 3 \cdot 7 + 4 \equiv 1 \ (\text{mod } 12)$, $x_3 = 7$, since $x_3 \equiv 3 \cdot 1 + 4 \equiv 7 \ (\text{mod } 12)$, and so on. Hence, the generator produces just three different integers before repeating. The sequence of pseudo-random numbers obtained is 5,7,1,7,1,7,1,.... □

Example 8.26. When we take $m = 9$, $a = 7$, $c = 4$, and $x_0 = 3$ in the linear congruential generator, we obtain the sequence 3,7,8,6,1,2,0,4,5,3,... (as should be verified by the reader). This sequence contains 9 different numbers before repeating. □

Remark: For computer simulations it is often necessary to generate pseudo-random numbers between 0 and 1. We can obtain such numbers by using a linear congruential generator to produce pseudo-random numbers x_i, $i = 1,2,3,...$ between 0 and m and then dividing each number by m, obtaining the sequence x_i/m, $i = 1,2,3,....$

The following theorem tells us how to find the terms of a sequence of pseudo-random numbers generated by the linear congruential method directly from the multiplier, the increment, and the seed.

Theorem 8.25. The terms of the sequence generated by the linear congruential method previously described are given by

$$x_k \equiv a^k x_0 + c(a^k - 1)/(a - 1) \ (\text{mod } m), \quad 0 \le x_k < m.$$

Proof. We prove this result using mathematical induction. For $k = 1$, the formula is obviously true, since $x_1 \equiv ax_0 + c \ (\text{mod } m)$, $0 \le x_1 < m$. Assume that the formula is valid for the kth term, so that

$$x_k \equiv a^k x_0 + c(a^k - 1)/(a - 1) \ (\text{mod } m), \quad 0 \le x_k < m.$$

Since

$$x_{k+1} \equiv ax_k + c \ (\text{mod } m), \quad 0 \le x_{k+1} < m,$$

we have

$$
\begin{aligned}
x_{k+1} &\equiv a(a^k x_0 + c(a^k - 1)/(a - 1)) + c \\
&\equiv a^{k+1} x_0 + c(a(a^k - 1)/(a - 1) + 1) \\
&\equiv a^{k+1} x_0 + c(a^{k+1} - 1)/(a - 1) \ (\text{mod } m),
\end{aligned}
$$

which is the correct formula for the $(k+1)$st term. This demonstrates that the formula is correct for all positive integers k. ∎

The *period length* of a linear-congruential pseudo-random number generator is the maximum length of the sequence obtained without repetition. We note that the longest possible period length for a linear congruential generator is the modulus m. The following theorem tells us when this maximum length is obtained.

Theorem 8.26. The linear congruential generator produces a sequence of period length m if and only if $(c,m) = 1$, $a \equiv 1 \ (\text{mod } p)$ for all primes p dividing m, and $a \equiv 1 \ (\text{mod } 4)$ if $4 \mid m$.

Because the proof of Theorem 8.26 is complicated and quite lengthy we omit it. The reader is referred to Knuth [91] for a proof.

The case of the linear congruential generator with $c = 0$ is of special interest because of its simplicity. In this case, the method is called the *pure multiplicative congruential method*. We specify the modulus m, multiplier a, and seed x_0. The sequence of pseudo-random numbers is defined recursively by

$$x_{n+1} \equiv ax_n \ (\text{mod } m), \quad 0 < x_{n+1} < m.$$

In general, we can express the pseudo-random numbers generated in terms of the multiplier and seed:

$$x_n \equiv a^n x_0 \ (\text{mod } m), \quad 0 < x_{n+1} < m.$$

If l is the period length of the sequence obtained using this pure multiplicative generator, then l is the smallest positive integer such that

$$x_0 \equiv a^l x_0 \ (\text{mod } m).$$

If $(x_0, m) = 1$, using Corollary 3.4.1 we have

$$a^l \equiv 1 \ (\text{mod } m).$$

From this congruence, we know that the largest possible period length is $\lambda(m)$, where $\lambda(m)$ is the minimal universal exponent modulo m.

For many applications, the pure multiplicative generator is used with the modulus m equal to the Mersenne prime $M_{31} = 2^{31} - 1$. When the modulus m is a prime, the maximum period length is $m - 1$, and this is obtained when a is a primitive root of m. To find a primitive root of M_{31} that can be used with good results, we first demonstrate that 7 is a primitive root of M_{31}.

Theorem 8.27. The integer 7 is a primitive root of $M_{31} = 2^{31} - 1$.

Proof. To show that 7 is a primitive root of $M_{31} = 2^{31} - 1$, it is sufficient to show that

$$7^{(M_{31}-1)/q} \not\equiv 1 \pmod{M_{31}}$$

for all prime divisors q of $M_{31} - 1$. With this information, we can conclude that $\operatorname{ord}_{M_{31}} 7 = M_{31} - 1$. To find the factorization of $M_{31} - 1$, we note that

$$
\begin{aligned}
M_{31} - 1 &= 2^{31} - 2 = 2(2^{30} - 1) = 2(2^{15} - 1)(2^{15} + 1) \\
&= 2(2^5 - 1)(2^{10} + 2^5 + 1)(2^5 + 1)(2^{10} - 2^5 + 1) \\
&= 2 \cdot 3^2 \cdot 7 \cdot 11 \cdot 31 \cdot 151 \cdot 331.
\end{aligned}
$$

If we show that

$$7^{(M_{31}-1)/q} \not\equiv 1 \pmod{M_{31}}$$

for $q = 2, 3, 7, 11, 31, 151$, and 331, then we know that 7 is a primitive root of $M_{31} = 2147483647$. Since

$$
\begin{aligned}
7^{(M_{31}-1)/2} &\equiv 2147483646 \not\equiv 1 \pmod{M_{31}} \\
7^{(M_{31}-1)/3} &\equiv 1513477735 \not\equiv 1 \pmod{M_{31}} \\
7^{(M_{31}-1)/7} &\equiv 120536285 \not\equiv 1 \pmod{M_{31}} \\
7^{(M_{31}-1)/11} &\equiv 1969212174 \not\equiv 1 \pmod{M_{31}} \\
7^{(M_{31}-1)/31} &\equiv 512 \not\equiv 1 \pmod{M_{31}} \\
7^{(M_{31}-1)/151} &\equiv 535044134 \not\equiv 1 \pmod{M_{31}} \\
7^{(M_{31}-1)/331} &\equiv 1761885083 \not\equiv 1 \pmod{M_{31}},
\end{aligned}
$$

we see that 7 is a primitive root of M_{31}. ∎

In practice we do not want to use the primitive root 7 as the generator, since the first few integers generated are small. Instead, we find a larger primitive root using Corollary 8.4.1. We use 7^k where $(k, M_{31} - 1) = 1$. For instance, since $(5, M_{31} - 1) = 1$, we know that $7^5 = 16807$ is a primitive root. Since $(13, M_{31} - 1) = 1$, another possibility is to use $7^{13} \equiv 252246292 \pmod{M_{31}}$ as the multiplier.

We have only briefly touched the subject of pseudo-random numbers. For a thorough discussion of pseudo-random numbers see Knuth [91] and for a survey of the relationships between pseudo-random number generators and cryptography, see the chapter by Lagarias in [84].

8.7 Exercises

1. Find the sequence of two-digit pseudo-random numbers generated using the middle-square method, taking 69 as the seed.

2. Find the first ten terms of the sequence of pseudo-random numbers generated by the linear congruential method with $x_0 = 6$ and $x_{n+1} \equiv 5x_n + 2 \pmod{19}$. What is the period length of this generator?

3. Find the period length of the sequence of pseudo-random numbers generated by the linear congruential method with $x_0 = 2$ and $x_{n+1} \equiv 4x_n + 7 \pmod{25}$.

4. Show that if either $a = 0$ or $a = 1$ is used for the multiplier in the generation of pseudo-random numbers by the linear congruential method, the resulting sequence would not be a good choice for a sequence of pseudo-random numbers.

5. Using Theorem 8.26, find those integers a which give period length m, where $(c, m) = 1$, for the linear congruential generator $x_{n+1} \equiv ax_n + c \pmod{m}$, for each of the following moduli.

   a) $m = 1000$ c) $m = 10^6 - 1$
   b) $m = 30030$ d) $m = 2^{25} - 1$

* 6. Show that every linear congruential pseudo-random number generator can be simply expressed in terms of a linear congruential generator with increment $c = 1$ and seed 0, by showing that the terms generated by the linear congruential generator $x_{n+1} \equiv ax_n + c \pmod{m}$, with seed x_0, can be expressed as $x_n \equiv b \cdot y_n + x_0 \pmod{m}$, where $b \equiv (a - 1)x_0 + c \pmod{m}$, $y_0 = 0$, and $y_{n+1} \equiv ay_n + 1 \pmod{m}$.

7. Find the period length of the pure multiplicative pseudo-random number generator $x_n \equiv cx_{n-1} \pmod{2^{31} - 1}$ for each of the following multipliers c.

   a) 2   c) 4   e) 13
   b) 3   d) 5   f) 17

8. Show that the maximal possible period length for a pure multiplicative generator of the form $x_{n+1} \equiv ax_n \pmod{2^e}$, $e \geq 3$, is 2^{e-2}. Show that this is obtained when $a \equiv \pm 3 \pmod 8$.

Another way to generate pseudo-random numbers is to use the *Fibonacci generator*. Let m be a positive integer. Two initial integers x_0 and x_1 both less than m are specified and the rest of the sequence is generated recursively by the congruence $x_{n+1} \equiv x_n + x_{n-1} \pmod{m}$, $0 \leq x_{n+1} < m$.

9. Find the first eight pseudo-random numbers generated by the Fibonacci generator with modulus $m = 31$ and initial values $x_0 = 1$ and $x_1 = 24$.

10. Find a good choice for the multiplier a in the pure multiplicative pseudo-random number generator $x_{n+1} \equiv ax_n \pmod{101}$. (*Hint:* Find a primitive root of 101 that is not too small.)

11. Find a good choice for the multiplier a in the pure multiplicative pseudo-random number generator $x_n \equiv ax_{n-1} \pmod{2^{25} - 1}$. (*Hint*: Find a primitive root of $2^{25} - 1$ and then take an appropriate power of this root.)

12. Find the multiplier a and increment c of the linear congruential pseudo-random number generator $x_{n+1} \equiv ax_n + c \pmod{1003}$, $0 \le x_{n+1} < 1003$, if $x_0 = 1$, $x_2 = 402$, and $x_3 = 361$.

13. Find the multiplier a of the pure multiplicative pseudo-random number generator $x_{n+1} \equiv ax_n \pmod{1000}$, $0 \le x_{n+1} < 1000$, if 313 and 145 are consecutive terms generated.

14. The *discrete exponential generator* takes a positive integer x_0 as its seed and generates pseudo-random numbers $x_1, x_2, x_3,...$ using the recursive definition $x_{n+1} \equiv g^{x_n} \pmod{p}$, $0 < x_{n+1} < p$, for $n = 0,1,2,...$ where p is an odd prime and g is a primitive root modulo p

    a) Find the sequence of pseudo-random numbers generated using the discrete exponential generator with $p = 17$, $g = 3$, and $x_0 = 2$.
    b) Find the sequence of pseudo-random numbers generated using the discrete exponential generator with $p = 47$, $g = 5$, and $x_0 = 3$.
    c) Given a term of a sequence of pseudo-random numbers generated using a discrete exponential generator, can the previous term be found easily when the prime p and primitive root g are known?

15. Another method of generating pseudo-random numbers is to use the *power generator* with parameters m, d. Here m is a positive integer and d is a positive integer relatively prime to $\phi(m)$. The generator starts with a positive integer x_0 as its seed and generates pseudo-random numbers $x_1, x_2, x_3,...$ using the recursive definition $x_{n+1} \equiv x_n^d \pmod{m}$, $0 < x_{n+1} < m$.

    a) Find the sequence of pseudo-random numbers generated using a power generator with $m = 15$, $d = 3$, and seed $x_0 = 2$.
    b) Find the sequence of pseudo-random numbers generated using a power generator with $m = 23$, $d = 2$, and seed $x_0 = 3$.

8.7 Computer Projects

Programming Projects

Write programs to generate pseudo-random numbers using the following generators:

1. The middle-sequence generator.

2. The linear congruential generator.

3. The pure multiplicative generator.

4. The Fibonacci generator (see the preamble to Exercise 9).

5. The discrete exponential generator (see Exercise 14).

6. The power generator (see Exercise 15).

Computations and Explorations

Using the programs you have written or a computation program, carry out the following computations and explorations:

1. Examine the behavior of the sequence of five digit pseudo-random numbers produced by the middle square method starting with different choices of the initial term.

2. Find the period length of different linear congruential pseudo-random generators of your choice.

3. Find the period length of different *quadratic congruential* pseudo-random number generators, that is, generators of the form $x_{n+1} \equiv (ax_n^2 + bx_n + c) \pmod{m}$, $0 \le x_{n+1} < m$, where a, b, and c are integers. Can you find conditions that guarantee that the period of this generator is m?

4. Determine the length of the period of the Fibonacci generator described in Exercise 9 for various choices of the modulus m. Do you think this is a good generator of pseudo-random numbers?

5. There are a variety of empirical tests to measure the randomness of pseudo-random number generators. Ten such tests are described in Knuth [91]. Look up these tests and apply some of them to different pseudo-random number generators.

8.8 An Application to the Splicing of Telephone Cables

An interesting application of the preceding material involves the splicing of telephone cables. We base our discussion on the exposition of Ore [31], who relates the contents of an original article by Lawther [115], reporting on work done for the Southwestern Bell Telephone Company.

To develop the application, we first make the following definition.

Definition. Let m be a positive integer and let a be an integer relatively prime to m. The ± 1-*exponent of a modulo m* is the smallest positive integer x such that

$$a^x \equiv \pm 1 \pmod{m}.$$

We are interested in determining the largest possible ± 1-exponent of an integer modulo m; we denote this by $\lambda_0(m)$. The following two theorems relate the value of the maximal ± 1-exponent $\lambda_0(m)$ to $\lambda(m)$, the minimal universal exponent modulo m.

First, we consider positive integers that possess primitive roots.

Theorem 8.28. If m is a positive integer, $m > 2$, with a primitive root, then the maximal ± 1-exponent $\lambda_0(m)$ equals $\phi(m)/2 = \lambda(m)/2$.

Proof. We first note that if m has a primitive root, then $\lambda(m) = \phi(m)$. By

Theorem 6.6 we know that $\phi(m)$ is even, so that $\phi(m)/2$ is an integer, if $m > 2$. Euler's Theorem tells us that

$$a^{\phi(m)} = (a^{\phi(m)/2})^2 \equiv 1 \pmod{m},$$

for all integers a with $(a,m) = 1$. By Exercise 13 of Section 8.3, we know that when m has a primitive root, the only solutions of $x^2 \equiv 1 \pmod{m}$ are $x \equiv \pm 1 \pmod{m}$. Hence,

$$a^{\phi(m)/2} \equiv \pm 1 \pmod{m}.$$

This implies that

$$\lambda_0(m) \leq \phi(m)/2.$$

Now let r be a primitive root of modulo m with ± 1-exponent e. Then

$$r^e \equiv \pm 1 \pmod{m},$$

so that

$$r^{2e} \equiv 1 \pmod{m}.$$

Since $\operatorname{ord}_m r = \phi(m)$, Theorem 8.1 tells us that $\phi(m) \mid 2e$, or equivalently, that $(\phi(m)/2) \mid e$. Hence, the maximum ± 1-exponent $\lambda_0(m)$ is at least $\phi(m)/2$. However, we know that $\lambda(m) \leq \phi(m)/2$. Consequently, $\lambda_0(m) = \phi(m)/2 = \lambda(m)/2$. ∎

We now will find the maximal ± 1-exponent of integers without primitive roots.

Theorem 8.29. If m is a positive integer without a primitive root, then the maximal ± 1-exponent $\lambda_0(m)$ equals $\lambda(m)$, the minimal universal exponent of m.

Proof. We first show that if a is an integer of order $\lambda(m)$ modulo m with ± 1-exponent e such that

$$a^{\lambda(m)/2} \not\equiv -1 \pmod{m},$$

then $e = \lambda(m)$. Consequently, once we have found such an integer a, we will have shown that $\lambda_0(m) = \lambda(m)$.

Assume that a is an integer of order $\lambda(m)$ modulo m with ± 1-exponent e such that

$$a^{\lambda(m)/2} \not\equiv -1 \pmod{m}.$$

Since $a^e \equiv \pm 1 \pmod{m}$, it follows that $a^{2e} \equiv 1 \pmod{m}$. By Theorem 8.1 we know that $\lambda(m) \mid 2e$. Since $\lambda(m) \mid 2e$ and $e \leq \lambda(m)$, either $e = \lambda(m)/2$ or $e = \lambda(m)$. To see that $e \neq \lambda(m)/2$, note that $a^e \equiv \pm 1 \pmod{m}$, but $a^{\lambda(m)/2} \not\equiv 1 \pmod{m}$, since $\operatorname{ord}_m a = \lambda(m)$, and $a^{\lambda(m)/2} \not\equiv -1 \pmod{m}$, by hypothesis. Therefore, we can conclude that if $\operatorname{ord}_m a = \lambda(m)$, a has ± 1-exponent e, and $a^e \equiv -1 \pmod{m}$, then $e = \lambda(m)$.

We now find an integer a with the desired properties. Let the prime-power factorization of m be $m = 2^{t_0} p_1^{t_1} p_2^{t_2} \cdots p_s^{t_s}$. We consider several cases.

We first consider those m with at least two different odd prime factors. Among the prime-powers $p_i^{t_i}$ dividing m, let $p_j^{t_j}$ be one with the smallest power of 2 dividing $\phi(p_j^{t_j})$. Let r_i be a primitive root of $p_i^{t_i}$ for $i = 1, 2,...,s$. Let a be an integer satisfying the simultaneous congruences

$$a \equiv 3 \pmod{2^{t_0}},$$
$$a \equiv r_i \pmod{p_i^{t_i}} \quad \text{for all } i \text{ with } i \neq j,$$
$$a \equiv r_j^2 \pmod{p_j^{t_j}}.$$

Such an integer a is guaranteed to exist by the Chinese remainder theorem. Note that

$$\text{ord}_m a = [\lambda(2^{t_0}), \phi(p_i^{t_2}),..., \phi(p_j^{t_j})/2 ,..., \phi(p_s^{t_s})],$$

and, by our choice of $p_j^{t_j}$, we know that this least common multiple equals $\lambda(m)$. Since $a \equiv r_j^2 \pmod{p_j^{t_j}}$, it follows that $a^{\phi(p_j^{t_j})/2} \equiv r_j^{\phi(p_j^{t_j})} \equiv 1 \pmod{p_j^{t_j}}$. Because $\phi(p_j^{t_j})/2 \mid \lambda(m)/2$, we know that

$$a^{\lambda(m)/2} \equiv 1 \pmod{p_j^{t_j}},$$

so that

$$a^{\lambda(m)/2} \not\equiv -1 \pmod{m}.$$

Consequently, the ±1-exponent of a is $\lambda(m)$.

The next case we consider deals with integers of the form $m = 2^{t_0}p^{t_1}$, where p is an odd prime, $t_1 \geq 1$ and $t_0 \geq 2$, since m has no primitive roots. When $t_0 = 2$ or 3, we have

$$\lambda(m) = [2, \phi(p_1^{t_1})] = \phi(p_1^{t_1}).$$

Let a be a solution of the simultaneous congruences

$$a \equiv 1 \pmod{4}$$
$$a \equiv r \pmod{p_1^{t_1}},$$

where r is a primitive root of $p_1^{t_1}$. We see that $\text{ord}_m a = \lambda(m)$. Because

$$a^{\lambda(m)/2} \equiv 1 \pmod{4},$$

we know that

$$a^{\lambda(m)/2} \not\equiv -1 \pmod{m}.$$

Consequently, the ±1-exponent of a is $\lambda(m)$.

When $t_0 \geq 4$, let a be a solution of the simultaneous congruences

$$a \equiv 3 \pmod{2^{t_0}}$$
$$a \equiv r \pmod{p_1^{t_1}};$$

the Chinese remainder theorem tells us that such an integer exists. We see that $\text{ord}_m a = \lambda(m)$. Since $4 \mid \lambda(2^{t_0})$, we know that $4 \mid \lambda(m)$. Hence,

$$a^{\lambda(m)/2} \equiv 3^{\lambda(m)/2} \equiv (3^2)^{\lambda(m)/4} \equiv 1 \pmod{8}.$$

Thus,

$$a^{\lambda(m)/2} \not\equiv -1 \pmod{m},$$

so that the ± 1-exponent of a is $\lambda(m)$.

Finally, when $m = 2^{t_0}$ with $t_0 \geq 3$, from Theorem 8.12 we know that $\operatorname{ord}_m 5 = \lambda(m)$, but

$$5^{\lambda(m)/2} \equiv (5^2)^{(\lambda(m)/4)} \equiv 1 \pmod{8}.$$

Therefore, we see that

$$5^{\lambda(m)/2} \not\equiv -1 \pmod{m};$$

we conclude that the ± 1-exponent of 5 is $\lambda(m)$.

This finishes the argument since we have dealt with all cases where m does not have a primitive root. ∎

We now develop a system for splicing telephone cables. Telephone cables are made up of concentric layers of insulated copper wire, as illustrated in Figure 8.1, and are produced in sections of specified length.

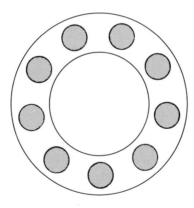

Figure 8.1. A cross-section of one layer of a telephone cable.

Telephone lines are constructed by splicing together sections of cable. When two wires are adjacent in the same layer in multiple sections of the cable, there are often problems with interference and crosstaik. Consequently, two wires adjacent in the same layer in one section should not be adjacent in the same layer in any nearby sections. For practical purpose, the splicing system should be simple. We use the following rules to describe the system. Wires in concentric layers are spliced to wires in the corresponding layers of the next section, following identical splicing direction at each connection. In a layer with m wires, we connect the wire in position j in one section, where $1 \leq j \leq m$ to the wire in position $S(j)$ in the next section, where $S(j)$ is the least positive residue of $1 + (j-1)s$ modulo m. Here, s is called the *spread* of the splicing system. We see that when a wire in one section is spliced to a wire in the next section, the adjacent wire in the first section is spliced to the wire in the next section in the position obtained by counting forward s modulo m from the position of the last wire spliced in this section. To have a one-to-one correspondence between wires of adjacent sections, we require that the spread s be relatively prime to the number of wires m. This shows that if wires in positions j and k are sent to the same wire in the next section, then $S(j) = S(k)$ and

$$1 + (j-1)s \equiv 1 + (k-1)s \pmod{m},$$

so that $js \equiv ks \pmod{m}$. Since $(m, s) = 1$, from Corollary 3.4.1 we see that $j \equiv k \pmod{m}$, which is impossible.

Example 8.27. Let us connect 9 wires with a spread of 2. We have the correspondence

$1 \rightarrow 1$	$2 \rightarrow 3$	$3 \rightarrow 5$
$4 \rightarrow 7$	$5 \rightarrow 9$	$6 \rightarrow 2$
$7 \rightarrow 4$	$8 \rightarrow 6$	$9 \rightarrow 8.$

$\square$

This is illustrated in Figure 8.2.

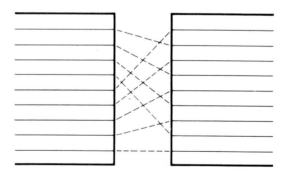

Figure 8.2. Splicing of 9 wires with spread of 2.

The following result tells us the correspondence of wires in the first section of cable to the wires in the nth section.

Theorem 8.30. Let $S_n(j)$ denote the position of the wire in the nth section spliced to the jth wire of the first section. Then

$$S_n(j) \equiv 1 + (j-1)s^{n-1} \pmod{m}.$$

Proof. For $n = 2$, by the rules for the splicing system, we have

$$S_2(j) \equiv 1 + (j-1)s \pmod{m},$$

so the proposition is true for $n = 2$. Now assume that

(8.4) $$S_n(j) \equiv 1 + (j-1)s^{n-1} \pmod{m}.$$

Then, in the next section, we have the wire in position $S_n(j)$ spliced to the wire in position

$$\begin{aligned} S_{n+1}(j) &\equiv 1 + (S_n(j)-1)s \\ &\equiv 1 + ((j-1)s^{n-1})s \\ &\equiv 1 + (j-1)s^n \pmod{m}. \end{aligned}$$

This shows that the proposition is true. ∎

In a splicing system, we want to have wires adjacent in one section separated as long as possible in the following sections. After n splices, Theorem 8.30 tells us that the adjacent wires in the jth and $(j + 1)$th positions are connected to wires in positions $S_n(j) \equiv 1 + (j-1)s^n \pmod{m}$ and $S_n(j+1) = 1 + js^n \pmod{m}$, respectively. These wires are adjacent in the nth section if, and only if,

$$S_n(j) - S_n(j+1) \equiv \pm 1 \pmod{m},$$

or equivalently,

$$(1 + (j-1)s^n) - (1+js^n) \equiv \pm 1 \pmod{m},$$

which holds if and only if

$$s^n \equiv \pm 1 \pmod{m}.$$

We can now apply the material at the beginning of this section. To keep adjacent wires in the first section separated as long as possible, we should pick for the spread s an integer with maximal ± 1-exponent $\lambda_0(m)$.

Example 8.28. With 100 wires, we should choose a spread s so that the ± 1-exponent of s is $\lambda_0(100) = \lambda(100) = 20$. The appropriate computations show that $s = 3$ is such a spread. □

8.8 Exercises

1. Find the maximal ± 1-exponent of each of the following positive integers.

a)	17	d)	36
b)	22	e)	99
c)	24	f)	100

2. Find an integer with maximal ± 1-exponent modulo of the following positive integers.

   a)   13   d) 25
   b)   14   e) 36
   c)   15   f) 60

3. Devise a splicing scheme for telephone cables containing each of the following numbers of wires.

   a)   50 wires   b)   76 wires   c)   125 wires

* 4. Show that using any splicing system of telephone cables with m wires arranged in a concentric layer, adjacent wires in one section can be kept separated in at most $[(m-1)/2]$ successive sections of cable. Show that when m is prime this upper limit is achieved using the system developed in this section.

8.8 Computer Projects

Programming Projects

Write programs to do the following:

1. Given an integer m find the maximal ± 1-exponent of m.

2. Develop a scheme for splicing telephone cables as described in this section.

Computations and Explorations

Using the programs you have written or a computation program, carry out the following computations and explorations:

1. Find the maximal ± 1-exponent of each positive integer less than 1000.

9

Quadratic Residues

9.1 Quadratic Residues and Nonresidues

Let p be an odd prime and a an integer relatively prime to p. In this chapter we devote our attention to the question: Is a a perfect square modulo p? We begin with a definition.

Definition. If m is a positive integer, we say that the integer a is a *quadratic residue of* m if $(a,m) = 1$ and the congruence $x^2 \equiv a \pmod{m}$ has a solution. If the congruence $x^2 \equiv a \pmod{m}$ has no solution, we say that a is a *quadratic nonresidue of* m.

Example 9.1. To determine which integers are quadratic residues of 11, we compute the squares of the integers $1,2,3,...,10$. We find that $1^2 \equiv 10^2 \equiv 1 \pmod{11}$, $2^2 \equiv 9^2 \equiv 4 \pmod{11}$, $3^2 \equiv 8^2 \equiv 9 \pmod{11}$, $4^2 \equiv 7^2 \equiv 5 \pmod{11}$, and $5^2 \equiv 6^2 \equiv 3 \pmod{11}$. Hence, the quadratic residues of 11 are $1,3,4,5,9$; the integers $2,6,7,8,10$ are quadratic nonresidues of 11. □

Note that the quadratic residues of the positive integer m are just the kth power residues of m with $k=2$, as defined in Section 8.4. We will show that if p is an odd prime, then there are exactly as many quadratic residues as quadratic nonresidues of p among the integers $1,2,...,p-1$. To demonstrate this fact, we use the following lemma.

Lemma 9.1. Let p be an odd prime and a an integer not divisible by p. Then, the congruence

$$x^2 \equiv a \pmod{p}$$

has either no solutions or exactly two incongruent solutions modulo p.

Proof. If $x^2 \equiv a \pmod p$ has a solution, say $x = x_0$, then we can easily demonstrate that $x = -x_0$ is a second incongruent solution. Since $(-x_0)^2 = x_0^2 \equiv a \pmod p$, we see that $-x_0$ is a solution. We note that $x_0 \not\equiv -x_0 \pmod p$, for if $x_0 \equiv -x_0 \pmod p$, then we have $2x_0 \equiv 0 \pmod p$. This is impossible since p is odd and $p \nmid x_0$ (since $x_0^2 \equiv a \pmod p$ and $p \nmid a$).

To show that there are no more than two incongruent solutions, assume that $x = x_0$ and $x = x_1$ are both solutions of $x^2 \equiv a \pmod p$. Then we have $x_0^2 \equiv x_1^2 \equiv a \pmod p$, so that $x_0^2 - x_1^2 = (x_0 + x_1)(x_0 - x_1) \equiv 0 \pmod p$. Hence, $p \mid (x_0 + x_1)$ or $p \mid (x_0 - x_1)$, so that $x_1 \equiv -x_0 \pmod p$ or $x_1 \equiv x_0 \pmod p$. Therefore, if there is a solution of $x^2 \equiv a \pmod p$, there are exactly two incongruent solutions. ∎

This leads us to the following theorem.

Theorem 9.1. If p is an odd prime, then there are exactly $(p-1)/2$ quadratic residues of p and $(p-1)/2$ quadratic nonresidues of p among the integers $1, 2, \ldots, p-1$.

Proof. To find all the quadratic residues of p among the integers $1, 2, \ldots, p-1$ we compute the least positive residues modulo p of the squares of the integers $1, 2, \ldots, p-1$. Since there are $p-1$ squares to consider and since each congruence $x^2 \equiv a \pmod p$ has either zero or two solutions, there must be exactly $(p-1)/2$ quadratic residues of p among the integers $1, 2, \ldots, p-1$. The remaining $p-1 - (p-1)/2 = (p-1)/2$ positive integers less than $p-1$ are quadratic nonresidues of p. ∎

The special notation associated with quadratic residues is described in the following definition.

Definition. Let p be an odd prime and a an integer not divisible by p. The *Legendre symbol* $\left(\dfrac{a}{p}\right)$ is defined by

$$\left(\frac{a}{p}\right) = \begin{cases} 1 & \text{if } a \text{ is a quadratic residue of } p \\ -1 & \text{if } a \text{ is a quadratic nonresidue of } p. \end{cases}$$

This symbol is named after the French mathematician Adrien-Marie Legendre* who introduced the use of this notation.

Example 9.2. The previous example shows that the Legendre symbols $\left(\dfrac{a}{11}\right)$, $a = 1, 2,...,10$, have the following values:

$$\left(\frac{1}{11}\right) = \left(\frac{3}{11}\right) = \left(\frac{4}{11}\right) = \left(\frac{5}{11}\right) = \left(\frac{9}{11}\right) = 1,$$

$$\left(\frac{2}{11}\right) = \left(\frac{6}{11}\right) = \left(\frac{7}{11}\right) = \left(\frac{8}{11}\right) = \left(\frac{10}{11}\right) = -1.$$

□

We now present a criterion for deciding whether an integer is a quadratic residue of a prime. This criterion is useful in demonstrating properties of the Legendre symbol.

Theorem 9.2. Euler's Criterion. Let p be an odd prime and let a be a positive integer not divisible by p. Then

$$\left(\frac{a}{p}\right) \equiv a^{(p-1)/2} \ (\text{mod } p).$$

Proof. First, assume that $\left(\dfrac{a}{p}\right) = 1$. Then, the congruence $x^2 \equiv a \ (\text{mod } p)$ has a solution, say $x = x_0$. Using Fermat's little theorem, we see that

$$a^{(p-1)/2} = (x_0^2)^{(p-1)/2} = x_0^{p-1} \equiv 1 \ (\text{mod } p).$$

Hence, if $\left(\dfrac{a}{p}\right) = 1$, we know that $\left(\dfrac{a}{p}\right) \equiv a^{(p-1)/2} \ (\text{mod } p).$

---

* **ADRIEN-MARIE LEGENDRE** (1752-1833) was born into a well-to-do family. He was professor at the Ecole Militaire in Paris from 1775 to 1780. In 1795 he was appointed professor at the Ecole Normale. His memoir *Recherches d'Analyse Indeterineé* published in 1785 contains a discussion of the law of quadratic reciprocity, a statement of Dirichlet's theorem on primes in arithmetic progressions, and a discussion of the representation of positive integers as the sum of three squares. He established the $n=5$ case of Fermat's last theorem. Legendre wrote a textbook on geometry, *Eléments de geometrie* that was used for over 100 years, and served as a model for other textbooks. Legendre made fundamental discoveries in mathematical astronomy and geodesy. He gave the first treatment of the law of least squares.

Now consider the case where $\left(\dfrac{a}{p}\right) = -1$. Then, the congruence $x^2 \equiv a \pmod{p}$ has no solutions. By Theorem 3.10 for each integer i such that $1 \leq i \leq p-1$, there is a unique integer j with $1 \leq j \leq p-1$, such that $ij \equiv a \pmod{p}$. Furthermore, since the congruence $x^2 \equiv a \pmod{p}$ has no solutions, we know that $i \neq j$. Thus, we can group the integers $1, 2, \ldots, p-1$ into $(p-1)/2$ pairs each with product a. Multiplying these pairs together, we find that

$$(p-1)! \equiv a^{(p-1)/2} \pmod{p}.$$

Since Wilson's theorem tells us that $(p-1)! \equiv -1 \pmod{p}$, we see that

$$-1 \equiv a^{(p-1)/2} \pmod{p}.$$

In this case, we also have $\left(\dfrac{a}{p}\right) \equiv a^{(p-1)/2} \pmod{p}$.   ∎

Example 9.3. Let $p = 23$ and $a = 5$. Since $5^{11} \equiv -1 \pmod{23}$, Euler's criterion tells us that $\left(\dfrac{5}{23}\right) = -1$. Hence, 5 is a quadratic nonresidue of 23.   □

We now prove some properties of the Legendre symbol.

Theorem 9.3. Let p be an odd prime and a and b integers not divisible by p. Then

$$\text{(i)} \quad \text{if } a \equiv b \pmod{p}, \text{ then } \left(\frac{a}{p}\right) = \left(\frac{b}{p}\right).$$

$$\text{(ii)} \quad \left(\frac{a}{p}\right)\left(\frac{b}{p}\right) = \left(\frac{ab}{p}\right).$$

$$\text{(iii)} \quad \left(\frac{a^2}{p}\right) = 1.$$

Proof of (i). If $a \equiv b \pmod{p}$, then $x^2 \equiv a \pmod{p}$ has a solution if and only if $x^2 \equiv b \pmod{p}$ has a solution. Hence, $\left(\dfrac{a}{p}\right) = \left(\dfrac{b}{p}\right)$.

Proof of (ii). By Euler's criterion we know that

$$\left(\frac{a}{p}\right) \equiv a^{(p-1)/2} \pmod{p}, \quad \left(\frac{b}{p}\right) \equiv b^{(p-1)/2} \pmod{p},$$

and

$$\left(\frac{ab}{p}\right) \equiv (ab)^{(p-1)/2} \pmod{p}.$$

Hence,

$$\left(\frac{a}{b}\right)\left(\frac{b}{p}\right) \equiv a^{(p-1)/2}b^{(p-1)/2} = (ab)^{(p-1)/2} \equiv \left(\frac{ab}{p}\right) \pmod{p}.$$

Since the only possible values of a Legendre symbol are ± 1, we conclude that

$$\left(\frac{a}{p}\right)\left(\frac{b}{p}\right) = \left(\frac{ab}{p}\right).$$

Proof of (iii). Since $\left(\dfrac{a}{p}\right) = \pm 1$, from part (ii) it follows that

$$\left(\frac{a^2}{p}\right) = \left(\frac{a}{p}\right)\left(\frac{a}{p}\right) = 1.$$
∎

Part (ii) of Theorem 9.3 has the following interesting consequence. The product of two quadratic residues, or of two quadratic nonresidues, of a prime is a quadratic residue of that prime, whereas the product of a quadratic residue and a quadratic nonresidue is a quadratic nonresidue.

Using Euler's criterion, we can classify those primes having -1 as a quadratic residue.

Theorem 9.4. If p is an odd prime, then

$$\left(\frac{-1}{p}\right) = \begin{cases} 1 & \text{if } p \equiv 1 \pmod 4 \\ -1 & \text{if } p \equiv -1 \pmod 4. \end{cases}$$

Proof. By Euler's criterion we know that

$$\left(\frac{-1}{p}\right) \equiv (-1)^{(p-1)/2} \pmod{p}.$$

If $p \equiv 1 \pmod 4$, then $p = 4k + 1$ for some integer k. Thus,

$$(-1)^{(p-1)/2} = (-1)^{2k} = 1,$$

so that $\left(\dfrac{-1}{p}\right) = 1$. If $p \equiv 3 \pmod 4$, then $p = 4k + 3$ for some integer k. Thus,

$$(-1)^{(p-1)/2} = (-1)^{2k+1} = -1,$$

so that $\left(\dfrac{-1}{p}\right) = -1$.
∎

The following elegant result of Gauss provides another criterion to determine whether an integer a relatively prime to the prime p is a quadratic residue of p.

Lemma 9.2. Gauss' Lemma. Let p be an odd prime and a an integer with $(a,p) = 1$. If s is the number of least positive residues of the integers a, $2a$, $3a$,...,$((p-1)/2)a$ that are greater than $p/2$, then $\left(\dfrac{a}{p}\right) = (-1)^s$.

Proof. Consider the integers $a,2a,...,((p-1)/2)a$. Let $u_1,u_2,...,u_s$ be the least positive residues of those that are greater than $p/2$, and let v_1, $v_2,...,v_t$ be the least positive residues of those integers that are less than $p/2$. Since $(ja,p) = 1$ for all j with $1 \le j \le (p-1)/2$, these least positive residues are in the set $1,2,...,p-1$.

We will show that $p-u_1$, $p-u_2$, ..., $p-u_s$, v_1, $v_2,...,v_t$ comprise the set of integers 1, 2,...,$(p-1)/2$, in some order. To see this, we need only show that no two of these integers are congruent modulo p, since there are exactly $(p-1)/2$ numbers in the set, and all are positive integers not exceeding $(p-1)/2$.

Clearly no two of the u_i's are congruent modulo p and no two of the v_j's are congruent modulo p; if a congruence of either of these two sorts held, we would have $ma \equiv na \pmod{p}$ where m and n are both positive integers not exceeding $(p-1)/2$. Since $p \nmid a$, this implies that $m \equiv n \pmod{p}$ which is impossible.

In addition, one of the integers $p - u_i$ cannot be congruent to a v_j, for if such a congruence held, we would have $ma \equiv p - na \pmod{p}$, so that $ma \equiv -na \pmod{p}$. Since $p \nmid a$, this implies that $m \equiv -n \pmod{p}$. This is impossible because both m and n are in the set $1,2,...,(p-1)/2$.

Now that we know that $p - u_1, p - u_2,...,p - u_s, v_1, v_2, ..., v_t$ are the integers $1,2,...,(p-1)/2$, in some order, we conclude that

$$(p-u_1)(p-u_2) \cdots (p-u_s)v_1 v_2 \cdots v_t \equiv \left(\frac{p-1}{2}\right)! \pmod{p},$$

which implies that

(9.1) $$(-1)^s u_1 u_2 \cdots u_s v_1 v_2 \cdots v_t \equiv \left(\frac{p-1}{2}\right)! \pmod{p}.$$

But, since $u_1, u_2,...,u_s, v_1, v_2,...,v_t$ are the least positive residues of a, $2a$,...,$((p-1)/2)a$, we also know that

(9.2) $\qquad u_1 u_2 \cdots u_s v_1 v_2 \cdots v_t \equiv a \cdot 2a \cdots \left[\dfrac{p-1}{2}\right] a$

$$= a^{\frac{p-1}{2}} \left[\dfrac{p-1}{2}\right]! \ (\text{mod } p).$$

Hence, from (9.1) and (9.2), we see that

$$(-1)^s a^{\frac{p-1}{2}} \left[\dfrac{p-1}{2}\right]! \equiv \left[\dfrac{p-1}{2}\right]! \ (\text{mod } p).$$

Because $(p,((p-1)/2)!) = 1$, this congruence implies that

$$(-1)^s a^{\frac{p-1}{2}} \equiv 1 \ (\text{mod } p).$$

By multiplying both sides by $(-1)^s$, we obtain

$$a^{\frac{p-1}{2}} \equiv (-1)^s \ (\text{mod } p).$$

Since Euler's criterion tells us that $a^{\frac{p-1}{2}} \equiv \left[\dfrac{a}{p}\right] \ (\text{mod } p)$, it follows that

$$\left[\dfrac{a}{p}\right] \equiv (-1)^s \ (\text{mod } p),$$

establishing Gauss' lemma. ∎

Example 9.4. Let $a = 5$ and $p = 11$. To find $\left[\dfrac{5}{11}\right]$ by Gauss' lemma, we compute the least positive residues of $1 \cdot 5$, $2 \cdot 5$, $3 \cdot 5$, $4 \cdot 5$, and $5 \cdot 5$. These are 5, 10, 4, 9, and 3, respectively. Since exactly two of these are greater than 11/2, Gauss' lemma tells us that $\left[\dfrac{5}{11}\right] = (-1)^2 = 1$. □

Using Gauss' lemma, we can characterize all primes that have 2 as a quadratic residue.

Theorem 9.5. If p is an odd prime, then

$$\left[\dfrac{2}{p}\right] = (-1)^{(p^2-1)/8} .$$

Hence, 2 is a quadratic residue of all primes $p \equiv \pm1 \ (\text{mod } 8)$ and a quadratic nonresidue of all primes $p \equiv \pm3 \ (\text{mod } 8)$.

Proof. By Gauss' lemma we know that if s is the number of least positive residues of the integers

$$1 \cdot 2, \; 2 \cdot 2, \; 3 \cdot 2, \; ..., \; \left[\frac{p-1}{2}\right] \cdot 2$$

that are greater than $p/2$, then $\left[\dfrac{2}{p}\right] = (-1)^s$. Since all these integers are less than p, we only need to count those greater than $p/2$ to find how many have least positive residue greater than $p/2$.

The integer $2j$, where $1 \le j \le (p-1)/2$, is less than $p/2$ when $j \le p/4$. Hence, there are $[p/4]$ integers in the set less than $p/2$. Consequently, there are $s = \dfrac{p-1}{2} - [p/4]$ greater than $p/2$. Therefore, by Gauss' lemma we see that

$$\left[\frac{2}{p}\right] = (-1)^{\frac{p-1}{2} - [p/4]}.$$

To prove the theorem, we must show that

$$\frac{p-1}{2} - [p/4] \equiv (p^2-1)/8 \pmod{2}.$$

To establish this, we need to consider the congruence class of p modulo 8, since, as we will see, both sides of the above congruence depend only on the congruence class of p modulo 8.

We first consider $(p^2-1)/8$. If $p \equiv \pm 1 \pmod 8$, then $p = 8k \pm 1$ where k is an integer, so that

$$(p^2-1)/8 = ((8k \pm 1)^2 - 1)/8 = (64k^2 \pm 16k)/8 = 8k^2 \pm 2k \equiv 0 \pmod 2.$$

If $p \equiv \pm 3 \pmod 8$, then $p = 8k \pm 3$ where k is an integer, so that

$$(p^2-1)/8 = ((8k \pm 3)^2 - 1)/8 = (64k^2 \pm 48k + 8)/8 = 8k^2 + 6k + 1$$
$$\equiv 1 \pmod 2.$$

Now consider $\dfrac{p-1}{2} - [p/4]$. If $p \equiv 1 \pmod 8$, then $p = 8k + 1$ for some integer k and

$$\frac{p-1}{2} - [p/4] = 4k - [2k + 1/4] = 2k \equiv 0 \pmod 2;$$

if $p \equiv 3 \pmod 8$, then $p = 8k + 3$ for some integer k, and

$$\frac{p-1}{2} - [p/4] = 4k + 1 - [2k + 3/4] = 2k + 1 \equiv 1 \pmod 2;$$

if $p \equiv 5 \pmod 8$, then $p = 8k + 5$ for some integer k, and

$$\frac{p-1}{2} - [p/4] = 4k + 2 - [2k + 5/4] = 2k + 1 \equiv 1 \pmod 2;$$

if $p \equiv 7 \pmod 8$, then $p = 8k + 7$ for some integer k, and

$$\frac{p-1}{2} - [p/4] = 4k + 3 - [2k + 7/4] = 2k + 2 \equiv 0 \pmod 2.$$

Comparing the congruence classes modulo 2 of $\frac{p-1}{2} - [p/4]$ and $(p^2-1)/8$ for the four possible congruence classes of the odd prime p modulo 8, we see that we always have $\frac{p-1}{2} - [p/4] \equiv (p^2-1)/8 \pmod 2$. It follows that for every prime p we have $(\frac{2}{p}) = (-1)^{(p^2-1)/8}$.

From the computations of the congruence class of $(p^2-1)/8 \pmod 2$, we see that $\left[\frac{2}{p}\right] = 1$ if $p \equiv \pm 1 \pmod 8$, while $\left[\frac{2}{p}\right] = -1$ if $p \equiv \pm 3 \pmod 8$. ∎

Example 9.5. By Theorem 9.5 we see that

$$\left[\frac{2}{7}\right] = \left[\frac{2}{17}\right] = \left[\frac{2}{23}\right] = \left[\frac{2}{31}\right] = 1,$$

while

$$\left[\frac{2}{3}\right] = \left[\frac{2}{5}\right] = \left[\frac{2}{11}\right] = \left[\frac{2}{13}\right] = \left[\frac{2}{19}\right] = \left[\frac{2}{29}\right] = -1.$$ □

We now present an example to show how to evaluate Legendre symbols.

Example 9.6. To evaluate $\left[\frac{317}{11}\right]$, we use part (i) of Theorem 9.3 to obtain

$$\left[\frac{317}{11}\right] = \left[\frac{9}{11}\right] = \left[\frac{3}{11}\right]^2 = 1, \text{ since } 317 \equiv 9 \pmod{11}.$$

To evaluate $\left[\frac{89}{13}\right]$, since $89 \equiv -2 \pmod{13}$, we have

$$\left[\frac{89}{13}\right] = \left[\frac{-2}{13}\right] = \left[\frac{-1}{13}\right]\left[\frac{2}{13}\right].$$ Because $13 \equiv 1 \pmod 4$, Theorem 9.4 tells us

that $\left[\frac{-1}{13}\right] = 1$. Since $13 \equiv -3 \pmod 8$, we see from Theorem 9.5 that

$\left[\frac{2}{13}\right] = -1$. Consequently, $\left[\frac{89}{13}\right] = -1$. □

In the next section, we state and prove a theorem called the *law of quadratic reciprocity.* of fundamental importance for the evaluation of Legendre symbols.

Flipping Coins Electronically

An interesting and useful application of the properties of quadratic residues is a method to "flip coins" electronically invented by Blum [139]. This method takes advantage of the difference in the length of time needed to find primes and the time needed to factor integers which are the product of two primes, also the basis of the RSA cipher discussed in Chapter 7.

Suppose that $n = pq$, where p and q are distinct odd primes and suppose that the congruence $x^2 \equiv a \pmod{n}$, $0 < a < n$, has a solution $x = x_0$. We will show that there are exactly four incongruent solutions modulo n. In other words, we will show that a has four incongruent *square roots modulo n.* To see this, let $x_0 \equiv x_1 \pmod{p}$, $0 < x_1 < p$, and let $x_0 \equiv x_2 \pmod{q}$, $0 < x_2 < q$. Then the congruence $x^2 \equiv a \pmod{p}$ has exactly two incongruent solutions modulo p, namely $x \equiv x_1 \pmod{p}$ and $x \equiv p - x_1 \pmod{p}$. Similarly the congruence $x^2 \equiv a \pmod{q}$ has exactly two incongruent solutions modulo q, namely $x \equiv x_2 \pmod{q}$ and $x \equiv q - x_2 \pmod{q}$.

From the Chinese remainder theorem, there are exactly four incongruent solutions of the congruence $x^2 \equiv a \pmod{n}$; these four incongruent solutions are the unique solutions modulo pq of the four sets of simultaneous congruences

(i) $x \equiv x_1 \pmod{p}$ (iii) $x \equiv p - x_1 \pmod{p}$
       $x \equiv x_2 \pmod{q}$ $x \equiv x_2 \pmod{q}$

(ii) $x \equiv x_1 \pmod{p}$ (iv) $x \equiv p - x_1 \pmod{p}$
       $x \equiv q - x_2 \pmod{q}$ $x \equiv q - x_2 \pmod{q}$.

We denote solutions of (i) and (ii) by x and y, respectively. Solutions of (iii) and (iv) are easily seen to be $n - y$ and $n - x$, respectively.

We also note that when $p \equiv q \equiv 3 \pmod{4}$, the solutions of $x^2 \equiv a \pmod{p}$ and of $x^2 \equiv a \pmod{q}$ are $x \equiv \pm a^{(p+1)/4} \pmod{p}$ and $x \equiv \pm a^{(q+1)/4} \pmod{q}$, respectively. By Euler's criterion, we know that $a^{(p-1)/2} \equiv \left(\dfrac{a}{p}\right) = 1 \pmod{p}$

and $a^{(q-1)/2} \equiv \left(\dfrac{a}{q}\right) = 1 \pmod{q}$ (recall that we are assuming that $x^2 \equiv a \pmod{pq}$ has a solution, so that a is a quadratic residue of both p and q) . Hence,

$$(a^{(p+1)/4})^2 = a^{(p+1)/2} = a^{(p-1)/2} \cdot a \equiv a \pmod{p}$$

and

$$(a^{(q+1)/4})^2 = a^{(q+1)/2} = a^{(q-1)/2} \cdot a \equiv a \pmod{q}.$$

Using the Chinese remainder theorem, together with the explicit solutions just constructed, we can easily find the four incongruent solutions of $x^2 \equiv a \pmod{n}$. The following example illustrates this procedure.

Example 9.7. Suppose we know *a priori* that the congruence

$$x^2 \equiv 860 \pmod{11021}$$

has a solution. Since $11021 = 103 \cdot 107$, to find the four incongruent solutions we solve the congruences

$$x^2 \equiv 860 \equiv 36 \pmod{103}$$

and

$$x^2 \equiv 860 \equiv 4 \pmod{107}.$$

The solutions of these congruences are

$$x \equiv \pm 36^{(103+1)/4} \equiv \pm 36^{26} \equiv \pm 6 \pmod{103}$$

and

$$x \equiv \pm 4^{(107+1)/4} \equiv \pm 4^{27} \equiv \pm 2 \pmod{107},$$

respectively. Using the Chinese remainder theorem, we obtain $x \equiv \pm 212$, $\pm 109 \pmod{11021}$ as the solutions of the four systems of congruences described by the four possible choices of signs in the system of congruences $x \equiv \pm 6 \pmod{103}$, $x \equiv \pm 2 \pmod{107}$. □

We can now describe a method for electronically flipping coins. Suppose that Bob and Alice are communicating electronically. Alice picks two distinct large primes p and q, with $p \equiv q \equiv 3 \pmod 4$. Alice sends Bob the integer $n = pq$. Bob picks, at random, a positive integer x less than n and sends to Alice the integer a with $x^2 \equiv a \pmod{n}$, $0 < a < n$. Alice finds the four solutions of $x^2 \equiv a \pmod{n}$, namely x, y, $n-x$, and $n-y$. Alice picks one of these four solutions and sends it to Bob. Note that since $x + y \equiv 2x_1 \not\equiv 0 \pmod p$ and $x + y \equiv 0 \pmod q$, we have $(x+y,n) = q$, and similarly $(x+(n-y),n) = p$. Thus, if Bob receives either y or $n-y$, he can rapidly factor n by using the Euclidean algorithm to find one of the two prime factors of n. On the other hand, if Bob receives either x or $n-x$, he has no way to factor n in a reasonable length of time.

Consequently, Bob wins the coin flip if he can factor n, whereas Alice wins if Bob cannot factor n. From previous comments, we know that there is an equal chance for Bob to receive a solution of $x^2 \equiv a \pmod{n}$ that helps him rapidly factor n, or a solution of $x^2 \equiv a \pmod{n}$ that does not help him factor n. Hence, the coin flip is fair.

9.1 Exercises

1. Find all the quadratic residues of each of the following integers.

   a) 3   c) 13
   b) 5   d) 19

2. Find all the quadratic residues of each of the following integers.

   a) 7   c) 15
   b) 8   d) 18

3. Find the value of the Legendre symbols $\left(\dfrac{j}{5}\right)$ for $j = 1,2,3,4$.

4. Find the value of the Legendre symbols $\left(\dfrac{j}{7}\right)$ for $j = 1,2,3,4,5,6$.

5. Evaluate the Legendre symbol $\left(\dfrac{7}{11}\right)$

   a) using Euler's criterion.
   b) using Gauss' lemma.

6. Let a and b be integers not divisible by the prime p. Show that either one or all three of the integers a, b, and ab are quadratic residues of p.

7. Show that if p is an odd prime, then

$$\left(\frac{-2}{p}\right) = \begin{cases} 1 & \text{if } p \equiv 1 \text{ or } 3 \pmod 8 \\ -1 & \text{if } p \equiv -1 \text{ or } -3 \pmod 8. \end{cases}$$

8. Show that if the prime-power factorization of n is

$$n = p_1^{2t_1+1} p_2^{2t_2+1} \cdots p_k^{2t_k+1} p_{k+1}^{2t_{k+1}} \cdots p_m^{2t_m}$$

   and q is a prime not dividing n, then

$$\left(\frac{n}{q}\right) = \left(\frac{p_1}{q}\right)\left(\frac{p_2}{q}\right) \cdots \left(\frac{p_k}{q}\right).$$

9. Show that if p is prime and $p \equiv 3 \pmod 4$, then $[(p-1)/2]! \equiv (-1)^t \pmod p$, where t is the number of positive integers less than $p/2$ that are nonquadratic residues of p.

10. Show that if b is a positive integer not divisible by the prime p, then

$$\left(\frac{b}{p}\right) + \left(\frac{2b}{p}\right) + \left(\frac{3b}{p}\right) + \cdots + \left(\frac{(p-1)b}{p}\right) = 0.$$

11. Let p be prime and a a quadratic residue of p. Show that if $p \equiv 1 \pmod 4$, then $-a$ is also a quadratic residue of p, while if $p \equiv 3 \pmod 4$, then $-a$ is a quadratic nonresidue of p.

12. Consider the quadratic congruence $ax^2 + bx + c \equiv 0 \pmod{p}$, where p is prime and a, b, and c are integers with $p \nmid a$.

    a) Let $p = 2$. Determine which quadratic congruences (mod 2) have solutions.

    b) Let p be an odd prime and let $d = b^2 - 4ac$. Show that the congruence $ax^2 + bx + c \equiv 0 \pmod{p}$ is equivalent to the congruence $y^2 \equiv d \pmod{p}$, where $y = 2ax + b$. Conclude that if $d \equiv 0 \pmod{p}$, then there is exactly one solution x modulo p, if d is a quadratic residue of p, then there are two incongruent solutions, and if d is a quadratic nonresidue of p, then there are no solutions.

13. Find all solutions of the quadratic congruences.

    a) $x^2 + x + 1 \equiv 0 \pmod{7}$
    b) $x^2 + 5x + 1 \equiv 0 \pmod{7}$
    c) $x^2 + 3x + 1 \equiv 0 \pmod{7}$

14. Show that if p is prime and $p \geq 7$, then there are always two consecutive quadratic residues of p. (*Hint*: First show that at least one of $2, 5$, and 10 is a quadratic residue of p.)

* 15. Show that if p is prime and $p \geq 7$, then there are always two quadratic residues of p that differ by 2.

16. Show that if p is prime and $p \geq 7$, then there are always two quadratic residues of p that differ by 3.

17. Show that if a is a quadratic residue of the prime p, then the solutions of $x^2 \equiv a \pmod{p}$ are

    a) $x \equiv \pm a^{n+1} \pmod{p}$, if $p = 4n + 3$.

    b) $x \equiv \pm a^{n+1}$ or $\pm 2^{2n+1} a^{n+1} \pmod{p}$, if $p = 8n + 5$.

* 18. Show that if p is a prime and $p = 8n + 1$, and r is a primitive root modulo p, then the solutions of $x^2 \equiv \pm 2 \pmod{p}$ are given by

    $$x \equiv \pm (r^{7n} \pm r^n) \pmod{p},$$

    where the $\pm$ sign in the first congruence corresponds to the $\pm$ sign inside the parentheses in the second congruence.

19. Find all solutions of the congruence $x^2 \equiv 1 \pmod{15}$.

20. Find all solutions of the congruence $x^2 \equiv 58 \pmod{77}$.

21. Find all solutions of the congruence $x^2 \equiv 207 \pmod{1001}$.

22. Let p be an odd prime, e a positive integer, and a an integer relatively prime to p. Show that the congruence $x^2 \equiv a \pmod{p^e}$, has either no solutions or exactly two incongruent solutions.

* 23. Let p be an odd prime, e a positive integer, and a an integer relatively prime to p. Show that there is a solution to the congruence $x^2 \equiv a \pmod{p^{e+1}}$ if and only if there is a solution to the congruence $x^2 \equiv a \pmod{p^e}$. Use Exercise

22 to conclude that the congruence $x^2 \equiv a \pmod{p^e}$ has no solutions if a is a quadratic nonresidue of p, and exactly two incongruent solutions modulo p if a is a quadratic residue of p.

24. Let n be an odd integer. Find the number of incongruent solutions modulo n of the congruence $x^2 \equiv a \pmod{n}$, where n has prime-power factorization $n = p_1^{t_1} p_2^{t_2} \cdots p_m^{t_m}$, in terms of the Legendre symbols $\left(\dfrac{a}{p_1} \right), \ldots, \left(\dfrac{a}{p_m} \right)$.

    (*Hint:* Use Exercise 23.)

25. Find the number of incongruent solutions of each of the following congruences.

    a) $x^2 \equiv 31 \pmod{75}$
    b) $x^2 \equiv 16 \pmod{105}$
    c) $x^2 \equiv 46 \pmod{231}$
    d) $x^2 \equiv 1156 \pmod{3^2 5^3 7^5 11^6}$

* 26. Show that the congruence $x^2 \equiv a \pmod{2^e}$, where e is an integer, $e \geq 3$, has either no solutions or exactly four incongruent solutions. (*Hint:* Use the fact that $(\pm x)^2 \equiv (2^{e-1} \pm x)^2 \pmod{2^e}$.)

27. Show that there are infinitely many primes of the form $4k + 1$. (*Hint:* Assume that $p_1, p_2, \ldots, p_n$ are the only such primes. Form $N = 4(p_1 p_2 \cdots p_n)^2 + 1$, and show, using Theorem 9.4, that N has a prime factor of the form $4k + 1$ that is not one of $p_1, p_2, \ldots, p_n$.)

* 28. Show that there are infinitely many primes of each of the following forms.

    a) $8k + 3$ b) $8k + 5$ c) $8k + 7$

    (*Hint:* For each part, assume that there are only finitely many primes $p_1, p_2, \ldots, p_n$ of the particular form. For part (a) look at $(p_1 p_2 \cdots p_n)^2 + 2$, for part (b), look at $(p_1 p_2 \cdots p_n)^2 + 4$, and for part (c), look at $(4 p_1 p_2 \cdots p_n)^2 - 2$. In each part, show that there is a prime factor of this integer of the required form not among the primes $p_1, p_2, \ldots, p_n$. Use Theorems 9.4 and 9.5.)

29. Show that if p is an odd prime, then the congruence $x^2 \equiv a \pmod{p^n}$ has a solution for all positive integers n if and only if a is a quadratic residue of p.

30. Show that if p is an odd prime with primitive root r, and a is a positive integer not divisible by p, then a is a quadratic residue of p if and only if $\text{ind}_r a$ is even.

31. Show that every primitive root of an odd prime p is a quadratic nonresidue of p.

32. Let p be an odd prime. Show that there are $(p-1)/2 - \phi(p-1)$ quadratic nonresidues of p that are not primitive roots of p.

* 33. Let p and $q = 2p + 1$ both be odd primes. Show that the $p - 1$ primitive roots of q are the quadratic nonresidues of q, other than the nonresidue $2p$ of q.

* 34. Show that if p and $q = 4p + 1$ are both primes and if a is a quadratic nonresidue of q with $\text{ord}_q a \neq 4$, then a is a primitive root of q.

* **35.** Show that a prime p is a Fermat prime if and only if every quadratic nonresidue of p is also a primitive root of p.

* **36.** Show that a prime divisor p of the Fermat number $F_n = 2^{2^n} + 1$ must be of the form $2^{n+2}k + 1$. (*Hint*: Show that $\operatorname{ord}_p 2 = 2^{n+1}$. Then show that $2^{(p-1)/2} \equiv 1 \pmod{p}$ using Theorem 9.5. Conclude that $2^{n+1} \mid (p-1)/2$.)

* **37.** a) Show that if p is a prime of the form $4k + 3$ and $q = 2p + 1$ is prime, then q divides the Mersenne number $M_p = 2^p - 1$. (*Hint*: Consider the Legendre symbol $\left[\dfrac{2}{q}\right]$.)

   b) From part (a), show that $23 \mid M_{11}$, $47 \mid M_{23}$, and $503 \mid M_{251}$.

* **38.** Show that if n is a positive integer and $2n + 1$ is prime, and if $n \equiv 0$ or $3 \pmod 4$, then $2n + 1$ divides the Mersenne number $M_n = 2^n - 1$, while if $n \equiv 1$ or $2 \pmod 4$, then $2n + 1$ divides $M_n + 2 = 2^n + 1$. (*Hint*: Consider the Legendre symbol $\left[\dfrac{2}{2n+1}\right]$ and use Theorem 9.5.)

  39. Show that if p is an odd prime, then every prime divisor q of the Mersenne number M_p must be of the form $q = 8k \pm 1$, where k is a positive integer. (*Hint*: Use Exercise 38.)

  40. Show how Exercise 39, together with Theorem 6.12, can be used to help show that M_{17} is prime.

* **41.** Show that if p is an odd prime, then

$$\sum_{j=1}^{p-2} \left[\frac{j(j+1)}{p}\right] = -1.$$

(*Hint*: First show that $\left[\dfrac{j(j+1)}{p}\right] = \left[\dfrac{\bar{j}+1}{p}\right]$ where $\bar{j}$ is an inverse of j modulo p).

* **42.** Let p be an odd prime. Among pairs of consecutive positive integers less than p, let **(RR)**, **(RN)**, **(NR)**, and **(NN)** denote the number of pairs of two quadratic residues, of a quadratic residue followed by a quadratic nonresidue, of a quadratic nonresidue followed by a quadratic residue, and of two quadratic nonresidues, respectively.

   a) Show that

$$\mathbf{(RR)} + \mathbf{(RN)} = \frac{1}{2}(p - 2 - (-1)^{(p-1)/2})$$

$$\mathbf{(NR)} + \mathbf{(NN)} = \frac{1}{2}(p - 2 + (-1)^{(p-1)/2})$$

$$\mathbf{(RR)} + \mathbf{(NR)} = \frac{1}{2}(p - 1) - 1$$

$$\mathbf{(RN)} + \mathbf{(NN)} = \frac{1}{2}(p - 1).$$

b) Using Exercise 41 show that

$$\sum_{j=1}^{p-2} \left[\frac{j(j+1)}{p} \right] = (\mathbf{RR}) + (\mathbf{NN}) - (\mathbf{RN}) - (\mathbf{NR}) = -1.$$

c) From parts (a) and (b), find **(RR), (RN), (NR),** and **(NN)**.

43. Use Theorem 8.16 to prove Theorem 9.1.

* 44. Let p and q be odd primes. Show that 2 is a primitive root of q, if $q = 4p + 1$.

* 45. Let p and q be odd primes. Show that 2 is a primitive root of q if p is of the form $4k + 1$ and $q = 2p + 1$.

* 46. Let p and q be odd primes. Show that -2 is a primitive root of q if p is of the form $4k - 1$ and $q = 2p + 1$.

* 47. Let p and q be odd primes. Show that -4 is a primitive root of q, if $q = 2p + 1$.

48. Find the solutions of $x^2 \equiv 482 \pmod{2773}$ (note that $2773 = 47 \cdot 59$).

* 49. In this exercise we develop a method for deciphering messages enciphered using a Rabin cipher. Recall that the relationship between a ciphertext block C and the corresponding plaintext block P in a Rabin cipher is $C \equiv P\,(P+b) \pmod{n}$, where $n = pq$, p and q are distinct odd primes, and b is a positive integer less than n.

a) Show that $C + a \equiv (P+b)^2 \pmod{n}$, where $a \equiv (\bar{2}b)^2 \pmod{n}$, and $\bar{2}$ is an inverse of 2 modulo n.

b) Using the algorithm in the text for solving congruences of the type $x^2 \equiv a \pmod{n}$, together with part (a), show how to find a plaintext block P from the corresponding ciphertext block C. Explain why there are four possible plaintext messages. (This ambiguity is a disadvantage of Rabin ciphers.)

c) Decipher the ciphertext message 1819 0459 0803 that was enciphered using the Rabin cipher with $b = 3$ and $n = 47 \cdot 59 = 2773$.

50. Let p be an odd prime and let C be the ciphertext obtained by modular exponentiation, with exponent e and modulus p, from the plaintext P, *i.e.*, $C \equiv P^e \pmod{p}$, $0 < C < n$, where $(e, p-1) = 1$. Show that C is a quadratic residue of p if and only if P is a quadratic residue of p.

* 51. a) Show that the second player in a game of electronic poker (see Section 7.3) can obtain an advantage by noting which cards have numerical equivalents that are quadratic residues modulo p. (*Hint*: Use Exercise 50.)

b) Show that the advantage of the second player noted in part (a) can be eliminated if the numerical equivalents of cards that are quadratic nonresidues are all multiplied by a fixed quadratic nonresidue.

* 52. Show that if the probing sequence for resolving collisions in a hashing scheme is $h_j(K) \equiv h(K) + aj + bj^2 \pmod{m}$, where $h(K)$ is a hashing function, m is a positive integer, and a and b are integers with $(b,m) = 1$, then only half the possible file locations are probed. This is called the *quadratic search*.

9.1 Computer Projects

Programming Projects

Write programs to do the following:

1. Evaluate Legendre symbols using Euler's criterion.

2. Evaluate Legendre symbols using Gauss' lemma.

3. Given a positive integer n that is the product of two distinct primes both congruent to 3 modulo 4, find the four square roots of the least positive residue of x^2 where x is an integer relatively prime to n.

* 4. Flip coins electronically using the procedure described in this section.

** 5. Decipher messages that were enciphered using a Rabin cipher (see Exercise 49).

Computations and Explorations

Using the programs you have written or a computation program, carry out the following computations and explorations:

1. Find the smallest quadratic nonresidue of each prime less than 1000.

2. Find the smallest quadratic nonresidue of 100 randomly selected primes between 100,000 and 1,000,000 and 100 randomly selected primes between 100,000,000 and 1,000,000,000. Can you make any conjectures based on your evidence?

3. Use numerical evidence to determine for which odd primes p there are more quadratic residues a of p with $1 \leq a \leq (p-1)/2$ than there are with $(p+1)/2 \leq a \leq p-1$.

4. Let p be a prime with $p \equiv 3 \pmod{4}$. It has been proved that if R is the largest number of consecutive quadratic residues of p and N is the largest number of consecutive quadratic nonresidues of p, then $R = N < \sqrt{p}$. Verify this result for all primes of this type less than 1000.

5. Let p be a prime with $p \equiv 1 \pmod{4}$. It has been conjectures that if N is the largest number of consecutive quadratic nonresidues of p then $N < \sqrt{p}$ when p is sufficiently large. Find evidence for this conjecture. For which small primes does this inequality fail?

9.2 The Law of Quadratic Reciprocity

Suppose that p and q are distinct odd primes. Suppose further that we know whether q is a quadratic residue of p. Do we also know whether p is a quadratic residue of q? The answer to this question was found by Euler in the mid-1700s. He found the answer by examining numerical evidence, but he did not prove that his answer was correct. Later, in 1785, Legendre reformulated Euler's answer, in its modern, elegant form in a theorem known as the *law of quadratic reciprocity*. This theorem tells us whether the congruence $x^2 \equiv q \pmod{p}$ has solutions, once we know whether there are solutions of $x^2 \equiv p \pmod{q}$.

Theorem 9.6. The Law of Quadratic Reciprocity. Let p and q be odd primes. Then

$$\left(\frac{p}{q}\right)\left(\frac{q}{p}\right) = (-1)^{\frac{p-1}{2} \cdot \frac{q-1}{2}}.$$

Legendre published several proposed proofs of this theorem, but each of his proofs contained a serious gap. The first correct proof was provided by Gauss, who claimed to have rediscoverd this result when he was 18 years old. Gauss devoted considerable attention to his search for a proof. In fact, he wrote that "for an entire year this theorem tormented me and absorbed my greatest efforts until at last I obtained a proof." Once he found a proof, he devised seven more proofs, each based on a different approach. Finding new and different proofs of the law of quadratic reciprocity has long challenged mathematicians. Many surprising and amazing proofs have been found. To provide a rough idea on how many different proofs are known, an article published a few years ago offered what was facetiously called the "152nd proof" of the law of quadratic reciprocity.

Before we prove the law of quadratic reciprocity, we will discuss its consequences and how it is used to evaluate Legendre symbols. We first note that the quantity $(p-1)/2$ is even when $p \equiv 1 \pmod 4$ and odd when $p \equiv 3 \pmod 4$. Consequently, we see that $\dfrac{p-1}{2} \cdot \dfrac{q-1}{2}$ is even if $p \equiv 1 \pmod 4$ or $q \equiv 1 \pmod 4$, while $\dfrac{p-1}{2} \cdot \dfrac{q-1}{2}$ is odd if $p \equiv q \equiv 3 \pmod 4$. Hence, we have

$$\left(\frac{p}{q}\right)\left(\frac{q}{p}\right) = \begin{cases} 1 & \text{if } p \equiv 1 \pmod 4 \text{ or } q \equiv 1 \pmod 4 \quad (\text{or both}) \\ -1 & \text{if } p \equiv q \equiv 3 \pmod 4. \end{cases}$$

Since the only possible values of $\left(\dfrac{p}{q}\right)$ and $\left(\dfrac{q}{p}\right)$ are ± 1, we see that

$$\left(\frac{p}{q}\right) = \begin{cases} \left(\dfrac{q}{p}\right) & \text{if } p \equiv 1 \ (\text{mod } 4) \text{ or } q \equiv 1 \ (\text{mod } 4) \text{ (or both)} \\[3ex] -\left(\dfrac{q}{p}\right) & \text{if } p \equiv q \equiv 3 \ (\text{mod } 4). \end{cases}$$

This means that if p and q are odd primes, then $\left(\dfrac{p}{q}\right) = \left(\dfrac{q}{p}\right)$ unless both p and q are congruent to 3 modulo 4, and in that case, $\left(\dfrac{p}{q}\right) = -\left(\dfrac{q}{p}\right)$.

Example 9.8. Let $p = 13$ and $q = 17$. Since $p \equiv q \equiv 1 \ (\text{mod } 4)$, the law of quadratic reciprocity tells us that $\left(\dfrac{13}{17}\right) = \left(\dfrac{17}{13}\right)$. By part (i) of Theorem 9.3 we know that $\left(\dfrac{17}{13}\right) = \left(\dfrac{4}{13}\right)$, and from part (iii) of Theorem 9.3, it follows that $\left(\dfrac{4}{13}\right) = \left(\dfrac{2^2}{13}\right) = 1$. Combining these equalities, we conclude that $\left(\dfrac{13}{17}\right) = 1$. □

Example 9.9. Let $p = 7$ and $q = 19$. Since $p \equiv q \equiv 3 \ (\text{mod } 4)$, by the law of quadratic reciprocity, we know that $\left(\dfrac{7}{19}\right) = -\left(\dfrac{19}{7}\right)$. From part (i) of Theorem 9.3, we see that $\left(\dfrac{19}{7}\right) = \left(\dfrac{5}{7}\right)$. Again, using the law of quadratic reciprocity, since $5 \equiv 1 \ (\text{mod } 4)$ and $7 \equiv 3 \ (\text{mod } 4)$, we have $\left(\dfrac{5}{7}\right) = \left(\dfrac{7}{5}\right)$. By part (i) of Theorem 9.3 and Theorem 9.5, we know that $\left(\dfrac{7}{5}\right) = \left(\dfrac{2}{5}\right) = -1$. Hence $\left(\dfrac{7}{19}\right) = 1$. □

We can use the law of quadratic reciprocity and Theorems 9.3 and 9.5 to evaluate Legendre symbols. Unfortunately, prime factorizations must be computed to evaluate Legendre symbols in this way.

Example 9.10. We will calculate $\left(\dfrac{713}{1009}\right)$ (note that 1009 is prime). We factor $713 = 23 \cdot 31$, so that by part (ii) of Theorem 9.3, we have

$$\left(\frac{713}{1009}\right) = \left(\frac{23 \cdot 31}{1009}\right) = \left(\frac{23}{1009}\right)\left(\frac{31}{1009}\right).$$

To evaluate the two Legendre symbols on the right side of this equality, we use the law of quadratic reciprocity. Since $1009 \equiv 1 \pmod 4$, we see that

$$\left(\frac{23}{1009}\right) = \left(\frac{1009}{23}\right), \quad \left(\frac{31}{1009}\right) = \left(\frac{1009}{31}\right).$$

Using Theorem 9.3, part (i), we have

$$\left(\frac{1009}{23}\right) = \left(\frac{20}{23}\right), \quad \left(\frac{1009}{31}\right) = \left(\frac{17}{31}\right).$$

By parts (ii) and (iii) of Theorem 9.3, it follows that

$$\left(\frac{20}{23}\right) = \left(\frac{2^2 5}{23}\right) = \left(\frac{2^2}{23}\right)\left(\frac{5}{23}\right) = \left(\frac{5}{23}\right).$$

The law of quadratic reciprocity, part (i) of Theorem 9.3, and Theorem 9.5 tell us that

$$\left(\frac{5}{23}\right) = \left(\frac{23}{5}\right) = \left(\frac{3}{5}\right) = \left(\frac{5}{3}\right) = \left(\frac{2}{3}\right) = -1.$$

Thus, $\left(\dfrac{23}{1009}\right) = -1.$

Likewise, using the law of quadratic reciprocity, Theorem 9.3, and Theorem 9.5, we find that

$$\left(\frac{17}{31}\right) = \left(\frac{31}{17}\right) = \left(\frac{14}{17}\right) = \left(\frac{2}{17}\right)\left(\frac{7}{17}\right) = \left(\frac{7}{17}\right) = \left(\frac{17}{7}\right) = \left(\frac{3}{7}\right)$$

$$= -\left(\frac{7}{3}\right) = -\left(\frac{4}{3}\right) = -\left(\frac{2^2}{3}\right) = -1.$$

Consequently, $\left(\dfrac{31}{1009}\right) = -1.$

Therefore, $\left(\dfrac{713}{1009}\right) = (-1)(-1) = 1.$ □

We now present one of the many possible approaches for proving the law of quadratic reciprocity. Before presenting the proof, we give a somewhat technical lemma, which we use in the proof of this important law.

Lemma 9.3. If p is an odd prime and a is an odd integer not divisible by p, then

$$\left(\frac{a}{p}\right) = (-1)^{T(a,p)},$$

where

$$T(a,p) = \sum_{j=1}^{(p-1)/2} [ja/p].$$

Proof. Consider the least positive residues of the integers $a, 2a,...,((p-1)/2)a$; let $u_1, u_2,..., u_s$ be those greater than $p/2$ and let $v_1, v_2,..., v_t$ be those less than $p/2$. The division algorithm tells us that

$$ja = p[ja/p] + \text{remainder},$$

where the remainder is one of the u_j's or v_j's. By adding the $(p-1)/2$ equations of this sort, we obtain

(9.3) $$\sum_{j=1}^{(p-1)/2} ja = \sum_{j=1}^{(p-1)/2} p[ja/p] + \sum_{j=1}^{s} u_j + \sum_{j=1}^{t} v_j.$$

As we showed in the proof of Gauss' lemma, the integers $p - u_1,..., p - u_s$, $v_1,..., v_t$ are precisely the integers $1, 2,..., (p-1)/2$, in some order. Hence, summing all these integers, we obtain

(9.4) $$\sum_{j=1}^{(p-1)/2} j = \sum_{j=1}^{s} (p-u_j) + \sum_{j=1}^{t} v_j = ps - \sum_{j=1}^{s} u_j + \sum_{j=1}^{t} v_j.$$

Subtracting (9.4) from (9.3), we find that

$$\sum_{j=1}^{(p-1)/2} ja - \sum_{j=1}^{(p-1)/2} j = \sum_{j=1}^{(p-1)/2} p[ja/p] - ps + 2\sum_{j=1}^{s} u_j$$

or equivalently, since $T(a,p) = \sum_{j=1}^{(p-1)/2} [ja/p]$,

$$(a-1)\sum_{j=1}^{(p-1)/2} j = pT(a,p) - ps + 2\sum_{j=1}^{s} u_j.$$

Reducing this last equation modulo 2, since a and p are odd, yields

$$0 \equiv T(a,p) - s \pmod{2}.$$

Hence,

$$T(a,p) \equiv s \pmod{2}.$$

To finish the proof, we note that from Gauss' lemma

$$\left(\frac{a}{p}\right) = (-1)^s.$$

Consequently, since $(-1)^s = (-1)^{T(a,p)}$, it follows that

$$\left(\frac{a}{p}\right) = (-1)^{T(a,p)}.$$

∎

Although Lemma 9.3 is used primarily as a tool in the proof of the law of quadratic reciprocity, it can also be used to evaluate Legendre symbols.

Example 9.11. To find $\left(\dfrac{7}{11}\right)$ using Lemma 9.3 we evaluate the sum

$$\sum_{j=1}^{5} [7_j/11] = [7/11] + [14/11] + [21/11] + [28/11] + [35/11]$$
$$= 0 + 1 + 1 + 2 + 3 = 7.$$

Hence, $\left(\dfrac{7}{11}\right) = (-1)^7 = -1.$

Likewise, to find $\left(\dfrac{11}{7}\right)$, we note that

$$\sum_{j=1}^{3} [11j/7] = [11/7] + [22/7] + [33/7] = 1 + 3 + 4 = 8,$$

so that $\left(\dfrac{11}{7}\right) = (-1)^8 = 1.$

□

Before we present a proof of the law of quadratic reciprocity, we use an example to illustrate the method of proof.

Let $p = 7$ and $q = 11$. We consider pairs of integers (x,y) with $1 \le x \le \dfrac{7-1}{2} = 3$ and $1 \le y \le \dfrac{11-1}{2} = 5$. There are 15 such pairs. We note that none of these pairs satisfy $11x = 7y$, since the equality $11x = 7y$ implies that $11 \mid 7y$, so that either $11 \mid 7$, which is absurd, or $11 \mid y$, which is impossible because $1 \le y \le 5$.

We divide these 15 pairs into two groups, depending on the relative sizes of $11x$ and $7y$.

The pairs of integers (x,y) with $1 \le x \le 3$, $1 \le y \le 5$, and $11x > 7y$ are precisely those pairs satisfying $1 \le x \le 3$ and $1 \le y \le 11x/7$. For a fixed integer x with $1 \le x \le 3$, there are $[11x/7]$ allowable values of y. Hence, the total number of pairs satisfying $1 \le x \le 3$, $1 \le y \le 5$, and $11x > 7y$ is

$$\sum_{j=1}^{3} [11/7] = [11/7] + [22/7] + [33/7] = 1 + 3 + 4 = 8;$$

these eight pairs are $(1,1)$, $(2,1)$, $(2,2)$, $(2,3)$, $(3,1)$, $(3,2)$, $(3,3)$, $(3,4)$.

The pairs of integers (x,y) with $1 \le x \le 3$, $1 \le y \le 5$, and $11x < 7y$ are precisely those pairs satisfying $1 \le y \le 5$ and $1 \le x \le 7y/11$. For a fixed integer y with $1 \le y \le 5$, there are $[7y/11]$ allowable values of x. Hence, the total number of pairs satisfying $1 \le x \le 3$, $1 \le y \le 5$, and $11x < 7y$ is

$$\sum_{j=1}^{5} [7j/11] = [7/11] + [14/11] + [21/11] + [28/11] + [35/11]$$
$$= 0 + 1 + 1 + 2 + 3 = 7.$$

These seven pairs are $(1,2)$, $(1,3)$, $(1,4)$, $(1,5)$, $(2,4)$, $(2,5)$, and $(3,5)$.

Consequently, we see that

$$\frac{11-1}{2} \cdot \frac{7-1}{2} = 5 \cdot 3 = 15 = \sum_{j=1}^{3} [11j/7] + \sum_{j=1}^{5} [7j/11] = 8 + 7.$$

Hence,

$$(-1)^{\frac{11-1}{2} \cdot \frac{7-1}{2}} = (-1)^{\sum_{j=1}^{3}[11j/7] + \sum_{j=1}^{5}[7j/11]}$$
$$= (-1)^{\sum_{j=1}^{3}[11j/7]} (-1)^{\sum_{j=1}^{5}[7j/11]}.$$

Since Lemma 9.3 tells us that $\left(\dfrac{11}{7}\right) = (-1)^{\sum_{j=1}^{3}[11j/7]}$ and

$\left(\dfrac{7}{11}\right) = (-1)^{\sum_{j=1}^{5}[7j/11]}$, we see that $\left(\dfrac{7}{11}\right)\left(\dfrac{11}{7}\right) = (-1)^{\frac{7-1}{2} \cdot \frac{11-1}{2}}$.

This establishes the special case of the law of quadratic reciprocity when $p = 7$ and $q = 11$. ☐

We now prove the law of quadratic reciprocity, using the idea illustrated in the example.

Proof. We consider pairs of integers (x,y) with $1 \le x \le (p-1)/2$ and $1 \le y \le (q-1)/2$. There are $\dfrac{p-1}{2} \dfrac{q-1}{2}$ such pairs. We divide these pairs into two groups, depending on the relative sizes of qx and py.

First, we note that $qx \ne py$ for all of these pairs. For if $qx = py$, then $q \mid py$, which implies that $q \mid p$ or $q \mid y$. However, since q and p are distinct primes, we know that $q \nmid p$, and since $1 \le y \le (q-1)/2$, we know that $q \nmid y$.

To enumerate the pairs of integers (x,y) with $1 \le x \le (p-1)/2$, $1 \le y \le (q-1)/2$, and $qx > py$, we note that these pairs are precisely those where $1 \le x \le (p-1)/2$ and $1 \le y \le qx/p$. For each fixed value of the integer x, with $1 \le x \le (p-1)/2$, there are $[qx/p]$ integers satisfying $1 \le y \le qx/p$. Consequently, the total number of pairs of integers (x,y) with $1 \le x \le (p-1)/2$, $1 \le y \le (q-1)/2$, and $qx > py$ is $\displaystyle\sum_{j=1}^{(p-1)/2} [qj/p]$.

We now consider the pairs of integers (x,y) with $1 \le x \le (p-1)/2$, $1 \le y \le (q-1)/2$, and $qx < py$. These pairs are precisely the pairs of integers (x,y) with $1 \le y \le (q-1)/2$ and $1 \le x \le py/q$. Hence, for each fixed value of the integer y, where $1 \le y \le (q-1)/2$, there are exactly $[py/q]$ integers x satisfying $1 \le x \le py/q$. This shows that the total number of pairs of integers (x,y) with $1 \le x \le (p-1)/2$, $1 \le y \le (q-1)/2$, and $qx < py$ is $\displaystyle\sum_{j=1}^{(q-1)/2} [pj/q]$.

Adding the numbers of pairs in these classes, and recalling that the total number of such pairs is $\dfrac{p-1}{2} \cdot \dfrac{q-1}{2}$, we see that

$$\sum_{j=1}^{(p-1)/2} [qj/p] \; + \; \sum_{j=1}^{(q-1)/2} [pj/q] \; = \; \frac{p-1}{2} \cdot \frac{q-1}{2} \; ,$$

or using the notation of Lemma 9.3,

$$T(q,p) \; + \; T(p,q) \; = \; \frac{p-1}{2} \cdot \frac{q-1}{2} .$$

Hence,

$$(-1)^{T(q,p)+T(p,q)} \; = \; (-1)^{T(q,p)}(-1)^{T(p,q)} \; = \; (-1)^{\frac{p-1}{2} \cdot \frac{q-1}{2}} .$$

Lemma 9.2 tells us that $(-1)^{T(q,p)} = \left[\dfrac{q}{p}\right]$ and $(-1)^{T(p,q)} = \left[\dfrac{p}{q}\right]$. Hence

$$\left[\frac{p}{q}\right]\left[\frac{q}{p}\right] \; = \; (-1)^{\frac{p-1}{2} \cdot \frac{q-1}{2}} .$$

This concludes the proof of the law of quadratic reciprocity. ∎

The law of quadratic reciprocity has many applications. One use is to prove the validity of the following primality test for Fermat numbers.

Theorem 9.7. Pepin's Test. The Fermat number $F_m = 2^{2^m} + 1$ is prime if and only if

$$3^{(F_m-1)/2} \equiv -1 \; (\bmod \; F_m).$$

Proof. We will first show that F_m is prime if the congruence in the statement of the theorem holds. Assume that

$$3^{(F_m-1)/2} \equiv -1 \; (\bmod \; F_m).$$

Then, by squaring both sides, we obtain

$$3^{F_m-1} \equiv 1 \; (\bmod \; F_m).$$

From this congruence, we see that if p is a prime dividing F_m, then

$$3^{F_m-1} \equiv 1 \ (\mathrm{mod}\ p),$$

and hence,

$$\mathrm{ord}_p\, 3 \mid (F_m - 1) = 2^{2^m}.$$

Consequently, $\mathrm{ord}_p\, 3$ must be a power of 2. However,

$$\mathrm{ord}_p\, 3 \nmid 2^{2^{m}-1} = (F_m - 1)/2,$$

since $3^{(F_m-1)/2} \equiv -1 \ (\mathrm{mod}\ F_m)$. Hence, the only possibility is that $\mathrm{ord}_p\, 3 = 2^{2^m} = F_m - 1$. Since $\mathrm{ord}_p\, 3 = F_m - 1 \le p - 1$ and $p \mid F_m$, we see that $p = F_m$, and consequently, F_m must be prime.

Conversely, if $F_m = 2^{2^m} + 1$ is prime for $m \ge 1$, then the law of quadratic reciprocity tells us that

(9.5)
$$\left(\frac{3}{F_m}\right) = \left(\frac{F_m}{3}\right) = \left(\frac{2}{3}\right) = -1,$$

since $F_m \equiv 1 \ (\mathrm{mod}\ 4)$ and $F_m \equiv 2 \ (\mathrm{mod}\ 3)$.

Now, using Euler's criterion, we know that

(9.6)
$$\left(\frac{3}{F_m}\right) \equiv 3^{(F_m-1)/2} \ (\mathrm{mod}\ F_m).$$

By the two equations involving $\left(\dfrac{3}{F_m}\right)$, (9.5) and (9.6), we conclude that

$$3^{(F_m-1)/2} \equiv -1 \ (\mathrm{mod}\ F_m).$$

This finishes the proof. ∎

Example 9.12. Let $m = 2$. Then $F_2 = 2^{2^2} + 1 = 17$ and

$$3^{(F_2-1)/2} = 3^8 \equiv -1 \ (\mathrm{mod}\ 17).$$

By Pepin's test, we see that $F_2 = 17$ is prime.

Let $m = 5$. Then $F_5 = 2^{2^5} + 1 = 2^{32} + 1 = 4294967297$. We note that

$$3^{(F_5-1)/2} = 3^{2^{31}} = 3^{2147483648} \equiv 10324303 \neq -1 \ (\mathrm{mod}\ 4294967297).$$

Hence, by Pepin's test, we see that F_5 is composite. □

9.2 Exercises

1. Evaluate each of the following Legendre symbols.

   a) $\left[\dfrac{3}{53}\right]$ d) $\left[\dfrac{31}{641}\right]$

   b) $\left[\dfrac{7}{79}\right]$ e) $\left[\dfrac{111}{991}\right]$

   c) $\left[\dfrac{15}{101}\right]$ f) $\left[\dfrac{105}{1009}\right]$

2. Using the law of quadratic reciprocity, show that if p is an odd prime, then

$$\left[\frac{3}{p}\right] = \begin{cases} 1 & \text{if } p \equiv \pm 1 \pmod{12} \\ -1 & \text{if } p \equiv \pm 5 \pmod{12}. \end{cases}$$

3. Show that if p is an odd prime, then

$$\left[\frac{-3}{p}\right] = \begin{cases} 1 & \text{if } p \equiv 1 \pmod 6 \\ -1 & \text{if } p \equiv -1 \pmod 6. \end{cases}$$

4. Find a congruence describing all primes for which 5 is a quadratic residue.

5. Find a congruence describing all primes for which 7 is a quadratic residue.

6. Show that there are infinitely many primes of the form $5k + 4$. (*Hint:* Let n be a positive integer and form $Q = 5(n!)^2 - 1$. Show that Q has a prime divisor of the form $5k + 4$ greater than n. To do this, use the law of quadratic reciprocity to show that if a prime p divides Q, then $\left[\dfrac{p}{5}\right] = 1$.)

7. Use Pepin's test to show that the following Fermat numbers are primes.

   a) $F_1 = 5$ b) $F_3 = 257$ c) $F_4 = 65537$

* 8. Use Pepin's test to conclude that 3 is a primitive root of every Fermat prime.

* 9. In this exercise we give another proof of the law of quadratic reciprocity. Let p and q be distinct odd primes. Let **R** be the interior of the rectangle with vertices $\mathbf{O} = (0,0)$, $\mathbf{A} = (p/2,0)$, $\mathbf{B} = (q/2,0)$, and $\mathbf{C} = (p/2,q/2)$ as shown.

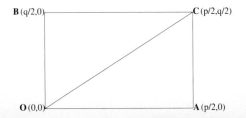

a) Show that the number of lattice points (points with integer coordinates) in **R** is $\dfrac{p-1}{2} \cdot \dfrac{q-1}{2}$.

b) Show that there are no lattice points on the diagonal connecting **O** and **C**.

c) Show that the number of lattice points in the triangle with vertices **O, A, C** is $\displaystyle\sum_{j=1}^{(p-1)/2} [\,jq/p\,]$.

d) Show that the number of lattice points in the triangle with vertices **O, B,** and **C** is $\displaystyle\sum_{j=1}^{(q-1)/2} [\,jp/q\,]$.

e) Conclude from parts (a), (b), (c), and (d) that
$$\sum_{j=1}^{(p-1)/2} [\,jq/p\,] + \sum_{j=1}^{(q-1)/2} [\,jp/q\,] = \frac{p-1}{2} \cdot \frac{q-1}{2}.$$

Derive the law of quadratic reciprocity using this equation and Lemma 9.2.

9.2 Computer Projects

Programming Projects

Write programs to do the following:

1. Evaluate Legendre symbols, using the law of quadratic reciprocity.

2. Given a positive integer n, determine whether the nth Fermat number F_n is prime using Pepin's test.

Computations and Explorations

Using the programs you have written or a computational program, perform the following computations and explorations.

1. Use Pepin's test to show that the Fermat numbers F_6, F_7, and F_8 are all composite. Can you go further?

9.3 The Jacobi Symbol

In this section we define the Jacobi symbol, named after the German mathematician Carl Jacobi* who introduced this symbol. The Jacobi symbol is a

---

* **CARL GUSTAV JACOB JACOBI (1804-1851)** was born into a well-to-do German banking family. Jacobi received an excellent early education at home. He studied at the Univeristy of Berlin, mastered mathematics through the texts of Euler, and obtained his doctorate in 1825. In 1826 he became a lecturer at the University of Königsberg; he was appointed a professor there in 1831. Besides his work in number theory, Jacobi made important contributions to analysis, geometry, and mechanics. He was also interested in the history of mathematics and was a catalyst in the publication of the collected works of Euler, a job still not completed although it started more than 125 years ago!

generalization of the Legendre symbol studied in the previous two sections. Jacobi symbols are useful in the evaluation of Legendre symbols and in the definition of a type of pseudoprime.

Definition. Let n be an odd positive integer with prime factorization $n = p_1^{t_1} p_2^{t_2} \cdots p_m^{t_m}$ and let a be an integer relatively prime to n. Then, the *Jacobi symbol* $\left(\dfrac{a}{n}\right)$ is defined by

$$\left(\frac{a}{n}\right) = \left(\frac{a}{p_1^{t_1} p_2^{t_2} \cdots p_m^{t_m}}\right) = \left(\frac{a}{p_1}\right)^{t_1} \left(\frac{a}{p_2}\right)^{t_2} \cdots \left(\frac{a}{p_m}\right)^{t_m},$$

where the symbols on the right-hand side of the equality are Legendre symbols.

Example 9.13. From the definition of the Jacobi symbol, we see that

$$\left(\frac{2}{45}\right) = \left(\frac{2}{3^2 \cdot 5}\right) = \left(\frac{2}{3}\right)^2 \left(\frac{2}{5}\right) = (-1)^2(-1) = -1,$$

and

$$\left(\frac{109}{385}\right) = \left(\frac{109}{5 \cdot 7 \cdot 11}\right) = \left(\frac{109}{5}\right)\left(\frac{109}{7}\right)\left(\frac{109}{11}\right) = \left(\frac{4}{5}\right)\left(\frac{4}{7}\right)\left(\frac{10}{11}\right)$$

$$= \left(\frac{2}{5}\right)^2 \left(\frac{2}{7}\right)^2 \left(\frac{-1}{11}\right) = (-1)^2 \, 1^2(-1) = -1.$$

□

When n is prime, the Jacobi symbol is the same as the Legendre symbol. However, when n is composite, the value of the Jacobi symbol $\left(\dfrac{a}{n}\right)$ does *not* tell us whether the congruence $x^2 \equiv a \pmod{n}$ has solutions. We do know that if the congruence $x^2 \equiv a \pmod{n}$ has solutions, then $\left(\dfrac{a}{n}\right) = 1$. To see this, note that if p is a prime divisor of n and if $x^2 \equiv a \pmod{n}$ has solutions, then the congruence $x^2 \equiv a \pmod{p}$ also has solutions. Thus, $\left(\dfrac{a}{p}\right) = 1$. Consequently, $\left(\dfrac{a}{n}\right) = \prod_{j=1}^{m} \left(\dfrac{a}{p_j}\right)^j = 1$. To see that it is possible that $\left(\dfrac{a}{n}\right) = 1$ when there are no solutions to $x^2 \equiv a \pmod{n}$, let $a = 2$ and $n = 15$. Note that $\left(\dfrac{2}{15}\right) = \left(\dfrac{2}{3}\right)\left(\dfrac{2}{5}\right) = (-1)(-1) = 1$. However, there are no solutions to $x^2 \equiv 2 \pmod{15}$, since the congruences $x^2 \equiv 2 \pmod{3}$ and $x^2 \equiv 2 \pmod{5}$ have no solutions.

We now show that the Jacobi symbol enjoys some properties similar to those of the Legendre symbol.

Theorem 9.8. Let n be an odd positive integer and let a and b be integers relatively prime to n. Then

$$
\text{(i)} \quad \text{if } a \equiv b \ (\text{mod } n), \text{ then } \left[\frac{a}{n}\right] = \left[\frac{b}{n}\right],
$$

$$
\text{(ii)} \quad \left[\frac{ab}{n}\right] = \left[\frac{a}{n}\right]\left[\frac{b}{n}\right],
$$

$$
\text{(iii)} \quad \left[\frac{-1}{n}\right] = (-1)^{(n-1)/2},
$$

$$
\text{(iv)} \quad \left[\frac{2}{n}\right] = (-1)^{(n^2-1)/8}.
$$

Proof. In the proof of all four parts of this theorem we use the prime factorization $n = p_1^{t_1} p_2^{t_2} \cdots p_m^{t_m}$.

Proof of (i). We know that if p is a prime dividing n, then $a \equiv b \ (\text{mod } p)$. Hence, from Theorem 9.3 (i), we have $\left[\frac{a}{b}\right] = \left[\frac{b}{p}\right]$. Consequently, we see that

$$
\left[\frac{a}{n}\right] = \left[\frac{a}{p_1}\right]^{t_1}\left[\frac{a}{p_2}\right]^{t_2} \cdots \left[\frac{a}{p_m}\right]^{t_m} = \left[\frac{b}{p_1}\right]^{t_1}\left[\frac{b}{p_2}\right]^{t_2} \cdots \left[\frac{b}{p_m}\right]^{t_m} = \left[\frac{b}{n}\right].
$$

Proof of (ii). From Theorem 9.3 (ii), we know that $\left[\frac{ab}{p_i}\right] = \left[\frac{a}{p_i}\right]\left[\frac{b}{p_i}\right]$. Hence,

$$
\left[\frac{ab}{n}\right] = \left[\frac{ab}{p_1}\right]^{t_1}\left[\frac{ab}{p_2}\right]^{t_2} \cdots \left[\frac{ab}{p_m}\right]^{t_m}
$$

$$
= \left[\frac{a}{p_1}\right]^{t_1}\left[\frac{b}{p_1}\right]^{t_1}\left[\frac{a}{p_2}\right]^{t_2}\left[\frac{b}{p_2}\right]^{t_2} \cdots \left[\frac{a}{p_m}\right]^{t_m}\left[\frac{b}{p_m}\right]^{t_m}
$$

$$
= \left[\frac{a}{n}\right]\left[\frac{b}{n}\right].
$$

Proof of (iii). Theorem 9.4 tells us that if p is prime, then $\left[\frac{-1}{p}\right] = (-1)^{(p-1)/2}$. Consequently,

$$\left(\frac{-1}{n}\right) = \left(\frac{-1}{p_1}\right)^{t_1}\left(\frac{-1}{p_2}\right)^{t_2}\cdots\left(\frac{-1}{p_m}\right)^{t_m}$$

$$= (-1)^{t_1(p_1-1)/2 + t_2(p_2-1)/2 + \cdots + t_m(p_m-1)/2}.$$

From the prime factorization of n, we have

$$n = (1 + (p_1-1))^{t_1}(1 + (p_2-1))^{t_2}\cdots(1 + (p_m-1))^{t_m}.$$

Since (p_i-1) is even, it follows that

$$(1 + (p_i-1))^{t_i} \equiv 1 + t_i(p_i-1) \pmod{4}$$

and

$$(1 + t_i(p_i-1))(1 + t_j(p_j-1)) \equiv 1 + t_i(p_j-1) + t_j(p_j-1) \pmod{4}.$$

Therefore,

$$n \equiv 1 + t_1(p_1-1) + t_2(p_2-1) + \cdots + t_m(p_m-1) \pmod{4}.$$

This implies that

$$(n-1)/2 \equiv t_1(p_1-1)/2 + t_2(p_2-1)/2 + \cdots + t_m(p_m-1)/2 \pmod{2}.$$

Combining this congruence for $(n-1)/2$ with the expression for $\left(\frac{-1}{n}\right)$ shows that

$$\left(\frac{-1}{n}\right) = (-1)^{\frac{n-1}{2}}.$$

Proof of (iv). If p is prime, then $\left(\frac{2}{p}\right) = (-1)^{(p^2-1)/8}$. Hence,

$$\left(\frac{2}{n}\right) = \left(\frac{2}{p_1}\right)^{t_1}\left(\frac{2}{p_2}\right)^{t_2}\cdots\left(\frac{2}{p_m}\right)^{t_m} = (-1)^{t_1(p_1^2-1)/8 + t_2(p_2^2-1)/8 + \cdots + t_m(p_m^2-1)/8}.$$

As in the proof of (iii), we note that

$$n^2 = (1 + (p_1^2-1))^{t_1}(1 + (p_2^2-1))^{t_2}\cdots(1 + (p_m^2-1))^{t_m}.$$

Since $p_j^2 - 1 \equiv 0 \pmod{8}$, we see that

$$(1 + (p_i^2-1))^{t_i} \equiv 1 + t_i(p_i^2-1) \pmod{64}$$

and

$$(1+t_i(p_i^2-1))(1 + t_j(p_j^2-1)) \equiv 1 + t_i(p_i^2-1) + t_j(p_j^2-1) \pmod{64}.$$

Hence,

$$n^2 \equiv 1 + t_1(p_1^2-1) + t_2(p_2^2-1) + \cdots + t_m(p_m^2-1) \pmod{64}.$$

This implies that

$$(n^2 - 1)/8 \equiv t_1(p_1^2 - 1)/8 + t_2(p_2^2 - 1)/8 + \cdots + t_m(p_m^2 - 1)/8 \pmod 8.$$

Combining this congruence for $(n^2 - 1)/8$ with the expression for $\left(\dfrac{2}{n}\right)$ tells us

that $\left(\dfrac{2}{n}\right) = (-1)^{(n^2 - 1)/8}$.

■

We now demonstrate that the reciprocity law holds for the Jacobi symbol as well as the Legendre symbol.

Theorem 9.9. The Reciprocity Law for Jacobi Symbols. Let n and m be relatively prime odd positive integers. Then

$$\left(\frac{n}{m}\right)\left(\frac{m}{n}\right) = (-1)^{\frac{m-1}{2}\frac{n-1}{2}}.$$

Proof. Let the prime factorizations of m and n be $m = p_1^{a_1} p_2^{a_2} \cdots p_s^{a_s}$ and $n = q_1^{b_1} q_2^{b_2} \cdots q_r^{b_r}$. We see that

$$\left(\frac{m}{n}\right) = \prod_{i=1}^{r}\left(\frac{m}{q_i}\right)^{b_i} = \prod_{i=1}^{r}\prod_{j=1}^{s}\left(\frac{p_j}{q_j}\right)^{b_i a_j}$$

and

$$\left(\frac{n}{m}\right) = \prod_{j=1}^{s}\left(\frac{n}{p_j}\right)^{a_j} = \prod_{j=1}^{s}\prod_{i=1}^{r}\left(\frac{q_i}{p_j}\right)^{a_j b_i}.$$

Thus,

$$\left(\frac{m}{n}\right)\left(\frac{n}{m}\right) = \prod_{i=1}^{r}\prod_{j=1}^{s}\left[\left(\frac{p_j}{q_i}\right)\left(\frac{q_i}{p_j}\right)\right]^{a_j b_i}.$$

By the law of quadratic reciprocity we know that

$$\left(\frac{p_j}{q_i}\right)\left(\frac{q_i}{p_j}\right) = (-1)^{\left[\frac{p_j - 1}{2}\right]\left[\frac{q_i - 1}{2}\right]}.$$

Hence,

$$(9.7) \qquad \left[\frac{m}{n}\right]\left[\frac{n}{m}\right] = \prod_{i=1}^{r}\prod_{j=1}^{s}(-1)^{a_j\left[\frac{p_j-1}{2}\right]b_i\left[\frac{q_i-1}{2}\right]} = (-1)^{\sum_{i=1}^{r}\sum_{j=1}^{s}a_j\left[\frac{p_j-1}{2}\right]b_i\left[\frac{q_i-1}{2}\right]}.$$

We note that

$$\sum_{i-1}^{r}\sum_{j=1}^{s}a_j\left[\frac{p_j-1}{2}\right]b_i\left[\frac{q_i-1}{2}\right] = \sum_{j=1}^{s}a_j\left[\frac{p_j-1}{2}\right]\sum_{i=1}^{r}b_i\left[\frac{q_i-1}{2}\right].$$

As we demonstrated in the proof of Theorem 9.8 (iii),

$$\sum_{j=1}^{s}a_j\left[\frac{p_j-1}{2}\right] \equiv \frac{m-1}{2} \pmod{2}$$

and

$$\sum_{i=1}^{r}b_i\left[\frac{q_i-1}{2}\right] \equiv \frac{n-1}{2} \pmod{2}.$$

Thus,

$$(9.8) \qquad \sum_{i=1}^{r}\sum_{j=1}^{s}a_j\left[\frac{p_j-1}{2}\right]b_i\left[\frac{q_i-1}{2}\right] \equiv \frac{m-1}{2}\cdot\frac{n-1}{2} \pmod{2}.$$

Therefore, by equations (9.7) and (9.8) we can conclude that

$$\left[\frac{m}{n}\right]\left[\frac{n}{m}\right] = (-1)^{\frac{m-1}{2}\cdot\frac{n-1}{2}}.$$

∎

We now develop an efficient algorithm for evaluating Jacobi symbols. Let a and b be relatively prime positive integers with $a < b$. Let $R_0 = a$ and $R_1 = b$. Using the division algorithm and factoring out the highest power of two dividing the remainder, we obtain

$$R_0 = R_1 q_1 + 2^{s_1}R_2,$$

where s_1 is a nonnegative integer and R_2 is an odd positive integer less than R_1. When we successively use the division algorithm, and factor out the highest power of two dividing remainders, we obtain

$$R_1 = R_2 q_2 + 2^{s_2}R_3$$
$$R_2 = R_3 q_3 + 2^{s_3}R_4$$
$$\vdots$$
$$\vdots$$
$$R_{n-3} = R_{n-2}q_{n-2} + 2^{s_{n-2}}R_{n-1}$$
$$R_{n-2} = R_{n-1}q_{n-1} + 2^{s_{n-1}}\cdot 1,$$

where s_j is a nonnegative integer and R_j is an odd positive integer less than R_{j-1} for $j = 2,3,...,n-1$. Note that the number of divisions required to reach the final equation does not exceed the number of divisions required to find the greatest common divisor of a and b using the Euclidean algorithm.

We illustrate this sequence of equations with the following example.

Example 9.14. Let $a = 401$ and $b = 111$. Then

$$401 = 111 \cdot 3 + 2^2 \cdot 17$$
$$111 = 17 \cdot 6 + 2^0 \cdot 9$$
$$17 = 9 \cdot 1 + 2^3 \cdot 1.$$

Using the sequence of equations we have described, together with the properties of the Jacobi symbol, we prove the following theorem, which gives an algorithm for evaluating Jacobi symbols.

Theorem 9.10. Let a and b be positive integers with $a > b$. Then

$$\left(\frac{a}{b}\right) = (-1)^{s_1 \frac{R_1^2-1}{8} + \cdots + s_{n-1} \frac{R_{n-1}^2-1}{8} + \frac{R_1-1}{2} \cdot \frac{R_2-1}{2} + \cdots + \frac{R_{n-2}-1}{2} \cdot \frac{R_{n-1}-1}{2}},$$

where the integers R_j and $s_j, j = 1, 2, \ldots, n-1$, are as previously described.

Proof. From the first equation and (i), (ii) and (iv) of Theorem 9.8, we have

$$\left(\frac{a}{b}\right) = \left(\frac{R_0}{R_1}\right) = \left(\frac{2^{s_1} R_2}{R_1}\right) = \left(\frac{2}{R_1}\right)^{s_1} \left(\frac{R_2}{R_1}\right) = (-1)^{s_1 \frac{R_1^2-1}{8}} \left(\frac{R_2}{R_1}\right).$$

Using Theorem 9.9, the reciprocity law for Jacobi symbols, we have

$$\left(\frac{R_2}{R_1}\right) = (-1)^{\frac{R_1-1}{2} \cdot \frac{R_2-1}{2}} \left(\frac{R_1}{R_2}\right),$$

so that

$$\left(\frac{a}{b}\right) = (-1)^{\frac{R_1-1}{2} \cdot \frac{R_2-1}{2} + s_1 \frac{R_1^2-1}{8}} \left(\frac{R_1}{R_2}\right).$$

Similarly, using the subsequent divisions, we find that

$$\left(\frac{R_{j-1}}{R_j}\right) = (-1)^{\frac{R_j-1}{2} \cdot \frac{R_{j+1}-1}{2} + s_j \cdot \frac{R_j^2-1}{8}} \left(\frac{R_j}{R_{j+1}}\right)$$

for $j = 2, 3, \ldots, n-1$. When we combine all the equalities, we obtain the desired expression for $\left(\frac{a}{b}\right)$. ∎

The following example illustrates the use of Theorem 9.10.

Example 9.15. To evaluate $\left(\frac{401}{111}\right)$, we use the sequence of divisions in the previous example and Theorem 9.10. This tells us that

$$\left(\frac{401}{111}\right) = (-1)^{2 \cdot \frac{111^2-1}{8} + 0 \cdot \frac{17^2-1}{8} + 3 \cdot \frac{9^2-1}{8} + \frac{111-1}{2} \cdot \frac{17-1}{2} + \frac{17-1}{2} \cdot \frac{9-1}{2}} = 1.$$

$\square$

The following corollary describes the computational complexity of the algorithm for evaluating Jacobi symbols given in Theorem 9.10.

Corollary 9.10.1. Let a and b be relatively prime positive integers with $a < b$. Then the Jacobi symbol $\left(\frac{a}{b}\right)$ can be evaluated using $O((\log_2 b)^3)$ bit operations.

Proof. To find $\left(\frac{a}{b}\right)$ using Theorem 9.10, we perform a sequence of $O(\log_2 b)$ divisions. To see this, note that the number of divisions does not exceed the number of divisions needed to find (a,b) using the Euclidean algorithm. Thus, by Lamé's theorem we know that $O(\log_2 b)$ divisions are needed. Each division can be done using $O((\log_2 b)^2)$ bit operations. Each pair of integers R_j and s_j can be found using $O(\log_2 b)$ bit operations once the appropriate division has been carried out.

Consequently, $O((\log_2 b)^3)$ bit operations are required to find the integers R_j, s_j, $j = 1,2,...,n-1$ from a and b. Finally, to evaluate the exponent of -1 in the expression for $\left(\frac{a}{b}\right)$ in Theorem 9.10, we use the last three bits in the binary expansions of $R_j, j = 1,2,...,n-1$ and the last bit in the binary expansions of $s_j, j = 1,2,...,n-1$. Therefore, we use $O(\log_2 b)$ additional bit operations to find $\left(\frac{a}{b}\right)$. Since $O((\log_2 b)^3) + O(\log_2 b) = O((\log_2 b)^3)$, the corollary holds.

∎

We can improve this corollary if we use more care when estimating the number of bit operations used by divisions. In particular, we can show that $O((\log_2 b)^2)$ bit operations suffice for evaluating $\left(\frac{a}{b}\right)$ We leave this as an exercise.

9.3 Exercises

1. Evaluate each of the following Jacobi symbols.

   a) $\left(\frac{5}{21}\right)$ d) $\left(\frac{1009}{2307}\right)$

   b) $\left(\frac{27}{101}\right)$ e) $\left(\frac{2663}{3299}\right)$

c) $\left[\dfrac{111}{1001}\right]$ f) $\left[\dfrac{10001}{20003}\right]$

2. For which positive integers n that are relatively prime to 15 does the Jacobi symbol $\left[\dfrac{15}{n}\right]$ equal 1?

3. For which positive integers n that are relatively prime to 30 does the Jacobi symbol $\left[\dfrac{30}{n}\right]$ equal 1?

4. Let a and b be relatively prime integers such that b is odd and positive and $a = (-1)^s 2^t q$ where q is odd. Show that

$$\left[\frac{a}{b}\right] = (-1)^{\frac{b-1}{2}\cdot s + \frac{b^2-1}{8}\cdot t}\left[\frac{q}{b}\right].$$

5. Let n be an odd square-free positive integer. Show that there is an integer a such that $(a,n) = 1$ and $\left[\dfrac{a}{n}\right] = -1$.

6. Let n be an odd square-free positive integer.

   a) Show that $\sum \left[\dfrac{k}{n}\right] = 0$, where the sum is taken over all k in a reduced set of residues modulo n. (*Hint*: Use Exercise 5.)

   b) From part (a), show that the number of integers in a reduced set of residues modulo n such that $\left[\dfrac{k}{n}\right] = 1$ is equal to the number with $\left[\dfrac{k}{n}\right] = -1$.

* 7. Let a and $b = r_0$ be relatively prime odd positive integers such that

$$a = r_0 q_1 + \varepsilon_1 r_1$$
$$r_0 = r_1 q_2 + \varepsilon_2 r_2$$
$$\cdot$$
$$\cdot$$
$$\cdot$$
$$r_{n-1} = r_{n-1} q_{n-1} + \varepsilon_n r_n$$

where q_i is a nonnegative even integer, $\varepsilon_i = \pm 1$, r_i is a positive integer with $r_i < r_{i-1}$, for $i = 1,2,...,n_j$, and $r_n = 1$. These equations are obtained by successively using the modified division algorithm given in Exercise 16 of Section 1.5.

   a) Show that the Jacobi symbol $\left[\dfrac{a}{b}\right]$ is given by

$$\left(\frac{a}{b}\right) = (-1)^{\left[\frac{r_0-1}{2}\cdot\frac{\varepsilon_1 r_1-1}{2} + \frac{r_1-1}{2}\cdot\frac{\varepsilon_2 r_2-1}{2} + \cdots + \frac{r_{n-1}-1}{2}\cdot\frac{\varepsilon_n r_n-1}{2}\right]}.$$

b) Show that the Jacobi symbol $\left(\dfrac{a}{b}\right)$ is given by

$$\left(\frac{a}{b}\right) = (-1)^T,$$

where T is the number of integers i, $1 \le i \le n$, with $r_{i-1} \equiv \varepsilon_i r_i \equiv 3$ (mod 4).

* 8. Show that if a and b are odd integers and $(a,b) = 1$, then the following reciprocity law holds for the Jacobi symbol:

$$\left(\frac{a}{|b|}\right)\left(\frac{b}{|a|}\right) = \begin{cases} -(-1)^{\frac{a-1}{2}\cdot\frac{b-1}{2}} & \text{if } a < 0 \text{ and } b < 0 \\ (-1)^{\frac{a-1}{2}\cdot\frac{b-1}{2}} & \text{otherwise.} \end{cases}$$

In Exercises 9-15 we deal with the *Kronecker symbol* which is defined as follows. Let a be a positive integer that is not a perfect square such that $a \equiv 0$ or 1 (mod 4). We define

$$\left(\frac{a}{2}\right) = \begin{cases} 1 & \text{if } a \equiv 1 \pmod 8 \\ -1 & \text{if } a \equiv 5 \pmod 8. \end{cases}$$

$$\left(\frac{a}{p}\right) = \text{the Legendre symbol } \left(\frac{a}{p}\right) \text{ if } p \text{ is an odd } \prime \text{ such that } p \nmid a.$$

$$\left(\frac{a}{n}\right) = \prod_{j=1}^{r}\left(\frac{a}{p_j}\right)^{t_j} \text{ if } (a,n) = 1 \text{ and } n = \prod_{j=1}^{r}p_j^{t_j} \text{ is the prime factorization of } n.$$

9. Evaluate each of the following Kronecker symbols,

a) $\left(\dfrac{5}{12}\right)$ b) $\left(\dfrac{13}{20}\right)$ c) $\left(\dfrac{101}{200}\right)$

For Exercises 10-15 let a be a positive integer that is not a perfect square such that $a \equiv 0$ or 1 (mod 4).

10. Show that $\left(\dfrac{a}{2}\right) = \left(\dfrac{2}{|a|}\right)$ if $2 \nmid a$, where the symbol on the right is a Jacobi symbol.

11. Show that if n_1 and n_2 are positive integers and if $(a_1, n_1 n_2) = 1$, then

$$\left(\frac{a}{n_1 n_2}\right) = \left(\frac{a}{n_1}\right)\cdot\left(\frac{a}{n_2}\right).$$

* 12. Show that if n is a positive integer relatively prime to a and if a is odd, then $\left(\dfrac{a}{n}\right) = \left(\dfrac{n}{|a|}\right)$, while if a is even, and $a = 2^s t$ where t is odd, then

$$\left(\frac{a}{n}\right) = \left(\frac{2}{n}\right)^s (-1)^{\frac{t-1}{2} \cdot \frac{n-1}{2}} \left(\frac{n}{|t|}\right).$$

* 13. Show that if n_1 and n_2 are positive integers relatively prime to a and $n_1 \equiv n_2 \pmod{|a|}$, then $\left(\dfrac{a}{n_1}\right) = \left(\dfrac{a}{n_2}\right)$.

* 14. Show that if $a \neq 0$, then there exists a positive integer n with $\left(\dfrac{a}{n}\right) = -1$.

* 15. Show that if $a \neq 0$, then $\left(\dfrac{a}{|a|-1}\right) = \begin{cases} 1 \text{ if } a > 0 \\ -1 \text{ if } a < 0. \end{cases}$

16. Show that if a and b are relatively prime positive integers with $a < b$ the Jacobi symbol $\left(\dfrac{a}{b}\right)$ can be evaluated using $O((\log_2 b)^2)$ bit operations.

9.3 Computer Projects

Programming Projects

Write programs to do the following:

1. Evaluate Jacobi symbols using the method of Theorem 9.7.
2. Evaluate Jacobi symbols using Exercises 4 and 7.
3. Evaluate Kronecker symbols (defined in the preamble to Exercise 9).

Computations and Explorations

Using the programs you have written or a computation program, carry out the following computations and explorations:

1. Find the value of the Legendre symbol $\left(\dfrac{1656169}{2355151}\right)$.

9.4 Euler Pseudoprimes

Let p be an odd prime number and let b be an integer not divisible by p. By Euler's criterion, we know that

$$b^{(p-1)/2} \equiv \left(\frac{b}{p}\right) \pmod{p}.$$

Hence, if we wish to test the positive integer n for primality, we can take an integer b, with $(b,n) = 1$, and determine whether

$$b^{(n-1)/2} \equiv \left[\frac{b}{n}\right] \pmod{n},$$

where the symbol on the right-hand side of the congruence is the Jacobi symbol. If we find that this congruence fails, then n is composite.

Example 9.16. Let $n = 341$ and $b = 2$. We calculate that $2^{170} \equiv 1 \pmod{341}$. Since $341 \equiv -3 \pmod 8$, using Theorem 9.8 (iv), we see that $\left[\dfrac{2}{341}\right] = -1$.

Consequently, $2^{170} \not\equiv \left[\dfrac{2}{341}\right] \pmod{341}$. This demonstrates that 341 is not prime.   □

Thus, we can define a type of pseudoprime based on Euler's criterion.

Definition. An odd, composite, positive integer n that satisfies the congruence

$$b^{(n-1)/2} \equiv \left[\frac{b}{n}\right] \pmod{n},$$

where b is a positive integer is called an *Euler pseudoprime to the base b.*

An Euler pseudoprime to the base b is a composite integer that masquerades as a prime by satisfying the congruence given in the definition.

Example 9.17. Let $n = 1105$ and $b = 2$. We calculate that $2^{552} \equiv 1 \pmod{1105}$. Since $1105 \equiv 1 \pmod 8$, we see that $\left[\dfrac{2}{1105}\right] = 1$.

Hence, $2^{552} \equiv \left[\dfrac{2}{1105}\right] \pmod{1105}$. Because 1105 is composite, it is an Euler pseudoprime to the base 2.   □

The following theorem shows that every Euler pseudoprime to the base b is a pseudoprime to this base.

Theorem 9.11. If n is an Euler pseudoprime to the base b, then n is a pseudoprime to the base b.

Proof. If n is an Euler pseudoprime to the base b, then

$$b^{(n-1)/2} \equiv \left[\frac{b}{n}\right] \pmod{n}.$$

Hence, by squaring both sides of this congruence, we find that

$$(b^{(n-1)/2})^2 \equiv \left(\frac{b}{n}\right)^2 \pmod{n}.$$

Since $\left(\dfrac{b}{n}\right) = \pm 1$, we see that $b^{n-1} \equiv 1 \pmod{n}$. This means that n is a pseudoprime to the base b. ∎

Not every pseudoprime is an Euler pseudoprime. For example, the integer 341 is not an Euler pseudoprime to the base 2, as we have shown, but is a pseudoprime to this base.

We know that every Euler pseudoprime is a pseudoprime. Next, we show that every strong pseudoprime is an Euler pseudoprime.

Theorem 9.12. If n is a strong pseudoprime to the base b, then n is an Euler pseudoprime to this base .

Proof. Let n be a strong pseudoprime to the base b. Then if $n - 1 = 2^s t$, where t is odd, either $b^t \equiv 1 \pmod{n}$ or $b^{2^r t} \equiv -1 \pmod{n}$ where $0 \le r \le s-1$. Let $n = \prod_{i=1}^{m} p_i^{a_i}$ be the prime-power factorization of n.

First, consider the case where $b^t \equiv 1 \pmod{n}$. Let p be a prime divisor of n. Since $b^t \equiv 1 \pmod{p}$, we know that $\mathrm{ord}_p b \,|\, t$. Because t is odd, we see that $\mathrm{ord}_p b$ is also odd. Hence, $\mathrm{ord}_p b \mid (p-1)/2$, since $\mathrm{ord}_p b$ is an odd divisor of the even integer $\phi(p) = p - 1$. Therefore,

$$b^{(p-1)/2} \equiv 1 \pmod{p}.$$

Consequently, by Euler's criterion, we have $\left(\dfrac{b}{p}\right) = 1$.

To compute the Jacobi symbol $\left(\dfrac{b}{n}\right)$, we note that $\left(\dfrac{b}{p}\right) = 1$ for all primes p dividing n. Hence,

$$\left(\frac{b}{n}\right) = \left(\frac{b}{\prod_{i=1}^{m} p_i^{a_i}}\right) = \prod_{i=1}^{m} \left(\frac{b}{p_i}\right)^{a_i} = 1.$$

Since $b^t \equiv 1 \pmod{n}$, we know that $b^{(n-1)/2} = (b^t)^{2^{s-1}} \equiv 1 \pmod{n}$. Therefore, we have

$$b^{(n-1)/2} \equiv \left(\frac{b}{n}\right) \equiv 1 \pmod{n}.$$

We conclude that n is an Euler pseudoprime to the base b.

Next, consider the case where

$$b^{2^r t} \equiv -1 \pmod{n}$$

for some r with $0 \le r \le s - 1$. If p is a prime divisor of n, then

$$b^{2^r t} \equiv -1 \pmod{p}.$$

Squaring both sides of this congruence, we obtain

$$b^{2^{r+1} t} \equiv 1 \pmod{p}.$$

This implies that $\text{ord}_p b \mid 2^{r+1} t$, but that $\text{ord}_p b \nmid 2^r t$. Hence,

$$\text{ord}_p b = 2^{r+1} c,$$

where c is an odd integer. Since $\text{ord}_p b \mid (p-1)$ and $2^{r+1} \mid \text{ord}_p b$, it follows that $2^{r+1} \mid (p-1)$.

Therefore, we have $p = 2^{r+1} d + 1$, where d is an integer. Since

$$b^{(\text{ord}_p b)/2} \equiv -1 \pmod{p},$$

we have

$$\left(\frac{b}{p} \right) \equiv b^{(p-1)/2} = b^{(\text{ord}_p b/2)((p-1)/\text{ord}_p b)}$$

$$\equiv (-1)^{(p-1)/\text{ord}_p b} = (-1)^{(p-1)/2^{r+1} c} \pmod{p}.$$

Because c is odd, we know that $(-1)^c = -1$. Hence,

(9.9)
$$\left(\frac{b}{p} \right) = (-1)^{(p-1)/2^{r+1}} = (-1)^d,$$

recalling that $d = (p-1)/2^{r+1}$. Since each prime p_i dividing n is of the form $p_i = 2^{r+1} d_i + 1$, it follows that

$$n = \prod_{i=1}^m p_i^{a_i}.$$

$$= \prod_{i=1}^m (2^{r+1} d_i + 1)^{a_i}$$

$$\equiv \prod_{i=1}^m (1 + 2^{r+1} a_i d_i)$$

$$\equiv 1 + 2^{r+1} \sum_{i=1}^m a_i d_i \pmod{2^{2r+2}}.$$

Therefore,

$$t2^{s-1} = (n-1)/2 \equiv 2^r \sum_{i=1}^m a_i d_i \pmod{2^{r+1}}.$$

This congruence implies that

$$t2^{s-1-r} \equiv \sum_{i=1}^{m} a_i d_i \pmod 2$$

and

$$(9.10) \qquad b^{(n-1)/2} = (b^{2^r t})^{2^{s-1-r}} \equiv (-1)^{2^{s-1-r}} = (-1)^{\sum_{i=1}^{m} a_i d_i} \pmod n.$$

On the other hand, from (9.9), we have

$$\left(\frac{b}{n} \right) = \prod_{i=1}^{m} \left(\frac{b}{p_i} \right)^{a_i} = \prod_{i=1}^{m} ((-1)^{d_i})^{a_i} = \prod_{i=1}^{m} (-1)^{a_i d_i} = (-1)^{\sum_{i=1}^{m} a_i d_i}.$$

Therefore, combining the previous equation with (9.10), we see that

$$b^{(n-1)/2} \equiv \left(\frac{b}{n} \right) \pmod n.$$

Consequently, n is an Euler pseudoprime to the base b. ∎

Although every strong pseudoprime to the base b is an Euler pseudoprime to this base, note that not every Euler pseudoprime to the base b is a strong pseudoprime to the base b, as the following example shows.

Example 9.18. We have previously shown that the integer 1105 is an Euler pseudoprime to the base 2. However, 1105 is not a strong pseudoprime to the base 2 since

$$2^{(1105-1)/2} = 2^{552} \equiv 1 \pmod{1105},$$

while

$$2^{(1105-1)/2^2} = 2^{276} \equiv 781 \not\equiv \pm 1 \pmod{1105}. \qquad \square$$

Although an Euler pseudoprime to the base b is not always a strong pseudoprime to this base, when certain extra conditions are met, an Euler pseudoprime to the base b is, in fact, a strong pseudoprime to this base. The following two theorems give results of this kind.

Theorem 9.13. If $n \equiv 3 \pmod 4$ and n is an Euler pseudoprime to the base b, then n is a strong pseudoprime to the base b.

Proof. From the congruence $n \equiv 3 \pmod 4$, we know that $n - 1 = 2 \cdot t$ where $t = (n-1)/2$ is odd. Since n is an Euler pseudoprime to the base b, it follows that

$$b^t = b^{(n-1)/2} \equiv \left(\frac{b}{n} \right) \pmod n.$$

Since $\left(\dfrac{b}{n} \right) = \pm 1$, we know that either $b^t \equiv 1 \pmod n$ or $b^t \equiv -1 \pmod n$.

Hence, one of the congruences in the definition of a strong pseudoprime to the base b must hold. Consequently, n is a strong pseudoprime to the base b.

Theorem 9.14. If n is an Euler pseudoprime to the base b and $\left(\dfrac{b}{n}\right) = -1$, then n is a strong pseudoprime to the base b.

Proof. We write $n - 1 = 2^s t$, where t is odd and s is a positive integer. Since n is an Euler pseudoprime to the base b, we have

$$b^{2^{s-1}t} = b^{(n-1)/2} \equiv \left(\frac{b}{n}\right) \pmod{n}.$$

But since $\left(\dfrac{b}{n}\right) = -1$, we see that

$$b^{t2^{s-1}} \equiv -1 \pmod{n}.$$

This is one of the congruences in the definition of a strong pseudoprime to the base b. Since n is composite, it is a strong pseudoprime to the base b. ∎

Using the concept of Euler pseudoprimality, we will develop a probabilistic primality test. This test was first suggested by Solovay and Strassen [128].

Before presenting the test, we give some helpful lemmata.

Lemma 9.4. If n is an odd positive integer that is not a perfect square, then there is at least one integer b with $1 < b < n, (b,n) = 1$, and $\left(\dfrac{b}{n}\right) = -1$, where $\left(\dfrac{b}{n}\right)$ is the Jacobi symbol.

Proof. If n is prime, the existence of such an integer b is guaranteed by Theorem 9.1. If n is composite, since n is not a perfect square, we can write $n = rs$ where $(r,s) = 1$ and $r = p^e$, with p an odd prime and e an odd positive integer.

Now let t be a quadratic nonresidue of the prime p; such a t exists by Theorem 9.1. We use the Chinese remainder theorem to find an integer b with $1 < b < n, (b,n) = 1$, and such that b satisfies the two congruences

$$b \equiv t \pmod{r}$$
$$b \equiv 1 \pmod{s}.$$

Then

$$\left(\frac{b}{r}\right) = \left(\frac{b}{p^e}\right) = \left(\frac{b}{p}\right)^e = (-1)^e = -1,$$

and $\left(\dfrac{b}{s}\right) = 1$. Since $\left(\dfrac{b}{n}\right) = \left(\dfrac{b}{r}\right)\left(\dfrac{b}{s}\right)$, it follows that $\left(\dfrac{b}{n}\right) = -1$. ∎

Lemma 9.5. Let n be an odd composite integer. Then there is at least one integer

b with $1 < b < n$, $(b,n) = 1$, and

$$b^{(n-1)/2} \not\equiv \left[\frac{b}{n}\right] \pmod{n}.$$

Proof. Assume that for all positive integers not exceeding n and relatively prime to n, that

(9.11) $$b^{(n-1)/2} \equiv \left[\frac{b}{n}\right] \pmod{n}.$$

Squaring both sides of this congruence tells us that

$$b^{n-1} \equiv \left[\frac{b}{n}\right]^2 \equiv (\pm 1)^2 = 1 \pmod{n},$$

if $(b,n) = 1$. Hence, n must be a Carmichael number. Therefore, by Theorem 8.22 we know that $n = q_1 q_2 \cdots q_r$, where $q_1, q_2, \ldots, q_r$ are distinct odd primes.

We will now show that

$$b^{(n-1)/2} \equiv 1 \pmod{n}$$

for all integers b with $1 \le b \le n$ and $(b,n) = 1$. Suppose that b is an integer such that

$$b^{(n-1)/2} \equiv -1 \pmod{n}.$$

We use the Chinese remainder theorem to find an integer a with $1 < a < n$, $(a,n) = 1$, and

$$a \equiv b \pmod{q_1}$$
$$a \equiv 1 \pmod{q_2 q_3 \cdots q_r}.$$

Then we observe that

(9.12) $$a^{(n-1)/2} \equiv b^{(n-1)/2} \equiv -1 \pmod{q_1},$$

while

(9.13) $$a^{(n-1)/2} \equiv 1 \pmod{q_2 q_3 \cdots q_r}.$$

From congruences (9.12) and (9.13), we see that

$$a^{(n-1)/2} \not\equiv \pm 1 \pmod{n},$$

contradicting congruence (9.11). Hence, we must have

$$b^{(n-1)/2} \equiv 1 \pmod{n},$$

for all b with $1 \le b \le n$ and $(b,n) = 1$. Consequently, from the definition of an Euler pseudoprime, we know that

$$b^{(n-1)/2} \equiv \left\lfloor \frac{b}{n} \right\rfloor = 1 \pmod{n}$$

for all b with $1 \leq b \leq n$ and $(b,n) = 1$. However, Lemma 9.4 tells us that this is impossible. Hence, the original assumption is false. There must be at least one integer b with $1 < b < n$, $(b,n) = 1$, and

$$b^{(n-1)/2} \not\equiv \left\lfloor \frac{b}{n} \right\rfloor \pmod{n}.$$ ∎

We can now state and prove the theorem that is the basis of the probabilistic primality test.

Theorem 9.15. Let n be an odd composite integer. Then the number of positive integers less than n and relatively prime to n that are bases to which n is an Euler pseudoprime, does not exceed $\phi(n)/2$.

Proof. By Lemma 9.5 we know that there is an integer b with $1 < b < n$, $(b,n) = 1$, and

(9.14) $$b^{(n-1)/2} \not\equiv \left\lfloor \frac{b}{n} \right\rfloor \pmod{2n}.$$

Now, let $a_1, a_2, \ldots, a_m$ denote the positive integers less than n satisfying $1 \leq a_j \leq n$, $(a_j, n) = 1$, and

(9.15) $$a_j^{(n-1)/2} \equiv \left\lfloor \frac{a_j}{n} \right\rfloor \pmod{n},$$

for $j = 1, 2, \ldots, m$.

Let $r_1, r_2, \ldots, r_m$ be the least positive residues of the integers $ba_1, ba_2, \ldots, ba_m$ modulo n. We note that the integers r_j are distinct and $(r_j, n) = 1$ for $j = 1, 2, \ldots, m$. Furthermore,

(9.16) $$r_j^{(n-1)/2} \not\equiv \left\lfloor \frac{r_j}{n} \right\rfloor \pmod{n}.$$

For, if it were true that

$$r_j^{(n-1)/2} \equiv \left\lfloor \frac{r_j}{n} \right\rfloor \pmod{n},$$

then we would have

$$(ba_j)^{(n-1)/2} \equiv \left\lfloor \frac{ba_j}{n} \right\rfloor \pmod{n}.$$

This would imply that,

$$b^{(n-1)/2}\, a_j^{(n-1)/2} \equiv \left[\frac{b}{n}\right]\left[\frac{a_j}{n}\right] \pmod{n},$$

and since (9.15) holds, we would have

$$b^{(n-1)/2} \equiv \left[\frac{b}{n}\right],$$

contradicting (9.14).

Since $a_j,\ j = 1,2,\ldots,m,$ satisfies the congruence (9.15) while $r_j,\ j = 1,2,\ldots,\ m,$ does not, as (9.16) shows, we know these two sets of integers share no common elements. Hence, looking at the two sets together, we have a total of $2m$ distinct positive integers less than n and relatively prime to n. Since there are $\phi(n)$ integers less than n that are relatively prime to n, we can conclude that $2m \le \phi(n)$, so that $m \le \phi(n)/2$. This proves the theorem. ■

By Theorem 9.15, we see that if n is an odd composite integer, when an integer b is selected at random from the integers $1,2,\ldots,n-1$, the probability that n is an Euler pseudoprime to the base b is less than 1/2. This leads to the following probabilistic primality test.

Theorem 9.16. The Solovay-Strassen Probabilistic Primality Test. Let n be a positive integer. Select, at random, k integers $b_1,b_2,\ldots,b_k$ from the integers $1,2,\ldots,n-1$. For each of these integers $b_j,\ j = 1,2,\ldots,k$, determine whether

$$b_j^{(n-1)/2} \equiv \left[\frac{b_j}{n}\right] \pmod{n}.$$

If any of these congruences fails, then n is composite. If n is prime then all these congruences hold. If n is composite, the probability that all k congruences hold is less than $1/2^k$. Therefore, if n passes this test when k is large, then n is "almost certainly prime".

Since every strong pseudoprime to the base b is an Euler pseudoprime to this base, more composite integers pass the Solovay-Strassen probabilistic primality test than the Rabin probabilistic primality test, although both require $O(k(\log_2 n)^3)$ bit operations.

9.4 Exercises

1. Show that the integer 561 is an Euler pseudoprime to the base 2.

2. Show that the integer 15841 is an Euler pseudoprime to the base 2, a strong pseudoprime to the base 2 and a Carmichael number.

3. Show that if n is an Euler pseudoprime to the bases a and b, then n is an Euler pseudoprime to the base ab.

4. Show that if n is an Euler pseudoprime to the base b, then n is also an Euler pseudoprime to the base $n - b$.

5. Show that if $n \equiv 5 \pmod 8$ and n is an Euler pseudoprime to the base 2, then n is a strong pseudoprime to the base 2.

6. Show that if $n \equiv 5 \pmod{12}$ and n is an Euler pseudoprime to the base 3, then n is a strong pseudoprime to the base 3.

7. Find a congruence condition for an Euler pseudoprime n to the base 5 that guarantees that n is a strong pseudoprime to the base 5.

** **8.** Let the composite positive integer n have prime-power factorization $n = p_1^{a_1} p_2^{a_2} \cdots p_m^{a_m}$, where $p_j = 1 + 2^{k_j} q_j$ for $j = 1,2,\ldots,m$, where $k_1 \le k_2 \le \cdots \le k_m$, and where $n = 1 + 2^k q$. Show that n is an Euler pseudoprime to exactly

$$\delta_n \prod_{j=1}^m ((n-1)/2, \, p_j - 1)$$

different bases b with $1 \le b < n$, where

$$\delta_n = \begin{cases} 2 & \text{if } k_1 = k \\ 1/2 & \text{if } k_j < k \text{ and } a_j \text{ is odd for some } j \\ 1 & \text{otherwise.} \end{cases}$$

9. For how many integers b, $1 \le b < 561$ is 561 an Euler pseudoprime to the base b?

10. For how many integers b, $1 \le b < 1729$ is 1729 an Euler pseudoprime to the base b?

9.4 Computer Projects

Programming Projects

Write programs to do the following:

1. Given an integer n and a positive integer b greater than one, determine whether n passes the test for Euler pseudoprimes to the base b.

2. Given an integer n, perform the Solovay-Strassen probabilistic primality test on n.

Computations and Explorations

Using the programs you have written or a computation program, carry out the following computations and explorations:

1. Find all Euler pseudoprimes to the base 2 less than 1000000. Do the same thing for the bases 3, 5, 7, and 11. Devise a primality test based on your results.

2. Find 10 integers each with between 50 and 60 decimal digits that are "probably prime" since they pass more than 20 iterations of the Solovay-Strassen probabilistic primality test.

9.5 Zero-Knowledge Proofs

Suppose you want to convince another person that you have some important private information without revealing this information. For example, you may want to convince someone that you know the prime factorization of a 200-digit positive integer without telling them the prime factors. Or you may have a proof of an important theorem and you want to convince the mathematical community you have such a proof without revealing it. In this section we will discuss methods, commonly known as *zero-knowledge* or *minimum-disclosure proofs*, that can be used to convince someone else that you have some private, verifiable information, without revealing it. Zero-knowledge proofs were first invented in the mid-1980s.

In a zero-knowledge proof, there are two parties, the *prover*, the person who has the secret information, and the *verifier*, who wants to be convinced that the prover has this secret information. When a zero-knowledge proof is used, the probability is extremely small that someone who does not have the information can successfully cheat the verifier by masquerading as the prover. Moreover, the verifier learns nothing, or almost nothing, about the information other than that the prover possesses it. In particular, the verifier cannot convince a third party that the verifier knows this information.

Remark: Since zero-knowledge proofs supply the verifier with a small amount of information, zero-knowledge proofs are more properly called minimum-disclosure proofs. Nevertheless, we will use the original terminology for such proofs.

We will illustrate the use of zero-knowledge proofs by describing several examples of such proofs, each based on the ease of finding square roots modulo products of two primes compared with the difficulty of finding square roots when the two primes are not known. (See Section 9.1 for a discussion of this topic.)

In our first example, suppose Paula, the *prover*, wants to convince Vince, the *verifier*, that she knows the prime factorization of n, where n is the product of two primes p and q. For simplicity in computation, we also suppose that both p and q are primes congruent to 3 modulo 4. She wants to convince Vince she knows p and q without helping him find them. The following interactive procedure achieves these goals.

i. Vince, who knows n, but not p and q, chooses, at random, an integer x. He computes y, the least nonnegative residue of x^4 modulo n and sends this to Paula.

ii. When Paula receives y, she computes its square root modulo n. (We will explain how she can do this after describing the steps of the procedure.) This square root is the least positive residue of x^2 modulo n. She sends this integer to Vince.

iii. Vince checks Paula's answer by finding the remainder of x^2 when it is divided by n.

To see why Paula can find the least positive residue of x^2 modulo n in step (ii), note that because she knows p and q, she can easily find the four square roots of x^4

modulo n Next, note that only one of the four square roots of x^4 modulo n is a quadratic residue modulo n (see Exercise 3). So, to find x^2, she can select the correct square root of the four square roots of x^4 modulo n by computing the value of the Legendre symbols of each of these square roots modulo p and modulo q. Note that someone not knowing p and q is unable to find the square root of y modulo n in a reasonable amount of time, unlike Paula who knows these primes.

We illustrate this procedure in the following example.

Example 9.19. Suppose that Paula's private information is her factorization of $n = 103 \cdot 239 = 24617$. She can use the procedure just described in the text to convince Vince she knows the primes $p = 103$ and $q = 239$ without revealing these primes to Vince. (In practice, primes p and q with hundreds of digits would be used rather than the small primes used in this example.)

To illustrate this procedure, suppose that in step (i) Vince selects 9134 when he chooses an integer at random. He computes the least positive residue of 9134^4 modulo 24617 which equals 20682. He sends the integer 20682 to Paula.

In step (ii) Paula determines the integer x^2 using the congruences

$$x^2 \equiv \pm 20682^{(103+1)/4} = \pm 20682^{26} \equiv \pm 59 \pmod{103}$$

$$x^2 \equiv \pm 20682^{(239+1)/4} = \pm 20682^{60} \equiv \pm 75 \pmod{239}.$$

She finds, using procedures described in Section 3.3, the four simultaneous solutions for x^2 modulo 24617, $x^2 \equiv 2943, 11786, 12831, 21674 \pmod{24617}$. However, of these four solutions, only one, 2943, is a quadratic residue modulo 24617. Consequently, Paula concludes that $x^2 \equiv 2943 \pmod{24617}$.

In step (iii) Vince checks Paula's answer by noting that $x^2 = 9134^2 \equiv 2943 \pmod{24617}$. $\square$

We now describe a method, based on zero-knowledge techniques, invented by Shamir in 1985 to verify the identity of the prover. We again suppose that $n = pq$ where p and q are two large primes both congruent to 3 modulo 4. Let I be a positive integer which represents some particular information, such as a personal identification number. The prover selects a small positive integer c that has the property that the integer v obtained by concatenating I with c (the number obtained by writing the digits of I followed by the digits of c) is a quadratic residue modulo n. (The number c can be found by trying different values of c until one is found for which the integer obtained by concatenating I with c is a quadratic residue, which happens with probability close to 1/2.) The prover can easily find u, a square root of v modulo n.

The prover convinces the verifier, that she knows the primes p and q using an interactive proof. Each cycle of the proof is based on the following steps.

i. The prover, Paula, chooses a random number r and sends to the verifier a message containing two values: x where $x \equiv r^2$ (mod n), $0 \leq x < n$, and y where $y \equiv v\bar{x}$ (mod n), $0 \leq y < n$. Here, as usual, $\bar{x}$ is an inverse of x modulo n.

ii. The verifier, Vince, checks that $xy \equiv v$ (mod n) and chooses, at random, a bit b which he sends to the prover.

iii. If the bit b sent by Vince is 0, Paula sends r to Vince. Otherwise, if the bit b is 1, Paula sends the least positive residue of $u\bar{r}$ modulo n to Vince, where $\bar{r}$ is an inverse of r modulo n.

iv. Vince computes the square of what Paula has sent back. If Vince sent a 0, he checks that this square is x, that is, that $r^2 \equiv x$ (mod n). If he sent 1 , he checks that this square is y, that is, that $s^2 \equiv y$ (mod n).

This procedure is also based on the fact that the prover can find u, a square root of v modulo n, whereas someone who does not know p and q will not be able to compute a square root modulo n in a reasonable amount of time.

The four steps of this procedure form one cycle. Cycles can be repeated sufficiently often to guarantee a high degree of security, as we will describe later.

We illustrate this type of zero-knowledge proof with the following example.

Example 9.20. Suppose that Paula wants to verify her identity to Vince by convincing him that she knows the prime factors of $n = 31 \cdot 61 = 1891$. Her identification number is $I = 391$. Note that 391 is a quadratic residue of 1891, since, as the reader can verify, it is a quadratic residue of both 31 and of 61, so she can take $v = 391$ (that is, since does not have to concatenate an integer c onto I). Paula finds that $u = 239$ is a square root of 391 modulo 1891. She can easily perform this calculation, since she knows the primes 31 and 61. (Note that we have selected small primes p and q in this example to illustrate the procedure. In practice primes with hundreds of digits should be used.)

We illustrate one cycle of this procedure. In step (i) Paula chooses a random number, say $r = 998$. She sends to Vince two numbers, $x \equiv r^2 \equiv 998^2 \equiv 1338$ (mod 1891) and $y \equiv v\bar{x} \equiv 391 \cdot 1296 \equiv 1839$ (mod 1891).

In step (ii), Vince checks that $xy \equiv 1338 \cdot 1839 \equiv 391$ (mod 1891). and chooses, at random, a bit b, say $b = 1$ which he sends to Paula.

In step (iii) Paula sends $s \equiv u\bar{r} \equiv 239 \cdot 1855 \equiv 851$ (mod 1891) to Vince. Finally, in step (iv), Vince checks that $s^2 \equiv 851^2 \equiv 1839 \equiv y$ (mod 1891). □

Note that if the prover sends the verifier both r and s, the verifier will know the private information $u = rs$, which is the secret information held by the prover. By passing the test with sufficiently many cycles, the prover has shown that on request that she can produce either r or s. It follows that she must know u, since, in each cycle, she knows both r and s. The choice of the random bit by the verifier makes it impossible for someone to fix the procedure by using numbers that have been rigged to pass the test. For example, someone could compute the square of a known number r and send $x = r^2$, instead of choosing a random number. Similarly, someone could select a number x such that $\sqrt{x}$ is a known square. However, it is impossible to do precalculations to make both x and y the squares of known numbers without knowing u.

Because the bit chosen by the verifier is chosen at random, the probability it is a 0 is 1/2 as is the probability it is a 1. If someone does not know u, the square root of v, the probability they pass one iteration of this test is almost exactly 1/2. Consequently, the probability that someone masquerading as the prover passes the test with 30 cycles is approximately $1/2^{30}$ which is less than one in a billion.

A variation of this procedure, known as the Fiat-Shamir method, is the basis for verification procedures used by smart cards, such as for verifying personal identification numbers.

Next, we describe a method that can be used to prove, using a zero-knowledge proof, that someone has some particular information. Suppose that the prover, Paula, has information represented by a sequence of numbers v_1, v_2, ..., v_m, where $1 \leq v_j < n$ for $j = 1,2,...,m$. Here, as before, n is the product of two primes p and q are both congruent to 3 modulo 4. Paula makes public the sequence of integers s_1, $s_2,...,s_m$ where $s_j \equiv \bar{v}_j^2$ (mod n), $1 \leq s_j < n$. Paula wants to convince the verifier, Vince, that she knows the private information v_1, v_2, ..., v_m, without revealing this information to Vince. What Vince knows is her public moduli n and her public information $s_1, s_2,..., s_m$.

The following procedure can be used to convince Vince she has this information. Each cycle of the procedure has the following steps.

i. Paula chooses a random number r and computes $x = r^2$ which she sends to Vince.

ii. Vince selects a subset S of the set $\{1,2,...,m\}$ and sends this subset to Paula.

iii. Paula computes y, the least positive residue modulo n of the product of r and the integers v_j with j in S, that is $y \equiv r \prod_{j \in S} v_j$ (mod n), $0 \leq y < n$.

iv. Vince verifies that $x \equiv y^2 z$ (mod n) where z is the product of the integers s_j with j in S, that is, $z \equiv \prod_{j \in S} s_j$ (mod n), $0 \leq z < n$.

Note that the congruence in step (iv) holds since

$$y^2 z \equiv r^2 \prod_{j \in S} v_j^2 \prod_{j \in S} s_j$$

$$\equiv r^2 \prod_{j \in S} v_j^2 \bar{v}_j^2$$

$$\equiv r^2 \text{ (mod } n).$$

The random number r is used so that the verifier cannot determine the value of the integer v_j, part of the secret information, by selecting the set $S = \{j\}$. When this procedure is carried out, the verifier is given no new information that will help him determine the private information $c_1,...,c_m$.

We illustrate one cycle of this interactive zero-knowledge proof in the following example.

Example 9.21. Suppose Paula wants to convince Vince she has some secret information, which is represented by the integers $v_1 = 1144$, $v_2 = 877$, $v_3 = 2001$, $v_4 = 1221$, $v_5 = 101$. Her secret modulus is $n = 47 \cdot 53 = 2491$. (In practice, primes with hundreds of digits are used rather than the small primes used in this example.)

Her public information consists of the integers s_j with $s_j \equiv \bar{v}_j^2 \pmod{2491}$, $0 < s_j < 2491$, $j = 1,2,3,4,5$. It follows, after routine calculation, that her public information consists of the integers $s_1 = 197$, $s_2 = 2453$, $s_3 = 1553$, $s_4 = 941$, $s_5 = 494$.

Paula can convince Vince that she has the secret information using the procedure described in the text. We describe one cycle of the procedure. In step (i) Paula chooses a random number, say $r = 1253$. Next, she sends $x = 679$, the least positive residue of r^2 modulo 2491, to Vince.

In step (ii) Vince selects a subset of $\{1,2,3,4,5\}$, say $S = \{1,3,4,5\}$ and informs Paula of this choice.

In step (iii) Paula computes the number y with $0 \le y < 2491$ and with

$$y \equiv rv_1 v_3 v_4 v_5$$
$$\equiv 1253 \cdot 1144 \cdot 2001 \cdot 1221 \cdot 101$$
$$\equiv 68 \pmod{2491}.$$

Consequently, she sends $y = 68$ to Vince.

Finally, in step (iv) Vince confirms that $x \equiv y^2 s_1 s_3 s_4 s_5 \pmod{2491}$ by verifying that $x = 679 \equiv 68^2 \cdot 197 \cdot 1553 \cdot 941 \cdot 494 \pmod{2491}$.

Vince can ask Paula to run through more cycles of this procedure to verify that she does have the secret information. He stops once he feels that the probability she is cheating is small enough to satisfy his needs. □

How can the prover cheat in this interactive procedure for zero-knowledge proofs of information? That is, how can the prover fool the verifier into thinking that she really knows the private information $c_1,...,c_m$ when she does not? The only obvious way is for the prover to guess the set S before the verifier supplies this, in step (i) to take $x = r^2 \prod_{j \in S} v_j^2$, and in step (iii) to take $y = r$. Since there are 2^m possible sets S (since there are that many subsets of $\{1,2,...,m\}$), the probability that someone not knowing the private information fools the verifier using this technique is $1/2^m$. Furthermore, when this cycle is iterated T times, the probability this procedure fools the verifier decreases to $1/2^{mT}$. For instance, if $m = 10$ and $T = 3$, the probability of the verifier being fooled is less than one in a billion.

In this section we have only briefly touched upon the subject of zero-knowledge proofs. The reader interested in learning more about this subject should refer to the chapter by Goldwasser in [84], as well as to the reference supplied in that chapter.

9.5 Exercises

1. Suppose that $n = 3149 = 47 \cdot 67$ and that $x^4 \equiv 2070 \pmod{3149}$. Find the least nonnegative residue of x^2 modulo 3149.

2. Suppose that $n = 10807 = 103 \cdot 107$ and that $x^4 = 188 \pmod{10807}$. Find the least nonnegative residue of x^2 modulo 10807.

3. Suppose that $n = pq$ where p and q are primes both congruent to 3 modulo 4 and that x is a integer relatively prime to n. Show that of the four square roots of x^4 modulo n, only one is the least nonnegative residue of a square of an integer.

4. Suppose that Paula has identification number 1760 and modulus $1961 = 37 \cdot 53$. Show how she verifies her identity to Vince in one cycle of the Shamir procedure if she selects the random number 1101 and he chooses 1 as his random bit.

5. Suppose that Paula has identification number 7 and modulus $1411 = 17 \cdot 83$. Show how she can verify her identity to Vince in one cycle of the Shamir procedure if she selects the random number 822 and he chooses 1 as his random bit.

6. Run through the steps used to verify that the prover has the secret information in Example 9.21 when the random number $r = 888$ is selected by the prover in step (i) and the verifier selects the subset $\{2,3,5\}$ of $\{1,2,3,4,5\}$.

7. Run through the steps used to verify that the prover has the secret information in Example 9.21 when the random number $r = 1403$ is selected by the prover in step (i) and the verifier selects the subset $\{1,5\}$ of $\{1,2,3,4,5\}$.

8. Let $n = 2491 = 47 \cdot 53$. Suppose that Paula's identification information consists of the sequence of six numbers $x_1 = 881$, $x_2 = 1199$, $x_3 = 2144$, $x_4 = 110$, $x_5 = 557$, $x_6 = 2200$.

   a) Find Paula's public identification information, $y_1, y_2, y_3, y_4, y_5, y_6$.
   b) Suppose that Paula selects at random the number $r = 1091$ and Vince chooses the subset $S = 2,3,5,6$ and sends this to Paula. Find the number that Paula computes and sends back to Vince.
   c) What computation does Vince make to verify Paula's knowledge of her secret information?

9. Let $n = 3953 = 59 \cdot 67$. Suppose that Paula's identification information consists of the sequence of six numbers $x_1 = 1001$, $x_2 = 21$, $x_3 = 3097$, $x_4 = 989$, $x_5 = 157$, $x_6 = 1039$.

   a) Find Paula's public identification information, $y_1, y_2, y_3, y_4, y_5, y_6$.
   b) Suppose that Paula selects at random the number $r = 403$ and Vince chooses the subset $S = 1,2,4,6$ and sends this to Paula. Find the number that Paula computes and sends back to Vince.
   c) What computation does Vince make to verify Paula's knowledge of her secret information?

9.5 Computer Projects

Programming Projects

Write programs to do the following:

1. Given n, the product of two distinct primes both congruent to 3 modulo 4, and the least positive residue of x^4 modulo n where x is an integer relatively prime to n, find the least positive residue of x^2 modulo n.

Computations and Explorations

Using the programs you have written or a computation program, carry out the following computations and explorations:

1. Give one of your classmates the integer n where $n = pq$ and p and q are primes with more than fifty decimal digits, both congruent to 3 modulo 4. Convince your classmate you know both p and q using a minimum disclosure proof.

2. Convince one of your classmates you know a secret in the form of a sequence of ten positive integers each less than 10000, using the minimal disclosure proof described in the text.

10

Decimal Fractions and Continued Fractions

10.1 Decimal Fractions

In this chapter we discuss the representation of rational and irrational numbers as decimal fractions and continued fractions. We first consider base b expansions of real numbers, where b is a positive integer, $b > 1$. Let α be a positive real number, and let $a = [\alpha]$ be the integer part of α, so that $\gamma = \alpha - [\alpha]$ is the fractional part of α and $\alpha = a + \gamma$ with $0 \le \gamma < 1$. By Theorem 1.10 the integer a has a unique base b expansion. We now show that the fractional part γ also has a unique base b expansion.

Theorem 10.1. Let γ be a real number with $0 \le \gamma < 1$, and let b be a positive integer, $b > 1$. Then γ can be uniquely written as

$$\gamma = \sum_{j=1}^{\infty} c_j/b^j,$$

where the coefficients c_j are integers with $0 \le c_j \le b-1$ for $j = 1,2,\ldots$, with the restriction that for every positive integer N there is an integer n with $n \ge N$ and $c_n \ne b-1$.

In the proof of Theorem 10.1 we deal with infinite series. We will use the following formula for the sum of the terms of an infinite geometric series.

Theorem 10.2. Let a and r be real numbers with $|r| < 1$. Then

$$\sum_{j=0}^{\infty} ar^j = a/(1-r).$$

Most books on calculus or mathematical analysis contain a proof of Theorem 10.2 (see [100] for instance).

We can now prove Theorem 10.1.

Proof. We first let

$$c_1 = [b\gamma],$$

so that $0 \le c_1 \le b-1$, since $0 \le b\gamma < b$. In addition, let

$$\gamma_1 = b\gamma - c_1 = b\gamma - [b\gamma],$$

so that $0 \le \gamma_1 < 1$ and

$$\gamma = \frac{c_1}{b} + \frac{\gamma_1}{b}.$$

We recursively define c_k and γ_k for $k = 2,3,...,$ by

$$c_k = [b\gamma_{k-1}]$$

and

$$\gamma_k = b\gamma_{k-1} - c_k$$

so that $0 \le c_k \le b-1$, since $0 \le b\gamma_{k-1} < b$, and $0 \le \gamma_k < 1$. Then, it follows that

$$\gamma = \frac{c_1}{b} + \frac{c_2}{b^2} + \cdots + \frac{c_n}{b^n} + \frac{\gamma_n}{b^n}.$$

Since $0 \le \gamma_n < 1$, we see that $0 \le \gamma_n/b^n < 1/b^n$. Consequently,

$$\lim_{n \to \infty} \gamma_n/b^n = 0.$$

Therefore, we can conclude that

$$\gamma = \lim_{n \to \infty} \left[\frac{c_1}{b} + \frac{c_2}{b^2} + \cdots + \frac{c_n}{b^n} \right]$$

$$= \sum_{j=1}^{\infty} c_j/b^j.$$

To show that this expansion is unique, assume that

$$\gamma = \sum_{j=1}^{\infty} c_j/b^j = \sum_{j=1}^{\infty} d_j/b^j,$$

where $0 \le c_j \le b-1$ and $0 \le d_j \le b-1$, and, for every positive integer N, there are integers n and m with $c_n \neq b-1$ and $d_m \neq b-1$. Assume that k is the smallest index for which $c_k \neq d_k$, and assume that $c_k > d_k$ (the case $c_k < d_k$ is handled by switching the roles of the two expansions). Then

$$0 = \sum_{j=1}^{\infty} (c_j - d_j)/b^j = (c_k - d_k)/b^k + \sum_{j=k+1}^{\infty} (c_j - d_j)/b^j,$$

so that

(10.1) $$(c_k - d_k)/b^k = \sum_{j=k+1}^{\infty} (d_j - c_j)/b^j.$$

Since $c_k > d_k$, we have

(10.2) $$(c_k - d_k)/b^k \geq 1/b^k,$$

while

(10.3) $$\sum_{j=k+1}^{\infty} (d_j - c_j)/b^j \leq \sum_{j=k+1}^{\infty} (b-1)/b^j$$
$$= (b-1) \frac{1/b^{k+1}}{1 - 1/b}$$
$$= 1/b^k,$$

where we have used Theorem 10.2 to evaluate the sum on the right-hand side of the inequality. Note that equality holds in (10.3) if and only if $d_j - c_j = b-1$ for all j with $j \geq k + 1$, and this occurs if and only if $d_j = b-1$ and $c_j = 0$ for $j \geq k+1$. However, such an instance is excluded by the hypotheses of the theorem. Hence, the inequality in (10.3) is strict, and therefore (10.2) and (10.3) contradict (10.1). This shows that the base b expansion of α is unique. ∎

The unique expansion of a real number in the form $\sum_{j=1}^{\infty} c_j/b^j$ is called the *base b expansion* of this number and is denoted by $(.c_1 c_2 c_3 ...)_b$.

To find the base b expansion $(.c_1 c_2 c_3 ...)_b$ of a real number γ, we can use the recursive formula for the digits given in the proof of Theorem 10.1, namely

$$c_k = [b\gamma_{k-1}], \quad \gamma_k = b\gamma_{k-1} - [b\gamma_{k-1}],$$

where $\gamma_0 = \gamma$, for $k = 1, 2, 3, ...$.

Example 10.1. Let $(.c_1 c_2 c_3 ...)_b$ be the base 8 expansion of $1/6$. Then

$$c_1 = [8 \cdot \frac{1}{6}] = 1, \qquad \gamma_1 = 8 \cdot \frac{1}{6} - 1 = \frac{1}{3},$$

$$c_2 = [8 \cdot \frac{1}{3}] = 2, \qquad \gamma_2 = 8 \cdot \frac{1}{3} - 2 = \frac{2}{3},$$

$$c_3 = [8 \cdot \frac{2}{3}] = 5, \qquad \gamma_3 = 8 \cdot \frac{2}{3} - 5 = \frac{1}{3},$$

$$c_4 = [8 \cdot \frac{1}{3}] = 2, \qquad \gamma_4 = 8 \cdot \frac{1}{3} - 2 = \frac{2}{3},$$

$$c_5 = [8 \cdot \frac{2}{3}] = 5, \qquad \gamma_5 = 8 \cdot \frac{2}{3} - 5 = \frac{1}{3},$$

and so on. We see that the expansion repeats and hence,

$$1/6 = (.1252525...)_8.$$

$\square$

We will now discuss base b expansions of rational numbers. We will show that a number is rational if and only if its base b expansion is periodic or terminates.

Definition. A base b expansion $(.c_1 c_2 c_3 ...)_b$ is said to *terminate* if there is a positive integer n such that $c_n = c_{n+1} = c_{n+2} = \cdots = 0$.

Example 10.2. The decimal expansion of $1/8$, $(.125000...)_{10} = (.125)_{10}$, terminates. Also, the base 6 expansion of $4/9$, $(.24000...)_6 = (.24)_6$, terminates. $\square$

To describe those real numbers with terminating base b expansion, we prove the following theorem.

Theorem 10.3. The real number α, $0 \le \alpha < 1$, has a terminating base b expansion if and only if α is rational and can be written as $\alpha = r/s$, where $0 \le r < s$ and every prime factor of s also divides b.

Proof. First, suppose that α has a terminating base b expansion,

$$\alpha = (.c_1 c_2 ... c_n)_b.$$

Then

$$\alpha = \frac{c_1}{b} + \frac{c_2}{b^2} + \cdots + \frac{c_n}{b^n}$$

$$= \frac{c_1 b^{n-1} + c_2 b^{n-2} + \cdots + c_n}{b^n},$$

so that α is rational, and can be written with a denominator divisible only by primes dividing b.

Conversely, suppose that $0 \le \alpha < 1$, and

$$\alpha = r/s,$$

where each prime dividing s also divides b. Hence, there is a power of b, say b^N, that is divisible by s (for instance, take N to be the largest exponent in the prime-power factorization of s). Then

$$b^N \alpha = b^N r/s = ar,$$

where $sa = b^N$, and a is a positive integer since $s \mid b^N$. Now let $(a_m a_{m-1} \ldots a_1 a_0)_b$ be the base b expansion of ar. Then

$$\alpha = ar/b^N = \frac{a_m b^m + a_{m-1} b^{m-1} + \cdots + a_1 b + a_0}{b^N}$$

$$= a_m b^{m-N} + a_{m-1} b^{m-1-N} + \cdots + a_1 b^{1-N} + a_0 b^{-N}$$

$$= (.00 \ldots a_m a_{m-1} \ldots a_1 a_0)_b.$$

Hence, α has a terminating base b expansion. ∎

Note that every terminating base b expansion can be written as a nonterminating base b expansion with a tail-end consisting entirely of the digit $b-1$, since $(.c_1 c_2 \ldots c_m)_b = (.c_1 c_2 \ldots c_m - 1 \ b - 1 \ b - 1 \ldots)_b$. For instance, $(.12)_{10} = (.11999 \ldots)_{10}$. This is why we require in Theorem 10.1 that for every integer N there is an integer n, such that $n > N$ and $c_n \neq b - 1$; without this restriction base b expansions would not be unique.

A base b expansion that does not terminate may be *periodic*, for instance

$$1/3 = (.333 \ldots)_{10},$$
$$1/6 = (.1666 \ldots)_{10},$$

and

$$1/7 = (.142857142857142857 \ldots)_{10}.$$

Definition. A base b expansion $(.c_1 c_2 c_3 \ldots)_b$ is called *periodic* if there are positive integers N and k such that $c_{n+k} = c_n$ for $n \geq N$.

We denote by $(.c_1 c_2 \ldots c_{N-1} \overline{c_N \ldots c_{N+k-1}})_b$ the periodic base b expansion $(.c_1 c_2 \ldots c_{N-1} c_N \ldots c_{N+k-1} c_N \ldots c_{N+k-1} c_N \ldots)_b$. For instance, we have

$$1/3 = (.\overline{3})_{10},$$
$$1/6 = (.1\overline{6})_{10},$$

and

$$1/7 = (.\overline{142857})_{10}.$$

Note that the periodic parts of the decimal expansions of 1/3 and 1/7 begin immediately, while in the decimal expansion of 1/6 the digit 1 precedes the periodic part of the expansion. We call the part of a periodic base b expansion preceding the periodic part the *pre-period*, and the periodic part the *period*, where we take the period to have minimal possible length.

Example 10.3. The base 3 expansion of $2/45$ is $(.00\overline{1012})_3$. The pre-period is $(00)_3$ and the period is $(1012)_3$. □

The next theorem tells us that the rational numbers are those real numbers with periodic or terminating base b expansions. Moreover, the theorem gives the lengths of the pre-period and periods of base b expansions of rational numbers.

Theorem 10.4. Let b be a positive integer. Then a periodic base b expansion represents a rational number. Conversely, the base b expansion of a rational number either terminates or is periodic. Further, if $0 < \alpha < 1$, $\alpha = r/s$, where r and s are relatively prime positive integers, and $s = TU$ where every prime factor of T divides b and $(U,b) = 1$, then the period length of the base b expansion of α is $\mathrm{ord}_U b$, and the pre-period length is N, where N is the smallest positive integer such that $T \mid b^N$.

Proof. First, suppose that the base b expansion of α is periodic, so that

$$\alpha = (.c_1 c_2 ... c_N \overline{c_{N+1} ... c_{N+k}})_b$$

$$= \frac{c_1}{b} + \frac{c_2}{b^2} + \cdots + \frac{c_N}{b^N} + \left(\sum_{j=0}^{\infty} \frac{1}{b^{jk}} \right) \left(\frac{c_{N+1}}{b^{N+1}} + \cdots + \frac{c_{N+k}}{b^{N+k}} \right)$$

$$= \frac{c_1}{b} + \frac{c_2}{b^2} + \cdots + \frac{c_N}{b^N} + \left(\frac{b^k}{b^k - 1} \right) \left(\frac{c_{N+1}}{b^{N+1}} + \cdots + \frac{c_{N+k}}{b^{N+k}} \right),$$

where we have used Theorem 10.2 to see that

$$\sum_{j=0}^{\infty} \frac{1}{b^{jk}} = \frac{1}{1 - \frac{1}{b^k}} = \frac{b^k}{b^k - 1}.$$

Since α is the sum of rational numbers, it is rational.

Conversely, suppose that $0 < \alpha < 1$, $\alpha = r/s$, where r and s are relatively prime positive integers, $s = TU$, where every prime factor of T divides b, $(U,b) = 1$, and N is the smallest integer such that $T \mid b^N$.

Since $T \mid b^N$, we have $aT = b^N$, where a is a positive integer. Hence

(10.4) $$b^N \alpha = b^N \frac{r}{TU} = \frac{ar}{U}.$$

Furthermore, we can write

(10.5) $$\frac{ar}{U} = A + \frac{C}{U},$$

where A and C are integers with

$$0 \le A < b^N, \qquad 0 < C < U,$$

and $(C,U) = 1$. (The inequality for A follows since $0 < b^N \alpha = \dfrac{ar}{U} < b^N$, which results from the inequality $0 < \alpha < 1$ when both sides are multiplied by b^N). The fact that $(C,U) = 1$ follows easily from the condition $(r,s) = 1$. By Theorem 1.10 A has a base b expansion $A = (a_n a_{n-1}...a_1 a_0)_b$.

If $U = 1$, then the base b expansion of α terminates as shown above. Otherwise, let $v = \text{ord}_U b$. Then,

$$(10.6) \qquad b^v \frac{C}{U} = \frac{(tU+1)C}{U} = tC + \frac{C}{U},$$

where t is an integer, since $b^v \equiv 1 \pmod{U}$. However, we also have

$$(10.7) \qquad b^v \frac{C}{U} = b^v \left[\frac{c_1}{b} + \frac{c_2}{b^2} + \cdots + \frac{c_v}{b^v} + \frac{\gamma_v}{b^v} \right],$$

where $(.c_1 c_2 c_3...)_b$ is the base b expansion of $\dfrac{C}{U}$, so that

$$c_k = [b\gamma_{k-1}], \qquad \gamma_k = b\gamma_{k-1} - [b\gamma_{k-1}]$$

where $\gamma_0 = \dfrac{C}{U}$, for $k = 1,2,3,...$. From (10.7) we see that

$$(10.8) \qquad b^v \frac{C}{U} = \left[c_1 b^{v-1} + c_2 b^{v-2} + \cdots + c_v \right] + \gamma_v.$$

Equating the fractional parts of (10.6) and (10.8), noting that $0 \le \gamma_v < 1$, we find that

$$\gamma_v = \frac{C}{U}.$$

Consequently, we see that

$$\gamma_v = \gamma_0 = \frac{C}{U},$$

so that from the recursive definition of $c_1, c_2,...$ we can conclude that $c_{k+v} = c_k$ for $k = 1,2,3,...$. Hence $\dfrac{C}{U}$ has a periodic base b expansion

$$\frac{C}{U} = (.\overline{c_1 c_2 ... c_v})_b.$$

Combining (10.4) and (10.5), and inserting the base b expansions of A and $\dfrac{C}{U}$, we have

$$(10.9) \qquad b^N \alpha = (a_n a_{n-1}...a_1 a_0 . \overline{c_1 c_2 ... c_v})_b.$$

Dividing both sides of (10.9) by b^N, we obtain

$$\alpha = (.00...a_n a_{n-1}...a_1 a_0 \overline{c_1 c_2 ... c_v})_b,$$

(where we have shifted the decimal point in the base b expansion of $b^N \alpha$ N spaces to the left to obtain the base b expansion of α). In this base b expansion of α, the pre-period $(.00...a_n a_{n-1}...a_1 a_0)_b$ is of length N, beginning with $N - (n+1)$ zeros, and the period length is v.

We have shown that there is a base b expansion of α with a pre-period of length N and a period of length v. To finish the proof, we must show that we cannot regroup the base b expansion of α, so that either the pre-period has length less than N, or the period has length less than v. To do this, suppose that

$$\alpha = (.c_1 c_2...c_M \overline{c_{M+1}...c_{M+k}})_b$$

$$= \frac{c_1}{b} + \frac{c_2}{b^2} + \cdots + \frac{c_M}{b^M} + \left[\frac{b^k}{b^k - 1}\right]\left[\frac{c_{M+1}}{b^{M+1}} + \cdots + \frac{c_{M+k}}{b^{M+k}}\right]$$

$$= \frac{(c_1 b^{M-1} + c_2 b^{M-2} + \cdots + c_M)(b^k - 1) + (c_{M+1} b^{k-1} + \cdots + c_{M+k})}{b^M (b^k - 1)}.$$

Since $\alpha = r/s$, with $(r,s) = 1$, we see that $s \mid b^M(b^k - 1)$. Consequently, $T \mid b^M$ and $U \mid (b^k - 1)$. Hence, $M \geq N$, and $v \mid k$ (by Theorem 8.1, since $b^k \equiv 1 \pmod{U}$ and $v = \text{ord}_U b$). Therefore, the pre-period length cannot be less than N and the period length cannot be less than v. ∎

We can use Theorem 10.4 to determine the lengths of the pre-period and period of decimal expansions. Let $\alpha = r/s$, $0 < \alpha < 1$, and $s = 2^{s_1} 5^{s_2} t$, where $(t, 10) = 1$. Then by Theorem 10.4 the pre-period has length $\max(s_1, s_2)$ and the period has length $\text{ord}_t 10$.

Example 10.4. Let $\alpha = 5/28$. Since $28 = 2^2 \cdot 7$, Theorem 10.4 tells us that the pre-period has length 2 and the period has length $\text{ord}_7 10 = 6$. Since $5/28 = (.17\overline{857142})$, we see that these lengths are correct. □

Note that the pre-period and period lengths of a rational number r/s, in lowest terms, depends only on the denominator s, and not on the numerator r.

We observe that by Theorem 10.4 a base b expansion that is not terminating and is not periodic represents an irrational number.

Example 10.5. The number with decimal expansion

$$\alpha = .10100100010000... ,$$

consisting of a one followed by a zero, a one followed by two zeros, a one followed by three zeroes, and so on, is irrational because this decimal expansion does not terminate and is not periodic. □

The number α in the above example is concocted so that its decimal expansion is clearly not periodic. To show that naturally occurring numbers such as e and π are irrational, we cannot use Theorem 10.4 because we do not have explicit formulae for the decimal digits of these numbers. No matter how many decimal digits of their expansions we compute, we still cannot conclude that they are irrational from this evidence, because the period could be longer than the number of digits we have computed.

10.1 Exercises

1. Find the decimal expansions of each of the following numbers.

   a) 2/5   d) 8/15
   b) 5/12   e) 1/111
   c) 12/13   f) 1/1001

2. Find the base 8 expansions of each of the following numbers.

   a) 1/3   d) 1/6
   b) 1/4   e) 1/12
   c) 1/5   f) 1/22

3. Find the fraction, in lowest terms, represented by each of the following expansions.

   a) .12   b) .1$\overline{2}$ c) .$\overline{12}$

4. Find the fraction, in lowest terms, represented by each of the following expansions.

   a) $(.\overline{123})_7$ c) $(.\overline{17})_{11}$
   b) $(.0\overline{13})_6$ d) $(.\overline{ABC})_{16}$

5. For which positive integers b does the base b expansion of 11/210 terminate?

6. Find the pre-period and period lengths of the decimal expansions of each of the following rational numbers.

   a) 7/12   d) 10/23
   b) 11/30   e) 13/56
   c) 1/75   f) 1/61

7. Find the pre-period and period lengths of the base 12 expansions of each of the following rational numbers.

   a) 1/4   d) 5/24
   b) 1/8   e) 17/132
   c) 7/10   f) 7/360

8. Let b be a positive integer. Show that the period length of the base b expansion of $1/m$ is $m - 1$ if and only if m is prime and b is a primitive root of m.

9. For which primes p does the decimal expansion of $1/p$ have period length equal to each of the following integers?

   a) 1   d) 4
   b) 2   e) 5
   c) 3   f) 6

10. Find the base b expansions of each of the following numbers.

    a) $1/(b-1)$ b) $1/(b+1)$

11. Let b be an integer with $b > 2$. Show that the base b expansion of $1/(b-1)^2$ is $(.0123...b-3 \ b-1)_b$.

12. Show that the real number with base b expansion

$$(.0123...b-1 \ 101112...)_b,$$

constructed by successively listing the base b expansions of the integers, is irrational.

13. Show that

$$\frac{1}{b} + \frac{1}{b^4} + \frac{1}{b^9} + \frac{1}{b^{16}} + \frac{1}{b^{25}} + \cdots$$

is irrational, whenever b is a positive integer larger than one.

14. Let $b_1, b_2, b_3, \ldots$ be an infinite sequence of positive integers greater than one. Show that every real number can be represented as

$$c_0 + \frac{c_1}{b_1} + \frac{c_2}{b_1 b_2} + \frac{c_3}{b_1 b_2 b_3} + \cdots,$$

where $c_0, c_1, c_2, c_3, \ldots$ are integers such that $0 \leq c_k < b_k$ for $k = 1, 2, 3, \ldots$.

15. Show that every real number has an expansion

$$c_0 + \frac{c_1}{1!} + \frac{c_2}{2!} + \frac{c_3}{3!} + \cdots$$

where $c_0, c_1, c_2, c_3, \ldots$ are integers and $0 \leq c_k < k$ for $k = 1, 2, 3, \ldots$.

16. Show that every rational number has a terminating expansion of the type described in Exercise 15.

* 17. Suppose that p is a prime and the base b expansion of $1/p$ is $(.\overline{c_1 c_2 \cdots c_{p-1}})_b$, so that the period length of the base b expansion of $1/p$ is $p - 1$. Show that if m is a positive integer with $1 \leq m < p$, then

$$m/p = (.\overline{c_{k+1} \cdots c_{p-1} c_1 c_2 \cdots c_{k-1} c_k})_b,$$

where $k = \text{ind}_b m$ modulo p.

* 18. Show that if p is prime and $1/p = (.\overline{c_1 c_2 \cdots c_k})_b$ has an even period length, $k = 2t$, then $c_j + c_{j+t} = b - 1$ for $j = 1, 2, \ldots, t$.

19. For which positive integers n is the length of the period of the binary expansion of $1/n$ equal to $n-1$?

20. For which positive integers n is the length of the period of the decimal expansion of $1/n$ equal to $n-1$?

* 21. Show that the number e is irrational.

22. Pseudo-random numbers can be generated using the base m expansion of $1/P$ where P is a positive integer relatively prime to m. We set $x_n = c_{j+n}$ where j, the position of the seed, is a positive integer and $1/P = (.c_1 c_2 c_3 ...)_m$. This is called the $1/P$ *generator*. First the first ten terms of the pseudo-random sequence generator with each of the following parameters.

   a) $m = 7, P = 19$, and $j = 6$
   b) $m = 8, P = 21$, and $j = 5$

10.1 Computer Projects

Programming Projects

Write computer programs to do the following:

1. Find the base b expansion of a rational number, where b is a positive integer.

2. Find the numerator and denominator of a rational number in lowest terms from its base b expansion.

3. Find the pre-period and period lengths of the base b expansion of a rational number, where b is a positive integer.

4. Generate pseudorandom numbers using the $1/P$ generator with modulus m and seed in position j where P and m are relatively prime positive integers greater than 1 and j is a positive integer.

Computations and Explorations

Using the programs you have written or a computation program, carry out the following computations and explorations:

1. Find as many positive integers n as you can such that the length of the period of the decimal expansion of $1/n$ is $n-1$.

2. Find the first 10000 terms of the decimal expansion of π. Can you find any patterns? Make some conjectures about this expansion.

3. Find the first 10000 terms of the decimal expansion of e. Can you find any patterns? Make some conjectures about this expansion.

10.2 Finite Continued Fractions

Using the Euclidean algorithm we can express rational numbers as *continued fractions*. For instance, the Euclidean algorithm produces the following sequence of equations:

$$62 = 2 \cdot 23 + 16$$
$$23 = 1 \cdot 16 + 7$$
$$16 = 2 \cdot 7 + 2$$
$$7 = 3 \cdot 2 + 1.$$

When we divide both sides of each equation by the divisor of that equation, we obtain

$$\frac{62}{23} = 2 + \frac{16}{23} = 2 + \frac{1}{23/16}$$

$$\frac{23}{16} = 1 + \frac{7}{16} = 1 + \frac{1}{16/7}$$

$$\frac{16}{7} = 2 + \frac{2}{7} = 2 + \frac{1}{7/2}$$

$$\frac{7}{2} = 3 + \frac{1}{2}.$$

By combining these equations, we find that

$$\frac{62}{23} = 2 + \frac{1}{23/16}$$

$$= 2 + \cfrac{1}{1 + \cfrac{1}{16/7}}$$

$$= 2 + \cfrac{1}{1 + \cfrac{1}{2 + 7/2}}$$

$$= 2 + \cfrac{1}{1 + \cfrac{1}{2 + \cfrac{1}{3 + \cfrac{1}{2}}}}.$$

The final expression in this string of equations is a continued fraction expansion of 62/23.

We now define continued fractions.

Definition. A *finite continued fraction* is an expression of the form

$$a_0 + \cfrac{1}{a_1 + \cfrac{1}{a_2 + \cfrac{}{\ddots + \cfrac{1}{a_{n-1} + \cfrac{1}{a_n}}}}},$$

where $a_0, a_1, a_2, ..., a_n$ are real numbers with $a_1, a_2, a_3, ..., a_n$ positive. The real numbers $a_1, a_2, ..., a_n$ are called the *partial quotients* of the continued fraction. The continued fraction is called *simple* if the real numbers $a_0, a_1, ..., a_n$ are all integers.

Because it is cumbersome to fully write out continued fractions, we use the notation $[a_0; a_1, a_2, ..., a_n]$ to represent the continued fraction in the definition of a finite continued fraction.

We will now show that every finite simple continued fraction represents a rational number. Later we will demonstrate that every rational number can be expressed as a finite simple continued fraction.

Theorem 10.5. Every finite simple continued fraction represents a rational number.

Proof. We will prove the theorem using mathematical induction. For $n = 1$ we have

$$[a_o; a_1] = a_0 + \frac{1}{a_1} = \frac{a_0 a_1 + 1}{a_0},$$

which is rational. Now assume that for the positive integer k the simple continued fraction $[a_0; a_1, a_2, ..., a_k]$ is rational whenever $a_0, a_1, ..., a_k$ are integers with $a_1, ..., a_k$ positive. Let $a_0, a_1, ..., a_{k+1}$ be integers with $a_1, ..., a_{k+1}$ positive. Note that

$$[a_0; a_1, ..., a_{k+1}] = a_0 + \frac{1}{[a_1; a_2, ..., a_k, a_{k+1}]}.$$

By the induction hypothesis, $[a_1; a_2, ..., a_k, a_{k+1}]$ is rational; hence, there are integers r and s, with $s \neq 0$, such that this continued fraction equals r/s. Then

$$[a_o; a_1, ..., a_k, a_{k+1}] = a_0 + \frac{1}{r/s} = \frac{a_0 r + s}{r},$$

which is again a rational number.   ∎

We now show, using the Euclidean algorithm, that every rational number can be written as a finite simple continued fraction.

Theorem 10.6. Every rational number can be expressed by a finite simple continued fraction.

Proof. Let $x = a/b$ where a and b are integers with $b > 0$. Let $r_0 = a$ and $r_1 = b$. Then the Euclidean algorithm produces the following sequence of equations:

$$
\begin{aligned}
r_0 &= r_1 q_1 + r_2 & 0 < r_2 < r_1, \\
r_1 &= r_2 q_2 + r_3 & 0 < r_3 < r_2, \\
r_2 &= r_3 q_3 + r_4 & 0 < r_4 < r_3,
\end{aligned}
$$

$$
\begin{aligned}
\cdot \\
\cdot \\
\cdot
\end{aligned}
$$

$$
\begin{aligned}
r_{n-3} &= r_{n-2} q_{n-2} + r_{n-1} & 0 < r_{n-1} < r_{n-2}, \\
r_{n-2} &= r_{n-1} q_{n-1} + r_n & 0 < r_n < r_{n-1} \\
r_{n-1} &= r_n q_n.
\end{aligned}
$$

In these equations $q_2, q_3, \ldots, q_n$ are positive integers. Writing these equations in fractional form we have

$$
\frac{a}{b} = \frac{r_0}{r_1} = q_1 + \frac{r_2}{r_1} = q_1 + \cfrac{1}{r_1/r_2}
$$

$$
\frac{r_1}{r_2} = q_2 + \frac{r_3}{r_2} = q_2 + \cfrac{1}{r_2/r_3}
$$

$$
\frac{r_2}{r_3} = q_3 + \frac{r_4}{r_3} = q_3 + \cfrac{1}{r_3/r_4}
$$

$$
\begin{aligned}
\cdot \\
\cdot \\
\cdot
\end{aligned}
$$

$$
\frac{r_{n-3}}{r_{n-2}} = \frac{r_{n-1}}{r_{n-2}} = q_{n-2} + \cfrac{1}{r_{n-2}/r_{n-1}}
$$

$$
\frac{r_{n-2}}{r_{n-1}} = q_{n-1} + \frac{r_n}{r_{n-1}} = q_{n-1} + \cfrac{1}{r_{n-1}/r_n}
$$

$$
\frac{r_{n-1}}{r_n} = q_n .
$$

Substituting the value of r_1/r_2 from the second equation into the first equation, we obtain

(10.10)
$$\frac{a}{b} = q_1 + \cfrac{1}{q_2 + \cfrac{1}{r_2/r_3}}.$$

Similarly, substituting the value of r_2/r_3 from the third equation into (10.10) we obtain

$$\frac{c}{b} = q_1 + \cfrac{1}{q_2 + \cfrac{1}{q_3 + \cfrac{1}{r_3/r_4}}}.$$

Continuing in this manner, we find that

$$\frac{a}{b} = q_1 + \cfrac{1}{q_2 + \cfrac{1}{q_3 + \cfrac{}{\cfrac{}{+ q_{n-1} + \cfrac{1}{q_n}}}}}.$$

Hence $\dfrac{a}{b} = [q_1; q_2, \ldots, q_n]$. This shows that every rational number can be written as a finite simple continued fraction. ∎

We note that continued fractions for rational numbers are not unique. From the identity

$$a_n = (a_n - 1) + \frac{1}{1},$$

we see that

$$[a_0; a_1, a_2, \ldots, a_{n-1}, a_n] = [a_0; a_1, a_2, \ldots, a_{n-1}, a_n - 1, 1]$$

whenever $a_n > 1$.

Example 10.6. We have

$$\frac{7}{11} = [0; 1, 1, 1, 3] = [0; 1, 1, 1, 2, 1].$$ ☐

In fact, it can be shown that every rational number can be written as a finite simple continued fraction in exactly two ways, one with an odd number of terms, the other with an even number (see Exercise 12 at the end of this section).

Next, we will discuss the numbers obtained from a finite continued fraction by cutting off the expression at various stages.

Definition. The continued fractions $[a_0;a_1,a_2,..., a_k]$, where k is a nonnegative integer less than or equal to n, is called the *kth convergent* of the continued fraction $[a_0;a_1,a_2,...,a_n]$. The kth convergent is denoted by C_k.

In our subsequent work, we will need some properties of the convergents of a continued fraction. We now develop these properties, starting with a formula for the convergents.

Theorem 10.7. Let $a_0,a_1,a_2,...,a_n$ be real numbers, with $a_1,a_2,...,a_n$ positive. Let the sequences $p_0,p_1,...,p_n$ and $q_0,q_1,...,q_n$ be defined recursively by

$$p_0 = a_0 \qquad\qquad q_0 = 1$$
$$p_1 = a_0a_1+1 \qquad\qquad q_1 = a_1$$

and

$$p_k = a_kp_{k-1} + p_{k-2} \qquad\qquad q_k = a_kq_{k-1} + q_{k-2}$$

for $k = 2,3,..., n$. Then the kth convergent $C_k = [a_0;a_1,..., a_k]$ is given by

$$C_k = p_k/q_k.$$

Proof. We will prove this theorem using mathematical induction. For $k = 0$ we have

$$C_0 = [a_0] = a_0/1 = p_0/q_0.$$

For $k = 1$, we see that

$$C_1 = [a_0;a_1] = a_0 + \frac{1}{a_1} = \frac{a_0a_1+1}{a_1} = \frac{p_1}{q_1}.$$

Hence, the theorem is valid for $k = 0$ and $k = 1$.

Now assume that the theorem is true for the positive integer k where $2 \le k < n$. This means that

$$(10.11) \qquad C_k = [a_o;a_1,..., a_k] = \frac{p_k}{q_k} = \frac{a_kp_{k-1} + p_{k-2}}{a_kq_{k-1} + q_{k-2}}.$$

Because of the way in which the p_j's and q_j's are defined, we see that the real numbers $p_{k-1},p_{k-2},q_{k-1},\ q_{k-2}$ depend only on the partial quotients $a_0,a_1,..., a_{k-1}$. Consequently, we can replace the real number a_k by $a_k + 1/a_{k+1}$ in (10.11), to obtain

$$C_{k+1} = [a_0;a_1,...,a_k,a_{k+1}] = [a_0;a_1,..., a_{k-1},a_k + \frac{1}{a_{k+1}}]$$

$$= \frac{\left(a_k + \dfrac{1}{a_{k+1}}\right) p_{k-1} + p_{k-2}}{\left(a_k + \dfrac{1}{a_{k+1}}\right) q_{k-1} + q_{k-2}}$$

$$= \frac{a_{k+1}(a_k p_{k-1} + p_{k-2}) + p_{k-1}}{a_{k+1}(a_k q_{k-1} + q_{k-1}) + q_{k-1}}$$

$$= \frac{a_{k+1} p_k + p_{k-1}}{a_{k+1} q_k + q_{k-1}}$$

$$= \frac{p_{k+1}}{q_{k+1}}.$$

This finishes the proof by induction. ∎

We illustrate how to use Theorem 10.7 with the following example.

Example 10.7. We have $173/55 = [3;6,1,7]$. We compute the sequences p_j and q_j for $j = 0,1,2,3$, by

$$\begin{array}{ll}
p_0 = 3 & q_0 = 1 \\
p_1 = 3 \cdot 6 + 1 = 19 & q_1 = 6 \\
p_2 = 1 \cdot 19 + 3 = 22 & q_2 = 1 \cdot 6 + 1 = 7 \\
p_3 = 7 \cdot 22 + 19 = 173 & q_3 = 7 \cdot 7 + 6 = 55.
\end{array}$$

Hence, the convergents of the above continued fraction are

$$\begin{array}{l}
C_0 = p_0/q_0 = 3/1 = 3 \\
C_1 = p_1/q_1 = 19/6 \\
C_2 = p_2/q_2 = 22/7 \\
C_3 = p_3/q_3 = 173/55.
\end{array}$$ □

We now state and prove another important property of the convergents of a continued fraction.

Theorem 10.8. Let $C_k = p_k/q_k$ be the kth convergent of the continued fraction $[a_0;a_1,...,a_n]$ where k is a positive integer, $1 \le k \le n$. If p_k and q_k are as defined in Theorem 10.7, then

$$p_k q_{k-1} - p_{k-1} q_k = (-1)^{k-1}.$$

Proof. We use mathematical induction to prove the theorem. For $k = 1$ we have

$$p_1 q_0 - p_0 q_1 = (a_0 a_1 + 1) \cdot 1 - a_0 a_1 = 1.$$

Assume the theorem is true for an integer k where $1 \leq k < n$, so that

$$p_k q_{k-1} - p_{k-1} q_k = (-1)^{k-1}.$$

Then, we have

$$
\begin{aligned}
p_{k+1} q_k - p_k q_{k+1} &= (a_{k+1} p_k + p_{k-1}) q_k - p_k (a_{k+1} q_k + q_{k-1}) \\
&= p_{k-1} q_k - p_k q_{k-1} = -(-1)^{k-1} = (-1)^k,
\end{aligned}
$$

so that the theorem is true for $k + 1$. This finishes the proof by induction. ∎

We illustrate this theorem with the example we used to illustrate Theorem 10.7.

Example 10.8. For the continued fraction $[3;6,1,7]$ we have

$$
\begin{aligned}
p_0 q_1 - p_1 q_0 &= 3 \cdot 6 - 19 \cdot 1 = -1 \\
p_1 q_2 - p_2 q_1 &= 19 \cdot 7 - 22 \cdot 6 = 1 \\
p_2 q_3 - p_3 q_2 &= 22 \cdot 55 - 173 \cdot 7 = -1.
\end{aligned}
$$
□

As a consequence of Theorem 10.8, we see that for $k = 1, 2, \ldots$ the convergents p_k/q_k of a simple continued fraction are in lowest terms. Corollary 10.8.1 demonstrates this.

Corollary 10.8.1. Let $C_k = p_k/q_k$ be the kth convergent of the simple continued fraction $[a_0; a_1, \ldots, a_n]$, where the integers p_k and q_k are as defined in Theorem 10.9. Then the integers p_k and q_k are relatively prime.

Proof. Let $d = (p_k, q_k)$. By Theorem 10.8 we know that

$$p_k q_{k-1} - q_k p_{k-1} = (-1)^{k-1}.$$

Hence

$$d \mid (-1)^{k-1}.$$

Therefore $d = 1$. ∎

We also have the following useful corollary of Theorem 10.8.

Corollary 10.8.2. Let $C_k = p_k/q_k$ be the kth convergent of the simple continued fraction $[a_0; a_1, a_2, \ldots, a_k]$. Then

$$C_k - C_{k-1} = \frac{(-1)^{k-1}}{q_k q_{k-1}}$$

for all integers k with $1 \leq k \leq n$. Also,

$$C_k - C_{k-2} = \frac{a_k(-1)^k}{q_k q_{k-2}}$$

for all integers k with $2 \leq k \leq n$.

Proof. By Theorem 10.8 we know that $p_k q_{k-1} - q_k p_{k-1} = (-1)^{k-1}$.

We obtain the first identity,

$$C_k - C_{k-1} = \frac{p_k}{q_k} - \frac{p_{k-1}}{q_{k-1}} = \frac{(-1)^{k-1}}{q_k q_{k-1}},$$

by dividing both sides by $q_k q_{k-1}$.

To obtain the second identity, note that

$$C_k - C_{k-2} = \frac{p_k}{q_k} - \frac{p_{k-2}}{q_{k-2}} = \frac{p_k q_{k-2} - p_{k-2} q_k}{q_k q_{k-2}}.$$

Since $p_k = a_k p_{k-1} + p_{k-2}$ and $q_k = a_k q_{k-1} + q_{k-2}$, we see that the numerator of the fraction on the right is

$$p_k q_{k-2} - p_{k-2} q_k = (a_k p_{k-1} + p_{k-2}) q_{k-2} - p_{k-2} (a_k q_{k-1} + q_{k-2})$$

$$= a_k (p_{k-1} q_{k-2} - p_{k-2} q_{k-1})$$

$$= a_k (-1)^{k-2},$$

using Theorem 10.8 to see that $p_{k-1} q_{k-2} - p_{k-2} q_{k-1} = (-1)^{k-2}$.

Therefore, we find that

$$C_k - C_{k-2} = \frac{a_k (-1)^k}{q_k q_{k-2}}.$$

This is the second identity of the corollary. ∎

Using Corollary 10.8.2 we can prove the following theorem which is useful when developing infinite continued fractions.

Theorem 10.9. Let C_k be the kth convergent of the finite simple continued fraction $[a_0; a_1, a_2, ..., a_n]$. Then

$$C_1 > C_3 > C_5 > \cdots ,$$
$$C_0 < C_2 < C_4 < \cdots ,$$

and every odd-numbered convergent C_{2j+1}, $j = 0, 1, 2, ...,$ is greater than every even-numbered convergent C_{2j}, $j = 0, 1, 2, ...$.

Proof. Since Corollary 10.8.2 tells us that, for $k = 2, 3, ..., n$,

$$C_k - C_{k-2} = \frac{a_k (-1)^k}{q_k q_{k-2}},$$

we know that

$$C_k < C_{k-2}$$

when k is odd, and

$$C_k > C_{k-2}$$

when k is even. Hence

$$C_1 > C_3 > C_5 > \cdots,$$

and

$$C_0 < C_2 < C_4 < \cdots.$$

To show that every odd-numbered convergent is greater than every even-numbered convergent, note that from Corollary 10.8.2 we have

$$C_{2m} - C_{2m-1} = \frac{(-1)^{2m-1}}{q_{2m}q_{2m-1}} < 0,$$

so that $C_{2m-1} > C_{2m}$. To compare C_{2k} and C_{2j-1}, we see that

$$C_{2j-1} > C_{2j+2k-1} > C_{2j+2k} > C_{2k}.$$

so that every odd-numbered convergent is greater than every even-numbered convergent. ∎

Example 10.9. Consider the finite simple continued fraction $[2;3,1,1,2,4]$. Then the convergents are

$$\begin{aligned}
C_0 &= & 2/1 &= 2 \\
C_1 &= & 7/3 &= 2.3333... \\
C_2 &= & 9/4 &= 2.25 \\
C_3 &= & 16/7 &= 2.2857... \\
C_4 &= & 41/18 &= 2.2777... \\
C_5 &= & 180/79 &= 2.2784... .
\end{aligned}$$

We see that

$$\begin{aligned}
C_0 &= 2 < C_2 = 2.25 < C_4 = 2.2777\cdots \\
&< C_5 = 2.2784... < C_3 = 2.2857... < C_1 = 2.3333... .
\end{aligned}$$ □

10.2 Exercises

1. Find the rational number, expressed in lowest terms, represented by each of the following simple continued fractions.

   a) $[2;7]$ e) $[1;1]$
   b) $[1;2,3]$ f) $[1;1,1]$
   c) $[0;5,6]$ g) $[1;1,1,1]$
   d) $[3;7,15,1]$ h) $[1;1,1,1,1]$

2. Find the rational number, expressed in lowest terms, represented by each of the following simple continued fractions.

a)	[10;3]	e)	[2;1,2,1,1,4]
b)	[3;2,1]	f)	[1;2,1,2]
c)	[0;1,2,3]	g)	[1;2,1,2,1]
d)	[2;1,2,1]	h)	[1;2,1,2,1,2]

3. Find the simple continued fraction expansion not terminating with the partial quotient one, of each of the following rational numbers.

a)	18/13	d)	310/99
b)	32/17	e)	−931/1005
c)	19/9	f)	831/8110

4. Find the simple continued fraction expansion not terminating with the partial quotient one, of each of the following rational numbers.

a)	6/5	d)	5/999
b)	22/7	e)	−943/1001
c)	19/29	f)	873/4867

5. Find the convergents of each of the continued fractions found in Exercise 3.

6. Find the convergents of each of the continued fractions found in Exercise 4.

7. Show that the convergents you found in Exercises 3 and 4 satisfy Theorem 10.9.

8. Let f_k denote the kth Fibonaccci number. Find the simple continued fraction, terminating with the partial quotient of one, of f_{k+1}/f_k, where k is a positive integer.

9. Show that if the simple continued fraction expression of the rational number α, $\alpha > 1$, is $[a_0;a_1,...,a_k]$, then the simple continued fraction expression of $1/\alpha$ is $[0;a_0,a_1,...,a_k]$.

10. Show that if $a_0 \neq 0$, then

$$p_k/p_{k-1} = [a_k;a_{k-1}, \ldots, a_1,a_0]$$

and

$$q_k/q_{k-1} = [a_k;a_{k-1},...,a_2,a_1],$$

where $C_{k-1} = p_{k-1}/q_{k-1}$ and $C_k = p_k/q_k, k \geq 1$, are successive convergents of the continued fraction $[a_0;a_1,...,a_n]$. (*Hint*: Use the relation $p_k = a_k p_{k-1} + p_{k-2}$ to show that $p_k/p_{k-1} = a_k + 1/(p_{k-1}/p_{k-2})$.

☞ 11. Show that $q_k \geq f_k$ for $k = 1,2,...$ where $C_k = p_k/q_k$ is the kth convergent of the simple continued fraction $[a_0;a_1,...,a_n]$ and f_k denotes the kth Fibonacci number.

12. Show that every rational number has exactly two finite simple continued fraction expansions.

* **13.** Let $[a_0;a_1,a_2,...,a_n]$ be the simple continued fraction expansion of r/s where $(r,s) = 1$ and $r \geq 1$. Show that this continued fraction is symmetric, *i.e.* $a_0 = a_n, a_1 = a_{n-1}, a_2 = a_{n-2},...,$ if and only if $s \mid (r^2+1)$ if n is odd and $s \mid (r^2-1)$ if n is even. (*Hint*: Use Exercise 10 and Theorem 10.8).

* **14.** Explain how finite continued fractions for rational numbers, with both plus and minus signs allowed, can be generated from the division algorithm given in Exercise 6 of Section 1.5.

15. Let $a_0,a_1,a_2,...,a_k$ be real numbers with $a_1,a_2,...$ positive and let x be a positive real number. Show that $[a_0;a_1,...,a_k] < [a_0;a_1,...,a_k+x]$ if k is odd and $[a_0;a_1,...,a_k] > [a_0;a_1,...,a_k+x]$ if k is even.

16. Determine whether n can be expressed as the sum of positive integers a and b where all the partial quotients of the finite simple continued fraction of a/b are either 1 or 2 where n is

a)	13	d)	23
b)	17	e)	27
c)	19	f)	29

10.2 Computer Projects

Programming Projects

Write programs to do the following:

1. Find the simple continued fraction expansion of a rational number

2. Find the convergents of a finite simple continued fraction and find the rational number that this continued fraction represents.

Computations and Explorations

Using the programs you have written or a computation program, carry out the following computations and explorations:

1. Find the finite continued fractions of x and $2x$ for twenty different rational numbers. Can you find a rule for finding the finite simple continued fraction of $2x$ from that of x?

2. Determine for each integer n with $n \leq 1000$ whether there are integers a and b with $n = a + b$ such that the partial quotients of the continued fraction of a/b are all either 1 or 2. Can you make any conjectures?

10.3 Infinite Continued Fractions

Suppose that we have an infinite sequence of positive integers $a_0,a_1,a_2,....$ How can we define the infinite continued fraction $[a_0,a_1,a_2,...]$? To make sense of infinite continued fractions, we need a result from mathematical analysis. We state the result below, and refer the reader to a mathematical analysis book, such as Rudin [100], for a proof.

Theorem 10.10. Let $x_0, x_1, x_2, \ldots$ be a sequence of real numbers such that $x_0 < x_1 < x_2 < \ldots$ and $x_k < U$ for $k = 0, 1, 2, \ldots$ for some real number U, or $x_0 > x_1 > x_2 > \ldots$ and $x_k > L$ for $k = 0, 1, 2, \ldots$ for some real number L. Then the terms of the sequence $x_0, x_1, x_2, \ldots$ tend to a limit x, *i.e.* there exists a real number x such that

$$\lim_{k \to \infty} x_k = x.$$

Theorem 10.10 tells us that the terms of an infinite sequence tend to a limit in two special situations, when the terms of the sequence are increasing and all less than an upper bound, and when the terms of the sequence are decreasing and all are greater than a lower bound.

We can now define infinite continued fractions as limits of finite continued fractions, as the following theorem shows.

Theorem 10.11. Let $a_0, a_1, a_2, \ldots$ be an infinite sequence of integers with $a_1, a_2, \ldots$ positive, and let $C_k = [a_0; a_1, a_2, \ldots, a_k]$. Then the convergents C_k tend to a limit α, *i.e*

$$\lim_{k \to \infty} C_k = \alpha.$$

Before proving Theorem 10.11 we note that the limit α described in the statement of the theorem is called the value of the *infinite simple continued fraction* $[a_0; a_1, a_2, \ldots]$.

To prove Theorem 10.11 we will show that the infinite sequence of even-numbered convergents is increasing and has an upper bound and that the infinite sequence of odd-numbered convergents is decreasing and has a lower bound. We then show that the limits of these two sequences, guaranteed to exist by Theorem 10.10, are in fact equal.

We now will prove Theorem 10.11.

Proof. Let m be an even positive integer. By Theorem 10.9 we see that

$$C_1 > C_3 > C_5 > \cdots > C_{m-1},$$
$$C_0 < C_2 < C_4 < \cdots < C_m,$$

and $C_{2j} < C_{2k+1}$ whenever $2j \le m$ and $2k + 1 < m$. By considering all possible values of m, we see that

$$C_1 > C_3 > C_5 > \cdots > C_{2n-1} > C_{2n+1} > \cdots,$$
$$C_0 < C_2 < C_4 < \cdots < C_{2n-2} < C_{2n} < \cdots,$$

and $C_{2j} > C_{2k+1}$ for all positive integers j and k. We see that the hypotheses of Theorem 10.10 are satisfied for each of the two sequences $C_1, C_3, C_2, \ldots$ and $C_0, C_2, C_4, \ldots$. Hence, the sequence $C_1, C_3, C_5, \ldots$ tends to a limit α_1 and the sequence $C_0, C_2, C_4, \ldots$ tends to a limit α_2, *i.e.*

$$\lim_{n \to \infty} C_{2n+1} = \alpha_1$$

and

$$\lim_{n \to \infty} C_{2n} = \alpha_2.$$

Our goal is to show that these two limits α_1 and α_2 are equal. Using Corollary 10.8.2 we have

$$C_{2n+1} - C_{2n} = \frac{p_{2n+1}}{q_{2n+1}} - \frac{p_{2n}}{q_{2n}} = \frac{(-1)^{(2n+1)-1}}{q_{2n+1}q_{2n}} = \frac{1}{q_{2n+1}q_{2n}}.$$

Since $q_k \geq k$ for all positive integers k (see Exercise 11 of Section 10.2), we know that

$$\frac{1}{q_{2n+1}q_{2n}} < \frac{1}{(2n+1)(2n)},$$

and hence

$$C_{2n+1} - C_{2n} = \frac{1}{q_{2n+1}q_{2n}}$$

tends to zero, *i.e.*

$$\lim_{n \to \infty} (C_{2n+1} - C_{2n}) = 0.$$

Hence, the sequences $C_1, C_3, C_5, \ldots$ and $C_0, C_2, C_4, \ldots$ have the same limit, since

$$\lim_{n \to \infty} (C_{2n+1} - C_{2n}) = \lim_{n \to \infty} C_{2n+1} - \lim_{n \to \infty} C_{2n} = 0.$$

Therefore $\alpha_1 = \alpha_2$, and we conclude that all the convergents tend to the limit $\alpha = \alpha_1 = \alpha_2$. This finishes the proof of the theorem. ∎

Previously, we showed that rational numbers have finite simple continued fractions. Next, we will show that the value of any infinite simple continued fraction is irrational.

Theorem 10.12. Let $a_0, a_1, a_2, \ldots$ be integers with $a_1, a_2, \ldots$ positive. Then $[a_0; a_1, a_2, \ldots]$ is irrational.

Proof. Let $\alpha = [a_0; a_1, a_2, \ldots]$ and let

$$C_k = p_k/q_k = [a_0; a_1, \ldots, a_k]$$

denote the kth convergent of α. When n is a positive integer, Theorem 10.11 shows that $C_{2n} < \alpha < C_{2n+1}$, so that

$$0 < \alpha - C_{2n} < C_{2n+1} - C_{2n}.$$

However, by Corollary 10.8.2 we know that

$$C_{2n+1} - C_{2n} = \frac{1}{q_{2n+1}q_{2n}},$$

this means that

$$0 < \alpha - C_{2n} = \alpha - \frac{p_{2n}}{q_{2n}} < \frac{1}{q_{2n+1}q_{2n}}.$$

and therefore we have

$$0 < \alpha q_{2n} - p_{2n} < 1/q_{2n+1}.$$

Assume that α is rational, so that $\alpha = a/b$ where a and b are integers with $b \neq 0$. Then

$$0 < \frac{aq_{2n}}{b} - p_{2n} < \frac{1}{q_{2n+1}},$$

and by multiplying this inequality by b we see that

$$0 < aq_{2n} - bp_{2n} < \frac{b}{q_{2n+1}}.$$

Note that $aq_{2n} - bp_{2n}$ is an integer for all positive integers n. However, since $q_{2n+1} > 2n+1$, for each integer n there is an integer n_0 such that $q_{2n_0+1} > b$, so that $b/q_{2n_0+1} < 1$. This is a contradiction, since the integer $aq_{2n_0} - bp_{2n_0}$ cannot be between 0 and 1. We conclude that α is irrational. ∎

We have demonstrated that every infinite simple continued fraction represents an irrational number. We will now show that every irrational number can be uniquely expressed by an infinite simple continued fraction, by first constructing such a continued fraction, and then by showing that it is unique.

Theorem 10.13. Let $\alpha = \alpha_0$ be an irrational number and define the sequence $a_0, a_1, a_2, \ldots$ recursively by

$$a_k = [\alpha_k], \quad \alpha_{k+1} = 1/(\alpha_k - a_k)$$

for $k = 0, 1, 2, \ldots$. Then α is the value of the infinite, simple continued fraction $[a_0; a_1, a_2, \ldots]$.

Proof. From the recursive definition of the integers a_k, we see that a_k is an integer for every k. Furthermore, using mathematical induction, we can show that α_k is irrational for every nonnegative integer k and that, as a consequence, α_{k+1} exists. First, note that $\alpha_0 = \alpha$ is irrational, so that $\alpha_0 \neq a_0 = [\alpha_0]$ and $\alpha_1 = 1/(\alpha_0 - a_0)$ exists.

Next, we assume that α_k is irrational. As a consequence, α_{k+1} exists. We can easily see that α_{k+1} is also irrational, since the relation

$$\alpha_{k+1} = 1/(\alpha_k - a_k)$$

implies that

(10.12)
$$\alpha_k = a_k + \frac{1}{\alpha_{k+1}},$$

and if α_{k+1} were rational, then α_k would also be rational. Now, since α_k is irrational and a_k is an integer, we know that $\alpha_k \neq a_k$, and

$$a_k < \alpha_k < a_k + 1,$$

so that

$$0 < \alpha_k - a_k < 1.$$

Hence,

$$\alpha_{k+1} = 1/(\alpha_k - a_k) > 1,$$

and consequently,

$$a_{k+1} = [\alpha_{k+1}] \geq 1$$

for $k = 0, 1, 2, \ldots$. This means that all the integers $a_1, a_2, \ldots$ are positive.

Note that by repeatedly using (10.12) we see that

$$\alpha = \alpha_0 = a_0 + \frac{1}{\alpha_1} = [a_0; \alpha_1]$$

$$= a_0 + \cfrac{1}{a_1 + \cfrac{1}{\alpha_2}} = [a_0; a_1, \alpha_2]$$

$$\cdot$$
$$\cdot$$
$$\cdot$$

$$= a_0 + \cfrac{1}{a_1 + \cfrac{1}{a_2 + \cfrac{\cdot}{\cdot \cdot + a_k + \cfrac{1}{\alpha_{k+1}}}}} = [a_0; a_1, a_2, \ldots, a_k, \alpha_{k+1}].$$

What we must now show is that the value of $[a_0; a_1, a_2, \ldots, a_k, \alpha_{k+1}]$ tends to α as k tends to infinity, *i.e.*, as k grows without bound. By Theorem 10.7 we see that

$$\alpha = [a_0; a_1, \ldots, a_k, \alpha_{k+1}] = \frac{\alpha_{k+1} p_k + p_{k+1}}{\alpha_{k+1} q_k + q_{k-1}},$$

where $C_j = p_j/q_j$ is the jth convergent of $[a_0; a_1 a_2, \ldots]$. Hence

$$\alpha - C_k = \frac{\alpha_{k+1}p_k + p_{k-1}}{\alpha_{k+1}q_k + q_{k-1}} - \frac{p_k}{q_k}$$

$$= \frac{-(p_k q_{k-1} - p_{k-1}q_k)}{(\alpha_{k+1}q_k + q_{k-1})q_k}$$

$$= \frac{(-1)^{k-1}}{(\alpha_{k+1}q_k + q_{k-1})q_k},$$

where we have used Theorem 10.8 to simplify the numerator on the right-hand side of the second equality. Since

$$\alpha_{k+1}q_k + q_{k-1} > a_{k+1}q_k + q_{k-1} = q_{k+1},$$

we see that

$$|\alpha - C_k| < \frac{1}{q_k q_{k+1}}.$$

Since $q_k > k$ (from Exercise 11 of Section 10.2), we note that $1/(q_k q_{k+1})$ tends to zero as k tends to infinity. Hence, C_k tends to α as k tends to infinity, or phrased differently, the value of the infinite simple continued fraction $[a_0;a_1,a_2,...]$ is α. ■

To show that the infinite simple continued fraction that represents an irrational number is unique, we prove the following theorem.

Theorem 10.14. If the two infinite simple continued fractions $[a_0;a_1,a_2,...]$ and $[b_0;b_1,b_2,...]$ represents the same irrational number, then $a_k = b_k$ for $k = 0,1,2,...$.

Proof. Suppose that $\alpha = [a_0;a_1,a_2,...]$. Then, since $C_0 = a_0$ and $C_1 = a_0 + 1/a_1$, Theorem 10.9 tells us that

$$a_0 < \alpha < a_0 + 1/a_1,$$

so that $a_0 = [\alpha]$. Further, we note that

$$[a_0;a_1,a_2,...] = a_0 + \frac{1}{[a_1;a_2,a_3,...]},$$

since

$$\alpha = [a_0;a_1,a_2,...] = \lim_{k \to \infty} [a_0;a_1,a_2,...,a_k]$$

$$= \lim_{k \to \infty} (a_0 + \frac{1}{[a_1;a_2,a_3,...,a_k]})$$

$$= a_0 + \frac{1}{\lim_{k \to \infty} [a_1;a_2,...,a_k]}$$

$$= a_0 + \frac{1}{[a_1;a_2,a_3,...]}.$$

Suppose that

$$[a_0;a_1,a_2,\ldots] = [b_0;b_1,b_2,\ldots].$$

Our remarks show that

$$a_0 = b_0 = [\alpha]$$

and that

$$a_0 + \cfrac{1}{[a_1;a_2,\ldots]} = b_0 + \cfrac{1}{[b_1;b_2,\ldots]},$$

so that

$$[a_1;a_2,\ldots] = [b_1;b_2,\ldots].$$

Now assume that $a_k = b_k$, and that $[a_{k+1};a_{k+2},\ldots] = [b_{k+1};b_{k+2},\ldots]$. Using the same argument, we see that $a_{k+1} = b_{k+1}$, and

$$a_{k+1} + \cfrac{1}{[a_{k+2};a_{k+3},\ldots]} = b_{k+1} + \cfrac{1}{[b_{k+1};b_{k+3},\ldots]},$$

which implies that

$$[a_{k+2};a_{k+3},\ldots] = [b_{k+2};b_{k+3},\ldots].$$

Hence, by mathematical induction we see that $a_k = b_k$ for $k = 0,1,2,\ldots$. ∎

To find the simple continued fraction expansion of a real number, we use the algorithm given in Theorem 10.13. We illustrate this procedure with the following example.

Example 10.10. Let $\alpha = \sqrt{6}$. We find that

$$a_0 = [\sqrt{6}] = 2, \quad \alpha_1 = \frac{1}{\sqrt{6}-2} = \frac{\sqrt{6}+2}{2},$$

$$a_1 = [\frac{\sqrt{6}+2}{2}] = 2, \quad \alpha_2 = \cfrac{1}{(\frac{\sqrt{6}+2}{2})-2} = \sqrt{6}+2,$$

$$a_2 = [\sqrt{6}+2] = 4, \quad \alpha_3 = \cfrac{1}{(\sqrt{6}+2)-4} = \frac{\sqrt{6}+2}{2} = \alpha_1.$$

Since $\alpha_3 = \alpha_1$, we see that $a_3 = a_1$, $a_4 = a_2,\ldots$, and so on. Hence

$$\sqrt{6} = [2;2,4,2,4,2,4,\ldots].$$

The simple continued fraction of $\sqrt{6}$ is periodic. We will discuss periodic simple continued fractions in the next section. □

The convergents of the infinite simple continued fraction of an irrational number are good approximations to α. In fact, if p_k/q_k is the kth convergent of this continued fraction, then from the proof of Theorem 10.13 we know that

$$|\alpha - p_k/q_k| < 1/(q_k q_{k+1}),$$

so that

$$|\alpha - p_k/q_k| < 1/q_k^2,$$

since $q_k < q_{k+1}$.

The next theorem and corollary show that the convergents of the simple continued fraction of α are the best rational approximations to α, in the sense that p_k/q_k is closer to α than any other rational number with a denominator less than q_k.

Theorem 10.15. Let α be an irrational number and let p_j/q_j, $j = 1,2,...$, be the convergents of the infinite simple continued fraction of α. If r and s are integers with $s > 0$ and if k is a positive integer such that

$$|s\alpha - r| < |q_k\alpha - p_k|,$$

then $s \geq q_{k+1}$.

Proof. Assume that $|s\alpha - r| < |q_k\alpha - p_k|$, but that $1 \leq s < q_{k+1}$. We consider the simultaneous equations

$$p_k x + p_{k+1} y = r$$
$$q_k x + q_{k+1} y = s.$$

By multiplying the first equation by q_k and the second by p_k, and then subtracting the second from the first, we find that

$$(p_{k+1} q_k - p_k q_{k+1}) y = r q_k - s p_k.$$

By Theorem 10.8 we know that $p_{k+1} q_k - p_k q_{k+1} = (-1)^k$, so that

$$y = (-1)^k (r q_k - s p_k).$$

Similarly, multiplying the first equation by q_{k+1} and the second by p_{k+1} and then subtracting the first from the second, we find that

$$x = (-1)^k (s p_{k+1} - r q_{k+1}).$$

We note that $x \neq 0$ and $y \neq 0$. If $x = 0$ then $s p_{k+1} = r q_{k+1}$. Since $(p_{k+1}, q_{k+1}) = 1$, Lemma 2.3 tells us that $q_{k+1} | s$, which implies that $q_{k+1} \leq s$, contrary to our assumption. If $y = 0$, then $r = p_k x$ and $s = q_k x$, so that

$$|s\alpha - r| = |x| |q_k\alpha - p_k| \geq |q_k\alpha - p_k|,$$

since $|x| \geq 1$, contrary to our assumption.

We will now show that x and y have opposite signs. First, suppose that $y < 0$. Since $q_k x = s - q_{k+1} y$, we know that $x > 0$, because $q_k x > 0$ and $q_k > 0$. When $y > 0$, since $q_{k+1} y \geq q_{k+1} > s$, we see that $q_k x = s - q_{k+1} y < 0$, so that $x < 0$.

By Theorem 10.9 we know that either $p_k/q_k < \alpha < p_{k+1}/q_{k+1}$ or that $p_{k+1}/q_{k+1} < \alpha < p_k/q_k$. In either case, we easily see that $q_k\alpha - p_k$ and $q_{k+1}\alpha - p_{k+1}$ have opposite signs.

From the simultaneous equations we started with, we see that

$$|s\alpha - r| = |(q_k x + q_{k+1} y)\alpha - (p_k x + p_{k+1} y)|$$
$$= |x(q_k\alpha - p_k) + y(q_{k+1}\alpha - p_{k+1})|.$$

Combining the conclusions of the previous two paragraphs, we see that $x(q_k\alpha - p_k)$ and $y(q_{k+1}\alpha - p_{k+1})$ have the same sign, so that

$$|s\alpha - r| = |x| \, |q_k\alpha - p_k| + |y| \, |q_{k+1}\alpha - p_{k+1}|$$
$$\geq |x| \, |q_k\alpha - p_k|$$
$$\geq |q_k\alpha - p_k|,$$

since $|x| \geq 1$. This contradicts our assumption.

We have shown that our assumption is false, and consequently, the proof is complete. ∎

Corollary 10.15.1. Let α be an irrational number and let p_j/q_j, $j = 1,2,...$ be the convergents of the infinite simple continued fraction of α. If r/s is a rational number, where r and s are integers with $s > 0$ and if k is a positive integer such that

$$|\alpha - r/s| < |\alpha - p_k/q_k|,$$

then $s > q_k$.

Proof. Suppose that $s \leq q_k$ and that

$$|\alpha - r/s| < |\alpha - p_k/q_k|.$$

By multiplying these two inequalities, we find that

$$s|\alpha - r/s| < q_k|\alpha - p_k/q_k|$$

so that

$$|s\alpha - r| < |q_k\alpha - p_k|,$$

violating the conclusion of Theorem 10.15. ∎

Example 10.11. The simple continued fraction of the real number π is $\pi = [3;7,15,1,292,1,1,1,2,1,3,...]$. Note that there is no discernible pattern in the sequence of partial quotients. The convergents of this continued fraction are the best rational approximations to π. The first five are 3, 22/7, 333/106, 355/113, and 103993/33102. We conclude from Corollary 10.15.1 that 22/7 is the best rational approximation of π with denominator less than or equal to 7, that 335/113 is the best rational approximation of π with denominator less than or equal to 113, and so on.

□

Finally, we conclude this section with a result that shows that any sufficiently close rational approximation to an irrational number must be a convergent of the infinite simple continued fraction expansion of this number.

Theorem 10.16. If α is an irrational number and if r/s is a rational number in lowest terms, where r and s are integers with $s > 0$, such that

$$|\alpha - r/s| < 1/(2s^2),$$

then r/s is a convergent of the simple continued fraction expansion of α.

Proof. Assume that r/s is not a convergent of the simple continued fraction expansion of α. Then, there are successive convergents p_k/q_k and p_{k+1}/q_{k+1} such that $q_k \le s < q_{k+1}$. By Theorem 10.15 we see that

$$|q_k\alpha - p_k| \le |s\alpha - r| = s|\alpha - r/s| < 1/(2s).$$

Dividing by q_k we obtain

$$|\alpha - p_k/q_k| < 1/(2sq_k).$$

Since we know that $|sp_k - rq_k| \ge 1$ (we know that $sp_k - rq_k$ is a nonzero integer since $r/s \; ! = p_k/q_k$), it follows that

$$\frac{1}{sq_k} \le \frac{|sp_k - rq_k|}{sq_k}$$

$$= \left| \frac{p_k}{q_k} - \frac{r}{s} \right|$$

$$\le \left| \alpha - \frac{p_k}{q_k} \right| + \left| \alpha - \frac{r}{s} \right|$$

$$< \frac{1}{2sq_k} + \frac{1}{2s^2}$$

(where we have used the triangle inequality to obtain the second inequality). Hence, we see that

$$1/2sq_k < 1/2s^2.$$

Consequently,

$$2sq_k > 2s^2,$$

which implies that $q_k > s$, contradicting the assumption.   ■

10.3 Exercises

1. Find the simple continued fractions of each of the following real numbers.

    a) $\sqrt{2}$ c) $\sqrt{5}$

    b) $\sqrt{3}$ d) $(1+\sqrt{5})/2$

2. Find the first five partial quotients of the simple continued fractions of each of the following real numbers.

    a) $\sqrt[3]{2}$ c) $(e-1)/(e+1)$

    b) 2π d) $(e^2-1)/(e^2+1)$

3. Find the best rational approximation to π with a denominator less than or equal to 100000.

4. The infinite simple continued fraction expansion of the number e is

    $$e = [2;1,2,1,1,4,1,1,6,1,1,8,...].$$

    a) Find the first eight convergents of the continued fraction of e.

    b) Find the best rational approximation to e having a denominator less than or equal to 536.

* 5. Let α be an irrational number with simple continued fraction expansion $\alpha = [a_0;a_1,a_2,...]$. Show that the simple continued fraction of $-\alpha$ is $[-a_0-1;1,a^1-1,a_2,a_3,...]$ if $a_1 > 1$ and $[-a_0-1;a_2+1,a_3, \cdots]$ if $a_1 = 1$.

* 6. Show that if p_k/q_k and p_{k+1}/q_{k+1} are consecutive convergents of the simple continued fraction of an irrational number α, then

    $$|\alpha - p_k/q_k| < 1/(2q_k^2)$$

    or

    $$|\alpha - p_{k+1}/q_{k+1}| < 1/(2q_{k+1}^2).$$

    (*Hint*: First show that $|\alpha - p_{k+1}/q_{k+1}| + |\alpha - p_k/q_k| = |p_{k+1}/q_{k+1} - p_k/q_k| = 1/(q_k q_{k+1})$.)

☞ 7. Let α be an irrational number, $\alpha > 1$. Show that the kth convergent of the simple continued fraction of $1/\alpha$ is the reciprocal of the $(k-1)$th convergent of the simple continued fraction of α.

* 8. Let α be an irrational number, and let p_j/q_j denote the jth convergent of the simple continued fraction expansion of α. Show that at least one of any three consecutive convergents satisfies the inequality

    $$|\alpha - p_j/q_j| < 1/(\sqrt{5}\, q_j^2).$$

    Conclude that there are infinitely many rational numbers p/q, where p and q are integers with $q \neq 0$, such that

$$| \alpha - p/q | < 1/(\sqrt{5}\, q^2).$$

* 9. Show that if $\alpha = (1 + \sqrt{5})/2$, and $c > \sqrt{5}$, then there are only a finite number of rational numbers p/q, where p and q are integers, $q \neq 0$, such that

$$| \alpha - p/q | < 1/(cq^2).$$

(*Hint*: Consider the convergents of the simple continued fraction expansion of $\sqrt{5}$.)

If α and β are two real numbers, we say that β is *equivalent* to α if there are integers a, b, c, and d such that $ad - bc = \pm 1$ and $\beta = \dfrac{a\alpha + b}{c\alpha + d}$.

10. Show that a real number α is equivalent to itself.

11. Show that if α and β are real numbers with β equivalent to α, then α is equivalent to β. Hence, we can say that two numbers α and β are equivalent.

12. Show that if α, β, and λ are real numbers such that α and β are equivalent and β and λ are equivalent, then α and λ are equivalent.

13. Show that any two rational numbers are equivalent.

* 14. Show that two irrational numbers α and β are equivalent if and only if the tails of their simple continued fractions agree, *i.e.*, $\alpha = [a_0 ; a_1 , a_2 ,..., a_j , c_1 , c_2 , c_3 ,...]$, $\beta = [b_0 ; b_1 , b_2 ,..., b_k , c_1 , c_2 , c_3 ,...]$, where a_i, $i = 0, 1, 2,..., j$, b_i, $i = 0, 1, 2,..., k$, and c_i, $i = 1, 2, 3,...$ are integers, all positive except perhaps a_0 and b_0.

Let α be an irrational number, and let the simple continued fraction expansion of α be $\alpha = [a_0 ; a_1 , a_2 ,...]$. Let p_k/q_k denote, as usual, the kth convergent of this continued fraction. We define the *pseudoconvergents* of this continued fraction to be

$$p_{k,t}/q_{k,t} = (tp_{k-1} + p_{k-2})/(tq_{k-1} + q_{k-2}),$$

where k is a positive integer, $k \geq 2$, and t is an integer with $0 < t < a_k$.

15. Show that each pseudoconvergent is in lowest terms

* 16. Show that the sequence of rational numbers $p_{k,2}/q_{k,2} ,..., p_{k,a_{k-1}}/q_{k,a_{k-1}} , p_k/q_k$ is increasing if k is even, and decreasing if k is odd.

* 17. Show that if r and s are integers with $s > 0$ such that

$$| \alpha - r/s | \leq | \alpha - p_{k,t}/q_{k,t} |$$

where k is a positive integer and $0 < t < a_k$, then $s > q_{k,t}$ or $r/s = p_{k-1}/q_{k-1}$. This shows that the closest rational approximations to a real number are the convergents and pseudoconvergents of its simple continued fraction.

18. Find the pseudoconvergents of the simple continued fraction of π for $k = 2$.

19. Find a rational number r/s which is closer to π than $22/7$ with denominator s less than 106. (*Hint*: Use Exercise 17.)

   20. Find the rational number r/s which is closest to e with denimator s less than
         100.

10.3 Computer Projects

Programming Projects

  Write programs to do the following:

   1. Given a real number x, find the simple continued fraction of x.

   2. Given an irrational number x and a positive integer n, find the best rational
         approximation to x with denominator not exceeding n.

Computations and Explorations

   1. Compute the first 100 partial quotients of the simple continued fraction of e^2.
         From this find the rule for the partial quotients of this simple continued
         fraction.

   2. Computer the first 1000 partial quotients of the simple continued fraction of π.
         What is the largest partial quotient that appears? How often does the integer
         one appear as a partial quotient?

10.4 Periodic Continued Fractions

  We call the infinite simple continued fraction $[a_0;a_1,a_2,...]$ *periodic* if there are
positive integers N and k such that $a_n = a_{n+k}$ for all positive integers n with $n \geq N$.
We use the notation

$$[a_0;a_1,a_2,...,a_{N-1},\overline{a_N,a_{N+1},...,a_{N+k-1}}]$$

to express the periodic infinite simple continued fraction

$$[a_0;a_1,a_2,...,a_{N-1},a_N,a_{N+1},...,a_{N+k-1},a_N,a_{N+1},...].$$

For instance, $[1;2,\overline{3,4}]$ denotes the infinite simple continued fraction
$[1;2,3,4,3,4,3,4,...]$.

  In Section 10.1 we showed that the base b expansion of a number is periodic if
and only if the number is rational. To characterize those irrational numbers with
periodic infinite simple continued fractions, we need the following definition.

Definition. The real number α is said to be a *quadratic irrational* if α is irrational
and if α is a root of a quadratic polynomial with integer coefficients, *i.e.*

$$A\alpha^2 + B\alpha + C = 0,$$

where A, B, and C are integers and $A \neq 0$.

Example 10.12. Let $\alpha = 2 + \sqrt{3}$. Then α is irrational, for if α were rational, then
by Exercise 22 of Section 1.1, $\alpha - 2 = \sqrt{3}$ would be rational, contradicting

Theorem 2.11. Next, note that

$$\alpha^2 - 4\alpha + 1 = (7+4\sqrt{3}) - 4(2+\sqrt{3}) + 1 = 0.$$

Hence α is a quadratic irrational.   □

We will show that the infinite simple continued fraction of an irrational number is periodic if and only if this number is a quadratic irrational. Before we do this, we first develop some useful results about quadratic irrationals.

Lemma 10.1. The real number α is a quadratic irrational if and only if there are integers a,b, and c with $b > 0$ and $c \neq 0$, such that b is not a perfect square and

$$\alpha = (a+\sqrt{b})/c.$$

Proof. If α is a quadratic irrational, then α is irrational, and there are integers A,B, and C such that $A\alpha^2 + B\alpha + C = 0$. From the quadratic formula, we know that

$$\alpha = \frac{-B \pm \sqrt{B^2 - 4AC}}{2A}.$$

Since α is a real number, we have $B^2 - 4AC > 0$, and since α is irrational, $B^2 - 4AC$ is not a perfect square and $A \neq 0$. By either taking $a = -B$, $b = B^2 - 4AC$, $c = 2A$ or $a = B$, $b = B^2 - 4AC$, $c = -2A$, we have our desired representation of α.

Conversely, if

$$\alpha = (a+\sqrt{b})/c,$$

where a,b, and c are integers with $b > 0$, $c \neq 0$, and b not a perfect square, then by Exercise 22 of Section 1.1 and Theorem 2.11, we can easily see that α is irrational. Furthermore, we note that

$$c^2\alpha^2 - 2ac\alpha + (a^2 - b) = 0,$$

so that α is a quadratic irrational.   ■

The following lemma will be used when we show that periodic simple continued fractions represent quadratic irrationals.

Lemma 10.2. If α is a quadratic irrational and if r,s,t, and u are integers, then $(r\alpha+s)/(t\alpha+u)$ is either rational or a quadratic irrational.

Proof. From Lemma 10.1, there are integers a,b, and c with $b > 0$, $c \neq 0$, and b not a perfect square such that

$$\alpha = (a+\sqrt{b})/c.$$

Thus

$$\frac{r\alpha+s}{t\alpha+u} = \left[\frac{r(a+\sqrt{b})}{c} + s\right] \Big/ \left[\frac{t(a+\sqrt{b})}{c} + u\right]$$

$$= \frac{(ar+cs)+r\sqrt{b}}{(at+cu)+t\sqrt{b}}$$

$$= \frac{[(ar+cs)+r\sqrt{b}][(at+cu)-t\sqrt{b}]}{[(at+cu)+t\sqrt{b}][(at+cu)-t\sqrt{b}]}$$

$$= \frac{[(ar+cs)(at+cu)-rtb]+[r(at+cu)-t(ar+cs)]\sqrt{b}}{(at+cu)^2-t^2b}.$$

Hence, by Lemma 10.1 $(r\alpha+s)/(t\alpha+u)$ is a quadratic irrational, unless the coefficient of $\sqrt{b}$ is zero, which would imply that this number is rational. ∎

In our subsequent discussions of simple continued fractions of quadratic irrationals we will use the notion of the conjugate of a quadratic irrational.

Definition. Let $\alpha = (a+\sqrt{b})/c$ be a quadratic irrational. Then the *conjugate* of α, denoted by α', is defined by $\alpha' = (a-\sqrt{b})/c$.

Lemma 10.3. If the quadratic irrational α is a root of the polynomial $Ax^2 + Bx + C = 0$, then the other root of this polynomial is α', the conjugate of α.

Proof. From the quadratic formula, we see that the two roots of $Ax^2 + Bx + C = 0$ are

$$\frac{-B\pm\sqrt{B^2-4AC}}{2A}.$$

If α is one of these roots, then α' is the other root, because the sign of $\sqrt{B^2-4AC}$ is reversed to obtain α' from α. ∎

The following lemma tells us how to find the conjugates of arithmetic expressions involving quadratic irrationals.

Lemma 10.4. If $\alpha_1 = (a_1+b_1\sqrt{d})/c_1$ and $\alpha_2 = (a_2+b_2\sqrt{d})/c_2$ are rational or quadratic irrationals, then

    (i) $(\alpha_1+\alpha_2)' = \alpha_1' + \alpha_2'$

    (ii) $(\alpha_1-\alpha_2)' = \alpha_1' - \alpha_2'$

    (iii) $(\alpha_1\alpha_2)' = \alpha_1'\alpha_2'$

    (iv) $(\alpha_1/\alpha_2)' = \alpha_1'/\alpha_2'.$

The proof of (iv) will be given here; the proofs of the other parts are easier. These appear at the end of this section as problems for the reader.

Proof of (iv). Note that

$$\alpha_1/\alpha_2 = \frac{(a_1+b_1\sqrt{d})/c_1}{(a_2+b_2\sqrt{d})/c_2}$$

$$= \frac{c_2(a_1+b_1\sqrt{d})(a_2-b_2\sqrt{d})}{c_1(a_2+b_2\sqrt{d})(a_2-b_2\sqrt{d})}$$

$$= \frac{(c_2a_1a_2-c_2b_1b_2d) + (c_2a_2b_1-c_2a_1b_2)\sqrt{d}}{c_1(a_2^2-b_2^2d)}.$$

While

$$\alpha_1'/\alpha_2' = \frac{(a_1-b_1\sqrt{d})/c_2}{(a_2-b_2\sqrt{d})/c_2}$$

$$= \frac{c_2(a_1-b_1\sqrt{d})(a_2+b_2\sqrt{d})}{c_1(a_2-b_2\sqrt{d})(a_2+b_2\sqrt{d})}$$

$$= \frac{(c_2a_1a_2-c_2b_1b_2d) - (c_2a_2b_1-c_2a_1b_2)\sqrt{d}}{c_1(a_2^2-b_2^2d)}.$$

Hence $(\alpha_1/\alpha_2)' = \alpha_1'/\alpha_2'$. $\qquad\qquad\square$

The fundamental result about periodic simple continued fractions is Lagrange's Theorem. (Note that this theorem is different than Lagrange's theorem on polynomial congruences discussed in Chapter 8. In this chapter we do not refer to that result.)

Theorem 10.17. Lagrange's Theorem. The infinite simple continued fraction of an irrational number is periodic if and only if this number is a quadratic irrational.

We first prove that a periodic continued fraction represents a quadratic irrational. The converse, that the simple continued fraction of a quadratic irrational is periodic, will be proved after a special algorithm for obtaining the continued fraction of a quadratic irrational is developed.

Proof. Let the simple continued fraction of α be periodic, so that

$$\alpha = [a_0;a_1,a_2,...,a_{N-1},\overline{a_N,a_{N+1},...,a_{N+k}}]$$

Now let

$$\beta = [\overline{a_N;a_{N+1},...,a_{N+k}}].$$

Then

$$\beta = [a_N;a_{N+1},...,a_{N+k},\beta],$$

and by Theorem 10.7 it follows that

(10.13)
$$\beta = \frac{\beta p_k + p_{k-1}}{\beta q_k + q_{k-1}},$$

where p_k/q_k and p_{k-1}/q_{k-1} are convergents of $[a_N;a_{N+1},...,a_{N+k}]$. Since the simple continued fraction of β is infinite, β is irrational, and by (10.13) we have

$$q_k \beta^2 + (q_{k-1} - p_k)\beta - p_{k-1} = 0,$$

so that β is a quadratic irrational. Now note that

$$\alpha = [a_0;a_1,a_2,...,a_{N-1},\beta],$$

so that from Theorem 10.9 we have

$$\alpha = \frac{\beta p_{N-1} + p_{N-2}}{\beta q_{N-1} + q_{N-2}},$$

where p_{N-1}/q_{N-1} and p_{N-2}/q_{N-2} are convergents of $[a_0;a_1,a_2,...,a_{N-1}]$. Since β is a quadratic irrational, Lemma 10.2 tells us that α is also a quadratic irrational (we know that α is irrational because it has an infinite simple continued fraction expansion). ∎

The following example shows how to use the proof of Theorem 10.17 to find the quadratic irrational represented by a periodic simple continued fraction.

Example 10.13. Let $x = [3;\overline{1,2}]$. By Theorem 10.17 we know that x is a quadratic irrational. To find the value of x, we let $x = [3;y]$ where $y = [\overline{1;2}]$, as in the proof of Theorem 10.17. We have $y = [1;2,y]$, so that

$$y = 1 + \cfrac{1}{2 + \cfrac{1}{y}} = \frac{3y + 1}{2y + 1}.$$

It follows that $2y^2 - 2y - 1 = 0$. Since y is positive, by the quadratic formula we have $y = \dfrac{1 + \sqrt{3}}{2}$. Since $x = 3 + \dfrac{1}{y}$, we have

$$x = 3 + \frac{2}{1 + \sqrt{3}} = 3 + \frac{2 - \sqrt{3}}{-2} = \frac{4 + \sqrt{3}}{2}.$$

□

To develop an algorithm for finding the simple continued fraction of a quadratic irrational, we need the following lemma.

Lemma 10.5. If α is a quadratic irrational, then α can be written as

$$\alpha = (P + \sqrt{d})/Q,$$

where P, Q, and d are integers, $Q \neq 0$, $d > 0$, d is not a perfect square, and $Q \mid (d - P^2)$.

Proof. Since α is a quadratic irrational, Lemma 10.1 tells us that

$$\alpha = (a + \sqrt{b})/c,$$

where a, b, and c are integers, $b > 0$, and $c \neq 0$. We multiply both the numerator and denominator of this expression for α by $|c|$ to obtain

$$\alpha = \frac{a|c| + \sqrt{bc^2}}{c|c|}$$

(where we have used the fact that $|c| = \sqrt{c^2}$). Now let $P = a|c|$, $Q = c|c|$, and $d = bc^2$. Then P, Q, and d are integers, $Q \neq 0$ since $c \neq 0$, $d > 0$ (since $b > 0$), d is not a perfect square since b is not a perfect square, and finally $Q | (d - P^2)$ since $d - P^2 = bc^2 - a^2 c^2 = c^2(b - a^2) = \pm Q(b - a^2)$. ∎

We now present an algorithm for finding the simple continued fractions of quadratic irrationals.

Theorem 10.18. Let α be a quadratic irrational, so that by Lemma 10.5 there are integers P_0, Q_0, and d such that

$$\alpha = (P_0 + \sqrt{d})/Q_0,$$

where $Q_0 \neq 0, d > 0$, d is not a perfect square, and $Q_0 | (d - P_0^2)$. Recursively define

$$\alpha_k = (P_k + \sqrt{d})/Q_k,$$
$$a_k = [\alpha_k],$$
$$P_{k+1} = a_k Q_k - P_k,$$
$$Q_{k+1} = (d - P_{k+1}^2)/Q_k,$$

for $k = 0, 1, 2, \dots$. Then $\alpha = [a_0; a_1, a_2, \dots]$.

Proof. Using mathematical induction, we will show that P_k and Q_k are integers with $Q_k \neq 0$ and $Q_k | (d - P_k^2)$, for $k = 0, 1, 2, \dots$. First, note that this assertion is true for $k = 0$ from the hypotheses of the theorem. Now assume that P_k and Q_k are integers with $Q_k \neq 0$ and $Q_k | (d - P_k^2)$. Then

$$P_{k+1} = a_k Q_k - P_k$$

is also an integer. Further,

$$Q_{k+1} = (d - P_{k+1}^2)/Q_k$$
$$= [d - (a_k Q_k - P_k)^2]/Q_k$$
$$= (d - P_k^2)/Q_k + (2a_k P_k - a_k^2 Q_k).$$

Since $Q_k | (d - P_k^2)$, by the induction hypothesis, we see that Q_{k+1} is an integer, and since d is not a perfect square, we see that $d \neq P_k^2$, so that $Q_{k+1} = (d - P_{k+1}^2)/Q_k \neq 0$. Since

$$Q_k = (d - P_{k+1}^2)/Q_{k+1},$$

we can conclude that $Q_{k+1} | (d - P_{k+1}^2)$. This finishes the inductive argument.

To demonstrate that the integers $a_0, a_1, a_2, \ldots$ are the partial quotients of the simple continued fraction of α, we use Theorem 10.13. If we can show that

$$\alpha_{k+1} = 1/(\alpha_k - a_k),$$

for $k = 0, 1, 2, \ldots$, then we know that $\alpha = [a_0; a_1, a_2, \ldots]$. Note that

$$\alpha_k - a_k = \frac{P_k + \sqrt{d}}{Q_k} - a_k$$

$$= [\sqrt{d} - (a_k Q_k - P_k)]/Q_k$$

$$= (\sqrt{d} - P_{k+1})/Q_k$$

$$= (\sqrt{d} - P_{k+1})(\sqrt{d} + P_{k+1})/Q_k(\sqrt{d} + P_{k+1})$$

$$= (d - P_{k+1}^2)/Q_k(\sqrt{d} + P_{k+1})$$

$$= Q_k Q_{k+1}/Q_k(\sqrt{d} + P_{k+1})$$

$$= Q_{k+1}/(\sqrt{d} + P_{k+1})$$

$$= 1/\alpha_{k+1},$$

where we have used the defining relation for Q_{k+1} to replace $d - P_{k+1}^2$ with $Q_k Q_{k+1}$. Hence, we can conclude that $\alpha = [a_0; a_1, a_2, \ldots]$. ∎

We illustrate the use of the algorithm given in Theorem 10.18 with the following example.

Example 10.14. Let $\alpha = (3 + \sqrt{7})/2$. Using Lemma 10.5, we write

$$\alpha = (6 + \sqrt{28})/4$$

where we set $P_0 = 6$, $Q_0 = 4$, and $d = 28$. Hence $a_0 = [\alpha] = 2$, and

$$
\begin{array}{llll}
P_1 & = & 2 \cdot 4 - 6 = 2, & \alpha_1 & = & (2 + \sqrt{28})/6, \\
Q_1 & = & (28 - 2^2)/4 = 6, & a_1 & = & [(2 + \sqrt{28})/6] = 1, \\
\\
P_2 & = & 1 \cdot 6 - 2 = 4, & \alpha_2 & = & (4 + \sqrt{28})/2, \\
Q_2 & = & (28 - 4^2)/6 = 2, & a_2 & = & [(4 + \sqrt{28})/2] = 4, \\
\\
P_3 & = & 4 \cdot 2 - 4 = 4, & \alpha_3 & = & (4 + \sqrt{28})/6, \\
Q_3 & = & (28 - 4^2)/2 = 6, & a_3 & = & [(4 + \sqrt{28})/6] = 1, \\
\end{array}
$$

$$P_4 = 1 \cdot 6 - 4 = 2, \qquad \alpha_4 = (2 + \sqrt{28})/4,$$
$$Q_4 = (28 - 2^2)/6 = 4, \qquad a_4 = [(2 + \sqrt{28})/4] = 1,$$

$$P_5 = 1 \cdot 4 - 2 = 2, \qquad \alpha_5 = (2 + \sqrt{28})/6,$$
$$Q_5 = (28 - 2^2)/4 = 6, \qquad a_5 = [(2 + \sqrt{28})/6] = 1,$$

and so, with repetition, since $P_1 = P_5$ and $Q_1 = Q_5$. Hence, we see that

$$(3 + \sqrt{7})/2 = [2; 1, 4, 1, 1, 1, 4, 1, 1, \dots]$$
$$= [2; \overline{1, 4, 1, 1}]. \qquad \square$$

We now finish the proof of Lagrange's Theorem by showing that the simple continued fraction expansion of a quadratic irrational is periodic.

Proof (continued). Let α be a quadratic irrational, so that by Lemma 10.5 we can write α as

$$\alpha = (P_0 + \sqrt{d})/Q_0.$$

Furthermore, by Theorem 10.18 we have $\alpha = [a_0; a_1, a_2, \dots]$ where

$$\alpha_k = (P_k + \sqrt{d})/Q_k,$$
$$a_k = [\alpha_k],$$
$$P_{k+1} = a_k Q_k - P_k,$$
$$Q_{k+1} = (d - P_{k+1}^2)/Q_k,$$

for $k = 0, 1, 2, \dots$.

Since $\alpha = [a_0; a_1, a_2, \dots, \alpha_k]$, Theorem 10.9 tells us that

$$\alpha = (p_{k-1} \alpha_k + p_{k-2})/(q_{k-1} \alpha_k + q_{k-2}).$$

Taking conjugates of both sides of this equation, and using Lemma 10.4, we see that

(10.14) $$\alpha' = (p_{k-1} \alpha_k' + p_{k-2})/(q_{k-1} \alpha_k' + q_{k-2}).$$

When we solve (10.14) for α_k', we find that

$$\alpha_k' = \frac{-q_{k-2}}{q_{k-1}} \left(\frac{\alpha' - \dfrac{p_{k-2}}{q_{k-2}}}{\alpha' - \dfrac{p_{k-1}}{q_{k-1}}} \right).$$

Note that the convergents p_{k-2}/q_{k-2} and p_{k-1}/q_{k-1} tend to α as k tends to infinity, so that

$$\left(\alpha' - \frac{p_{k-2}}{q_{k-2}} \right) \bigg/ \left(\alpha' - \frac{p_{k-1}}{q_{k-1}} \right)$$

tends to 1. Hence, there is an integer N such that $\alpha_k' < 0$ for $k \geq N$. Since $\alpha_k > 0$ for $k \geq 1$, we have

$$\alpha_k - \alpha'_k = \frac{P_k + \sqrt{d}}{Q_k} - \frac{P_k - \sqrt{d}}{Q_k} = \frac{2\sqrt{d}}{Q_k} > 0,$$

so that $Q_k > 0$ for $k \geq N$.

Since $Q_k Q_{k+1} = d - P_{k+1}^2$, we see that for $k \geq N$,

$$Q_k \leq Q_k Q_{k+1} = d - P_{k+1}^2 \leq d.$$

Also for $k \geq N$, we have

$$P_{k+1}^2 \leq d = P_{k+1}^2 - Q_k Q_{k+1},$$

so that

$$-\sqrt{d} < P_{k+1} < \sqrt{d}.$$

From the inequalities $0 \leq Q_k \leq d$ and $-\sqrt{d} < P_{k+1} < \sqrt{d}$, that hold for $k \geq N$, we see that there are only a finite number of possible values for the pair of integers P_k, Q_k for $k > N$. Since there are infinitely many integers k with $k \geq N$, there are two integers i and j such that $P_i = P_j$ and $Q_i = Q_j$ with $i < j$. Hence, from the defining relation for α_k, we see that $\alpha_i = \alpha_j$. Consequently, we can see that $a_i = a_j, a_{i+1} = a_{j+1}, a_{i+2} = a_{j+2}, \ldots$. Hence

$$\alpha = [a_0; a_1, a_2, \ldots, a_{i-1}, a_i, a_{i+1}, \ldots, a_{j-1}, a_i, a_{i+1}, \ldots, a_{j-1}, \ldots]$$

$$= [a_0; a_1, a_2, \ldots, a_{i-1}, \overline{a_j, a_{i+1}, \ldots, a_{j-1}}].$$

This shows that α has a periodic simple continued fraction. ∎

Next, we investigate those periodic simple continued fractions that are *purely periodic*, *i.e.* those without a pre-period.

Definition. The continued fraction $[a_0; a_1, a_2, \ldots]$ is called *purely periodic* if there is an integer n such that $a_k = a_{n+k}$, for $k = 0, 1, 2, \ldots$, so that

$$[a_0; a_1, a_2, \ldots] = [\overline{a_0; a_1, a_2, a_3, \ldots, a_{n-1}}].$$

Example 10.15. The continued fraction $[\overline{2;3}] = (1 + \sqrt{3})/2$ is purely periodic while $[2; \overline{2,4}] = \sqrt{6}$ is not. □

The next definition and theorem describe those quadratic irrationals with purely periodic simple continued fractions.

Definition. A quadratic irrational α if called *reduced* if $\alpha > 1$ and $-1 < \alpha' < 0$, where α' is the conjugate of α.

Theorem 10.19. The simple continued fraction of the quadratic irrational α is purely periodic if and only if α is reduced. Further, if α is reduced and $\alpha = [\overline{a_0; a_1, a_2, \ldots, a_n}]$ then the continued fraction of $-1/\alpha'$ is $[\overline{a_n; a_{n-1}, \ldots, a_0}]$.

Proof. First, assume that α is a reduced quadratic irrational. Recall from Theorem 10.15 that the partial fractions of the simple continued fraction of α are given by

$$a_k = [\alpha_k], \quad \alpha_{k+1} = 1/(\alpha_k - a_k),$$

for $k = 0,1,2,...$, where $\alpha_0 = \alpha$. We see that

$$1/\alpha_{k+1} = \alpha_k - a_k,$$

and by taking conjugates and using Lemma 10.4, we see that

(10.15) $$1/\alpha'_{k+1} = \alpha'_k - a_k.$$

We can prove, by mathematical induction, that $-1 < \alpha'_k < 0$ for $k = 0,1,2,....$ First, note that since $\alpha_0 = \alpha$ is reduced, $-1 < \alpha'_0 < 0$. Now assume that $-1 < \alpha'_k < 0$. Then, since $a_k \geq 1$ for $k = 0,1,2,...$ (note that $a_0 \geq 1$ since $\alpha > 1$), we see from (10.15) that

$$1/\alpha'_{k+1} < -1,$$

so that $-1 < \alpha'_{k+1} < 0$. Hence, $-1 < \alpha'_k < 0$ for $k = 0,1,2,...$.

Next, note that from (10.15) we have

$$\alpha'_k = a_k + 1/\alpha'_{k+1},$$

and since $-1 < \alpha'_k < 0$, it follows that

$$-1 < a_k + 1/\alpha'_{k+1} < 0.$$

Consequently,

$$-1 - 1/\alpha'_{k+1} < a_k < -1/\alpha'_{k+1},$$

so that

$$a_k = [-1/\alpha'_{k+1}].$$

Since α is a quadratic irrational, the proof of Lagrange's Theorem shows that there are nonnegative integers i and j, $i < j$, such that $\alpha_i = \alpha_j$, and hence with $-1/\alpha'_i = -1/\alpha'_j$. Since $a_{i-1} = [-1/\alpha'_i]$ and $a_{j-1} = [-1/\alpha'_j]$, we see that $a_{i-1} = a_{j-1}$. Furthermore, since $\alpha_{i-1} = a_{i-1} + 1/\alpha_i$ and , $\alpha_{j-1} = a_{j-1} + 1/\alpha_j$ we also see that $\alpha_{i-1} = \alpha_{j-1}$. Continuing this argument, we see that $\alpha_{i-2} = \alpha_{j-2}, \alpha_{i-3} = \alpha_{j-3},...$, and finally, that $\alpha_0 = \alpha_{j-i}$. Since

$$\alpha_0 = \alpha = [a_0;a_1,...,a_{j-i-1},\alpha_{j-i}]$$

$$= [a_o;a_1,...,a_{j-i-1},\alpha_0]$$

$$= \overline{[a_0;a_1,...,a_{j-i-1}]},$$

we see that the simple continued fraction of α is purely periodic.

To prove the converse, assume that α is a quadratic irrational with a purely periodic continued fraction $\alpha = \overline{[a_0;a_1,a_2,...,a_k]}$. Since $\alpha = [a_0;a_1,a_2,...,a_k,\alpha]$, Theorem 10.9 tells that

(10.16)
$$\alpha = \frac{\alpha p_k + p_{k-1}}{\alpha q_k + q_{k-1}},$$

where p_{k-1}/q_{k-1} and p_k/q_k are the $(k-1)$th and kth convergents of the continued fraction expansion of α. From (10.16), we see that

(10.17)
$$q_k\alpha^2 + (q_{k-1}-p_k)\alpha - p_{k-1} = 0.$$

Now, let β be the quadratic irrational such that $\beta = \overline{[a_k;a_{k-1},...,a_1,a_0]}$, i.e. with the period of the simple continued fraction for α reversed. Then $\beta = [a_k;a_{k-1},...,a_1,a_0,\beta]$, so that by Theorem 10.9, it follows that

(10.18)
$$\beta = \frac{\beta p'_k + p'_{k-1}}{\beta q'_k + q'_{k-1}},$$

where p'_{k-1}/q'_{k-1} and p'_k/q'_k are the $(k-1)$th and kth convergents of the continued fraction expansion of β. Note, however, from problem 6 of Section 10.2, that

$$p_k/p_{k-1} = [a_n;a_{n-1},...,a_1,a_0] = p'_k/q'_k$$

and

$$q_k/q_{k-1} = [a_n;a_{n-1},...,a_2,a_1] = p'_{k-1}/q'_{k-1}.$$

Since p'_{k-1}/q'_{k-1} and p'_k/q'_k are convergents, we know that they are in lowest terms. Also, p_k/p_{k-1} and q_k/q_{k-1} are in lowest terms, since Theorem 10.10 tells us that $p_kq_{k-1} - p_{k-1}q_k = (-1)^{k-1}$. Hence,

$$p'_k = p_k, \ q'_k = p_{k-1}$$

and

$$p'_{k-1} = q_k, \ q'_{k-1} = q_{k-1}.$$

Inserting these values into (10.18), we see that

$$\beta = \frac{\beta p_k + q_k}{\beta p_{k-1} + q_{k-1}}.$$

Therefore, we know that

$$p_{k-1}\beta^2 + (q_{k-1} - p_k)\beta - q_k = 0.$$

This implies that

(10.19)
$$q_k(-1/\beta)^2 + (q_{k-1} - p_k)(-1/\beta) - p_{k-1} = 0.$$

By (10.17) and (10.19) we see that the two roots of the quadratic equation

$$q_kx^2 + (q_{k-1} - p_k)x - p_{k-1} = 0$$

are α and $-1/\beta$, so that by the quadratic equation, we have $\alpha' = -1/\beta$. Since $\beta = \overline{[a_n;a_{n-1},...,a_1,a_0]}$, we see that $\beta > 1$, so that $-1 < \alpha' = -1/\beta < 0$. Hence, α is a reduced quadratic irrational.

Furthermore, note that since $\beta = -1/\alpha'$, it follows that

$$-1/\alpha' = \overline{[a_n;a_{n-1},...,a_1,a_0]}.$$ ∎

We now find the form of the periodic simple continued fraction of $\sqrt{D}$, where D is a positive integer that is not a perfect square. Although $\sqrt{D}$ is not reduced, since its conjugate $-\sqrt{D}$ is not between -1 and 0, the quadratic irrational $[\sqrt{D}] + \sqrt{D}$ is reduced, since its conjugate, $[\sqrt{D}] - \sqrt{D}$, does lie between -1 and 0. Therefore, from Theorem 10.21, we know that the continued fraction of $[\sqrt{D}] + \sqrt{D}$ is purely periodic. Since the initial partial quotient of the simple continued fraction of $[\sqrt{D}] + \sqrt{D}$ is $[[\sqrt{D}] + \sqrt{D}] = 2[\sqrt{D}] = 2a_0$, where $a_0 = [\sqrt{D}]$, we can write

$$[\sqrt{D}] + \sqrt{D} = \overline{[2a_0;a_1,a_2,...,a_n]}$$
$$= [2a_0;a_1,a_2,...,a_n,\overline{2a_0,a_1,...,a_n}].$$

Subtracting $a_0 = \sqrt{D}$ from both sides of this equality, we find that

$$\sqrt{D} = [a_0;a_1,a_2,...,2a_0,a_1,a_2,...2a_0,...]$$
$$= [a_0;\overline{a_1,a_2,...,a_n,2a_0}].$$

To obtain even more information about the partial quotients of the continued fraction of $\sqrt{D}$, we note that from Theorem 10.21, the simple continued fraction expansion of $-1/([\sqrt{D}] - \sqrt{D})$ can be obtained from that for $[\sqrt{D}] + \sqrt{D}$, by reversing the period, so that

$$1/(\sqrt{D} - [\sqrt{D}]) = \overline{[a_n;a_{n-1},...,a_1,2a_0]}.$$

But also note that

$$\sqrt{D} - [\sqrt{D}] = [0;\overline{a_1,a_2,...,a_n2a_0}],$$

so that by taking reciprocals, we find that

$$1/(\sqrt{D} - [\sqrt{D}]) = \overline{[a_1;a_2,...,a_n,2a_0]}.$$

Therefore, when we equate these two expressions for the simple continued fraction of $1/(\sqrt{D} - [\sqrt{D}])$, we obtain

$$a_1 = a_n, a_2 = a_{n-1},...,a_n = a_1,$$

so that the periodic part of the continued fraction for $\sqrt{D}$ is symmetric from the first to the penultimate term.

In conclusion, we see that the simple continued fraction of $\sqrt{D}$ has the form

$$\sqrt{D} = [a_0;\overline{a_1,a_2,...,a_2,a_1,2a_0}].$$

We illustrate this with some examples.

Example 10.16. Note that

$$
\begin{aligned}
\sqrt{23} &= [4;\overline{1,3,1,8}] \\
\sqrt{31} & [5,\overline{1,1,3,5,3,1,1,10}] \\
\sqrt{46} &= [6;\overline{1,2,1,1,2,6,2,1,1,2,1,12}] \\
\sqrt{76} &= [8;\overline{1,2,1,1,5,4,5,1,1,2,1,16}]
\end{aligned}
$$

and

$$
\sqrt{97} = [9;\overline{1,5,1,1,1,1,1,1,5,1,18}],
$$

where each continued fraction has a pre-period of length 1 and a period ending with twice the first partial quotient which is symmetric from the first to the next to the last term. □

The simple continued fraction expansions of $\sqrt{d}$ for positive integers d such that d is not a perfect square and $d < 100$ can be found in Table 5 of the Appendix.

10.4 Exercises

1. Find the simple continued fractions of each of the following numbers.

   a) $\sqrt{7}$ d) $\sqrt{47}$
   b) $\sqrt{11}$ e) $\sqrt{59}$
   c) $\sqrt{23}$ f) $\sqrt{94}$

2. Find the simple continued fractions of each of the following numbers.

   a) $\sqrt{101}$ d) $\sqrt{201}$
   b) $\sqrt{103}$ e) $\sqrt{203}$
   c) $\sqrt{107}$ f) $\sqrt{209}$

3. Find the simple continued fractions of each of the following numbers.

   a) $1+\sqrt{2}$

   b) $(2+\sqrt{5})/3$

   c) $(5-\sqrt{7})/4$

4. Find the simple continued fractions of each of the following numbers.

   a) $(1+\sqrt{3})/2$

   b) $(14+\sqrt{37})/3$

   c) $(13-\sqrt{2})/7$

5. Find the quadratic irrational with each of the following simple continued fraction expansions.

   a) $[2;1,\overline{5}]$

b) $[2;\overline{1,5}]$

c) $[\overline{2;1,5}]$

6. Find the quadratic irrational with each of the following simple continued fraction expansions.

a) $[1;2,\overline{3}]$

b) $[1;\overline{2,3}]$

c) $[\overline{1;2,3}]$

7. Find the quadratic irrational with each of the following simple continued fraction expansions.

a) $[3;\overline{6}]$

b) $[4;\overline{8}]$

c) $[5;\overline{10}]$

d) $[6;\overline{12}]$

8. a) Let d be a positive integer. Show that the simple continued fraction of $\sqrt{d^2+1}$ is $[d;\overline{2d}]$.

b) Use part (a) to find the simple continued fractions of $\sqrt{101}, \sqrt{290}$, and $\sqrt{2210}$.

9. Let d be a integer, $d \geq 2$.

a) Show that the simple continued fraction of $\sqrt{d^2-1}$ is $[d-1;\overline{1,2d-2}]$.

b) Show that the simple continued fraction of $\sqrt{d^2-d}$ is $[d-1;\overline{2,2d-2}]$.

c) Use parts (a) and (b) to find the simple continued fractions of $\sqrt{99}$, $\sqrt{110}, \sqrt{272}$, and $\sqrt{600}$.

10. a) Show that if d is an integer, $d \geq 3$, then the simple continued fraction of $\sqrt{d^2-2}$ is $[d-1;\overline{1,d-2,1,2d-2}]$.

b) Show that if d is a positive integer, then the simple continued fraction of $\sqrt{d^2+2}$ is $[d;\overline{d,2d}]$.

c) Find the simple continued fraction expansions of $\sqrt{47}$, $\sqrt{51}$, and $\sqrt{287}$.

11. Let d be an odd positive integer.

a) Show that the simple continued fraction of $\sqrt{d^2+4}$ is $[d;\overline{(d-1)/2,1,1,(d-1)/2,2d}]$, if $d > 1$.

b) Show that the simple continued fraction of $\sqrt{d^2-4}$ is $[d-1;\overline{1,(d-3)/2,2,(d-3)/2,1,2d-2}]$, if $d > 3$.

12. Show that the simple continued fraction of $\sqrt{d}$, where d is a positive integer, has period length one if and only if $d = a^2+1$ where a is a nonnegative integer.

13. Show that the simple continued fraction of $\sqrt{d}$, where d is a positive integer, has period length two if and only if $d = a^2 + b$ where a and b are integers, $b > 1$, and $b \mid 2a$.

14. Prove that if $\alpha_1 = (a_1 + b_1\sqrt{d})/c_1$ and $\alpha_2 = (a_2 + b_2\sqrt{d})/c_2$ are quadratic irrationals, then

    a) $(\alpha_1 + \alpha_2)' = \alpha_1' + \alpha_2'$

    b) $(\alpha_1 - \alpha_2)' = \alpha_1' - \alpha_2'$

    c) $(\alpha_1 \alpha_2)' = \alpha_1' \cdot \alpha_2'$.

15. Which of the following quadratic irrationals have purely periodic continued fractions?

    a) $1 + \sqrt{5}$ d) $(11 - \sqrt{10})/9$

    b) $2 + \sqrt{8}$ e) $(3 + \sqrt{23})/2$

    c) $4 + \sqrt{17}$ f) $(17 + \sqrt{188})/3$

16. Suppose that $\alpha = (a + \sqrt{b})/c$, where a, b, and c are integers, $b > 0$, and b is not a perfect square. Show that is a reduced quadratic irrational if and only if $0 < a < \sqrt{b}$ and $\sqrt{b} - a < c < \sqrt{b} + a < 2\sqrt{b}$.

17. Show that if α is a reduced quadratic irrational, then $-1/\alpha'$ is also a reduced quadratic irrational.

* 18. Let k be a positive integer. Show that there are infinitely many positive integers D, such that the simple continued fraction expansion of $\sqrt{D}$ has a period of length k. (*Hint:* Let $a_1 = 2$, $a_2 = 5$, and for $k \geq 3$ let $a_k = 2a_{k-1} + a_{k-2}$. Show that if $D = (ta_k + 1)^2 + 2ta_{k-1} + 1$, where t is a nonnegative integer, then $\sqrt{D}$ has a period of length $k + 1$.)

* 19. Let k be a positive integer. Let $D_k = (3^k + 1)^2 + 3$. Show that the simple continued frac.ion of $\sqrt{D_k}$ has a period of length $6k$.

10.4 Computer Projects

Programming Projects

Write computer programs to do the following:

* 1. Find the quadratic irrational that is the value of a periodic simple continued fraction.

* 2. Find the periodic simple continued fraction expansion of a quadratic irrational.

Computations and Explorations

Using the programs you have written or a computation program, carry out the following computations and explorations:

1. Find the smallest positive integer D such that the length of the period of the simple continued fraction of $\sqrt{D}$ is 10, 100, 1000, and 10000.

2. Find the length of the largest period of the simple continued fraction of $\sqrt{D}$ where D is a positive integer less than 1000, less than 10000, and less than 100000. Can you make any conjectures?

10.5 Factoring Using Continued Fractions

We can factor the positive integer n if we can find positive integers x and y such that $x^2 - y^2 = n$ and $x - y \neq 1$. This is the basis of the Fermat factorization method discussed in Section 2.4. However, it is possible to factor n if we can find positive integers x and y that satisfy the weaker condition

(10.20) $x^2 \equiv y^2 \pmod{n}, \ 0 < y < x < n, \text{ and } x + y \neq n.$

To see this, note that if (10.20) holds, then n divides $x^2 - y^2 = (x + y)(x - y)$, and n divides neither $x - y$ nor $x + y$. It follows that $(n, x - y)$ and $(n, x + y)$ are divisors of n that do not equal 1 or n. We can find these divisors rapidly using the Euclidean algorithm.

Example 10.17. Note that $29^2 - 17^2 = 841 - 289 = 552 \equiv 0 \pmod{69}$. Since $29^2 - 17^2 = (29 - 17)(29 + 17) \equiv 0 \pmod{69}$, both $(29 - 17, 69) = (12, 69)$ and $(29 + 17, 69) = (46, 69)$ are divisors of 69 not equal to either 1 or 69; using the Euclidean algorithm we find that these factors are $(12, 69) = 3$ and $(46, 69) = 23$. □

The continued fraction expansion of $\sqrt{n}$ can be used to find solutions of the congruence $x^2 \equiv y^2 \pmod{n}$. The following theorem is the basis for this.

Theorem 10.20. Let n be a positive integer that is not a perfect square. Define $\alpha_k = (P_k + \sqrt{n})/Q_k$, $a_k = [\alpha_k]$, $P_{k+1} = a_k Q_k - P_k$, and $Q_{k+1} = (n - P_{k+1}^2)/Q_k$, for $k = 0, 1, 2, \ldots$ where $\alpha_0 = \sqrt{n}$. Furthermore, let p_k/q_k denote the kth convergent of the simple continued fraction expansion of $\sqrt{n}$. Then

$$p_k^2 - nq_k^2 = (-1)^{k-1} Q_{k+1}.$$

The proof of Theorem 10.20 depends on the following useful lemma.

Lemma 10.6. Let $r + s\sqrt{n} = t + u\sqrt{n}$ where r, s, t, and u are rational numbers and n is a positive integer that is not a perfect square. Then $r = t$ and $s = u$.

Proof. Since $r + s\sqrt{n} = t + u\sqrt{n}$, we see that if $s \neq u$ then

$$\sqrt{n} = \frac{r - t}{u - s}.$$

Since $(r - t)/(u - s)$ is rational and $\sqrt{n}$ is irrational it follows that $s = u$, and consequently $r = t$. ∎

We can now prove Theorem 10.20.

Proof. Since $\sqrt{n} = \alpha_0 = [a_0; a_1, a_2, \ldots, a_k, \alpha_{k+1}]$, Theorem 10.7 tells us that

$$\sqrt{n} = \frac{\alpha_{k+1}p_k + p_{k-1}}{\alpha_{k+1}q_k + q_{k-1}}.$$

Since $\alpha_{k+1} = (P_{k+1} + \sqrt{n})/Q_{k+1}$ we have

$$\sqrt{n} = \frac{(P_{k+1} + \sqrt{n})p_k + Q_{k+1}p_{k-1}}{(P_{k+1} + \sqrt{n})q_k + Q_{k+1}q_{k-1}}.$$

Therefore, we see that

$$nq_k + (P_{k+1}q_k + Q_{k+1}q_{k-1})\sqrt{n} = (P_{k+1}p_k + Q_{k+1}p_{k-1}) + p_k\sqrt{n}.$$

By Lemma 10.6 we see that $nq_k = P_{k+1}p_k + Q_{k+1}p_{k-1}$ and $P_{k+1}q_k + Q_{k+1}q_{k-1} = p_k$. When we multiply the first of these two equations by q_k and the second by p_k, subtract the first from the second, and then simplify, we obtain

$$p_k^2 - nq_k^2 = (p_k q_{k-1} - p_{k-1}q_k)Q_{k+1} = (-1)^{k-1}Q_{k+1},$$

where we have used Theorem 10.8 to complete the proof. ∎

We now outline the technique known as the *continued fraction algorithm* for factoring an integer n. Suppose that the terms p_k, q_k, P_k, Q_k, a_k, and α_k have their usual meanings in the computation of the continued fraction expansion of $\sqrt{n}$. By Theorem 10.20 it follows that for every nonnegative integer k

$$p_k^2 \equiv (-1)^{k-1}Q_{k+1} \pmod{n}$$

where p_k and Q_{k+1} are as defined in the statement of the theorem. Now suppose that k is odd and that Q_{k+1} is a square, that is, $Q_{k+1} = s^2$ where s is a positive integer. Then $p_k^2 \equiv s^2 \pmod{n}$ and we may be able to use this congruence of two squares modulo n to find factors of n. Summarizing, to factor n we carry out the algorithm described in Theorem 10.8 to find the continued fraction expansion of $\sqrt{n}$. We look for squares among the terms with even index in the sequence $\{Q_k\}$. Each such occurrence may lead to a nonproper factor of n (or may just lead to the factorization $n = 1 \cdot n$). We illustrate this technique with several examples.

Example 10.18. We can factor 1037 using the continued fraction algorithm. Take $\alpha = \sqrt{1037} = (0 + \sqrt{1037})/1$ with $P_0 = 0$ and $Q_1 = 1$ and generate the terms P_k, Q_k, α_k, and a_k. We look for squares among the terms with even index in the sequence $\{Q_k\}$. We find that $Q_1 = 13$ and $Q_2 = 49$. Since $49 = 7^2$ is a square, and the index of Q_2 is even, we examine the congruence $p_1^2 \equiv (-1)^2 Q_2 \pmod{1037}$. Computing the terms of the sequence $\{p_k\}$, we find that $p_1 = 129$. This gives the congruence $129^2 \equiv 49 \pmod{1037}$. Hence $129^2 - 7^2 = (129 - 7)(129 + 7) \equiv 0 \pmod{1037}$. This produces the factors $(129 - 7, 1037) = (122, 1037) = 61$ and $(129 + 7, 1037) = (136, 1037) = 17$ of 1037. □

Example 10.19. We can use the continued fraction algorithm to find factors of 1000009. (We follow computations of Riesel [58].) We have $Q_1 = 9$, $Q_2 = 445$, $Q_3 = 873$, and $Q_4 = 81$. Since $81 = 9^2$ is square, we examine the congruence $p_3^2 \equiv (-1)^4 Q_4 \pmod{1000009}$. However, $p_3 = 2000009 \equiv -9 \pmod{1000009}$, so that $p_3 + 9$ is divisible by 1000009. It follows that we do not get any propers factors of 1000009 from this.

We continue until we reach another square in the sequence $\{Q_k\}$ with k even. This happens when $k = 18$ with $Q_{18} = 16$. Calculating p_{17} gives $p_{17} = 494881$. From the congruence $p_{17}^2 \equiv (-1)^{18} Q_{18} \pmod{1000009}$ we have $494881^2 \equiv 4^2 \pmod{1000009}$. It follows that $(494881 - 4, 1000009) = (494877, 1000009) = 293$ and $(494881 + 4, 1000009) = (494885, 1000009) = 3413$ are factors of 1000009. □

More powerful techniques based on continued fraction expansions are known. These are described in Dixon [107], Guy [110], Riesel [58] and Wagstaff and Smith [132]. We describe one such generalization in the exercises.

10.5 Exercises

1. Find factors of 119 using the congruence $19^2 \equiv 2^2 \pmod{119}$.

2. Factor 1537 using the continued fraction algorithm.

3. Factor the integer 13290059 using the continued fraction algorithm. (*Hint:* Use a computer program to generate the integers Q_k for the continued fraction for $\sqrt{13290059}$. You will need more than 50 terms.)

4. Let n be a positive integer and let $p_1, p_2, \ldots,$ and p_m be primes. Suppose that there exist integers $x_1, x_2, \ldots x_r$ such that

$$
\begin{aligned}
x_1^2 &\equiv (-1)^{e_{01}} p_1^{e_{11}} \cdots p_m^{e_{m1}} \pmod{n}, \\
x_2^2 &\equiv (-1)^{e_{02}} p_1^{e_{12}} \cdots p_m^{e_{m2}} \pmod{n}, \\
&\;\;\vdots \\
x_r^2 &\equiv (-1)^{e_{0r}} p_1^{e_{1r}} \cdots p_m^{e_{mr}} \pmod{n},
\end{aligned}
$$

where

$$
\begin{aligned}
e_{01} + e_{02} + \cdots + e_{0r} &= 2e_0 \\
e_{11} + e_{12} + \cdots + e_{1r} &= 2e_1 \\
&\;\;\vdots \\
e_{m1} + e_{m2} + \cdots + e_{mr} &= 2e_m.
\end{aligned}
$$

Show that $x^2 \equiv y^2 \pmod{n}$ where $x = x_1 x_2 \cdots x_r$ and $y = (-1)^{e_0} p_1^{e_1} \cdots p_r^{e_r}$. Explain how to factor n using this information. Here the primes $p_1,...,p_r$, together with -1, are called the *factor base*.

5. Show that 143 can be factored by setting $x_1 = 17$ and $x_2 = 19$, taking the factor base to be $\{3,5\}$.

6. Let n be a positive integer and let p_1, $p_2,...,p_r$ be primes. Suppose that $Q_{k_i} = \prod_{j=1}^{r} p_j^{k_{ij}}$ for $i = 1,...,t$ where the integers Q_j have their usual meaning with respect to the continued fraction of $\sqrt{n}$. Explain how n can be factored if $\sum_{i=1}^{t} k_i$ is even and $\sum_{i=1}^{t} k_{ij}$ is even for $j = 1,2,...,r$.

7. Show that 12007001 can be factoring using the continued fraction expansion of $\sqrt{12007001}$ with factor base $-1,2,31,71,97$. (*Hint:* Use the factorizations $Q_1 = 2^3 \cdot 97$, $Q_{12} = 2^4 \cdot 71$, $Q_{28} = 2^{11}$. $Q_{34} = 31 \cdot 97$, and $Q_{41} = 31 \cdot 71$, and show that $p_0 p_{11} p_{27} p_{33} p_{40} = 9815310$.)

8. Factor 197209 using the continued fraction expansion of $\sqrt{197209}$ and factor base 2,3,5.

10.5 Computer Projects

Programming Projects

Write programs to do the following things.

* **1.** Factor positive integers using the continued fraction algorithm.

** **2.** Factor positive integers using factor bases and continued fraction expansions (see Exercise 6).

Computations and Explorations

Using the programs you have written or a computational program, carry out the following computations and explorations.

1. Use the continued fraction algorithm to factor $F_7 = 2^{2^7} + 1$.

* **2.** Use the continued fraction algorithm to find the prime factorization of N_{11}, where N_j is the jth term of the sequence defined by $N_1 = 2$, $N_{j+1} = p_1 p_2 \cdots p_j + 1$ where p_j is the largest prime factor of N_j. (For example, $N_2 = 3, N_3 = 7, N_4 = 43, N_5 = 1807$, and so on.)

11

Some Nonlinear Diophantine Equations

11.1 Pythagorean Triples

The Pythagorean theorem tells us that the sum of the squares of the lengths of the legs of a right triangle equals the square of the length of the hypotenuse. Conversely, any triangle for which the sum of the squares of the lengths of the two shortest sides equals the square of the third side is a right triangle. Consequently, to find all right triangles with integral side lengths, we need to find all triples of positive integers x, y, z satisfying the diophantine equation

(11.1) $$x^2 + y^2 = z^2.$$

Triples of positive integers satisfying this equation are called *Pythagorean triples* after the ancient Greek mathematician Pythagoras*.

---

* **PYTHAGORAS (c. 572 - c. 500 B.C.E.)** was born on the Greek island of Samos. After extensive travels and studies, Pythagoras founded his famous school at the Greek port of Crotona in what is now southern Italy. Besides being an academy devoted to the study of mathematics, philosophy, and science, the school was the site of a brotherhood sharing secret rites. The Pythagoreans, as the members of this brotherhood were called, published nothing and ascribed all their discoveries to Pythagoras himself. However, it is believed that Pythagoras himself discovered what is now called the Pythagorean Theorem, namely that $a^2 + b^2 = c^2$ where a, b, and c are the lengths of the two legs and of the hypotenuse of a right triangle, respectively. The Pythagoreans believed that the key to understanding the world lay with natural numbers and form. Their central tenet was "Everything is Number." Because of their fascination with the natural numbers, the Pythagoreans made many discoveries in number theory. In particular they studied perfect numbers and amicable numbers for the mystical properties they felt these numbers possessed.

Example 11.1. The triples $3,4,5$; $6,8,10$; and $5,12,13$ are Pythagorean triples because $3^2 + 4^2 = 5^2$, $6^2 + 8^2 = 10^2$, and $5^2 + 12^2 = 13^2$. ☐

Unlike most nonlinear diophantine equations, it is possible to explicitly describe all the integral solutions of (11.1). Before developing the result describing all Pythagorean triples, we need a definition.

Definition. A Pythagorean triple x,y,z is called *primitive* if $(x,y,z) = 1$.

Example 11.2. The Pythagorean triples $3,4,5$ and $5,12,13$ are primitive, whereas the Pythagorean triple $6,8,10$ is not. ☐

Let x,y,z be a Pythagorean triple with $(x,y,z) = d$. Then there are integers x_1,y_1,z_1 with $x = dx_1$, $y = dy_1$, $z = dz_1$, and $(x_1,y_1,z_1) = 1$. Furthermore, because

$$x^2 + y^2 = z^2,$$

we have

$$(x/d)^2 + (y/d)^2 = (z/d)^2,$$

so that

$$x_1^2 + y_1^2 = z_1^2.$$

Hence, x_1,y_1,z_1 is a primitive Pythagorean triple, and the original triple x,y,z is simply an integral multiple of this primitive Pythagorean triple.

Also note that any integral multiple of a primitive (or for that matter any) Pythagorean triple is again a Pythagorean triple. If x_1,y_1,z_1 is a primitive Pythagorean triple, then we have

$$x_1^2 + y_1^2 = z_1^2,$$

and hence,

$$(dx_1)^2 + (dy_1)^2 = (dz_1)^2,$$

so that dx_1,dy_1,dz_1 is a Pythagorean triple.

Consequently, all Pythagorean triples can be found by forming integral multiples of primitive Pythagorean triples. To find all primitive Pythagorean triples, we need some lemmata. The first lemma tells us that any two integers of a primitive Pythagorean triple are relatively prime.

Lemma 11.1. If x,y,z is a primitive Pythagorean triple, then $(x,y) = (x,z) = (y,z) = 1$.

Proof. Suppose that x,y,z is a primitive Pythagorean triple and $(x,y) > 1$. Then, there is a prime p such that $p \mid (x,y)$, so that $p \mid x$ and $p \mid y$. Since $p \mid x$ and $p \mid y$, we know that $p \mid (x^2 + y^2) = z^2$. Because $p \mid z^2$, we can conclude that $p \mid z$.

This is a contradiction since $(x,y,z) = 1$. Therefore, $(x,y) = 1$. In a similar manner we can easily show that $(x,z) = (y,z) = 1$. ∎

Next, we establish a lemma about the parity of the integers of a primitive Pythagorean triple.

Lemma 11.2. If x,y,z is a primitive Pythagorean triple, then x is even and y is odd or x is odd and y is even.

Proof. Let x,y,z be a primitive Pythagorean triple. By Lemma 11.1 we know that $(x,y) = 1$, so that x and y cannot both be even. Also x and y cannot both be odd. If x and y were both odd, then we would have

$$x^2 \equiv y^2 \equiv 1 \pmod{4},$$

so that

$$z^2 = x^2 + y^2 \equiv 2 \pmod{4}.$$

This is impossible. Therefore, x is even and y is odd, or *vice versa*. ∎

The final lemma that we need is a consequence of the fundamental theorem of arithmetic. It tells us that two relatively prime integers that multiply together to give a square must both be squares.

Lemma 11.3. If r, s, and t are positive integers such that $(r,s) = 1$ and $rs = t^2$, then there are integers m and n such that $r = m^2$ and $s = n^2$.

Proof. If $r = 1$ or $s = 1$, then the lemma is obviously true, so we may suppose that $r > 1$ and $s > 1$. Let the prime-power factorizations of r, s, and t be

$$r = p_1^{a_1} p_2^{a_2} \cdots p_u^{a_u},$$
$$s = p_{u+1}^{a_{u+1}} p_{u+2}^{a_{u+2}} \cdots p_v^{a_v},$$

and

$$t = q_1^{b_1} q_2^{b_2} \cdots q_k^{b_k}.$$

Since $(r,s) = 1$, the primes occurring in the factorizations of r and s are distinct. Since $rs = t^2$, we have

$$p_1^{a_1} p_2^{a_2} \cdots p_u^{a_u} p_{u+1}^{a_{u+1}} p_{u+2}^{a_{u+2}} \cdots p_v^{a_v} = q_1^{2b_1} q_2^{2b_2} \cdots q_k^{2b_k}.$$

From the fundamental theorem of arithmetic, the prime-powers occurring on the two sides of the above equation are the same. Hence, each p_i must be equal to q_j for some j with matching exponents, so that $a_i = 2b_j$. Consequently, every exponent a_i is even, and therefore $a_i/2$ is an integer. We see that $r = m^2$ and $s = n^2$, where m and n are the integers

$$m = p_1^{a_1/2} p_2^{a_2/2} \cdots p_u^{a_u/2}$$

and

$$n = p_{u+1}^{a_{u+1}/2} p_{u+2}^{a_{u+2}/2} \cdots p_v^{a_v/2}.$$ ∎

We can now prove the desired result that describes all primitive Pythagorean triples.

Theorem 11.1. The positive integers x, y, z form a primitive Pythagorean triple, with y even, if and only if there are relatively prime positive integers m and n, $m > n$, with m odd and n even or m even and n odd, such that

$$x = m^2 - n^2$$
$$y = 2mn$$
$$z = m^2 + n^2.$$

Proof. Let x, y, z be a primitive Pythagorean triple. Lemma 11.2 tells us that x is odd and y is even, or *vice versa*. Since we have assumed that y is even, x and z are both odd. Hence, $z + x$ and $z - x$ are both even, so that there are positive integers r and s with $r = (z + x)/2$ and $s = (z - x)/2$.

Since $x^2 + y^2 = z^2$, we have $y^2 = z^2 - x^2 = (z + x)(z - x)$. Hence,

$$\left[\frac{y}{2}\right]^2 = \left[\frac{z+x}{2}\right]\left[\frac{z-x}{2}\right] = rs.$$

We note that $(r, s) = 1$. To see this, let $(r, s) = d$. Since $d \mid r$ and $d \mid s$, $d \mid (r + s) = z$ and $d \mid (r - s) = x$. This means that $d \mid (x, z) = 1$, so that $d = 1$.

Using Lemma 11.3, we see that there are positive integers m and n such that $r = m^2$ and $s = n^2$. Writing $x, y,$ and z in terms of m and n, we have

$$x = r - s = m^2 - n^2,$$
$$y = \sqrt{4rs} = \sqrt{4m^2n^2} = 2mn,$$
$$z = r + s = m^2 + n^2.$$

We also see that $(m, n) = 1$, since any common divisor of m and n must also divide $x = m^2 - n^2$, $y = 2mn$, and $z = m^2 + n^2$, and we know that $(x, y, z) = 1$. We also note that m and n cannot both be odd, for if they were, then x, y, and z would all be even, contradicting the condition $(x, y, z) = 1$. Since $(m, n) = 1$ and m and n cannot both be odd, we see m is even and n is odd, or *vice versa*. This shows that every primitive Pythagorean triple has the appropriate form.

To see that every triple

$$x = m^2 - n^2$$
$$y = 2mn$$
$$z = m^2 + n^2,$$

where m and n are positive integers, $m > n$, $(m, n) = 1$, and $m \not\equiv n \pmod 2$, forms a primitive Pythagorean triple, first note that

$$x^2 + y^2 = (m^2 - n^2)^2 + (2mn)^2$$
$$= (m^4 - 2m^2 n^2 + n^4) + 4m^2 n^2$$
$$= m^4 + 2m^2 n^2 + n^4$$
$$= (m^2 + n^2)^2$$
$$= z^2.$$

To see that these values of x, y, and z are mutually relatively prime, assume that $(x,y,z) = d > 1$. Then, there is a prime p such that $p \mid (x,y,z)$. We note that $p \neq 2$, since x is odd (because $x = m^2 - n^2$ where m^2 and n^2 have opposite parity). Also, note that because $p \mid x$ and $p \mid z$, $p \mid (z+x) = 2m^2$ and $p \mid (z-x) = 2n^2$. Hence $p \mid m$ and $p \mid n$, contradicting the fact that $(m,n) = 1$. Therefore, $(x,y,z) = 1$, and x, y, z is a primitive Pythagorean triple, concluding the proof. ∎

The following example illustrates the use of Theorem 11.1 to produce Pythagorean triples.

Example 11.3. Let $m = 5$ and $n = 2$, so that $(m,n) = 1$, $m \not\equiv n \pmod 2$, and $m > n$. Hence, Theorem 11.1 tells us that

$$x = m^2 - n^2 = 5^2 - 2^2 = 21$$
$$y = 2mn = 2 \cdot 5 \cdot 2 = 20$$
$$z = m^2 + n^2 = 5^2 + 2^2 = 29$$

is a primitive Pythagorean triple.   □

We list the primitive Pythagorean triples generated using Theorem 11.1 with $m \leq 6$ in Table 11.1.

m	n	$x = m^2 - n^2$	$y = 2mn$	$z = m^2 + n^2$
2	1	3	4	5
3	2	5	12	13
4	1	15	8	17
4	3	7	24	25
5	2	21	20	29
5	4	9	40	41
6	1	35	12	37
6	5	11	60	61

Table 11.1. Some Primitive Pythagorean Triples.

11.1 Exercises

1. Find all

   a) primitive Pythagorean triples x, y, z with $z \leq 40$.

   b) Pythagorean triples x, y, z with $z \leq 40$.

2. Show that if x, y, z is a primitive Pythagorean triple, then either x or y is divisible by 3.

3. Show that if x, y, z is a primitive Pythagorean triple, then exactly one of x, y, and z is divisible by 5.

4. Show that if x, y, z is a primitive Pythagorean triple, then at least one of x, y, and z is divisible by 4.

5. Show that every positive integer greater than three is part of at least one Pythagorean triple.

6. Let $x_1 = 3$, $y_1 = 4$, $z_1 = 5$, and let x_n, y_n, z_n, for $n = 2, 3, 4, \ldots$, be defined recursively by

$$
\begin{aligned}
x_{n+1} &= 3x_n + 2z_n + 1 \\
y_{n+1} &= 3x_n + 2z_n + 2 \\
z_{n+1} &= 4x_n + 3z_n + 2.
\end{aligned}
$$

Show that x_n, y_n, z_n is a Pythagorean triple.

7. Show that if x, y, z is a Pythagorean triple with $y = x + 1$, then x, y, z is one of the Pythagorean triples given in Exercise 6.

8. Find all solutions in positive integers of the diophantine equation $x^2 + 2y^2 = z^2$.

9. Find all solutions in positive integers of the diophantine equation $x^2 + 3y^2 = z^2$

* 10. Find all solutions in positive integers of the diophantine equation $w^2 + x^2 + y^2 = z^2$.

11. Find all Pythagorean triples containing the integer 12.

12. Find formulae for the integers of all Pythagorean triples x, y, z with $z = y + 1$.

13. Find formulae for the integers of all Pythagorean triples x, y, z with $z = y + 2$.

* 14. Show that the number of Pythagorean triples x, y, z (with $x^2 + y^2 = z^2$) with a fixed integer x is $(\tau(x^2) - 1)/2$ if x is odd, and $(\tau(x^2/4) - 1)/2$ if x is even.

* 15. Find all solutions in positive integers of the diophantine equation $x^2 + py^2 = z^2$, where p is a prime.

16. Find all solutions in positive integers of the diophantine equation $1/x^2 + 1/y^2 = 1/z^2$.

11.1 Computer Projects

Programming Projects

Write programs to do the following:

1. Find all Pythagorean triples x, y, z with x, y, and z less than a given bound.

2. Find all Pythagorean triples containing a given integer.

Computations and Explorations

Using the programs you have written or a computation program, carry out the following computations and explorations:

1. Find as many Pythagorean triples x, y, z as you can where each of x, y, and z is one less than the square of an integer. Do you think that there are infinitely many such triples?

11.2 Fermat's Last Theorem

In the previous section, we showed that the diophantine equation $x^2 + y^2 = z^2$ has infinitely many solutions in nonzero integers x, y, z. What happens when we replace the exponent two in this equation with an integer greater than two? Next to the discussion of the equation $x^2 + y^2 = z^2$ in his copy of the works of Diophantus, Fermat wrote in the margin:

"However, it is impossible to write a cube as the sum of two cubes, a fourth power as the sum of two fourth powers and in general any power the sum of two similar powers. For this I have discovered a truly wonderful proof, but the margin is too small to contain it."

Since Fermat made this statement many people have searched for a proof of this assertion without success. Even though no correct proof has yet been discovered, the following conjecture is known as *Fermat's last theorem*.

Fermat's Last Theorem. The diophantine equation

$$x^n + y^n = z^n$$

has no solutions in nonzero integers x, y, z when n is an integer with $n \geq 3$.

Currently, we know that Fermat's last theorem is true for all positive integers n with $3 \leq n \leq 1,000,000$. In this section, we will show that the special case of Fermat's last theorem with $n = 4$ is true. That is, we will show that the diophantine equation

$$x^4 + y^4 = z^4$$

has no solutions in nonzero integers x, y, z. Note that if we could also show that the diophantine equations

$$x^p + y^p = z^p$$

has no solutions in nonzero integers x,y,z whenever p is an odd prime, then we would know that Fermat's last theorem is true (see Exercise 2 at the end of this section).

The proof we will give of the case when $n = 4$ uses the *method of infinite descent* devised by Fermat. This method is an offshoot of the well-ordering property, and shows that a diophantine equation has no solutions by showing that for every solution there is a "smaller" solution, contradicting the well-ordering property.

Using the method of infinite descent we will show that the diophantine equation $x^4 + y^4 = z^2$. has no solutions in nonzero integers x, y, and z. This is stronger than showing that Fermat's last theorem is true for $n = 4$, because any solution of $x^4 + y^4 = z^4 = (z^2)^2$ gives a solution of $x^4 + y^4 = z^2$.

Theorem 11.2. The diophantine equation

$$x^4 + y^4 = z^2$$

has no solutions in nonzero integers x,y,z.

Proof. Assume that the above equation has a solution in nonzero integers x,y,z. Since we may replace any number of the variables with their negatives without changing the validity of the equation, we may assume that x,y,z are positive integers.

We may also suppose that $(x,y) = 1$. To see this, let $(x,y) = d$. Then $x = dx_1$ and $y = dy_1$, with $(x_1,y_1) = 1$, where x_1 and y_1 are positive integers. Since $x^4 + y^4 = z^2$, we have

$$(dx_1)^4 + (dy_1)^4 = z^2,$$

so that

$$d^4(x_1^4 + y_1^4) = z^2.$$

Hence $d^4 \mid z^2$, and, by problem 50 of Section 2.3, we know that $d^2 \mid z$. Therefore, $z = d^2z_1$, where z_1 is a positive integer. Thus,

$$d^4(x_1^4 + y_1^4) = (d^2z_1)^2 = d^4z_1^2,$$

so that

$$x_1^4 + y_1^4 = z_1^2.$$

This gives a solution of $x^4 + y^4 = z^2$ in positive integers $x = x_1, y = y_1, z = z_1$ with $(x_1,y_1) = 1$.

So suppose that $x = x_0, y = y_0, z = z_0$ is a solution of $x^4 + y^4 = z^2$, where x_0, y_0, and z_0 are positive integers with $(x_0,y_0) = 1$. We will show that there is another solution in positive integers $x = x_1$, $y = y_1$, $z = z_1$ with $(x_1,y_1) = 1$, such that $z_1 < z_0$.

Since $x_0^4 + y_0^4 = z_0^2$, we have

$$(x_0^2)^2 + (y_0^2)^2 = z_0^2,$$

so that x_0^2, y_0^2, z_0 is a Pythagorean triple. Furthermore, we have $(x_0^2, y_0^2) = 1$, for if p is a prime such that $p \mid x_0^2$ and $p \mid y_0^2$, then $p \mid x_0$ and $p \mid y_0$, contradicting the fact that $(x_0,y_0) = 1$. Hence, x_0^2, y_0^2, z_0 is a primitive Pythagorean triple, and by Theorem 11.1, we know that there are positive integers m and n with $(m,n) = 1$, $m \not\equiv n \pmod 2$, and

$$\begin{aligned} x_0^2 &= m^2 - n^2 \\ y_0^2 &= 2mn \\ z_0 &= m^2 + n^2, \end{aligned}$$

where we have interchanged x_0^2 and y_0^2, if necessary, to make y_0^2 the even integer of this pair.

From the equation for x_0^2, we see that

$$x_0^2 + n^2 = m^2.$$

Since $(m,n) = 1$, it follows that x_0, n, m is a primitive Pythagorean triple, m is odd and n is even. Again using Theorem 11.1, we see that there are positive integers r and s with $(r,s) = 1$, $r \not\equiv s \pmod 2$, and

$$\begin{aligned} x_0 &= r^2 - s^2 \\ n &= 2rs \\ m &= r^2 + s^2. \end{aligned}$$

Since m is odd and $(m,n) = 1$, we know that $(m,2n) = 1$. We note that because $y_0^2 = (2n)m$, Lemma 11.3 tells us that there are positive integers z_1 and w with $m = z_1^2$ and $2n = w^2$. Since w is even, $w = 2v$ where v is a positive integer, so that

$$v^2 = n/2 = rs.$$

Since $(r,s) = 1$, Lemma 11.3 tells us that there are positive integers x_1 and y_1 such that $r = x_1^2$ and $s = y_1^2$. Note that since $(r,s) = 1$, it easily follows that $(x_1,y_1) = 1$. Hence,

$$x_1^4 + y_1^4 = r^2 + s^2 = m = z_1^2$$

where x_1,y_1,z_1 are positive integers with $(x_1,y_1) = 1$. Moreover, we have $z_1 < z_0$, because

$$z_1 \le z_1^4 = m^2 < m^2 + n^2 = z_0.$$

To complete the proof, assume that $x^4 + y^4 = z^2$ has at least one integral solution. By the well-ordering property, we know that among the solutions in positive integers, there is a solution with the smallest value z_0 of the variable z. However, we have shown that from this solution we can find another solution with a smaller value of the variable z, leading to a contradiction. This completes the proof by the method of infinite descent. ∎

Readers interested in the history of Fermat's last theorem and how investigations relating to this conjecture led to the genesis of the theory of algebraic numbers are encouraged to consult the books of Edwards [14] and Ribenboim [35]. A great deal of research relating to Fermat's last theorem is underway. Recently, the German mathematician Faltings established a result that shows that for any fixed positive integer n, $n \geq 3$, the diophantine equation $x^n + y^n = z^n$ has at most a finite number of solutions where x, y, and z are integers and $(x, y) = 1$.

11.2 Exercises

1. Show that if x, y, z is a Pythagorean triple and n is an integer $n > 2$, then $x^n + y^n \neq z^n$.

2. Show that Fermat's last theorem is a consequence of Theorem 11.2, and the assertion that $x^p + y^p = z^p$ has no solutions in nonzero integers when p is an odd prime.

3. Using Fermat's little theorem, show that if p is prime and
   a) if $x^{p-1} + y^{p-1} = z^{p-1}$, then $p \mid xyz$.
   b) if $x^p + y^p = z^p$, then $p \mid (x + y - z)$.

4. Show that the diophantine equation $x^4 - y^4 = z^2$ has no solutions in nonzero integers using the method of infinite descent.

5. Using Exercise 4, show that the area of a right triangle with integer sides is never a perfect square.

* 6. Show that the diophantine equation $x^4 + 4y^4 = z^2$ has no solutions in nonzero integers.

* 7. Show that the diophantine equation $x^4 - 8y^4 = z^2$ has no solutions in nonzero integers.

* 8. Show that the diophantine equation $x^4 + 3y^4 = z^4$ has infinitely many solutions.

9. Show that in a Pythagorean triple there is at most one perfect square.

10. Show that the diophantine equation $x^2 + y^2 = z^3$ has infinitely many integer solutions by showing that for each positive integer k the integers $x = 3k^2 - 1$, $y = k(k^2 - 3)$, $z = k^2 + 1$ form a solution.

11. This exercise asks for a proof of a theorem proved by the French mathematician Sophie Germain* in 1805. Suppose that n is a prime greater than 2 and that p is an odd prime such that $p \mid xyz$ whenever x, y, and z are integers such that $x^n + y^n + z^n \equiv 0 \pmod{p}$. Further suppose that there are no solutions of the congruence $w^n \equiv n \pmod{p}$. Show that if x, y, and z are integers such that $x^n + y^n + z^n = 0$, then $n \mid xyz$.

12. Show that the diophantine equation $w^3 + x^3 + y^3 = z^3$ has infinitely many nontrivial solutions. (*Hint:* Take $w = 9zk^4$, $x = z(1 - 9k^3)$ $y = 3zk(1 - 3k^3)$, where z and k are nonzero integers.)

13. Can you find four consecutive positive integers such that the sum of the cubes of the first three is the cube of the fourth integer?

14. Prove that the diophantine equation $w^4 + x^4 = y^4 + z^4$ has infinitely many nontrivial solutions. (*Hint:* Follow Euler by taking $w = m^7 + m^5n^2 - 2m^3n^4 + 3m^2n^5 + mn^6$, $x = m^6n - 3m^5n^2 - 2m^4n^3 + m^2n^5 + n^7$, $y = m^7 + m^5n^2 - 2m^3n^4 - 3m^2n^5 + mn^6$, $z = m^6n + 3m^5n^2 - 2m^4n^3 + m^2n^5 + n^7$, where m and n are positive integers.)

11.2 Computer Projects

Programming Projects

Write programs to do the following:

1. Given a positive integer n search for solutions of the diophantine equation $x^n + y^n = z^n$.

2. Generate solutions of the diophantine equation $x^2 + y^2 = z^3$. (See Exercise 10 at the end of this section.)

Computations and Explorations

Using the programs you have written or a computation program, carry out the following computation and explorations:

---

* **SOPHIE GERMAIN (1776-1831)** was born in Paris and educated at home, using her father's extensive library as a resource. She decided, as a young teenager, to study mathematics when she discovered that Archimedes was murdered by the Romans. She started by reading the works of Euler and Newton. Although Germain did not attend classes, she learned from university course notes that she managed to obtain. After reading the notes from Lagrange's lectures, she send him a letter under the pseudonym Mr. LeBlanc. Lagrange, impressed with the insights displayed in this letter, decided to meet Mr. LeBlanc; he was surprised to find that its author was a young woman. Germain corresponded under the pseudonym Mr. LeBlanc with many mathematicians, including Legrende who included many of her discoveries in his book *Theorie des Nombres*. She also made important contributions to the mathematical theories of elasticity and acoustics. Gauss was impressed by her work and recommended she receive a doctorate from the University of Göttingen. Unfortunately she died just before she was to receive this degree.

1. Euler conjectured that no sum of fewer than n nth powers of nonzero integers is equal to the nth power of an integer. Show that this conjecture is false (as was shown in 1966 by Lander and Parkin) by finding four fifth powers of integers that is also the fifth power of an integer. Can you find other counterexamples to Euler's claim?

11.3 Sums of Squares

Mathematicians throughout history have been interested in problems about the representation of integers as sums of squares. Diophantus, Fermat, Euler, and Lagrange are some of the mathematicians who made important contributions to the solution of such problems. In this section we discuss two questions of this kind: Which integers are the sum of two squares? and What is the least integer n such that every positive integer is the sum of n squares?

We begin by considering the first question. Not every positive integer is the sum of two squares. In fact, n is not the sum of two squares if it is of the form $4k+3$. To see this, note that since $a^2 \equiv 0$ or $1 \pmod 4$ for every integer a, $x^2 + y^2 \equiv 0, 1,$ or $2 \pmod 4$.

To conjecture which integers are the sum of two squares we first examine some small positive integers.

Example 11.4. Among the first 20 positive integers note that

$1 = 0^2 + 1^2$,	11 is not the sum of two squares,
$2 = 1^2 + 1^2$,	12 is not the sum of two squares,
3 is not the sum of two squares,	$13 = 3^2 + 2^2$,
$4 = 2^2 + 0^2$,	14 is not the sum of two squares,
$5 = 1^2 + 2^2$,	15 is not the sum of two squares,
6 is not the sum of two squares,	$16 = 4^2 + 0^2$,
7 is not the sum of two squares,	$17 = 4^2 + 1^2$,
$8 = 2^2 + 2^2$,	$18 = 3^2 + 3^2$,
$9 = 3^2 + 0^2$,	19 is not the sum of two squares,
$10 = 3^2 + 1^2$,	$20 = 2^2 + 4^2$.

□

It is not immediately obvious from the evidence in Example 11.4 which integers are the sum of two squares. (Can you see anything in common among those positive integer not representable as the sum of two squares?)

We now begin a discussion which will show that the prime factorization of an integer determines whether this integer is the sum of two squares. There are two reasons for this. The first is that the product of two integers that are sums of two squares is again the sum of two squares. The second reason is that a prime is representable as the sum of two squares if and only if it is not of the form $4k + 3$.

We will prove both of these results. Then we will state and prove the theorem that specifies which integers are the sum of two squares.

The proof that the product of sums of two squares is again the sum of two squares relies on an important algebraic identity which we will use several times in this section.

Theorem 11.3. If m and n are both sums of two squares then mn is also the sum of two squares.

Proof. Let $m = a^2 + b^2$ and $n = c^2 + d^2$. Then

$$(11.2) \qquad mn = (a^2 + b^2)(c^2 + d^2) = (ac + bd)^2 + (ad - bc)^2.$$

The reader can easily verify this identity by expanding all the terms. ∎

Example 11.5. Since $5 = 2^2 + 1^2$ and $13 = 3^2 + 2^2$, it follows from (11.2) that

$$65 = 5 \cdot 13 = (2^2 + 1^2)(3^2 + 2^2) = \\ (2 \cdot 3 + 1 \cdot 2)^2 + (2 \cdot 2 - 1 \cdot 3)^2 = 8^2 + 1^2$$

□

One crucial result is that every prime of the form $4k + 1$ is the sum of two squares. To prove this result we will need the following lemma.

Lemma 11.4. If p is a prime of the form $4m + 1$ where m is an integer, then there exist integers x and y such that $x^2 + y^2 = kp$ for some positive integer k with $k < p$.

Proof. By Theorem 9.4 we know that -1 is a quadratic residue of p. Hence there is an integer a, $a < p$ such that $a^2 \equiv -1 \pmod{p}$. It follows that $a^2 + 1 = kp$ for some positive integer k. Hence $x^2 + y^2 = kp$ where $x = a$ and $y = 1$. From the inequality $kp = x^2 + y^2 < (p-1)^2 + 1 < p^2$ we see that $k < p$. ∎

We can now prove the following theorem which tells us that all primes not of the form $4k + 3$ are the sum of two squares.

Theorem 11.4. If p is a prime not of the form $4k + 3$ then there are integers x and y such that $x^2 + y^2 = p$.

Proof. Note that 2 is the sum of two squares since $1^2 + 1^2 = 2$. Now suppose that p is a prime of the form $4k + 1$. Let m be the smallest positive integer such that $x^2 + y^2 = mp$ has a solution in integers x and y. By Lemma 11.4 there is such an integer less than p; by the well-ordering property a least such integer exists. We will show that $m = 1$.

Assume that $m > 1$. Let a and b be defined by

$$a \equiv x \ (\text{mod } m), \ b \equiv y \ (\text{mod } m)$$

and

$$-m/2 < a \leq m/2, \ -m/2 < b \leq m/2.$$

It follows that $a^2 + b^2 \equiv x^2 + y^2 = mp \equiv 0 \ (\text{mod } m)$. Hence there is an integer k such that

$$a^2 + b^2 = km.$$

We have

$$(a^2 + b^2)(x^2 + y^2) = (km)(mp) = km^2 p.$$

By equation (11.2) we have

$$(a^2 + b^2)(x^2 + y^2) = (ax + by)^2 + (ay - bx)^2$$

Furthermore, since $a \equiv x \ (\text{mod } m)$ and $b \equiv y \ (\text{mod } m)$ we have

$$ax + by \equiv x^2 + y^2 \equiv 0 \ (\text{mod } m)$$
$$ay - bx \equiv xy - yx \equiv 0 \ (\text{mod } m).$$

Hence $(ax+by)/m$ and $(ay-bx)/m$ are integers, so that

$$\left[\frac{ax + by}{m}\right]^2 + \left[\frac{ax - by}{m}\right]^2 = km^2 p/m^2 = kp$$

is the sum of two squares. If we show that $0 < k < m$ this will contradict the choice of m as the minimum positive integer such that $x^2 + y^2 = mp$ has a solution in integers. We know that $a^2 + b^2 = km$, $-m/2 < a \leq m/2$, and $-m/2 < b \leq m/2$. Hence $a^2 \leq m^2/4$ and $b^2 \leq m^2/4$. We have

$$0 \leq km = a^2 + b^2 \leq 2(m^2/4) = m^2/2.$$

Consequently $0 \leq k \leq m/2$. It follows that $k < m$. All that remains to do is show that $k \neq 0$. If $k = 0$ we have $a^2 + b^2 = 0$. This implies that $a = b = 0$, so that $x \equiv y \equiv 0 \ (\text{mod } m)$, which shows that $m \mid x$ and $m \mid y$. Since $x^2 + y^2 = mp$ this implies that $m^2 \mid mp$ which implies that $m \mid p$. Since m is less than p, this implies that $m = 1$ which is what we wanted to prove. ∎

We can now put all the pieces together and prove the fundamental result that classifies the positive integers which are representable as the sum of two squares.

Theorem 11.5. The positive integer n is the sum of two squares if and only if each prime factor of n of the form $4k+3$ occurs to an even power in the prime factorization of n.

Proof. Suppose that in the prime factorization of n there are no primes of the form $4k+3$ that appear to an odd power. Write $n = t^2 u$ where u is the product of primes. No primes of the form $4k+3$ appear in u. By Theorem 11.4 each prime in u

can be written as the sum of two squares. Applying Theorem 11.3 one fewer time than the number of different primes in u shows that u is also the sum of two squares, say

$$u = x^2 + y^2.$$

It then follows that n is also the sum of two squares, namely

$$n = (tx)^2 + (ty)^2.$$

Now suppose that there is a prime p, $p \equiv 3 \pmod 4$, that occurs in the prime factorization of n to an odd power, say the $(2j+1)$th power. Furthermore, suppose that n is the sum of two squares, that is,

$$n = x^2 + y^2.$$

Let $(x,y)=d$, $a = x/d$, $b = y/d$, and $m = n/d^2$. It follows that $(a,b) = 1$ and

$$a^2 + b^2 = m.$$

Suppose that p^k is the largest power of p dividing d. Then m is divisible by $p^{2j-2k+1}$ and $2j-2k+1$ is at least 1 since it is nonnegative. Hence $p \mid m$. We know that p does not divide a, for if $p \mid a$, then $p \mid b$ since $b^2 = m - a^2$ and $(a,b) = 1$.

Thus there is an integer z such that $az \equiv b \pmod p$. It follows that

$$a^2 + b^2 \equiv a^2 + (az)^2 = a^2(1 + z^2)\pmod p.$$

Since $a^2 + b^2 = m$ and $p \mid m$, we see that

$$a^2(1 + z^2) \equiv 0 \pmod p.$$

Because $(a,p) = 1$, it follows that $1 + z^2 \equiv 0 \pmod p$. This implies that $z^2 \equiv -1 \pmod p$ which is impossible since -1 is not a quadratic residue of p since $p \equiv 3 \pmod 4$. This contradiction shows that n could not have been the sum of two squares. ∎

Since there are positive integers not representable as the sum of two squares, we can ask whether every positive integer is the sum of three squares. The answer is no, since it is impossible to write 7 as the sum of three squares (as the reader should show). Since three squares do not suffice, we ask whether four squares do. The answer to this is yes, as we will show. Fermat wrote that he had a proof of this fact, although he never published it (and most historians of mathematics believe he actually had such a proof). Euler was unable to find a proof, although he made substantial progress towards a solution. It was in 1770 when Lagrange presented the first published solution.

The proof that every positive integer is the sum of four squares depends on the following theorem which shows that the product of two integers both representable as the sum of four squares can also be so represented. Just as for the analogous result for two squares there is an important algebraic identity used in the proof.

Theorem 11.6. If m and n are positive integers which are both the sum of four squares, then mn is also the sum of four squares.

Proof. Let $m = a^2 + b^2 + c^2 + d^2$ and $n = e^2 + f^2 + g^2 + h^2$. That mn is also the sum of four squares follows from the following algebraic identity:

$$
\begin{aligned}
(11.3) \quad mn &= (a^2 + b^2 + c^2 + d^2)(e^2 + f^2 + g^2 + h^2) \\
&= (ae + bf + cg + dh)^2 + (af - be + ch - dg)^2 + \\
&\quad (ag - bh - ce + df)^2 + (ah + bg - cf - de)^2.
\end{aligned}
$$

The reader can easily verify this identity by multiplying all the terms. ∎

We illustrate the use of Theorem 11.6 with an example.

Example 11.6. Since $7 = 2^2 + 1^2 + 1^2 + 1^2$ and $10 = 3^2 + 1^2 + 0^2 + 0^2$ from (11.3) it follows that

$$
\begin{aligned}
70 = 7 \cdot 10 &= (2^2 + 1^2 + 1^2 + 1^2)(3^2 + 1^2 + 0^2 + 0^2) \\
&= (2 \cdot 3 + 1 \cdot 1 + 1 \cdot 0 + 1 \cdot 0)^2 + (2 \cdot 1 - 1 \cdot 3 + 1 \cdot 0 - 1 \cdot 0)^2 + \\
&\quad (2 \cdot 0 - 1 \cdot 0 - 1 \cdot 3 + 1 \cdot 1)^2 + (2 \cdot 0 + 1 \cdot 0 - 1 \cdot 1 - 1 \cdot 3)^2 \\
&= 7^2 + 1^2 + 2^2 + 4^2.
\end{aligned}
$$

□

We will now begin our work to show that every prime is the sum of four squares. We begin with a lemma.

Lemma 11.5. If p is an odd prime then there exists an integer k, $k < p$, such that

$$ kp = x^2 + y^2 + z^2 + w^2 $$

has a solution in integers x, y, z, and w.

Proof. We will first show that there are integers x and y such that

$$ x^2 + y^2 + 1 \equiv 0 \pmod{p} $$

with $0 \le x < p/2$ and $0 \le y < p/2$.

Let

$$ S = \{0^2, 1^2, \ldots, (\frac{p-1}{2})^2\} $$

and

$$ T = \{-1 - 0^2, -1 - 1^2, \ldots, -1 - (\frac{p-1}{2})^2\}. $$

No two elements of S are congruent modulo p (since $x^2 \equiv y^2 \pmod{p}$ implies that

$x \equiv \pm y \pmod{p}$). Likewise, no two elements of T are congruent modulo p. It is easy to see that the set $S \cup T$ contains $p+1$ distinct integers. By the pigeonhole principle there are two integers in this union that are congruent modulo p. It follows that there are integers x and y with $x^2 \equiv -1-y^2 \pmod{p}$ with $0 \le x \le (p-1)/2$ and $0 \le y < (p-1)/2$. We have

$$x^2 + y^2 + 1 \equiv 0 \pmod{p}.$$

Hence, $x^2 + y^2 + 1^2 + 0^2 = kp$ for some integer k. Since $x^2 + y^2 + 1 < 2((p-1)/2)^2 + 1 < p^2$, it follows that $k < p$. ∎

We can now prove that every prime is the sum of four squares.

Theorem 11.7. Let p be a prime. Then the equation $x^2 + y^2 + z^2 + w^2 = p$ has a solution where $x,y,z,$ and w are integers.

Proof. The result is true when $p = 2$, since $2 = 1^2 + 1^2 + 0^2 + 0^2$. Now assume that p is an odd prime. Let m be the smallest integer such that $x^2 + y^2 + z^2 + w^2 = mp$ has a solution where $x,y,z,$ and w are integers. (By Lemma 11.5 such integers exist and by the well-ordering property there is a minimal such integer.) The theorem will follow if we can show that $m=1$. To do this we assume that $m > 1$ and find a smaller such integer.

If m is even, then either all of x, y, z, and w are odd, all are even, or two are odd and two are even. In all these cases we can rearrange these integers (if necessary) so that $x \equiv y \pmod{2}$ and $z \equiv w \pmod{2}$. It then follows that $(x-y)/2$, $(x+y)/2$, $(z-w)/2$, and $(z+w)/2$ are integers and

$$\left[\frac{x-y}{2}\right]^2 + \left[\frac{x+y}{2}\right]^2 + \left[\frac{z-w}{2}\right]^2 + \left[\frac{z+w}{2}\right]^2 = (m/2)p.$$

This contradicts the minimality of m.

Now suppose that m is odd and $m > 1$. Let $a, b, c,$ and d be integers such that

$$a \equiv x \pmod{m}, \ b \equiv y \pmod{m}, \ c \equiv z \pmod{m}, \ d \equiv w \pmod{m},$$

and

$$-m/2 < a < m/2, \ -m/2 < b < m/2, \ -m/2 < c < m/2, \ -m/2 < d < m/2.$$

We have

$$a^2 + b^2 + c^2 + d^2 \equiv x^2 + y^2 + z^2 + w^2 \pmod{m}.$$

Hence

$$a^2 + b^2 + c^2 + d^2 = km.$$

for some integer k and

$$0 \leq a^2 + b^2 + c^2 + d^2 < 4(m/2)^2 = m^2.$$

Consequently, $0 \leq k < m$. If $k = 0$ we have $a = b = c = d = 0$ so that $x \equiv y \equiv z \equiv w \equiv 0 \pmod{m}$. From this it follows that $m^2 \mid mp$ which is impossible since $1 < m < p$. It follows that $k > 0$.

We have

$$(x^2 + y^2 + z^2 + w^2)(a^2 + b^2 + c^2 + d^2) = mp \cdot km = m^2 kp.$$

But by the identity in the proof of Theorem 11.6 we have

$$(ax + by + cz + dw)^2 + (bx - ay + dz - cw)^2 + (cx - dy - az + bw)^2 + (dx + cy - bz - aw)^2 = m^2 kp.$$

Each of the four terms being squared is divisible by m since

$$ax + by + cz + dw \equiv x^2 + y^2 + z^2 + w^2 \equiv 0 \pmod{m}$$
$$bx - ay + dz - cw \equiv yx - xy + wz - zw \equiv 0 \pmod{m}$$
$$cx - dy - az + bw \equiv zx - wy - xz + yw \equiv 0 \pmod{m}$$
$$dx + cy - bz - aw \equiv wx + zy - yz - xw \equiv 0 \pmod{m}.$$

Let $X, Y, Z,$ and W be the integers obtained by dividing these quantities by m, that is

$$X = (ax + by + cz + dw)/m$$
$$Y = (bx - ay + dz - cw)/m$$
$$Z = (cx - dy - az + bw)/m$$
$$W = (dx + cy - bz - aw)/m.$$

It then follows that

$$X^2 + Y^2 + Z^2 + W^2 = m^2 kp/m^2 = kp.$$

But this contradicts the choice of m. Hence m must be 1. ∎

We now can state and prove the fundamental theorem about representations of integers as sums of four squares.

Theorem 11.8. Every positive integer is the sum of the squares of four integers.

Proof. Suppose that n is a positive integer. Then by the fundamental theorem of arithmetic n is the product of primes. By Theorem 11.7 each of these prime factors can be written as the sum of four squares. Applying Theorem 11.6 a sufficient number of times it follows that n is also the sum of four squares. ∎

We have shown that every positive integer can be written as the sum of four squares. As we have mentioned this theorem was originally proved by Lagrange in 1770. Around the same time the English mathematician Waring generalized this problem. He stated, but did not prove, that every positive integer is the sum of 9 cubes of nonnegative integers, the sum of 19 fourth powers of nonnegative integers, and so on. We can phrase this conjecture in the following way.

Waring's Problem. If k is a positive integer is there an integer $g(k)$ such that every positive integer can be written as the sum of $g(k)$ kth powers of nonnegative integers, and no smaller number of kth powers will suffice?

Lagrange's theorem shows that we can take $g(2) = 4$ (since there are integers that are not the sum of three squares). In the nineteenth century mathematicians showed that such an integer $g(k)$ exists for $3 \leq k \leq 8$ and $k = 10$. But it was not until 1906 that the famous German mathematician David Hilbert showed that for every positive integer k there is a constant $g(k)$ such that every positive integer may be expressed as the sum of $g(k)$ kth powers of nonnegative integers. Hilbert's proof is extremely complicated and is not constructive, so that it gives no formula for $g(k)$. It is now known that $g(3) = 9, g(4) \geq 19, g(5) = 37,$ and

$$g(k) = [(3/2)^k] + 2^k - 2$$

for $6 \leq k \leq 471600000$. Proofs of these results rely on nonelementary results from analytical number theory. There are still many unanswered questions about the values of $g(k)$. The interested reader can learn about recent results on Waring's problem by consulting the numerous articles on this problem described in [26]. The 1990 paper of Wunderlich and Kubina [137] established the upper limit of the range for which it has been verified that $g(k)$ is given by this formula.

11.3 Exercises

1. Given that $13 = 3^2 + 2^2$, $29 = 5^2 + 2^2$, and $50 = 7^2 + 1^2$, write each of the following integers as the sum of two squares.

    a) $377 = 13 \cdot 29$ c) $1450 = 29 \cdot 50$
    b) $650 = 13 \cdot 50$ d) $18850 = 13 \cdot 29 \cdot 50$

2. Determine whether each of the following integers can be written as the sum of two squares.

    a) 19   d) 45   g) 99

b) 25   e) 65   h) 999
c) 29   f) 80   i) 1000

3. Represent each of the following integers as the sum of two squares.

   a) 34   d) 490
   b) 90   e) 21658
   c) 100   f) 324608

4. Show that a positive integer is the difference of two squares if and only if it is not of the form $4k + 2$ where k is an integer.

5. Represent each of the following integers as the sum of three squares if possible.

   a) 3   d) 18
   b) 7   e) 23
   c) 11   f) 28

6. Show that the positive integer n is not the sum of three squares of integers if n is of the form $8k + 7$ where k is an integer.

7. Show that the positive integer n is not the sum of three squares of integers if n is of the form $4^m(8k + 7)$ where m and k are nonnegative integers.

8. Prove or disprove that the sum of two integers representable as the sum of three squares of integers is also so representable.

9. Given that $7 = 2^2 + 1^2 + 1^2 + 1^2$, $15 = 3^2 + 2^2 + 1^2 + 1^2$, and $34 = 4^2 + 4^2 + 1^2 + 1^2$ write each of the following integers as the sum of four squares.

   a) $105 = 7 \cdot 15$ c) $238 = 7 \cdot 34$
   b) $510 = 15 \cdot 34$ d) $3570 = 7 \cdot 15 \cdot 34$

10. Write each of the following positive integers as the sum of four squares.

    a) 6   d) 89
    b) 12   e) 99
    c) 21   f) 555

11. Show that every integer n, $n \geq 170$, is the sum of the squares of five positive integers. (*Hint*: Write $m = n - 169$ as the sum of the squares of four integers, and use the fact that $169 = 13^2 = 12^2 + 5^2 = 12^2 + 4^2 + 3^2 = 10^2 + 8^2 + 2^2 + 1^2$.)

12. Show that the only positive integers that are not expressible as the sum of five squares of positive integers are 1,2,3,4,6,7,9,10,12,15,18,33. (*Hint:* Use Exercise 11, show that each of these integers cannot be expressed as stated, and then show all remaining positive integers less than 170 can be expressed as stated.)

* 13. Show that there are arbitrarily large integers that are not the sum of the squares of four positive integers.

We outline a second proof for Theorem 11.8 in Exercises 14-15.

* 14. Show that if p is prime and a is an integer not divisible by p, then there exist integers x and y such that $ax \equiv y \pmod{p}$ with $0 < |x| < \sqrt{p}$ and $0 < |y| < \sqrt{p}$. This result is called *Thue's Lemma*. (*Hint:* Use the pigeonhole principle to show that there are two integers of the form $au - v$ with $0 \le u \le [\sqrt{p}\,]$ and $0 \le v \le [\sqrt{p}\,]$ that are congruent modulo p. Construct x and y from the two values of u and the two values of v, respectively.)

15. Use Exercise 14 to prove Theorem 11.8. (*Hint:* Show that there is an integer a with $a^2 \equiv -1 \pmod{p}$. Then apply Thue's Lemma with this value of a.)

16. Show that 23 is the sum of nine cubes of nonnegative integers but not the sum of eight cubes of nonnegative integers.

Exercises 17-21 give an elementary proof that $g(4) \le 50$.

17. Show that

$$\sum_{1 \le i < j \le 4} \left[(x_i + x_j)^4 + (x_i - x_j)^4 \right] = 6 \left[\sum_{k=1}^{4} x_k^2 \right]^2.$$

(*Hint:* Start with the identity $(x_i + x_j)^4 + (x_i - x_j)^4 = 2x_i^4 + 12x_i^2 x_j^2 + 2x_j^4$.)

18. Show from Exercise 17 that every integer of the form $6n^2$, where n is a positive integer, is the sum of 12 fourth powers.

19. Use Exercise 18 and the fact that every positive integer is the sum of four squares to show that every positive integer of the form $6m$, where m is a positive integer, can be written as the sum of 48 fourth powers.

20. Show that the integers $0,1,2,81,16,17$ form a complete system of residues modulo 6 each of which is the sum of at most two fourth powers. Show from this that every integer n with $n > 81$ can be written as $6m + k$ where m is a positive integer and k comes from this complete system of residues. Conclude from this that every integer n with $n > 81$ is the sum of 50 fourth powers.

21. Show that every positive integer n with $n \le 81$ is the sum of at most 50 fourth powers. (*Hint:* For $51 \le n \le 81$ start by using three terms equal to 2^4.) Conclude from this exercise and Exercise 20 that $g(4) \le 50$.

11.3 Computer Projects

Programming Projects

Write programs to do the following:

* **1.** Determine whether a positive integer n can be represented as the sum of two squares and so represent it if possible.

* **2.** Given a positive integer n, represent n as the sum of four squares.

Computations and Exploration

Using the programs you have written or a computation program, carry out the following computations and explorations:

1. Find the number of ways that each integer less than 100 can be written as the sum of two squares. (Count the sum $(\pm x^2) + (\pm y^2)$ four times, once for each choice of signs.)

2. Using numerical evidence, make a conjecture concerning which positive integers can be expressed as the sum of three squares. (Be sure to consult Exercise 7.)

3. Explore which positive integers can be written as the sum of n cubes of nonnegative integers for $n = 2,3,4,5,6,7,8$.

11.4 Pell's Equation

In this section, we study diophantine equations of the form

(11.4) $$x^2 - dy^2 = n,$$

where d and n are fixed integers. When $d < 0$ and $n < 0$, there are no solutions of (11.4). When $d < 0$ and $n > 0$, there can be at most a finite number of solutions, since the equation $x^2 - dy^2 = n$ implies that $|x| \leq \sqrt{n}$ and $|y| \leq \sqrt{n/|d|}$. Also, note that when d is a perfect square, say $d = D^2$, then

$$x^2 - dy^2 = x^2 - D^2y = (x+Dy)(x-Dy) = n .$$

Hence, any solution of (11.4), when d is a perfect square, corresponds to a simultaneous solution of the equations

$$x + Dy = a$$
$$x - Dy = b,$$

where a and b are integers such that $n = ab$. In this case, there are only a finite number of solutions, since there is at most one solution in integers of these two equations for each factorization $n = ab$.

For the rest of this section, we are interested in the diophantine equation $x^2 - dy^2 = n$, where d and n are integers and d is a positive integer which is not a perfect square. As the following theorem shows, the simple continued fraction of $\sqrt{d}$ is very useful for the study of this equation.

Theorem 11.9. Let d and n be integers such that $d > 0$, d is not a perfect square, and $|n| < \sqrt{d}$. If $x^2 - dy^2 = n$, then x/y is a convergent of the simple continued fraction of $\sqrt{d}$.

Proof. First consider the case where $n > 0$. Since $x^2 - dy^2 = n$, we see that

(11.5) $(x + y\sqrt{d})(x - y\sqrt{d}) = n.$

From (11.5), we see that $x - y\sqrt{d} > 0$, so that $x > y\sqrt{d}$. Consequently,

$$\frac{x}{y} - \sqrt{d} > 0,$$

and since $0 < n < \sqrt{d}$, we see that

$$\frac{x}{y} - \sqrt{d} = \frac{(x - \sqrt{d}\,y)}{y}$$

$$= \frac{x^2 - dy^2}{y(x + y\sqrt{d})}$$

$$< \frac{n}{y(2y\sqrt{d})}$$

$$< \frac{\sqrt{d}}{2y^2\sqrt{d}}$$

$$= \frac{1}{2y^2}.$$

Since $0 < \dfrac{x}{y} - \sqrt{d} < \dfrac{1}{2y^2}$, Theorem 10.16 tells us that x/y must be a convergent of the simple continued fraction of $\sqrt{d}$.

When $n < 0$, we divide both sides of $x^2 - dy^2 = n$ by $-d$, to obtain

$$y^2 - (1/d)x^2 = -n/d.$$

By a similar argument to that given when $n > 0$, we see that y/x is a convergent of the simple continued fraction expansion of $1/\sqrt{d}$. Therefore, from Exercise 7 of Section 10.3, we know that $x/y = 1/(y/x)$ must be a convergent of the simple continued fraction of $\sqrt{d} = 1/(1/\sqrt{d})$.   ∎

We have shown that solutions of the diophantine equation $x^2 - dy^2 = n$, where $|n| < \sqrt{d}$, are given by the convergents of the simple continued fraction expansion of $\sqrt{d}$. We will restate Theorem 10.20 here since it will help us use these convergents to find solutions of this diophantine equation.

Theorem 10.20. Let d be a positive integer that is not a perfect square. Define $\alpha_k = (P_k + \sqrt{d})/Q_k$, $a_k = [\alpha_k]$, $P_{k+1} = a_k Q_k - P_k$, $Q_{k+1} = (d - P_{k+1}^2)/Q_k$, for $k = 0,1,2,...$ where $\alpha_0 = \sqrt{d}$. Furthermore, let p_k/q_k denote the kth convergent of the simple continued fraction expansion of $\sqrt{d}$. Then

$$p_k^2 - dq_k^2 = (-1)^{k-1} Q_{k+1}.$$

The special case of the diophantine equation $x^2 - dy^2 = n$ with $n = 1$ is called *Pell's equation*, after John Pell*. Although Pell played an important role in the mathematical community of his day, he played only a minor part in solving the equation named in his honor. The problem of finding the solutions of this equation has a long history. Special cases of Pell's equation are discussed in ancient works by Archimedes and Diophantus. Moreover, the twelfth century Indian mathematician Bhaskara described a method for finding the solutions of Pell's equation. In more recent times, in a letter written in 1657 Fermat posed to the "mathematicians of Europe" the problem of showing that there are infinitely many integral solutions of the equation $x^2 - dy^2 = 1$ when d is a positive integer greater than 1 which is not a square. Soon afterwards, the English mathematicians Wallis and Brouncker developed a method to find these solutions, but did not provide a proof that their method works. Euler provided all the theory needed for a proof in a paper published in 1767 and Lagrange published such a proof in 1768. The methods of Wallis and Brouncker, Euler, and Lagrange all are related to the use of the continued fraction of $\sqrt{d}$. We will show how this continued fraction is used to find the solutions of Pell's equation. In particular, we will use Theorems 11.9 and 10.20 to find all solutions of Pell's equation and the related equation $x^2 - dy^2 = -1$.

Theorem 11.10. Let d be a positive integer that is not a perfect square. Let p_k/q_k denote the kth convergent of the simple continued fraction of $\sqrt{d}$, $k = 1,2,3,...$ and let n be the period length of this continued fraction. Then, when n is even, the positive solutions of the diophantine equation $x^2 - dy^2 = 1$ are $x = p_{jn-1}$, $y = q_{jn-1}$, $j = 1,2,3,...$, and the diophantine equation $x^2 - dy^2 = -1$ has no solutions. When n is odd, the positive solutions of $x^2 - dy^2 = 1$ are $x = p_{2jn-1}, y = q_{2jn-1}$, $j = 1,2,3,...$ and the solutions of $x^2 - dy^2 = -1$ are $x = p_{(2j-1)n-1}$, $y = q_{(2j-1)n-1}, j = 1,2,3,....$

---

* **JOHN PELL (1611-1683)**, the son of a clergyman, was born in Sussex, England and was educated at Trinity College, Cambridge. He became a schoolmaster instead of following his father's wishes that he enter the clergy. After developing a reputation for scholarship in both mathematics and languages, he took a position at the University of Amsterdam. He remained there until, at the request of the Prince of Orange, he joined the faculty of a new college at Breda. Among Pell's writings in mathematics are a book *Idea of Mathematics*, as well as many pamphlets and articles. He corresponded and discussed mathematics with the leading mathematicians of his day, including Leibniz and Newton, the inventors of calculus. Euler may have called $x^2 - dy^2 = 1$ Pell's equation because he was familiar with a book in which Pell augmented the work of other mathematicians on the solutions of the equation $x^2 - 12y^2 = n$.

Pell was involved with diplomacy; he served in Switzerland as an agent of Oliver Cromwell and he joined the English diplomatic service in 1654. He finally decided to join the clergy in 1661, when he took his holy orders and became chaplain to the Bishop of London. Unfortunately, at the time of his death Pell was living in abject poverty.

Proof. Theorem 11.9 tells us that if x_0, y_0 is a positive solution of $x^2 - dy^2 = \pm 1$, then $x_0 = p_k$, $y_0 = q_k$ where p_k/q_k is a convergent of the simple continued fraction of $\sqrt{d}$. On the other hand, from Theorem 10.20 we know that

$$p_k^2 - dq_k^2 = (-1)^{k-1} Q_{k+1},$$

where Q_{k+1} is as defined in the statement of Theorem 10.20.

Because the period of the continued expansion of $\sqrt{d}$ is n, we know that $Q_{jn} = Q_0 = 1$ for $j = 1, 2, 3, \ldots$, (since $\sqrt{d} = \dfrac{P_0 + \sqrt{d}}{Q_0}$). Hence,

$$p_{jn-1}^2 - d\, q_{jn-1}^2 = (-1)^{jn} Q_{nj} = (-1)^{jn}.$$

This equation shows that when n is even p_{jn-1}, q_{jn-1} is a solution of $x^2 - dy^2 = 1$ for $j = 1, 2, 3, \ldots$, and when n is odd, p_{2jn-1}, q_{2jn-1} is a solution of $x^2 - dy^2 = 1$ and $p_{2(j-1)n-1}, q_{2(j-1)n-1}$ is a solution of $x^2 - dy^2 = -1$ for $j = 1, 2, 3, \ldots$.

To show that the diophantine equations $x^2 - dy^2 = 1$ and $x^2 - dy^2 = -1$ have no solutions other than those already found, we will show that $Q_{k+1} = 1$ implies that $n \mid k$ and that $Q_j \neq -1$ for $j = 1, 2, 3 \ldots$.

We first note that if $Q_{k+1} = 1$, then

$$\alpha_{k+1} = P_{k+1} + \sqrt{d}.$$

Since $\alpha_{k+1} = [a_{k+1}; a_{k+2}, \ldots]$, the continued fraction expansion of α_{k+1} is purely periodic. Hence, Theorem 10.19 tells us that $-1 < \alpha_{k+1}' = P_{k+1} - \sqrt{d} < 0$. This implies that $P_{k+1} = [\sqrt{d}]$, so that $\alpha_k = \alpha_0$, and $n \mid k$.

To see that $Q_j \neq -1$ for $j = 1, 2, 3, \ldots$, note that $Q_j = -1$ implies that $\alpha_j = -P_j - \sqrt{d}$. Since α_j has a purely periodic simple continued fraction expansion, we know that

$$-1 < \alpha_j' = -P_j + \sqrt{d} < 0$$

and

$$\alpha_j = -P_j - \sqrt{d} > 1.$$

From the first of these inequalities, we see that $P_j > -\sqrt{d}$ and, from the second, we see that $P_j < -1 - \sqrt{d}$. Since these two inequalities for p_j are contradictory, we see that $Q_j \neq -1$.

Since we have found all solutions of $x^2 - dy^2 = 1$ and $x^2 - dy^2 = -1$, where x and y are positive integers, we have completed the proof. ∎

We illustrate the use of Theorem 11.5 with the following examples.

Example 11.7. Since the simple continued fraction of $\sqrt{13}$ is $[3; \overline{1,1,1,1,6}]$ the positive solutions of the diophantine equation $x^2 - 13y^2 = 1$ are p_{10j-1}, q_{10j-1}, $j = 1, 2, 3, \ldots$ where p_{10j-1}/q_{10j-1} is the $(10j-1)$th convergent of the simple continued fraction expansion of $\sqrt{13}$. The least positive solution is $p_9 = 649$, $q_9 = 180$. The positive solutions of the diophantine equation $x^2 - 13y^2 = -1$ are $p_{10j-6}, q_{10j-6}, j = 1,2,3,...$; the least positive solution is $p_4 = 18, q_4 = 5$. □

Example 11.8. Since the continued fraction of $\sqrt{14}$ is $[3;\overline{1,2,1,6}]$, the positive solutions of $x^2 - 14y^2 = 1$ are p_{4j-1}, q_{4j-1}, $j = 1,2,3,...$ where p_{4j-1}/q_{4j-1} is the jth convergent of the simple continued fraction expansion of $\sqrt{14}$. The least positive solution is $p_3 = 15$, $q_3 = 4$. The diophantine equation $x^2 - 14y^2 = -1$ has no solutions, since the period length of the simple continued fraction expansion of $\sqrt{14}$ is even. □

We conclude this section with the following theorem that shows how to find all the positive solutions of Pell's equation $x^2 - dy^2 = 1$ from the least positive solution, without finding subsequent convergents of the continued fraction expansion of $\sqrt{d}$.

Theorem 11.11. Let x_1, y_1 be the least positive solution of the diophantine equation $x^2 - dy^2 = 1$, where d is a positive integer that is not a perfect square. Then all positive solutions x_k, y_k are given by

$$x_k + y_k\sqrt{d} = (x_1 + y_1\sqrt{d})^k$$

for $k = 1,2,3,...$. (Note that x_k and y_k are determined by the use of Lemma 11.4).

Proof. We need to show that x_k, y_k is a solution for $k = 1,2,3,...$ and that every solution is of this form.

To show that x_k, y_k is a solution, first note that by taking conjugates, it follows that $x_k - y_k\sqrt{d} = (x_1 - y_1\sqrt{d})^k$, because from Lemma 10.4, the conjugate of a power is the power of the conjugate. Now, note that

$$\begin{aligned}
x_k^2 - dy_k^2 &= (x_k + y_k\sqrt{d})(x_k - y_k\sqrt{d}) \\
&= (x_1 + y_1\sqrt{d})^k(x_1 - y_1\sqrt{d})^k \\
&= (x_1^2 - dy_1^2)^k \\
&= 1.
\end{aligned}$$

Hence x_k, y_k is a solution for $k = 1,2,3,...$.

To show that every positive solution is equal to x_k, y_k for some positive integer k, assume that X, Y is a positive solution different from x_k, y_k for $k = 1,2,3,...$. Then there is an integer n such that

$$(x_1 + y_1\sqrt{d})^n < X + Y\sqrt{d} < (x_1 + y_1\sqrt{d})^{n+1}.$$

When we multiply this inequality by $(x_1 + y_1\sqrt{d})^{-n}$, we obtain

$$1 < (x_1 - y_1\sqrt{d})^n(X + Y\sqrt{d}) < x_1 + y_1\sqrt{d},$$

since $x_1^2 - dy_1^2 = 1$ implies that $x_1 - y_1\sqrt{d} = (x_1 + y_1\sqrt{d})^{-1}$.

Now let

$$s + t\sqrt{d} = (x_1 - y_1\sqrt{d})^n(X + Y\sqrt{d}),$$

and note that

$$
\begin{aligned}
s^2 - dt^2 &= (s - t\sqrt{d})(s + t\sqrt{d}) \\
&= (x_1 + y_1\sqrt{d})^n(X - Y\sqrt{d})(x_1 - y_1\sqrt{d})^n(X + Y\sqrt{d}) \\
&= (x_1^2 - dy_1^2)^n(X^2 - dY^2) \\
&= 1.
\end{aligned}
$$

We see that s, t is a solution of $x^2 - dy^2 = 1$, and furthermore, we know that $1 < s + t\sqrt{d} < x_1 + y_1\sqrt{d}$. Moreover, since we know that $s + t\sqrt{d} > 1$, we see that $0 < (s + t\sqrt{d})^{-1} < 1$. Hence

$$s = \frac{1}{2}[(s + t\sqrt{d}) + (s - t\sqrt{d})] > 0$$

and

$$t = \frac{1}{2\sqrt{d}}[(s + t\sqrt{d}) - (s - t\sqrt{d})] > 0.$$

This means that s, t is a positive solution, so that $s \geq x_1$, and $t \geq y_1$, by the choice of x_1, y_1 as the smallest positive solution. But this contradicts the inequality $s + t\sqrt{d} < x_1 + y_1\sqrt{d}$. Therefore X, Y must be x_k, y_k for some choice of k. ∎

The following example illustrates the use of Theorem 11.11.

Example 11.9. From a previous example we know that the least positive solution of the diophantine equation $x^2 - 13y^2 = 1$ is $x_1 = 649$, $y_1 = 180$. Hence, all positive solutions are given by x_k, y_k where

$$x_k + y_k\sqrt{13} = (649 + 180\sqrt{13})^k.$$

For instance, we have

$$x_2 + y_2\sqrt{13} = 842401 + 233640\sqrt{13}$$

Hence $x_2 = 842401$, $y_2 = 233640$ is the least positive solution of $x^2 - 13y^2 = 1$, other than $x_1 = 649$, $y_1 = 180$. □

11.4 Exercises

1. Find all the solutions where x and y are integers of each of the following equations.

   a) $x^2 + 3y^2 = 4$

   b) $x^2 + 5y^2 = 7$

c) $2x^2 + 7y^2 = 30$

2. Find all the solutions where x and y are integers of each of the following equations.

a) $x^2 - y^2 = 8$

b) $x^2 - 4y^2 = 40$

c) $4x^2 - 9y^2 = 100$

3. For which of the following values of n does the diophantine equation $x^2 - 31y^2 = n$ have a solution?

a)   1 d) -3
b) -1 e)   4
c)   2 f) -45

4. Find the least positive solution in integers of each of the following diophantine equations.

a) $x^2 - 29y^2 = -1$ b) $x^2 - 29y^2 = 1$

5. Find the three smallest positive solutions of the diophantine equation $x^2 - 37y^2 = 1$.

6. For each of the following values of d determine whether the diophantine equation $x^2 - dy^2 = -1$ has solutions in integers.

a)   2 e)   17
b)   3 f)   31
c)   6 g)   41
d)   13 h)   50

7. The least positive solution of the diophantine equation $x^2 - 61y^2 = 1$ is $x_1 = 1766319049$, $y_1 = 226153980$. Find the least positive solution other than x_1, y_1.

* 8. Show that if p_k/q_k is a convergent of the simple continued fraction expansion of $\sqrt{d}$ then $|p_k^2 - dq_k^2| < 1 + 2\sqrt{d}$.

9. Show that if d is a positive integer divisible by a prime of the form $4k + 3$, then the diophantine equation $x^2 - dy^2 = -1$ has no solutions.

10. Let d and n be positive integers.

a)   Show that if r,s is a solution of the diophantine equation $x^2 - dy^2 = 1$ and X,Y is a solution of the diophantine equation $x^2 - dy^2 = n$ then $Xr \pm dYs$, $Xs \pm Yr$ is also a solution of $x^2 - dy^2 = n$.

b)   Show that the diophantine equation $x^2 - dy^2 = n$ either has no solutions or has infinitely many solutions.

11. Find those right triangles having legs with lengths that are consecutive integers. (*Hint*: Use Theorem 11.1 to write the lengths of the legs as $x = s^2 - t^2$ and

$y = 2st$, where s and t are positive integers such that $(s,t) = 1$, $s > t$ and s and t have opposite parity. Then $x - y = \pm 1$ implies that $(s - t)^2 - 2t^2 = \pm 1$.)

12. Show that the diophantine equation $x^4 - 2y^4 = 1$ has no nontrivial solutions.

13. Show that the diophantine equation $x^4 - 2y^2 = -1$ has no nontrivial solutions.

11.4 Computer Projects

Programming Projects

Write programs to do the following:

1. Find those integers n with $|n| < \sqrt{d}$ such that the diophantine equation $x^2 - dy^2 = n$ has no solutions.

2. Find the least positive solutions of the diophantine equations $x^2 - dy^2 = 1$ and $x^2 - dy^2 = -1$.

3. Find the solutions of Pell's equation from the least positive solution (see Theorem 11.11).

Computations and Explorations

Using the programs you have written or a computation program, carry out the following computations and explorations:

1. Find the least positive solution of the diophantine equation $x^2 - 109y^2 = 1$. (This problem was posed by Fermat to English mathematicians in the mid-1600s.)

2. Find the least positive solution of the diophantine equation $x^2 - 991y^2 = 1$.

3. Find the least positive solution of the diophantine equation $x^2 - 1000099y^2 = 1$.

Appendix

Table 1. Factor Table.

The least prime factor of each odd positive integer less than 10000 and not divisible by five is given in the table. The initial digits of the integer are listed to the side and the last digit is at the top of the column. Primes are indicated with a dash.

	1	3	7	9		1	3	7	9		1	3	7	9		1	3	7	9
0	—	—	—	3	40	—	13	11	—	80	3	11	3	—	120	—	3	17	3
1	—	—	—	—	41	3	7	3	—	81	—	3	19	3	121	7	—	—	23
2	3	—	3	—	42	—	3	7	3	82	—	—	—	—	122	3	—	3	—
3	—	3	—	3	43	—	—	19	—	83	3	7	3	—	123	—	3	—	3
4	—	—	—	7	44	3	—	3	—	84	29	3	7	3	124	17	11	29	—
5	3	—	3	—	45	11	3	—	3	85	23	—	—	—	125	3	7	3	—
6	—	3	—	3	46	—	—	—	7	86	3	—	3	11	126	13	3	7	3
7	—	—	7	—	47	3	11	3	—	87	13	3	—	3	127	31	19	—	—
8	3	—	3	—	48	13	3	—	3	88	—	—	—	7	128	3	—	3	—
9	7	3	—	3	49	—	17	7	—	89	3	19	3	29	129	—	3	—	3
10	—	—	—	—	50	3	—	3	—	90	17	3	—	3	130	—	—	—	7
11	3	—	3	7	51	7	3	11	3	91	—	11	7	—	131	3	13	3	—
12	11	3	—	3	52	—	—	17	23	92	3	13	3	—	132	—	3	—	3
13	—	7	—	—	53	3	13	3	7	93	7	3	—	3	133	11	31	7	13
14	3	11	3	—	54	—	3	—	3	94	—	23	—	13	134	3	17	3	19
15	—	3	—	3	55	19	7	—	13	95	3	8	3	7	135	7	3	23	3
16	7	—	—	13	56	3	—	3	—	96	31	3	—	3	136	—	29	—	37
17	3	—	3	—	57	—	3	—	3	97	—	7	—	11	137	3	—	3	7
18	—	3	11	3	58	7	11	—	19	98	3	—	3	23	138	—	3	19	3
19	—	—	—	—	59	3	—	3	—	99	—	3	—	3	139	13	7	11	—
20	3	7	3	11	60	—	3	—	3	100	7	17	19	—	140	3	23	3	—
21	—	3	7	3	61	13	—	—	—	101	3	—	3	—	141	17	3	13	3
22	13	—	—	—	62	3	7	3	17	102	—	3	13	3	142	7	—	—	—
23	3	—	3	—	63	—	3	7	3	103	—	—	17	—	143	3	—	3	—
24	—	3	13	3	64	—	—	—	11	104	3	7	3	—	144	11	3	—	3
25	—	11	—	7	65	3	—	3	—	105	—	3	7	3	145	—	—	31	—
26	3	—	3	—	66	—	3	23	3	106	—	—	11	—	146	3	7	3	13
27	—	3	—	3	67	11	—	—	7	107	3	29	3	13	147	—	3	7	3
28	—	—	7	17	68	3	—	3	13	108	23	3	—	3	148	—	—	—	—
29	3	—	3	13	69	—	3	17	3	109	—	—	—	7	149	3	—	3	—
30	7	3	—	3	70	—	19	7	—	110	3	—	3	—	150	19	3	11	3
31	—	—	—	11	71	3	23	3	—	111	11	3	—	3	151	—	17	37	7
32	3	17	3	7	72	7	3	—	3	112	19	—	7	—	152	3	—	3	11
33	—	3	—	3	73	17	—	11	—	113	3	11	3	17	153	—	3	29	3
34	11	7	—	—	74	3	—	3	7	114	7	3	31	3	154	23	—	7	—
35	3	—	3	—	75	—	3	—	3	115	—	—	13	19	155	3	—	3	—
36	19	3	—	3	76	—	7	13	—	116	3	—	3	7	156	7	3	—	3
37	7	—	13	—	77	3	—	3	19	117	—	3	11	3	157	—	11	19	—
38	3	—	3	—	78	11	3	—	3	118	—	7	—	29	158	3	—	3	7
39	17	3	—	3	79	7	13	—	17	119	3	—	3	11	159	37	3	—	3

Table 1. (Continued).

	1	3	7	9		1	3	7	9		1	3	7	9		1	3	7	9
160	—	7	—	—	200	3	—	3	7	240	7	3	29	3	280	—	—	7	53
161	3	—	3	—	201	—	3	—	3	241	—	19	—	41	281	3	29	3	—
162	—	3	—	3	202	43	7	—	—	242	3	—	3	7	282	7	3	11	3
163	7	23	—	11	203	3	19	3	—	243	11	3	—	3	283	19	—	—	17
164	3	31	3	17	204	13	3	23	3	244	—	7	—	31	284	3	—	3	7
165	13	3	—	3	205	7	—	11	29	245	3	11	3	—	285	—	3	—	3
166	11	—	—	—	206	3	—	3	—	246	23	3	—	3	286	—	7	47	19
167	3	7	3	23	207	19	3	31	3	247	7	—	—	37	287	3	13	3	—
168	41	3	7	3	208	—	—	—	—	248	3	13	3	19	288	43	3	—	3
169	19	—	—	—	209	3	7	3	—	249	47	3	11	3	289	7	11	—	13
170	3	13	3	—	210	11	3	7	3	250	41	—	23	13	290	3	—	3	—
171	29	3	17	3	211	—	—	29	13	251	3	7	3	11	291	41	3	—	3
172	—	—	11	7	212	3	11	3	—	252	—	3	7	3	292	23	37	—	29
173	3	—	3	37	213	—	3	—	3	253	—	17	43	—	293	3	7	3	—
174	—	3	—	3	214	—	—	19	7	254	3	—	3	—	294	17	3	7	3
175	17	—	7	—	215	3	—	3	17	255	—	3	—	3	295	13	—	—	11
176	3	41	3	29	216	—	3	11	3	256	13	11	17	7	296	3	—	3	—
177	7	3	—	3	217	13	41	7	—	257	3	31	3	—	297	—	3	13	3
178	13	—	—	—	218	3	37	3	11	258	29	3	13	3	298	11	19	29	7
179	3	11	3	7	219	7	3	13	3	259	—	—	7	23	299	3	41	3	—
180	—	3	13	3	220	31	—	—	47	260	3	19	3	—	300	—	3	31	3
181	—	7	23	17	221	3	—	3	7	261	7	3	—	3	301	—	23	7	—
182	3	—	3	31	222	—	3	17	3	262	—	43	37	11	302	3	—	3	13
183	—	3	11	3	223	23	7	—	—	263	3	—	3	7	303	7	3	—	3
184	7	19	—	43	224	3	—	3	13	264	19	3	—	3	304	—	17	11	—
185	3	17	3	11	225	—	3	37	3	265	11	7	—	—	305	3	43	3	7
186	—	3	—	3	226	7	31	—	—	266	3	—	3	17	306	—	3	—	3
187	—	—	—	—	227	3	—	3	43	267	—	3	—	3	307	37	7	17	—
188	3	7	3	—	228	—	3	—	3	268	7	—	—	—	308	3	—	3	—
189	31	3	7	3	229	29	—	—	11	269	3	—	3	—	309	11	3	19	3
190	—	11	—	23	230	3	7	3	—	270	37	3	—	3	310	7	29	13	—
191	3	—	3	19	231	—	3	7	3	271	—	—	11	—	311	3	11	3	—
192	17	3	41	3	232	11	23	13	17	272	3	7	3	—	312	—	3	53	3
193	—	—	13	7	233	3	—	3	—	273	—	3	7	3	313	31	13	—	43
194	3	29	3	—	234	—	3	—	3	274	—	13	41	—	314	3	7	3	47
195	—	3	19	3	235	—	13	—	7	275	3	—	3	31	315	23	3	7	3
196	37	13	7	11	236	3	17	3	23	276	11	3	—	3	316	29	—	—	—
197	3	—	3	—	237	—	3	—	3	277	17	47	—	7	317	3	19	3	11
198	7	3	—	3	238	—	—	7	—	278	3	11	3	—	318	—	3	—	3
199	11	—	—	—	239	3	—	3	—	279	—	3	—	3	319	—	31	23	7

Table 1. (Continued).

	1	3	7	9		1	3	7	9		1	3	7	9		1	3	7	9
320	3	—	3	—	360	13	3	—	3	400	—	—	—	19	440	3	7	3	—
321	13	3	—	3	361	23	—	—	7	401	3	—	3	—	441	11	3	7	3
322	—	11	7	—	362	3	—	3	19	402	—	3	—	3	442	—	—	19	43
323	3	53	3	41	363	—	3	—	3	403	29	37	11	7	443	3	11	3	23
324	7	3	17	3	364	11	—	7	41	404	3	13	3	—	444	—	3	—	3
325	—	—	—	—	365	3	13	3	—	405	—	3	—	3	445	—	61	—	7
326	3	13	3	7	366	7	3	19	3	406	31	17	7	13	446	3	—	3	41
327	—	3	29	3	367	—	—	—	13	407	3	—	3	—	447	17	3	11	3
328	17	7	19	11	368	3	29	3	7	408	7	3	61	3	448	—	—	7	67
329	3	37	3	—	369	—	3	—	3	409	—	—	17	—	449	3	—	3	11
330	—	3	—	3	370	—	7	11	—	410	3	11	3	7	450	7	3	—	3
331	7	—	31	—	371	3	47	3	—	411	—	3	23	3	451	13	—	—	—
332	3	—	3	—	372	61	3	—	3	412	13	7	—	—	452	3	—	3	7
333	—	3	47	3	373	7	—	37	—	413	3	—	3	—	453	23	3	13	3
334	13	—	—	17	374	3	19	3	23	414	41	3	11	3	454	19	7	—	—
335	3	7	3	—	375	11	3	13	3	415	7	—	—	—	455	3	29	3	47
336	—	3	7	3	376	—	53	—	—	416	3	23	3	11	456	—	3	—	3
337	—	—	11	31	377	3	7	3	—	417	43	3	—	3	457	7	17	23	19
338	3	17	3	—	378	19	3	7	3	418	37	47	53	59	458	3	—	3	13
339	—	3	43	3	379	17	—	—	29	419	3	7	3	13	459	—	3	—	3
340	19	41	—	7	380	3	—	3	31	420	—	3	7	3	460	43	—	17	11
341	3	—	3	13	381	37	3	11	3	421	—	11	—	—	461	3	7	3	31
342	11	3	23	3	382	—	—	43	7	422	3	41	3	—	462	—	3	7	3
343	47	—	7	19	383	3	—	3	11	423	—	3	19	3	463	11	41	—	—
344	3	11	3	—	384	23	3	—	3	424	—	—	31	7	464	3	—	3	—
345	7	3	—	3	385	—	—	7	17	425	3	—	3	—	465	—	3	—	3
346	—	—	—	—	386	3	—	3	53	426	—	3	17	3	466	59	—	13	7
347	3	23	3	7	387	7	3	—	3	427	—	—	7	11	467	3	—	3	—
348	59	3	11	3	388	—	11	13	—	428	3	—	3	—	468	31	3	43	3
349	—	7	13	—	389	3	17	3	7	429	7	3	—	3	469	—	13	7	37
350	3	31	3	11	390	47	3	—	3	430	11	13	59	31	470	3	—	3	17
351	—	3	—	3	391	—	7	—	—	431	3	19	3	7	471	7	3	53	3
352	7	13	—	—	392	3	—	3	—	432	29	3	—	3	472	—	—	29	—
353	3	—	3	—	393	—	3	31	3	433	61	7	—	—	473	3	—	3	7
354	—	3	—	3	394	7	—	—	11	434	3	43	3	—	474	11	3	47	3
355	53	11	—	—	395	3	59	3	37	435	19	3	—	3	475	—	7	67	—
356	3	7	3	43	396	17	3	—	3	436	7	—	11	17	476	3	11	3	19
357	—	3	7	3	397	11	29	41	23	437	3	—	3	29	477	13	3	17	3
358	—	—	17	37	398	3	7	3	—	438	13	3	41	3	478	7	—	—	—
359	3	—	3	59	399	13	3	7	3	439	—	23	—	53	479	3	—	3	—

Table 1. (Continued).

	1	3	7	9		1	3	7	9		1	3	7	9		1	3	7	9
480	—	3	11	3	520	7	11	41	—	560	3	13	3	71	600	17	3	—	3
481	17	—	—	61	521	3	13	3	17	561	31	3	41	3	601	—	7	11	13
482	3	7	3	11	522	23	3	—	3	562	7	—	17	13	602	3	19	3	—
483	—	3	7	3	523	—	—	—	13	563	3	43	3	—	603	37	3	—	3
484	47	29	37	13	524	3	7	3	29	564	—	3	—	3	604	7	—	—	23
485	3	23	3	43	525	59	3	7	3	565	—	—	—	—	605	3	—	3	73
486	—	3	31	3	526	—	19	23	11	566	3	7	3	—	606	11	3	—	3
487	—	11	—	7	527	3	—	3	—	567	53	3	7	3	607	13	—	59	—
488	3	19	3	—	528	—	3	17	3	568	13	—	11	—	608	3	7	3	—
489	67	3	59	3	529	11	67	—	7	569	3	—	3	41	609	—	3	7	3
490	13	—	7	—	530	3	—	3	—	570	—	3	13	3	610	—	17	31	41
491	3	17	3	—	531	47	3	13	3	571	—	29	—	7	611	3	—	3	29
492	7	3	13	3	532	17	—	7	73	572	3	59	3	17	612	—	3	11	3
493	—	—	—	11	533	3	—	3	19	573	11	3	—	3	613	—	—	17	7
494	3	—	3	7	534	7	3	—	3	574	—	—	7	—	614	3	—	3	11
495	—	3	—	3	535	—	53	11	23	575	3	11	3	13	615	—	3	47	3
496	11	7	—	—	536	3	31	3	7	576	7	3	73	3	616	61	—	7	31
497	3	—	3	13	537	41	3	19	3	577	29	23	53	—	617	3	—	3	37
498	17	3	—	3	538	—	7	—	17	578	3	—	3	7	618	7	3	23	3
499	7	—	19	—	539	3	—	3	—	579	—	3	11	3	619	41	11	—	—
500	3	—	3	—	540	11	3	—	3	580	—	7	—	37	620	3	—	3	7
501	—	3	29	3	541	7	—	—	—	581	3	—	3	11	621	—	3	—	3
502	—	—	11	47	542	3	11	3	61	582	—	3	—	3	622	—	7	13	—
503	3	7	3	—	543	—	3	—	3	583	7	19	13	—	623	3	23	3	17
504	71	3	7	3	544	—	—	13	—	584	3	—	3	—	624	79	3	—	3
505	—	31	13	—	545	3	7	3	53	585	—	3	—	3	625	.7	13	—	11
506	3	61	3	37	546	43	3	7	3	586	—	11	—	—	626	3	—	3	—
507	11	3	—	3	547	—	13	—	—	587	3	7	3	—	627	—	3	—	3
508	—	13	—	7	548	3	—	3	11	588	—	3	7	3	628	11	61	—	19
509	3	11	3	—	549	17	3	23	3	589	43	71	—	17	629	3	7	3	—
510	—	3	—	3	550	—	—	—	7	590	3	—	3	19	630	—	3	7	3
511	19	—	7	—	551	3	37	3	—	591	23	3	61	3	631	—	59	—	71
512	3	47	3	23	552	—	3	—	3	592	31	—	—	7	632	3	—	3	—
513	7	3	11	3	553	—	11	7	29	593	3	17	3	—	633	13	3	—	3
514	53	37	—	19	554	3	23	3	31	594	13	3	19	3	634	17	—	11	7
515	3	—	3	7	555	7	3	—	3	595	11	—	7	59	635	3	—	3	—
516	13	3	—	3	556	67	—	19	—	596	3	67	3	47	636	—	3	—	3
517	—	7	31	—	557	3	—	3	7	597	7	3	43	3	637	23	—	7	—
518	3	71	3	—	558	—	3	37	3	598	—	31	—	53	638	3	13	3	—
519	29	3	—	3	559	—	7	29	11	599	3	13	3	7	639	7	3	—	3

Table 1. (Continued).

	1	3	7	9		1	3	7	9		1	3	7	9		1	3	7	9
640	37	19	43	13	680	3	—	3	11	720	19	3	—	3	760	11	—	—	7
641	3	11	3	7	681	7	3	17	3	721	—	—	7	—	761	3	23	3	19
642	—	3	—	3	682	19	—	—	—	722	3	31	3	—	762	—	3	29	3
643	59	7	41	47	683	3	—	3	7	723	7	3	—	3	763	13	17	7	—
644	3	17	3	—	684	—	3	41	3	724	13	—	—	11	764	3	—	3	—
645	—	3	11	3	685	13	7	—	19	725	3	—	3	7	765	7	3	13	3
646	7	23	29	—	686	3	—	3	—	726	53	3	13	3	766	47	79	11	—
647	3	—	3	11	687	—	3	13	3	727	11	7	19	29	767	3	—	3	7
648	—	3	13	3	688	7	—	71	83	728	3	—	3	37	768	—	3	—	3
649	—	43	73	67	689	3	61	3	—	729	23	3	—	3	769	—	7	43	—
650	3	7	3	23	690	67	3	—	3	730	7	67	—	—	770	3	—	3	13
651	17	3	7	3	691	—	31	—	11	731	3	71	3	13	771	11	3	—	3
652	—	11	61	—	692	3	7	3	13	732	—	3	17	3	772	7	—	—	59
653	3	47	3	13	693	29	3	7	3	733	—	—	11	41	773	3	11	3	71
654	31	3	—	3	694	11	53	—	—	734	3	7	3	—	774	—	3	61	3
655	—	—	79	7	695	3	17	3	—	735	—	3	7	3	775	23	—	—	—
656	3	—	3	—	696	—	3	—	3	736	17	37	53	—	776	3	7	3	17
657	—	3	—	3	697	—	19	—	7	737	3	73	3	47	777	19	3	7	3
658	—	29	7	11	698	3	—	3	29	738	11	3	83	3	778	31	43	13	—
659	3	19	3	—	699	—	3	—	3	739	19	—	13	7	779	3	—	3	11
660	7	3	—	3	700	—	47	7	43	740	3	11	3	31	780	29	3	37	3
661	11	17	13	—	701	3	—	3	—	741	—	3	—	3	781	73	13	—	7
662	3	37	3	7	702	7	3	—	3	742	41	13	7	17	782	3	—	3	—
663	19	3	—	3	703	79	13	31	—	743	3	—	3	43	783	41	3	17	3
664	29	7	17	61	704	3	—	3	7	744	7	3	11	3	784	—	11	7	47
665	3	—	3	—	705	11	3	—	3	745	—	29	—	—	785	3	—	3	29
666	—	3	59	3	706	23	7	37	—	746	3	17	3	7	786	7	3	—	3
667	7	—	11	—	707	3	11	3	—	747	31	3	—	3	787	17	—	—	—
668	3	41	3	—	708	73	3	19	3	748	—	7	—	—	788	3	—	3	7
669	—	3	37	3	709	7	41	47	31	749	3	59	3	—	789	13	3	53	3
670	—	—	19	—	710	3	—	3	—	750	13	3	—	3	790	—	7	—	11
671	3	7	3	—	711	13	3	11	3	751	7	11	—	73	791	3	41	3	—
672	11	3	7	3	712	—	17	—	—	752	3	—	3	—	792	89	3	—	3
673	53	—	—	23	713	3	7	3	11	753	17	3	—	3	793	7	—	—	17
674	3	11	3	17	714	37	3	7	3	754	—	19	—	—	794	3	13	3	—
675	43	3	29	3	715	—	23	17	—	755	3	7	3	—	795	—	3	73	3
676	—	—	67	7	716	3	13	3	67	756	—	3	7	3	796	19	—	31	13
677	3	13	3	—	717	71	3	—	3	757	67	—	—	11	797	3	7	3	79
678	—	3	11	3	718	43	11	—	7	758	3	—	3	—	798	23	3	7	3
679	—	—	7	13	719	3	—	3	23	759	—	3	71	3	799	61	—	11	19

Table 1. (Continued).

	1 3 7 9		1 3 7 9		1 3 7 9		1 3 7 9
800	3 53 3 —	840	31 3 7 3	880	13 — — 23	920	3 — 3 —
801	— 3 — 3	841	13 47 19 —	881	3 7 3 —	921	61 3 13 3
802	13 71 23 7	842	3 — 3 —	882	— 3 7 3	922	— 23 — 11
803	3 29 3 —	843	— 3 11 3	883	— 11 — —	923	3 7 3 —
804	11 3 13 3	844	23 — — 7	884	3 37 3 —	924	— 3 7 3
805	83 — 7 —	845	3 79 3 11	885	53 3 17 3	925	11 19 — 47
806	3 11 3 —	846	— 3 — 3	886	— — — 7	926	3 59 3 13
807	7 3 41 3	847	43 37 7 61	887	3 19 3 13	927	73 3 — 3
808	— 59 — —	848	3 17 3 13	888	83 3 — 3	928	— — 37 7
809	3 — 3 7	849	7 3 29 3	889	17 — 7 11	929	3 — 3 17
810	— 3 11 3	850	— 11 47 67	890	3 29 3 59	930	71 3 41 3
811	— 7 — 23	851	3 — 3 7	891	7 3 37 3	931	— 67 7 —
812	3 — 3 11	852	— 3 — 3	892	11 — 79 —	932	3 — 3 19
813	47 3 79 3	853	19 7 — —	893	3 — 3 7	933	7 3 — 3
814	7 17 — 29	854	3 — 3 83	894	— 3 23 3	934	— — 13 —
815	3 31 3 41	855	17 3 43 3	895	— 7 13 17	935	3 47 3 7
816	— 3 — 3	856	7 — 13 11	896	3 — 3 —	936	11 3 17 3
817	— 11 13 —	857	3 — 3 23	897	— 3 47 3	937	— 7 — 83
818	3 7 3 19	858	— 3 31 3	898	7 13 11 89	938	3 11 3 41
819	— 3 7 3	859	11 13 — —	899	3 17 3 —	939	— 3 — 3
820	59 13 29 —	860	3 7 3 —	900	— 3 — 3	940	7 — 23 97
821	3 43 3 —	861	79 3 7 3	901	— — 71 29	941	3 — 3 —
822	— 3 19 3	862	37 — — —	902	3 7 3 —	942	— 3 11 3
823	— — — 7	863	3 89 3 53	903	11 3 7 3	943	— — — —
824	3 — 3 73	864	— 3 — 3	904	— — 83 —	944	3 7 3 11
825	37 3 23 3	865	41 17 11 7	905	3 11 3 —	945	13 3 7 3
826	11 — 7 —	866	3 — 3 —	906	13 3 — 3	946	— — — 17
827	3 — 3 17	867	13 3 — 3	907	47 43 29 7	947	3 — 3 —
828	7 3 — 3	868	— 19 7 —	908	3 31 3 61	948	19 3 53 3
829	— — — 43	869	3 — 3 —	909	— 3 11 3	949	— 11 — 7
830	3 19 3 7	870	7 3 — 3	910	19 — 7 —	950	3 13 3 37
831	— 3 — 3	871	31 — 23 —	911	3 31 3 11	951	— 3 31 3
832	53 7 11 —	872	3 11 3 7	912	7 3 — 3	952	— 89 7 13
833	3 13 3 31	873	— 3 — 3	913	23 — — 13	953	3 — 3 —
834	19 3 17 3	874	— 7 — 13	914	3 41 3 7	954	7 3 — 3
835	7 — 61 13	875	3 — 3 19	915	— 3 — 3	955	— 41 19 11
836	3 — 3 —	876	— 3 11 3	916	— 7 89 53	956	3 73 3 7
837	11 3 — 3	877	7 31 67 —	917	3 — 3 67	957	17 3 61 3
838	17 83 — —	878	3 — 3 11	918	— 3 — 3	958	11 7 — 43
839	3 7 3 37	879	59 3 19 3	919	7 29 17 —	959	3 53 3 29

Table 1. (Continued).

	1	3	7	9		1	3	7	9		1	3	7	9		1	3	7	9
960	—	3	13	3	970	89	31	18	7	980	3	—	3	17	990	—	3	—	3
961	7	—	59	—	971	3	11	3	—	981	—	3	—	3	991	11	23	47	7
962	3	—	3	—	972	—	3	71	3	982	7	11	31	—	992	3	—	3	—
963	—	3	23	3	973	37	—	7	—	983	3	—	3	—	993	—	3	19	3
964	31	—	11	—	974	3	—	3	—	984	13	3	43	3	994	—	61	7	—
965	3	7	3	13	975	7	3	11	3	985	—	59	—	—	995	3	37	3	23
966	—	3	7	3	976	43	13	—	—	986	3	7	3	71	996	7	3	—	3
967	19	17	—	—	977	3	29	3	7	987	—	3	7	3	997	13	—	11	17
968	3	23	3	—	978	—	3	—	3	988	41	—	—	11	998	3	67	3	7
969	11	3	—	3	979	—	7	97	41	989	3	13	3	19	999	97	3	13	3

Table 2. Values of Some Arithmetic Functions.

n	$\phi(n)$	$\tau(n)$	$\sigma(n)$
1	1	1	1
2	1	2	3
3	2	2	4
4	2	3	7
5	4	2	6
6	2	4	12
7	6	2	8
8	4	4	15
9	6	3	13
10	4	4	18
11	10	2	12
12	4	6	28
13	12	2	14
14	6	4	24
15	8	4	24
16	8	5	31
17	16	2	18
18	6	6	39
19	18	2	20
20	8	6	42
21	12	4	32
22	10	4	36
23	22	2	24
24	8	8	60
25	20	3	31
26	12	4	42
27	18	4	40
28	12	6	56
29	28	2	30
30	8	8	72
31	30	2	32
32	16	6	63
33	20	4	48
34	16	4	54
35	24	4	48
36	12	9	91
37	36	2	38
38	18	4	60
39	24	4	56
40	16	8	90
41	40	2	42
42	12	8	96
43	42	2	44
44	20	6	84
45	24	6	78
46	22	4	72
47	46	2	48
48	16	10	124
49	42	3	57

Table 2. (Continued).

n	$\phi(n)$	$\tau(n)$	$\sigma(n)$
50	20	6	93
51	32	4	72
52	24	6	98
53	52	2	54
54	18	8	120
55	40	4	72
56	24	8	120
57	36	4	80
58	28	4	90
59	58	2	60
60	16	12	168
61	60	2	62
62	30	4	96
63	36	6	104
64	32	7	127
65	48	4	84
66	20	8	144
67	66	2	68
68	32	6	126
69	44	4	96
70	24	8	144
71	70	2	72
72	24	12	195
73	72	2	74
74	36	4	114
75	40	6	124
76	36	6	140
77	60	4	96
78	24	8	168
79	78	2	80
80	32	10	186
81	54	5	121
82	40	4	126
83	82	2	84
84	24	12	224
85	64	4	108
86	42	4	132
87	56	4	120
88	40	8	180
89	88	2	90
90	24	12	234
91	72	4	112
92	44	6	168
93	60	4	128
94	46	4	144
95	72	4	120
96	32	12	252
97	96	2	98
98	42	6	171
99	60	6	156
100	40	9	217

Table 3. Primitive Roots Modulo Primes

The least primitive root r modulo p for each prime p, $p < 1000$ is given in the table.

p	r	p	r	p	r	p	r
2	1	191	19	439	15	709	2
3	2	193	5	443	2	719	11
5	2	197	2	449	3	727	5
7	3	199	3	457	13	733	6
11	2	211	2	461	2	739	3
13	2	223	3	463	3	743	5
17	3	227	2	467	2	751	3
19	2	229	6	479	13	757	2
23	5	233	3	487	3	761	6
29	2	239	7	491	2	769	11
31	3	241	7	499	7	773	2
37	2	251	6	503	5	787	2
41	6	257	3	509	2	797	2
43	3	263	5	521	3	809	3
47	5	269	2	523	2	811	3
53	2	271	6	541	2	821	2
59	2	277	5	547	2	823	3
61	2	281	3	557	2	827	2
67	2	283	3	563	2	829	2
71	7	293	2	569	3	839	11
73	5	307	5	571	3	853	2
79	3	311	17	577	5	857	3
83	2	313	10	587	2	859	2
89	3	317	2	593	3	863	5
97	5	331	3	599	7	877	2
101	2	337	10	601	7	881	3
103	5	347	2	607	3	883	2
107	2	349	2	613	2	887	5
109	6	353	3	617	3	907	2
113	3	359	7	619	2	911	17
127	3	367	6	631	3	919	7
131	2	373	2	641	3	929	3
137	3	379	2	643	11	937	5
139	2	383	5	647	5	941	2
149	2	389	2	653	2	947	2
151	6	397	5	659	2	953	3
157	5	401	3	601	2	967	5
163	2	409	21	673	5	971	6
167	5	419	2	677	2	977	3
173	2	421	2	683	5	983	5
179	2	431	7	691	3	991	6
181	2	433	5	701	2	997	7

Table 4. Indices

p	1	2	3	4	5	6	7	8	9	10	11	12	13	14	15	16
3	2	1														
5	4	1	3	2							Indices					
7	6	2	1	4	5	3										
11	10	1	8	2	4	9	7	3	6	5						
13	12	1	4	2	9	5	11	3	8	10	7	6				
17	16	14	1	12	5	15	11	10	2	3	7	13	4	9	6	8
19	18	1	13	2	16	14	6	3	8	17	12	15	5	7	11	4
23	22	2	16	4	1	18	19	6	10	3	9	20	14	21	17	8
29	28	1	5	2	22	6	12	3	10	23	25	7	18	13	27	4
31	30	24	1	18	20	25	28	12	2	14	23	19	11	22	21	0
37	36	1	26	2	23	27	32	3	16	24	30	28	11	33	13	4
41	40	26	15	12	22	1	39	38	30	8	3	27	31	25	37	24
43	42	27	1	12	25	28	35	39	2	10	30	13	32	20	26	24
47	46	18	20	36	1	38	32	8	40	19	7	10	11	4	21	26
53	52	1	17	2	47	18	14	3	34	48	6	19	24	15	12	4
59	58	1	50	2	6	51	18	3	42	7	25	52	45	19	56	4
61	60	1	6	2	22	7	49	3	12	23	15	8	40	50	28	4
67	66	1	39	2	15	40	23	3	12	16	59	41	19	24	54	4
71	70	6	26	12	28	32	1	18	52	34	31	38	39	7	54	24
73	72	8	6	16	1	14	33	24	12	9	55	22	59	41	7	32
79	78	4	1	8	62	5	53	12	2	66	68	9	34	57	63	16
83	82	1	72	2	27	73	8	3	62	28	24	74	77	9	17	4
89	88	16	1	32	70	17	81	48	2	86	84	33	23	9	71	64
97	96	34	70	68	1	8	31	6	44	35	86	42	25	65	71	40

p	17	18	19	20	21	22	23	24	25	26	27	28	29	30	31	32	33
19	10	9															
23	7	12	15	5	13	11					Indices						
29	21	11	9	24	17	26	20	8	16	19	15	14					
31	7	26	4	8	29	17	27	13	10	5	3	16	9	15			
37	7	17	35	25	22	31	15	29	10	12	6	34	21	14	9	5	20
41	33	16	9	34	14	29	36	13	4	17	5	11	7	23	28	10	18
43	38	29	19	37	36	15	16	40	8	17	3	5	41	11	34	9	31
47	16	12	45	37	6	25	5	28	2	29	14	22	35	39	3	44	27
53	10	35	37	49	31	7	39	20	42	25	51	16	46	13	33	5	23
59	40	43	38	8	10	26	15	53	12	46	34	20	28	57	49	5	17
61	47	13	26	24	55	16	57	9	44	41	18	51	35	29	59	5	21
67	64	13	10	17	62	60	28	42	30	20	51	25	44	55	47	5	32
71	49	58	16	40	27	37	15	44	56	45	8	13	68	60	11	30	57
73	21	20	62	17	39	63	46	30	2	67	18	49	35	15	11	40	61
79	21	6	32	70	54	72	26	13	46	38	3	61	11	67	56	20	69
83	56	63	47	29	80	25	60	75	54	78	52	10	12	18	38	5	14
89	6	18	35	14	82	12	57	49	52	39	3	25	59	87	31	80	85
97	89	78	81	69	5	24	77	76	2	59	18	3	13	9	46	74	60

Table 4. (Continued).

p	Numbers															
	34	35	36	37	38	39	40	41	42	43	44	45	46	47	48	49
37	8	19	18													
41	19	21	2	32	35	6	20				Indices					
43	23	18	14	7	4	33	22	6	21							
47	34	33	30	42	17	31	9	15	24	13	43	41	23			
53	11	9	36	30	38	41	50	45	32	22	8	29	40	44	21	23
59	41	24	44	55	39	37	9	14	11	33	27	48	16	23	54	36
61	48	11	14	39	27	46	25	54	56	43	17	34	58	20	10	38
67	65	38	14	22	11	58	18	53	63	9	61	27	29	50	43	46
71	55	29	64	20	22	65	46	25	33	48	43	10	21	9	50	2
78	29	34	28	64	70	65	25	4	47	51	71	13	54	31	38	66
79	25	37	10	19	36	35	74	75	58	49	76	64	30	59	17	28
83	57	35	64	20	48	67	30	40	81	71	26	7	61	23	76	16
89	22	63	34	11	51	24	30	21	10	29	28	72	73	54	65	74
97	27	32	16	91	19	95	7	85	39	4	58	45	15	84	14	62

p	Numbers															
	50	51	52	53	54	55	56	57	58	59	60	61	62	63	64	65
53	43	27	26								Indices					
59	13	32	47	22	35	31	21	30	29							
61	45	53	42	33	19	37	52	32	36	31	30					
67	31	37	21	57	52	8	26	49	45	36	56	7	48	35	6	34
71	62	5	51	23	14	59	19	42	4	3	66	69	17	53	36	67
73	10	27	3	53	26	56	57	68	43	5	23	58	19	45	48	60
79	50	22	42	77	7	52	65	33	15	31	71	45	60	55	24	18
83	55	46	79	59	53	51	11	37	13	34	19	66	39	70	6	22
89	68	7	55	78	19	66	41	36	75	43	15	69	47	83	8	5
97	36	63	93	10	52	87	37	55	47	67	43	64	80	75	12	26

p	Numbers															
	66	67	68	69	70	71	72	73	74	75	76	77	78	79	80	81
67	33															
71	63	47	61	41	35								Indices			
78	69	50	37	52	42	44	36									
79	73	48	29	27	41	51	14	44	23	47	40	43	39			
83	15	45	58	50	36	33	65	69	21	44	49	32	68	43	31	42
89	13	56	38	58	79	62	50	20	27	53	67	77	40	42	46	4
97	94	57	61	51	66	11	50	28	29	72	53	21	33	30	41	88

p	Numbers															
	82	83	84	85	86	87	88	89	90	91	92	93	94	95	96	
83	41											Indices				
89	37	61	26	76	45	60	44									
97	23	17	73	90	38	83	92	54	79	56	49	20	22	82	48	

Table 4. (Continued).

p	Indices															
	1	2	3	4	5	6	7	8	9	10	11	12	13	14	15	16
3	2	1														
5	2	4	3	1							Numbers					
7	3	2	6	4	5	1										
11	2	4	8	5	10	9	7	3	6	1						
13	2	4	8	3	6	12	11	9	5	10	7	1				
17	3	9	10	13	5	15	11	16	14	8	7	4	12	2	6	1
19	2	4	8	16	13	7	14	9	18	17	15	11	3	6	12	5
23	5	2	10	4	20	8	17	16	11	9	22	18	21	13	19	3
29	2	4	8	16	3	6	12	24	19	9	18	7	14	28	27	25
31	3	9	27	19	26	16	17	20	29	25	13	8	24	10	30	28
37	2	4	8	16	32	27	17	34	31	25	13	26	15	30	23	9
41	6	36	11	25	27	39	29	10	19	32	28	4	24	21	3	18
43	3	9	27	38	28	41	37	25	32	10	30	4	12	36	22	23
47	5	25	31	14	23	21	11	8	40	12	13	18	43	27	41	17
53	2	4	8	16	32	11	22	44	35	17	34	15	30	7	14	28
59	2	4	8	16	32	5	10	20	40	21	42	25	50	41	23	46
61	2	4	8	16	32	3	6	12	24	48	35	9	18	36	11	22
67	2	4	8	16	32	64	61	55	43	19	38	9	18	36	5	10
71	7	49	59	58	51	2	14	27	47	45	31	4	28	54	23	19
73	5	25	52	41	59	3	15	2	10	50	31	9	45	6	30	4
79	3	9	27	2	6	18	54	4	12	36	29	8	24	72	58	16
83	2	4	8	16	32	64	45	7	14	28	56	29	58	33	66	49
89	3	9	27	81	65	17	51	64	14	42	37	22	66	20	60	2
97	5	25	28	43	21	8	40	6	30	53	71	64	29	48	46	36

p	Indices															
	17	18	19	20	21	22	23	24	25	26	27	28	29	30	31	32	33
19	10	1									Numbers					
23	15	6	7	12	14	1										
29	21	13	26	23	17	5	10	20	11	22	15	1				
31	22	4	12	5	15	14	11	2	6	18	23	7	21	1		
37	18	36	35	33	29	21	5	10	20	3	6	12	24	11	22	7	14
41	26	33	34	40	35	5	30	16	14	2	12	31	22	9	13	37	17
43	26	35	19	14	42	40	34	16	5	15	2	6	18	11	33	13	39
47	38	2	10	3	15	28	46	42	22	16	33	24	26	36	39	7	35
53	3	6	12	24	48	43	33	13	26	52	51	49	45	37	21	42	31
59	33	7	14	28	56	53	47	35	11	22	44	29	58	57	55	51	43
61	44	27	54	47	33	5	10	20	40	19	38	15	30	60	59	57	53
67	20	40	13	26	52	37	7	14	28	56	45	23	46	25	50	33	66
71	62	8	56	37	46	38	53	16	41	3	21	5	35	32	11	6	42
73	20	27	62	18	17	12	60	8	40	54	51	36	34	24	47	16	7
79	48	65	37	32	17	51	74	64	34	23	69	49	68	46	59	19	57
83	15	30	60	37	74	65	47	11	22	44	5	10	20	40	80	77	71
89	6	18	54	73	41	34	13	39	28	84	74	44	43	40	31	4	12
97	83	27	38	93	77	94	82	22	13	65	34	73	74	79	7	35	78

Table 4. (Continued).

| p | \multicolumn{16}{c}{Indices} |
	34	35	36	37	38	39	40	41	42	43	44	45	46	47	48	49
37	28	19	1													
41	20	38	23	15	8	7	1					Numbers				
43	31	7	21	20	17	8	24	29	1							
47	34	29	4	20	6	30	9	45	37	44	32	19	1			
53	9	18	36	19	38	23	46	39	25	50	47	41	29	5	10	20
59	27	54	49	39	19	38	17	34	9	18	36	13	26	52	45	31
61	45	29	58	55	49	37	13	26	52	43	25	50	39	17	34	7
67	65	63	59	51	35	3	6	12	24	48	29	58	49	31	62	57
71	10	70	64	22	12	13	20	69	57	44	24	26	40	67	43	17
73	35	29	72	68	48	21	32	14	70	58	71	63	23	42	64	28
79	13	39	38	35	26	78	76	70	52	77	73	61	25	75	67	43
83	59	35	70	57	31	62	41	82	81	79	75	67	51	19	38	76
89	36	19	57	82	68	26	78	56	79	59	88	86	80	62	8	24
97	2	10	50	56	86	42	16	80	12	60	9	45	31	58	96	92

| p | \multicolumn{16}{c}{Indices} |
	50	51	52	53	54	55	56	57	58	59	60	61	62	63	64	65
53	40	27	1													
59	3	6	12	24	48	37	15	30	1			Numbers				
61	14	28	56	51	41	21	42	23	46	31	1					
67	47	27	54	41	15	30	60	53	39	11	22	44	21	42	17	34
71	48	52	9	63	15	34	25	33	18	55	30	68	50	66	36	39
73	67	43	69	53	46	11	55	56	61	13	65	33	19	22	37	39
79	50	71	55	7	21	63	31	14	42	47	62	28	5	15	45	56
83	69	55	27	54	25	50	17	34	68	53	23	46	9	18	36	72
89	72	38	25	75	47	52	67	23	69	29	87	83	71	35	16	48
97	72	69	54	76	89	57	91	67	44	26	33	68	49	51	61	14

| p | \multicolumn{16}{c}{Indices} |
	66	67	68	69	70	71	72	73	74	75	76	77	78	79	80	81
67	1															
71	60	65	29	61	1									Numbers		
73	49	26	57	66	38	44	1									
79	10	30	11	33	20	60	22	66	40	41	44	53	1			
83	61	39	78	73	63	43	3	6	12	24	48	13	26	52	21	42
89	55	76	50	61	5	15	45	46	49	58	85	77	53	70	32	7
97	70	59	4	20	3	15	75	84	32	63	24	23	18	90	62	19

| p | \multicolumn{15}{c}{Indices} |
	82	83	84	85	86	87	88	89	90	91	92	93	94	95	96
83	1														
89	21	63	11	33	10	30	1						Numbers		
97	95	87	47	41	11	55	81	17	85	37	88	52	66	39	1

Table 5. Simple Continued Fractions
for Square Roots of Positive Integers

d	$\sqrt{d}$	d	$\sqrt{d}$
2	$[1;\overline{2}]$	53	$[7;\overline{3,1,1,3,14}]$
3	$[1;\overline{1,2}]$	54	$[7;\overline{2,1,6,1,2,14}]$
5	$[2;\overline{4}]$	55	$[7;\overline{2,2,2,14}]$
6	$[2;\overline{2,4}]$	56	$[7;\overline{2,14}]$
7	$[2;\overline{1,1,1,4}]$	57	$[7;\overline{1,1,4,1,1,14}]$
8	$[2;\overline{1,4}]$	58	$[7;\overline{1,1,1,1,1,1,14}]$
10	$[3;\overline{6}]$	59	$[7;\overline{1,2,7,2,1,14}]$
11	$[3;\overline{3,6}]$	60	$[7;\overline{1,2,1,14}]$
12	$[3;\overline{2,6}]$	61	$[7;\overline{1,4,3,1,2,2,1,3,4,1,14}]$
13	$[3;\overline{1,1,1,1,6}]$	62	$[7;\overline{1,6,1,14}]$
14	$[3;\overline{1,2,1,6}]$	63	$[7;\overline{1,14}]$
15	$[3;\overline{1,6}]$	65	$[8;\overline{16}]$
17	$[4;\overline{8}]$	66	$[8;\overline{8,16}]$
18	$[4;\overline{4,8}]$	67	$[8;\overline{5,2,1,1,7,1,1,2,5,16}]$
19	$[4;\overline{2,1,3,1,2,8}]$	68	$[8;\overline{4,16}]$
20	$[4;\overline{2,8}]$	69	$[8;\overline{3,3,1,4,1,3,3,16}]$
21	$[4;\overline{1,1,2,1,1,8}]$	70	$[8;\overline{2,1,2,1,2,16}]$
22	$[4;\overline{1,2,4,2,1,8}]$	71	$[8;\overline{2,2,1,7,1,2,2,16}]$
23	$[4;\overline{1,3,1,8}]$	72	$[8;\overline{2,16}]$
24	$[4;\overline{1,8}]$	73	$[8;\overline{1,1,5,5,1,1,16}]$
26	$[5;\overline{10}]$	74	$[8;\overline{1,1,1,1,16}]$
27	$[5;\overline{5,10}]$	75	$[8;\overline{1,1,1,16}]$
28	$[5;\overline{3,2,3,10}]$	76	$[8;\overline{1,2,1,1,5,4,5,1,1,2,1,16}]$
29	$[5;\overline{2,1,1,2,10}]$	77	$[8;\overline{1,3,2,3,1,16}]$
30	$[5;\overline{2,10}]$	78	$[8;\overline{1,4,1,16}]$
31	$[5;\overline{1,1,3,5,3,1,1,10}]$	79	$[8;\overline{1,7,1,16}]$
32	$[5;\overline{1,1,1,10}]$	80	$[8;\overline{1,16}]$
33	$[5;\overline{1,2,1,10}]$	82	$[9;\overline{18}]$
34	$[5;\overline{1,4,1,10}]$	83	$[9;\overline{9,18}]$
35	$[5;\overline{1,10}]$	84	$[9;\overline{6,18}]$
37	$[6;\overline{12}]$	85	$[9;\overline{4,1,1,4,18}]$
38	$[6;\overline{6,12}]$	86	$[9;\overline{3,1,1,1,8,1,1,1,3,18}]$
39	$[6;\overline{4,12}]$	87	$[9;\overline{3,18}]$
40	$[6;\overline{3,12}]$	88	$[9;\overline{2,1,1,1,2,18}]$
41	$[6;\overline{2,2,12}]$	89	$[9;\overline{2,3,3,2,18}]$
42	$[6;\overline{2,12}]$	90	$[9;\overline{2,18}]$
43	$[6;\overline{1,1,3,1,5,1,3,1,1,12}]$	91	$[9;\overline{1,1,5,1,5,1,1,18}]$
44	$[6;\overline{1,1,1,2,1,1,1,12}]$	92	$[9;\overline{1,1,2,4,2,1,1,18}]$
45	$[6;\overline{1,2,2,2,1,12}]$	93	$[9;\overline{1,1,1,4,6,4,1,1,1,18}]$
46	$[6;\overline{1,3,1,1,2,6,2,1,1,3,1,12}]$	94	$[9;\overline{1,2,3,1,1,5,1,8,1,5,1,1,3,2,1,18}]$
47	$[6;\overline{1,5,1,12}]$	95	$[9;\overline{1,2,1,18}]$
48	$[6;\overline{1,12}]$	96	$[9;\overline{1,3,1,18}]$
50	$[7;\overline{14}]$	97	$[9;\overline{1,5,1,1,1,1,1,1,5,18}]$
51	$[7;\overline{7,14}]$	98	$[9;\overline{1,8,1,18}]$
52	$[7;\overline{4,1,2,1,4,14}]$	99	$[9;\overline{1,18}]$

Answers to Odd-Numbered Exercises

Section 1.1

1. **a.** $a(b + c) = (b + c)a = ba + ca = ab + ac.$
   b. $(a + b)^2 = (a + b)(a + b) = a(a + b) + b(a + b) = a^2 + ab + ba + b^2$
   $= a^2 + 2ab + b^2.$
   c. $a + (b + c) = a + (c + b) = (a + c) + b = (c + a) + b.$
   d. $(b - a) + (c - b) + (a - c) = (-a + b) + (-b + c) + (-c + a)$
   $= -a + (b - b) + (c - c) + a = 0.$

3. By the property of the additive identity element, $0 + (-0) = -0$. On the other hand, by the property of additive inverses, $0 + (-0) = 0$. Therefore $-0 = 0$.

5. If $x = x - 0$ is positive, $x > 0$. If $x > 0$, then $x - 0 = x$ is positive.

7. $a - c = a + (-b + b) - c = (a - b) + (b - c)$, which is positive from our hypothesis and the closure of the positive integers.

9. **a.** not well-ordered.   **b.** well-ordered.   **c.** well-ordered.   **d.** not well-ordered.
   e. well-ordered.   **f.** not well ordered.

11. Let $A = \{a, b, c, \dots\}$ be a nonempty set of negative integers. Then $B = \{-a, -b, -c, \dots\}$ is a set of positive integers which, by the well-ordering principle, has a smallest member, say z. Since $z < -x, -z > x$ for every x in A, hence $-z$ is the largest element of A.

13. If x is an integer, then $[x] + [-x] = x - x = 0$. Otherwise, $x = z + r$, where z is an integer and r is a real number with $0 < r < 1$. In this case, $[x] + [-x] = [z + r] + [-z - r] = z + (-z - 1) = -1$

15. We have $[x] \leq x$ and $[y] \leq y$. Adding these two inequalities gives $[x] + [y] \leq x + y$. Hence $[x + y] \geq [[x] + [y]] = [x] + [y]$.

17. Let $x = a + r$ and $y = b + s$, where a and b are integers and r and s are real numbers such that $0 \leq r, s < 1$. $[xy] = [ab + as + br + sr] = ab + [as + br + sr]$, whereas $[x][y] = ab$. Whence $[xy] \geq [x][y]$. If x and y are both negative, then $[xy] \leq [x][y]$. If one of x and y is positive and the other negative, then $[xy] \geq [x][y]$.

19. Let $x = [x] + r$. Since $0 \leq r < 1, x + \frac{1}{2} = [x] + r + \frac{1}{2}$. If $r < \frac{1}{2}$, then $[x]$ is the integer nearest to x and $[x + \frac{1}{2}] = [x]$ since $[x] \leq x + \frac{1}{2} = [x] + r + \frac{1}{2} < [x] + 1$. If $r \geq \frac{1}{2}$, then $[x] + 1$ is the integer nearest to x (choosing this integer if x is midway between $[x]$ and $[x + 1]$) and $[x + \frac{1}{2}] = [x] + 1$ since $[x] + 1 \leq x + r + \frac{1}{2} < [x] + 2$.

21. Let S be the set of all positive integers of the form $a - bk$. Now use the well-ordering principle on S.

23. **a.** true   **b.** false   **c.** false   **d.** false

25. Reduce to the case $1 < a < 2$, and hence $b > 2$. Let n be an integer such that $[ka] < n$ $< n + 1 \leq [(k + 1)a]$. Since $a = \frac{b}{b-1}$, we have $k \left(\frac{b}{b-1} \right) < n < n + 1 < k \left(\frac{b}{b-1} \right) + \frac{b}{b-1}$, from which we get $0 < nb - kb - n < nb - kb - n + b - 1 < b$. The first inequality implies $n < (n - k)b$. The last inequaltiy implies $(n - k)b - 1 < n$,

therefore, $[(n - k)b] = n$. For uniqueness, suppose $[ka] = [kb]$. Then, if $k > 0$, Exercise 20 tells us that $[a] = [b]$. But since $\frac{1}{a} + \frac{1}{b} = 1$, one of a and b must be greater than 2, and the other less than 2.

27. Assume that there are only finitely many Ulam numbers. Let the two largest Ulam numbers be u_{n-1} and u_n. Then the integer $u_n + u_{n-1}$ is an Ulam number larger than u_n, since it is the unique sum of two Ulam numbers since $u_i + u_j < u_n + u_{n-1}$ if $j < n$ or $j = n$ and $i < n - 1$, a contradiction.

Section1.2

1. **a.** 20   **b.** 55   **c.** 385   **d.** 2046
3. **a.** 243   **b.** 720   **c.** 14400   **d.** 32768
5. **a.** $5(\sum_{i=m}^{n} a_i)$ **b.** $-(\sum_{i=m}^{n} b_i)$ **c.** $(\sum_{i=m}^{n} a_i) - (\sum_{i=m}^{n} b_i)$
   d. $3(\sum_{i=m}^{n} a_i) + 4(\sum_{i=m}^{n} b_i)$.
7. $100^{100} > 100! > (50!)^2 > 2^{100}$
9. **a.** 315   **b.** 540   **c.** 90   **d.** 195
11. **a.** 2352   **b.** 6000   **c.** 1320   **d.** 5544
13. $\sum_{k=1}^{n}(\frac{1}{k(k+1)}) = \sum_{k=1}^{n}(\frac{1}{k} - \frac{1}{k+1})$. Let $a_j = \frac{1}{j+1}$. Using Exercise 12 we see that
   $\sum_{k=1}^{n}(\frac{1}{k(k+1)}) = \sum_{j=1}^{n}(a_{j-1} - a_j) = -(a_n - a_0) = 1 - \frac{1}{n+1}$.
15. Using Exercise 12, we see that $\sum_{k=1}^{n}((k+1)^2 - k^2) = (n+1)^2 - 1$, and $\sum_{k=1}^{n}(2k + 1) = 2\sum_{k=1}^{n} k + n$. Setting these two expressions equal and solving for $\sum_{k=1}^{n} k$, we find that $\sum_{k=1}^{n} k = \frac{n(n+1)}{2}$.
17. We sum both sides of the identity $(k+1)^4 - k^4 = 4k^3 + 6k^2 + 4k + 1$ from $k = 1$ to $k = n$. Using Exercises 12, 15, and 16, we find that $\sum_{k=1}^{n} k^3 = \frac{n^2(n+1)^2}{4}$.
19. **a.** $10! = (7!)(8 \cdot 9 \cdot 10) = (7!)(720) = (7!)(6!)$.
   b. $10! = (7!)(6!) = (7!)(5!) \cdot 6 = (7!)(5!)(3!)$.
   c. $16! = (14!)(15 \cdot 16) = (14!)(240) = (14!)(5!)(2!)$.
   d. $9! = (7!)(8 \cdot 9) = (7!)(6 \cdot 6 \cdot 2) = (7!)(3!)(3!)(2!)$.
21. Assume that $x \leq y$. Then $z! = x! + y! \leq y! + y! = 2(y!)$. Since $z > y$ we have $z! \geq (y + 1)y!$. This implies that $y + 1 = 2$ and $z = y + 1 = 2$. Hence the only solution with x, y, and z positive integers is $x = y = 1$ and $z = 2$.

Section1.3

1. $1 < 2^1$. Assume $n < 2^n$. We then have $n + 1 < 2^n + 1 < 2^n + 2^n = 2^{n+1}$.
3. $\sum_{k=1}^{1} \frac{1}{k^2} = 1 \leq 2 - \frac{1}{1}$. Assume that $\sum_{k=1}^{n} \frac{1}{k^2} \leq 2 - \frac{1}{n}$. Then, $\sum_{k=1}^{n+1} \frac{1}{k^2} = \sum_{k=1}^{n} \frac{1}{k^2} + \frac{1}{(n+1)^2} \leq 2 - \frac{1}{n+1}(1 - \frac{1}{n+1}) \leq 2 - \frac{1}{n+1}$.
5. $A^n = \begin{pmatrix} 1 & n \\ 0 & 1 \end{pmatrix}$.
7. $\sum_{j=1}^{1} j^2 = 1 = \frac{1(2)(3)}{6}$. Assume that $\sum_{j=1}^{n} j^2 = \frac{n(n+1)(2n+1)}{6}$. Then, $\sum_{j=1}^{n+1} j^2 = \sum_{j=1}^{n} j^2 + (n+1)^2 = \frac{n(n+1)(2n+1)}{6} + (n+1)^2 = \frac{(n+1)(n+2)[2(n+1)+1]}{6}$.
9. $\sum_{j=1}^{1} j(j+1) = 2 = \frac{1(2)(3)}{3}$. Assume it is true for n. Then $\sum_{j=1}^{n+1} j(j+1) = n(n+1)(n+2)/3 + (n+1)(n+2) = (n+1)(n+2)(n/3+1) = (n+1)(n+2)(n+3)/3$.
11. We have $\prod_{j=1}^{n} 2^j = 2^{\sum_{j=1}^{n} j} = 2^{n(n+1)/2}$ since $\sum_{j=1}^{n} j = \frac{n(n+1)}{2}$.
13. $3 + 3 \cdot 5^2 + 3 \cdot 5^4 + \cdots + 3 \cdot 5^{1000} = \sum_{k=0}^{500} 3 \cdot (5^2)^k = \frac{5^{1002}-1}{8}$.

15. $1 + \frac{1}{2} + \frac{1}{2^2} + \cdots + \frac{1}{2^{100}} = \sum_{k=0}^{100} (\frac{1}{2})^k = \frac{2^{101}-1}{2^{100}}$, since $\sum_{k=0}^{n} r^k = \frac{r^{n+1}-1}{r-1}$.

17. We see that $12 = 4 \cdot 3$. Now assume that postage of n cents can be formed, with $n = 4a + 5b$. To form $n + 1$ cents postage, if $a > 0$ we can replace a 4-cent stamp with a 5-cent stamp; that is, $n + 1 = 4(a - 1) + 5(b + 1)$. If no 4-cent stamps are present, then all 5-cent stamps were used. It follows that there must be at least three 5-cent stamps and these can be replaced by four 4-cent stamps; that is, $n + 1 = 4(a + 4) + 5(b - 3)$.

19. Note that $H_{2^1} = \sum_{j=1}^{2} 1/j = 3/2 \geq 1 + 1/2$, so the basis step holds. Assume that the inequality is true for n, that is, $H_{2^n} \geq 1 + \frac{n}{2}$. Then $H_{2^{n+1}} = \sum_{j=1}^{2^n} \frac{1}{j} + \sum_{j=2^n+1}^{2^{n+1}} \frac{1}{j} \geq H_{2^n} + \sum_{j=2^n+1}^{2^{n+1}} \frac{1}{2} \geq 1 + \frac{n}{2} + 2^n \cdot \frac{1}{2^{n+1}} = 1 + \frac{n}{2} + \frac{1}{2} = 1 + \frac{n+1}{2}$.

21. $(2 \cdot 1)! = 2 < 2^{2 \cdot 1}(1!)^2 = 4$. Assume that $(2n)! < 2^{2n}(n!)^2$. Then, $[2(n + 1)]! = (2n)!(2n + 1)(2n + 2) < 2^{2n}(n!)^2(2n + 1)(2n + 2) < 2^{2n}(n!)^2(2n + 2)^2 = 2^{2(n+1)}[(n + 1)!]^2$, as desired.

23. Let A be such a set. Define $B = \{x - k + 1 \mid x \in A \text{ and } x \geq k\}$. Then $A = \{x + k - 1 \mid x \in B\}$. Since $x \geq k$, B is a set of positive integers. Since $k \in A$ and $k \geq k, k - k + 1 = 1$ is in B. Since $n + 1$ is in A whenever n is, $n + 1 - k + 1$ is in B whenever $n - k + 1$ is. Thus B satisfies the hypothesis for mathematical induction, i.e. B is the set of positive integers. Then A is the set of integers greater than or equal to k.

25. For $n = 4$ we have $4^2 = 16 < 24 = 4!$. Assume that $n^2 < n!$. Then, $(n + 1)^2 = n^2 + 2n + 1 < n! + 2n + 1 < n \cdot n! + n! = (n + 1)n! = (n + 1)!$.

27. A puzzle of two pieces can be assembled in one move. Assume that a puzzle of n pieces requires $n - 1$ moves. Now, to assemble a puzzle of $n + 1$ pieces, first assemble n pieces, using $n - 1$ moves. Now make the move consisting of putting the two blocks, one of n pieces, the other of 1 piece, together. Thus, assembling a puzzle of $n + 1$ pieces takes $(n + 1) - 1$ moves.

29. Suppose that $f(n)$ is defined recursively by specifying the value of $f(1)$ and a rule for finding $f(n + 1)$ from $f(n)$. First, note that $f(1)$ is well-defined since this value is explicitly stated. Now assume that $f(n)$ is well-defined. Then $f(n + 1)$ also is well-defined since a rule is given for determining this value from $f(n)$.

31. $g(1) = 2, g(2) = 2^{g(1)} = 4, g(3) = 2^{g(2)} = 2^4 = 16$, and $g(4) = 2^{g(3)} = 2^{16}$.

33. The basis step consists of verifying the formula for $n = 1$ and $n = 2$. For $n = 1$ we have $f(1) = 1 = 2^1 + (-1)^1$ and for $n = 2$ we have $f(2) = 5 = 2^2 + (-1)^2$. Now assume that $f(k) = 2^k + (-1)^k$ for all positive integers k with $k < n$ where $n > 2$. By the inductive hypothiesis we have $f(n) = f(n - 1) + 2f(n - 2) = (2^{n-1} + (-1)^{n-1}) + 2(2^{n-2} + (-1)^{n-2}) = (2^{n-1} + 2^{n-1}) + (-1)^{n-2}(-1 + 2) = 2^n + (-1)^n$.

35. From Exercise 34, we see that $t_n = \sum_{j=1}^{n} j$, and $t_{n-1} = \sum_{j=1}^{n-1} j = \sum_{j=1}^{n-1}(n - j)$. Now, $t_{n-1} + t_n = \sum_{j=1}^{n-1}(n - j + j) + n = n \cdot n = n^2$.

37. **a.** 55   **b.** 233   **c.** 610   **d.** 2584   **e.** 6765   **f.** 75025

39. $\sum_{j=1}^{n} f_{2j-1} = f_{2n}$.

41. For $n = 1, f_2 f_0 - f_1^2 = 1 \cdot 0 - 1^2 = -1^1$, so the basis step holds. Assume that $f_{n+1}f_{n-1} - f_n^2 = (-1)^n$. Then $f_{n+2}f_n - f_{n+1}^2 = (f_{n+1} + f_n)f_n - f_{n+1}(f_n + f_{n-1}) = f_n^2 - f_{n+1}f_{n-1} = -(-1)^n = (-1)^{n+1}$.

43. $f_1 f_2 = 1 \cdot 1 = 1^2 = f_2^2$, so the basis step holds. By the induction hypothesis we have
$f_1 f_2 + \cdots + f_{2n-1}f_{2n} + f_{2n}f_{2n+1} + f_{2n+1}f_{2(n+1)} = f_{2n}^2 + f_{2n}f_{2n+1} + f_{2n+1}f_{2(n+1)} = f_{2n}(f_{2n} + f_{2n+1}) + f_{2n+1}f_{2(n+1)} = f_{2n}f_{2(n+1)} + f_{2n+1}f_{2(n+1)} = (f_{2n} + f_{2n+1})f_{2(n+1)} = f_{2(n+1)}^2$.

45. $f_1 = 1 < \alpha = 1.6\ldots$. Assume the statement is true for $n = 2, 3, \ldots, k$. Since α satisfies $x^2 - x - 1 = 0$, we have $f_{k+1} = f_k + f_{k-1} \leq \alpha^{n-1} + \alpha^{n-2} = \alpha^{n-2}(\alpha + 1) = \alpha^{n-2}\alpha^2 = \alpha^n$ as desired.

47. Let A be the set of positive integers which cannot be written as the sum of distinct Fibonacci numbers. By the well-ordering principle, A has a least element, say z. Write $z - 1$ as the sum of distinct Fibonacci numbers. If this sum does not include f_1 or f_2, then simply add f_1 or f_2 into the sum to get z as a sum of distinct Fibonacci numbers. Otherwise, suppose f_i is the least Fibonacci number not in this sum, then we can replace $f_1 + \cdots + f_{i-1}$ with f_i and get z by Example 1.15.

49. On one hand, $\det(F^n) = \det(F)^n = (-1)^n$. On the other hand, $\det\begin{pmatrix} f_{n+1} & f_n \\ f_n & f_{n-1} \end{pmatrix} = f_{n+1}f_{n-1} - f_n^2$.

51. By definition, $a_i = 3^i$ for $i = 0, 1, 2$. Assume that $a_k \leq 3^k$ for all integers k with $0 \leq k < n$ and $n \geq 2$. It follows that $a_n = a_{n-1} + a_{n-2} + a_{n-3} \leq 3^{n-1} + 3^{n-2} + 3^{n-3} = 3^{n-3}(1 + 3 + 9) = 13 \cdot 3^{n-3} < 27 \cdot 3^{n-3} = 3^n$.

53. Let P_n be the statement for n. Then P_2 is true, since we have $\left(\frac{a_1 + a_2}{2}\right)^2 - a_1 a_2 = \left(\frac{a_1 - a_2}{2}\right)^2 \geq 0$. Assume P_k is true. Then for $2k$ positive real numbers $a_1, \ldots, a_{2k}$ we have $a_1 + \cdots + a_{2k} \geq 2(\sqrt{a_1 a_2} + \sqrt{a_3 a_4} + \cdots + \sqrt{a_{2k-1} a_{2k}})$ by P_2. Apply P_k to this last expression to get $\geq 2n(a_1 a_2 \cdots a_{2n})^{1/2n}$ which establishes P_n for $n = 2^k$ for all k. Again, assume P_k is true. Let $g = (a_1 a_2 \cdots a_{n-1})^{1/(n-1)}$. Applying P_k, we have $a_1 + a_2 + \cdots + a_{k-1} + g \geq k(a_1 a_2 \cdots a_{k-1} g)^{1/k} = k(g^{k-1}g)^{1/k} = kg$. Therefore, $a_1 + a_2 + \cdots + a_{k-1} \geq (k-1)g$ which establishes P_{k-1}. Thus P_{2k} is true and P_n implies P_{n-1}. This estabishes P_n for all n.

55. $f_n = f_{n+2} - f_{n+1}$. Thus, $f_0 = 0, f_{-1} = 1, f_{-2} = -1, f_{-3} = 2, f_{-4} = -3, f_{-5} = 5, f_{-6} = -8, f_{-7} = 13, f_{-8} = -21, f_{-9} = 34, f_{-10} = -55$.

Section1.4

1. a. 1 **b.** 50 **c.** 1140 **d.** 462 **e.** 120 **f.** 1

3. a. $(a + b)^5 = a^5 + 5a^4 b + 10a^3 b^2 + 10a^2 b^3 + 5ab^4 + b^5$

 b. $(x + y)^{10} = x^{10} + 10x^9 y + 45x^8 y^2 + 120x^7 y^3 + 210x^6 y^4 + 252x^5 y^5 + 210x^4 y^6 + 120x^3 y^7 + 45x^2 y^8 + 10xy^9 + y^{10}$

 c. $(m - n)^7 = m^7 - 7m^6 n + 21m^5 n^2 - 35m^4 n^3 + 35m^3 n^4 - 21m^2 n^5 + 7mn^6 - n^7$

 d. $(2a + 3b)^4 = 16a^4 + 96a^3 b + 216a^2 b^2 + 216ab^3 + 81b^4$

 e. $(3x - 4y)^5 = 243x^5 - 1620x^4 y + 4320x^3 y^2 - 5760x^2 y^3 + 3840xy^4 - 1024y^5$

 f. $(5x + 7)^8 = 390625x^8 + 4375000x^7 + 21437500x^6 + 60025000x^5 + 105043750x^4 + 117649000x^3 + 82354300x^2 + 32941720x + 5764801$

5. On the one hand, $[1 + (-1)]^n = 0^n = 0$. On the other hand, by the binomial theorem, $\sum_{k=0}^{n}(-1)^k \binom{n}{k} = [1 + (-1)]^n$.

7. $\binom{n}{r}\binom{r}{k} = \frac{n!}{r!(n-r)!} \cdot \frac{r!}{k!(r-k)!} = \frac{n!(n-k)!}{k!(n-k)!(n-r)!(n-k-n+r)} = \binom{n}{k}\binom{n-k}{n-r}$.

9. The basis step is omitted. Assume that $\binom{r}{r} + \binom{r+1}{r} + \cdots + \binom{n}{r} = \binom{n+1}{r+1}$ is true whenever r is an integer with $1 \leq r \leq n$. We will now examine the formula with $n + 1$ in the place of n. If $r < n + 1$, then $\binom{r}{r} + \binom{r+1}{r} + \cdots + \binom{n}{r} + \binom{n+1}{r} = \binom{n+1}{r+1} + \binom{n+1}{r} = \binom{n+2}{r+2}$ by Theorem 1.5, so the formula holds in this case. If $r = n + 1$, then $\binom{r}{r} + \cdots + \binom{n+1}{r} = \binom{n+1}{n+1} = 1 = \binom{n+2}{n+2}$

11. We use Exercise 44 in Section 1.3. $\alpha^2 = \alpha + 1$ and $\beta^2 = \beta + 1$, since they are roots of $x^2 - x - 1 = 0$. Then we have $f_{2n} = (\alpha^{2n} - \beta^{2n})/\sqrt{5} = (1/\sqrt{5})((\alpha+1)^n - (\beta + 1)^n) = (1/\sqrt{5}) \left(\sum_{j=0}^{n} \binom{n}{j} \alpha^j - \sum_{j=0}^{n} \binom{n}{j} \beta^j \right) = (1/\sqrt{5}) \sum_{j=0}^{n} \binom{n}{j} (\alpha^j - \beta^j) = \sum_{j=1}^{n} \binom{n}{j} f_j$ since the first term is zero in the second to last sum.

13. Using Exercise 12, $\binom{x}{n} + \binom{x}{n+1} = \frac{x!}{n!(x-n)!} + \frac{x!}{(n+1)!(x-n-1)!} = \frac{x!(n+1)}{(n+1)!(x-n)!} + \frac{x!(x-n)}{(n+1)!(x-n)!} = \frac{x!(x-n+n+1)}{(n+1)!(x-n)!} = \frac{(x+1)!}{(n+1)!(x-n)!} = \binom{x+1}{n+1}$.

15. Let S be a set of n copies of $x + y$. Consider the ocefficient of $x^k y^{n-k}$ in the expansion of $(x + y)^n$. Choosing the x from each element of a k element subset of S, we notice that the coefficient of $x^k y^{n-k}$ is the number of k element subsets of S, $\binom{n}{k}$.

17. By counting elements with exactly 0, 1, 2, and 3 properties, we see that only elements with exactly 0 properties are counted in $n - [n(P_1) + n(P_2) + n(P_3)] + [n(P_1 P_2) + n(P_1 P_3) + n(P_2 P_3)] - [n(P_1 P_2 P_3)]$, and those only once.

19. A term of the sum is of the form $ax_1^{k_1} x_2^{k_2} \cdots x_m^{k_m}$ where $k_1 + k_2 + \cdots + k_m = n$ and $a = \frac{n!}{k_1! k_2! \cdots k_m!}$.

21. 27720

Section1.5

1. We find that $3 \mid 99$ since $99 = 3 \cdot 33$, $5 \mid 145$ since $145 = 5 \cdot 29$, $7 \mid 343$ since $343 = 49 \cdot 7$, and $888 \mid 0$ since $0 = 0 \cdot 888$.

3. **a.** $0 \mid 7$ **b.** $7 \mid 707$ **c.** $7 \nmid 1717$ **d.** $7 \mid 123321$ **e.** $7 \nmid -285714$ **f.** $7 \nmid -430597$

5. **a.** $100 = 5 \cdot 17 + 15$, **b.** $289 = 17 \cdot 17$ **c.** $-44 = -3 \cdot 17 + 7$ **d.** $-100 = -6 \cdot 17 + 2$

7. By hypothesis $b = ra$ and $d = sc$, for some r and s. Thus $bd = rs(ac)$ and $ac \mid bd$.

9. If $a \mid b$, then $b = na$ for some integer n, and $bc = n(ca)$, i.e. $ac \mid bc$. Now, suppose $ac \mid bc$. Thus $bc = nac$ and, as $c \neq 0$, $b = na$, i.e. $a \mid b$.

11. $a \mid b$ if and only if $b = na$, for some integer n, if and only if $b^k = n^k a^k$ if and only if $a^k \mid b^k$.

13. Let a and b be odd, and c even. Then $ab = (2x + 1)(2y + 1) = 4xy + 2x + 2y + 1 = 2(2xy + x + y) + 1$, so ab is odd. On the other hand $ac = (2x + 1)(2z) = 2(2xz + z)$ so ac is even.

15. By the division algorithm, $a = bq + r$, with $0 \leq r < b$. Thus $-a = -bq - r = -(q+1)b + b - r$. If $0 \leq b - r < b$ then we are done. Otherwise $b - r = b$, or $r = 0$ and $-a = -qb + 0$.

17. **a.** Follow the proof of the division algorithm.
    b. $17 = -7(-2) + 3$. Here $r = 3$.

19. If $m = kn - 1$, then $\left[\frac{m+1}{n}\right] = \left[\frac{kn}{n}\right] = k$, whereas $\left[\frac{m}{n}\right] = \left[\frac{kn-1}{n}\right] = k + \left[\frac{-1}{n}\right] = k - 1$. Thus $\left[\frac{m+1}{n}\right] = \left[\frac{m}{n}\right] + 1$, if $m = kn - 1$ for some integer k. By the division algorithm, $m = kn + r$ for some k, r. Then $\left[\frac{m+1}{n}\right] = \left[\frac{kn+r+1}{n}\right] = k + \left[\frac{r+1}{n}\right] = k$. And $\left[\frac{m}{n}\right] = \left[\frac{kn+r}{n}\right] = k$.

21. The positive integers divisible by the positive integer d are those integers of the form kd where k is a positive integer. The number of these that are less than x is the number of positive integers k with $kd \leq x$, or equivalently with $k \leq \frac{x}{d}$. There are $\left[\frac{x}{d}\right]$ such integers.

23. 128, 18

25. 462

27. When x is the weight of a letter in ounces, and c is the cost of mailing the letter in cents, we have $c = 22 + 17[x + \frac{1}{2}]$, \$ 1.08 is not a possible cost, whereas an 8 ounce letter will cost \$ 1.58 to mail.

29. $(4n + 1)(4m + 1) = 16mn + 4m + 4n + 1 = 4(4mn + m + n) + 1$, $(4n + 3)(4m + 3) = 16mn + 12m + 12n + 9 = 4(4mn + 3m + 3n + 2) + 1$.

31. Every odd integer may be written in the form $4k + 1$ or $4k + 3$. Observe that $(4k + 1)^4 = 16^2 k^4 + 4(4k)^3 + 6(4k)^2 + 4(4k) + 1 = 16(16k^4 + 16k^3 + 6k^2 + k) + 1$. Proceeding further, $(4k + 3)^4 = (4k)^4 + 12(4k)^3 + 36(4k)^2 + 108(4k) + 3^4 = 16(16k^4 + 48k^3 + 54k^2 + 27k + 5) + 1$.

33. Of any consecutive three integers, one is a multiple of three. Also, at least one is even. Therefore, the product is a multiple of $2 \cdot 3 = 6$.

35. Note that $(a - 1)a(a + 1)$ is a multiple of three by Exercise 26, and then use Exercise 9.

37. The Fibonacci sequence reduced modulo 3 is $1, 1, 2, 0, 2, 2, 1, 0, 1, 1, 0, \ldots$, which is periodic with period 8, since the 9th and 10th entries are the same as the 1st and 2nd. Since only the Fibonacci numbers corresponding to 0 in the list are divisible by 3, and these 0's are only in positions $4n$, with n an integer, the statement is proved.

39. Repeatedly use the recursive definition of the Fibonacci numbers. Use induction as in Exercises 34, 35, and 36 to show that $5 \mid f_{5k}$.

41. $T(39) = 59; T(59) = 89; T(89) = 134; T(134) = 67; T(67) = 101; T(101) = 152; T(152) = 76; T(76) = 38; T(38) = 19; T(19) = 29; T(29) = 44; T(44) = 22; T(22) = 11; T(11) = 17; T(17) = 26; T(26) = 13; T(13) = 20; T(20) = 10; T(10) = 5; T(5) = 8; T(8) = 4; T(4) = 2; T(2) = 1$.

43. The basis step is omitted. Assume that the conjecture holds for all integers less that n. By assumption there is an integer k such that k iterations of the transfromation T, starting at n, produces an integer m less than n. By the inductive hypothesis there is an integer l such that iterating T l times produces the integer 1. Hence iterating T $k + l$ times starting with n leads to 1.

45. We first show that $(2 + \sqrt{3})^n + (2 - \sqrt{3})^n$ is an even integer. By the binomial theorem it follows that $(2 + \sqrt{3})^n + (2 - \sqrt{3})^n = \sum_{j=0}^{n} 2^j \sqrt{3}^{n-j} + \sum_{j=0}^{n} 2^j (-1)^{n-j} \sqrt{3}^{n-j} = 2(2^n + 3 \cdot 2^{n-2} + 3^2 \cdot 2^{n-4} + \cdots) = 2l$ where l is an integer. Next, note that $(2 - \sqrt{3})^n < 1$. We see that $[(2 + \sqrt{3})^2] = (2 + \sqrt{3})^n + (2 - \sqrt{3})^n - 1$. It follows that $[(2 + \sqrt{3})^2]$ is odd.

Section1.6

1. $(1999)_{10} = (5554)_7, (6105)_7 = (2112)_{10}$.

3. $(1010111)_2 = (175)_{10}, (999)_{10} = (1111100111)_2$.

5. $(100011110101)_2 = (8F5)_{16}$.

7. Because we are using the blocks of three digits as one 'digit,' which has 1000 possible values.

9. $(101001)_{-2} = -39$ and $(12012)_{-3} = 26$.

11. This follows directly from Theorem 1.10.

13. Let w be the weight to be measured. By Exercise 10, w has a unique balanced ternary expansion. Place the object in pan 1. If $e_i = 1$ then place a weight of 3^i into pan 2. If $e_i = -1$ then place a weight of 3^i in pan 1. If $e_i = 0$ then do not use the weight of 3^i. Now the pans will be balanced.

15. To convert a number from base r to base r^n, take the number in blocks of size n. To go the other way, convert each digit of a base r^n number to base r, and concatenate the results.

17. Multiplying n by b^m gives $b^m n = b^m (a_k b^k + a_{k-1} b^{k-1} + \cdots + a_1 b + a_0) = (a_k b^{k+m} + a_{k-1} b^{k+m-1} + \cdots + a_1 b^{m+1} + a_0 b^m + 0 \cdot b^{m-1} + \cdots + 0) = (a_k a_{k-1} \ldots a_1 a_0 00 \ldots 00)_b$, where we have placed m zeroes at the end of the base b expansion of n.

19. a. -6 b. 13 c. -14 d. 0

21. Since 4 bits are required for every decimal digit, $4n$ bits are required to store the number in this manner.

23. To find a Cantor expansion of the positive integer n, let m be the unique positive integer such that $m! \leq n < (N+1)!$. By the division algorithm there is an integer a_m such that $n = m! \cdot a_m + r_m$ where $0 \leq a_m \leq m$ and $0 \leq r_m < m!$. We iterate, finding that $r_m = (m-1)! \cdot a_{m-1} + r_{m-1}$ where $0 \leq a_{m-1} \leq m - 1$ and $0 \leq r_{m-1} < (m-1)!$. We iterate $m - 2$ more times, where we have $r_i = (i-1)! \cdot a_{i-1} + r_{i-1}$ where $0 \leq a_{i-1} \leq i - 1$ and $0 \leq r_{i-1} < (i-1)!$ for $i = m+1, m, m-1, \ldots, 2$ with $r_{m+1} = n$. At the last stage we have $r_2 = 1! \cdot a_1 + 0$ where $r_1 = 0$ or 1 and $r_2 = a_1$. Now we must show that this expansion is unique. So suppose that n has two different Cantor expansions $n = a_m m! + a_{m-1}(m-1)! + \cdots + a_2 2! + a_1 1! = b_m m! + b_{m-1}(m-1)! + \cdots + b_2 2! + b_1 1!$, where a_j and b_j are integers, and $0 \leq a_j \leq j$ and $0 \leq b_j \leq j$ for $j = 1, 2, \ldots, m$. Suppose that k is the largest integer such that $a_k \neq b_k$, and without loss of generality, assume $a_k > b_k$, which implies that $a_k \geq b_k + 1$. Then $a_k k! + a_{k-1}(k-1)! + \cdots + a_1 1! = b_k k! + b_{k-1}(k-1)! + \cdots + b_1 1!$. Using the identity $\sum_{j=1}^{k} j \cdot j! = (k+1)! - 1$, proved in Exercise 16 of Section 1.3, we see that $b_k k! + b_{k-1}(k-1)! + \cdots + b_1 1! \leq b_k k! + (k-1) \cdot (k-1)! + \cdots + 1 \cdot 1! \leq b_k k! + k! - 1 = (b_k + 1)k! - 1 < a_k k!$. This is a contradiction, so the expansion is unique.

25. Call a position good if the number of ones in each column is even, and bad otherwise. Since a player can only affect one row, he or she must affect (adversely) some column sums. Thus any move from a good position produces a bad position. To find a move from a bad position to a good one, construct a binary number by putting a 1 in the place of column with odd sum, and a 0 in the place of a column with even sum. Subtracting this number of matches from the largest pile will arrive at a good position.

27. **a.** First show that the result of the operation must yield a multiple of 9. Then, it suffices to check only multiples of 9 with decreasing digits. There are only 79 of these. If we perform the operation on each of these 79 numbers and reorder the digits, we will have one of the following 23 numbers: 7551, 9954, 5553, 9990, 9981, 8820, 9810, 9620, 8532, 8550, 9720, 9972, 7731, 6543, 8730, 8640, 8721, 7443, 9963, 7632, 6552, 6642, or 6174. It will suffice to check only 9810, 7551, 9990, 8550, 9720, 8640, and 7632.

b. From the solution in part (a), construct a tree from the last seven numbers. The longest branch is 6 steps. Every number will reach the tree in two steps. The maximum is given by 8500 (for instance) which takes 8 steps.

29. Consider $a_0 = (3542)_6$. Then, T_6 repeats with period 2. Hence, there is no Kaprekar's constant for the base 6.

Section 1.7

1. $(10010110110)_2$.

3. $(10110001101)_2$.

5. $(1234321)_5 + (2030104)_5 = (3314430)_5$.

7. $(4320023)_5$.

9. $(16665)_{16}$.

11. $(FACE)_{16} \cdot (BAD)_{16} = (B705736)_{16}$.

13. We represent the integer $(18235187)_{10}$ using three words: $((018)(235)(187))_{1000}$ and the integer $(22135674)_{10}$ using three words: $((022)(135)(674))_{1000}$, where each base 1000 digit is represented by three base 10 digits in parentheses. To find the sum, difference, and product of these integers from their base 1000 we carry out the algorithms for such computations for base 1000.

15. To add numbers using the one's complement representation, first decide if the answer will be negative or positive. To do this is easy if both numbers have the same lead (sign) bit; otherwise conduct a bit-by-bit comparison of a positive summand's digits and the complement of the negative's. Now, add the other digits (all but the initial (sign) bit) as an ordinary binary number. If either summand (or both) is negative, then add $(1)_2$ to this answer. Also, add an appropriate sign bit to the front of the number.

17. Let $a = (a_m a_{m-1} \ldots a_2 a_1)_!$ and $b = (b_m b_{m-1} \ldots b_2 b_1)_!$. Then $a + b$ is obtained by adding the digits from right to left with the following rule for producing carries. If $a_j + b_j + c_{j-1}$, where c is the carry from adding a_{j-1} and b_{j-1}, is greater than or equal to j, then $c_j = 1$, and the resulting jth digit is $a_j + b_j + c_{j-1} - j$. Otherwise, $c_j = 0$. To subtract b from a, assuming $a > b$, we let $d_i = a_i - b_i + c_{i-1}$ and set $c_i = 0$ if $a_i - b_i + c_{i-1}$ is between 0 and j. Otherwise, $d_j = a_j - b_j + c_{j-1} + j + 1$ and set $c_j = -1$. In this manner, $a - b = (d_m d_{m-1} \ldots d_2 d_1)_!$.

19. We have $(a_n \ldots a_1 5)_{10}^2 = (10(a_n \ldots a_1)_{10} + 5)^2 = 100(a_n \ldots a_1)_{10}^2 + 100(a_n \ldots a_1)_{10} + 25 = 100(a_n \ldots a_1)_{10}((a_n \ldots a_1)_{10} + 1) + 25$. The decimal digits of this number consist of the decimal digits of $(a_n \ldots a_1)_{10}(a_n \ldots a_1)_{10} + 1$ followed by 25 since this first product is multiplied by 100 which shifts its decimal expansion two digits.

Section 1.8

1. **a.** $2n + 7 = O(n)$ **b.** $n^2/3$ is not $O(n)$ **c.** $10 = O(n)$ **d.** $\log(n^2 + 1) = O(n)$.
   e. $\sqrt{n^2 + 1} = O(n)$.   **f.** $\frac{n^2 + 1}{n + 1} = O(n)$.

3. First note that $(n^3 + 4n^2 \log n + 101n^2)$ is $O(n^3)$, and that $(14n \log n + 8n)$ is $O(n \log n)$ as in Example 1.36. Now applying Theorem 1.12 yields the result.

5. Use Exercise 4 and follow Example 1.34, noting that $(\log n)^3 \leq n^3$ whenever n is a positive integer.

7. Using Exercise 4, $\log n! \leq \log n^n = n \log n$. Hence, $\log n! = O(n \log n)$.

9. Suppose that $f = O(g)$ where $f(n)$ and $g(n)$ are positive integers for every integer n. Then there is an integer C such that $f(n) < Cg(n)$ for all $x \in S$. Then $f^k(n) < c^k g^k(n)$ for all $x \in S$. Hence $f^k = O(g^k)$.

11. The number of digits in the base b expansion of n is $1 + k$ where k is largest such that $b^k \leq n < b^{k+1}$ since there is a digit for each of the powers of $b^0, b^1, \ldots b^k$. Note that this inequality is equivalent to $k \leq \log_b n < k + 1$, so that $k = \lfloor \log_b n \rfloor$. Hence there are $\lfloor \log_b n \rfloor + 1$ digits in the base b expansion of n.

13. To multiply an n digit integer by an m digit integer in the conventional manner, one must multiply every digit of the first number by every digit of the second number. There are nm such pairs.

15. **a.** The number of bits in $n!$ does not exceed $\log n! + 1 = 1 + \sum \log j < \sum \log n = n \log n$. Then we have $n - 1$ multiplications, which gives $O((n - 1)(n \log n + 1)) = O(n^2 \log n)$ bit operations.
    b. We need to find three factorials, which will have the same big-O value as in part (a). We will also need to perform one subtraction (which will not affect the big-O value), one multiplication and one division. The factorials have at most $n \log n$ bits, so by Theorem 1.14, the multiplication will take at most $O((n \log n)^{1+\epsilon})$ bit operations. By Theorem 1.16, the division will take $O(M(n)) = O((n \log n)^{1+\epsilon})$, so in total we have the number of bit operations is $3O(n^2 \log n) + O((n \log n)^{1+\epsilon}) + O((n \log n)^{1+\epsilon}) = O(n^2 \log n)$.

17. $(1100011)_2$

19. **a.** $ab = (10^{2n} + 10^n)A_1 B_1 + 10^n(A_1 - A_0)(B_0 - B_1) + (10^n + 1)A_0 B_0$ where A_i and B_i are defined as in identity (1.9).
    b. $73 \cdot 87 = (10^2 + 10)7 \cdot 8 + 10(7 - 3)(7 - 8) + (11)3 \cdot 7 = 6160 - 40 + 231 = 6351$

  c. $4216 \cdot 2733 = (10100)42 \cdot 27 + (100)(42 - 16)(33 - 27) + (101)16 \cdot 33$. Then, $42 \cdot 27 = (10^2 + 10)4 \cdot 2 + 10(4 - 2)(7 - 2) + (11)2 \cdot 7 = 1134$, and, $26 \cdot 06 = (10^2 + 10)2 \cdot 0 + 10(2 - 6)(6 - 0) + (11)6 \cdot 6 = 156$, and $16 \cdot 33 = (10^2 + 10)1 \cdot 3 + 10(6 - 1)(3 - 3) + (11)6 \cdot 3 = 528$. Then $4216 \cdot 2733 = (10100)1134 + (100)156 + (101)528 = 11522328$.

21. That the given equation is an identity may be seen by direct calculation. The seven multiplications necessary to use this identity are: $a_{11}b_{11}, a_{12}b_{21}, (a_{11} - a_{21} - a_{22})(b_{11} - b_{21} - b_{22}), (a_{21} + a_{22})(b_{12} - b_{11}), (a_{11} + a_{12} - a_{21} - a_{22})b_{22}, (a_{11} - a_{21})(b_{22} - b_{12}), a_{22}(b_{11} - b_{21} - b_{12} + b_{22})$.

23. Let $k = \lceil \log_2 n \rceil + 1$, then the number of multiplications for a $2^k \times 2^k$ matrices is $O(7^k)$. But, $7^k = 2^{(\log_2 7)(\lceil \log_2 n \rceil + 1)} = O(2^{\log_2 n \, \log_2 7} 2^{\log_2 7}) = O(n^{\log_2 7})$. The other operations are absorbed into this term.

Section 1.9

1. a. 101 prime   **b.** 103 prime   **c.** 107 prime   **d.** 111 not prime   **e.** 113 prime   **f.** 121 not prime.

3. 2, 3, 5, 7, 11, 13, 17, 19, 23, 29, 31, 37, 41, 43, 47, 53, 59, 61, 67, 71, 73, 79, 83, 89, 97, 101, 103, 107, 109, 113, 127, 131, 137, 139, 149

5. Suppose that $n = x^4 - y^4 = (x - y)(x + y)(x^2 + y^2)$, where $x > y$. The integer n cannot be prime since it divisible by $x + y$ which cannot be 1 or n.

7. Using the identity given in the hint with some k such that $1 < k < n$ and $k \mid n$, then $a^k - 1 \mid a^n - 1$. Since $a^n - 1$ is prime by hypothesis, $a^k - 1 = 1$. From this, we see that $a = 2$ and $k = 1$, a contradiction. Thus $a = 2$ and n is prime.

9. $Q_1 = 3, Q_2 = 7, Q_3 = 31, Q_4 = 211, Q_5 = 2311, Q_6 = 30031$. The smallest prime factors are 3, 7, 31, 211, 2311, and 59.

11. If n is prime, we are done. Otherwise $n/p < (\sqrt[3]{n})^2$. If n/p is prime, then we are done. Otherwise, by Theorem 1.18, n/p has a prime factor less than $\sqrt{n/p} < \sqrt[3]{n}$.

13. Suppose that $p, p + 2$, and $p + 4$ were all prime. We consider three cases. First, suppose that p is of the form $3k$. Then p cannot be prime unless $k = 1$, and the prime triplet is 3, 5, and 7. Next, suppose that p is of the form $3k + 1$. Then $p + 2 = 3k + 3 = 3(k + 1)$, is not prime. We obtain no prime triplets in this case. Finally, suppose that p is of the form $3k + 2$. Then $p + 4 = 3k + 6 = 3(k + 2)$ is not prime. We obtain no prime triplet in this case either. Since the three cases are exhaustive, we have only 1 prime triplet of this kind, 3, 5, and 7.

15. a. 7 = 3+2+2.   **b.** 17 = 11+3+3.   **c.** 27 = 23+2+2.   **d.** 97 = 89+5+3.   **e.** 101 = 97+2+2.   **f.** 199 = 191+5+3.

17. If n is prime the statement is true for n. Otherwise, n is composite, so that n is the product of two integers a and b such that $1 < a \le b < n$. Since $n = ab$ and by the inductive hypothesis both a and b are the product of primes, we conclude that n is also the product of primes.

19. $\pi(250) = (\pi(\sqrt{250}) - 1) + 250 - ([\frac{250}{2}] + [\frac{250}{3}] + [\frac{250}{5}] + [\frac{250}{7}] + [\frac{250}{11}] + [\frac{250}{13}]) + ([\frac{250}{6}] + [\frac{250}{10}] + [\frac{250}{14}] + [\frac{250}{22}] + [\frac{250}{26}] + [\frac{250}{15}] + [\frac{250}{21}] + [\frac{250}{33}] + [\frac{250}{39}] + [\frac{250}{35}] + [\frac{250}{55}] + [\frac{250}{65}] + [\frac{250}{77}] + [\frac{250}{91}] + [\frac{250}{143}]) - ([\frac{250}{30}] + [\frac{250}{42}] + [\frac{250}{66}] + [\frac{250}{78}] + [\frac{250}{105}] + [\frac{250}{165}] + [\frac{250}{195}]) + ([\frac{250}{210}]) = 53$.

21. Let x_0 be a positive integer. It follows that $f(x_0) = p$ where p is prime. Let k be an integer. We have $f(x_0 + kp) = a_n(x_0 + kp)^n + \cdots + a_1(x_0 + kp) + a_0$. Note that by the binomial theorem, $(x_0 + kp)^j = \sum_{i=1}^{j} \binom{j}{i} x_0^{j-i}(kp)^i$. It follows that $f(x_0 + kp) = \sum_{j=0}^{n} a_j x_0^j + Np = f(x_0) + Np$, where $N = \sum_{j=1}^{n} N_j$ is an integer. Since $p \mid f(x_0)$ it follows that $p \mid (f(x_0) + Np) = f(x_0 + kp)$. Since $f(x_0 + kp)$ is supposed to be prime, it follows that $f(x_0 + kp) = p$ for all integers

k. This contradicts the fact that a polynomial of degree n takes on each value no more that n times. Hence $f(y)$ is composite for at least one integer y.

23. At each stage of the procedure for generating the lucky numbers the smallest number left, say k, is designated to be a lucky number and infinitely many numbers are left after the deletion of every kth integer left. It follows that there are infinitely many steps, and at each step a new lucky number is added to the sequence. Hence there are infinitely many lucky numbers.

25. If $p^{\alpha} - q^{\beta} = 1$, with p, q primes, then p or q is even, so p or q is 2. If $p = 2$, there are several cases: we have $2^{\alpha} - q^{\beta} = 1$. If α is even, say $\alpha = 2k$, $(2^{2k} - 1) = (2^k - 1)(2^k + 1) = q^{\beta}$. So $q|(2^k - 1)$ and $q|(2^k + 1)$, hence $q = 1$, a contradiction. If α is odd and β is odd, $2^{\alpha} = 1 + q^{\beta} = (1 + q)(q^{\beta - 1} - q^{\beta - 2} + \cdots + 1)$. So $1 + q = 2^n$ for some n. Then $2^{\alpha} = (2^n - 1)^{\beta} + 1 = 2^n(\text{odd number})$. So $2^{\alpha - n} = $ odd number and so $\alpha = n$. Therefore $2^{\alpha} = 1 + (2^{\alpha} - 1)^{\beta}$ and so $\beta = 1$ which is not allowed. If $\alpha = 2k + 1$ and $\beta = 2n$ we have $2^{2k+1} = 1 + q^{2n}$. Since q is odd, q^2 is of the form $4m + 1$, and by the Binomial Theorem, so is q^{2n}. Thus the right hand side of the last equation is of the form $4m + 2$, but this forces $k = 0$, a contradiction. If $q = 2$, we have $p^{\alpha} - 2^{\beta} = 1$. Whence $2^{\beta} = (p - 1)(p^{\alpha - 1} + p^{\alpha - 2} + \cdots + p + 1)$, where the last factor is the sum of α odd terms but must be a power of 2, therefore, $\alpha = 2k$ for some k. Then $2^{\beta} = (p^k - 1)(p^k + 1)$. These last two factors are powers of 2 which differ by 2 which forces $k = 1, \alpha = 2, \beta = 3, p = 3$, and $q = 2$ as the only solution.

27. Since $3p > 2n$, p, and $2p$ are the only multiples of p that appear as factors in $(2n)!$. So $p \mid (2n)!$ exactly twice. Since $2p > n$, p is the only multiple of p that appears as a factor in $n!$. So $p \mid n!$ exactly once. Then since $\binom{2n}{n} = \frac{2n!}{n!n!}$, the two factors of p in the numerator are cancelled by the two in the denominator.

29. By Bertrand's conjecture, there must be a prime in each interval of the form $[2^{k-1}, 2^k]$, for $k = 2, 3, 4, \ldots$. Thus, there are at least $k - 1$ primes less than 2^k. Since the prime 2 isn't counted here, we have at least k primes less than 2^k.

Section 2.1

1. a. 5. b. 111 c. 6 d. 1 e. 11 f. 2

3. $(a, 2a) = a$.

5. $(a + 1, a) = 1$.

7. By Theorem 2.2, $(ca, cb) = cma + cnb = |c| \cdot |ma + nb|$, where $cma + cnb$ is as small as possible. Therefore, $|ma + nb|$ is as small as possible, i.e. equal to (a, b).

9. 2 if a and b are both odd, 1 otherwise.

11. Let $a = 2k$. Since $(a, b) \mid b$, and b is odd, (a, b) is odd. Thus if $(a, b) \mid a = 2k$, $(a, b) \mid k$ because (a, b) is odd. So $(a, b) = (k, b) = (\frac{a}{2}, b)$.

13. Let $d = (a, b), e = (a, c)$. Then $1 = (a/d, b/d)(a/e, c/e) = (a/de, bc/de) = 1$. Now multiply by de.

15. Let p, q, r, be primes. Then $(pq, qr, pr) = 1$, but no pair of the numbers is relatively prime.

17. a. 2. b. 5 c. 99 d. 3 e. 7 f. 1001

19. Let $d \mid a_i, 1 \leq i \leq n$. Then clearly $dc \mid ca_i, 1 \leq i \leq n$. So $dc \mid (ca_1, ca_2, \ldots, ca_n)$. To see the other direction, note that $c \mid ca_i$ for all i, so $c \mid (ca_1, ca_2, \ldots, ca_n) = d$. Express d as $d = cd'$, where d' is as large as possible. But since $cd' \mid ca_i, d' \mid a_i$ and $d' = (a_1, a_2, \ldots, a_n)$.

21. Suppose that $(6k + a, 6k + b) = d$. Then $d \mid b - a$. We have $a, b \in \{-1, 1, 2, 3, 4\}$, so if $a < b$ it follows that $b - a \in \{1, 2, 3, 4, 6\}$. Hence $d \in \{1, 2, 3, 4, 5\}$. To show that $d = 1$ it is sufficient to show that neither 2 nor 3 divides $(6k + a, 6k + b)$. If $p = 2$ or $p = 3$ and $p \mid (6k + a, 6k + b)$ then $p \mid a$ and $p \mid b$. However, there are no such pairs a, b in the set $\{-1, 1, 2, 3, 5\}$.

23. From Exercise 21, we know that $6k - 1, 6k + 1, 6k + 2, 6k + 3$, and $6k + 5$ are pairwise relatively prime. To represent n as the sum of two relatively prime integers greater than one, let $n = 12k + h, 0 \leq h < 12$. Then h can be written as the sum of two elements, r and s from the set $\{-1, 1, 2, 3, 5\}$. Then $n = (6k + r) + (6k + s)$, which are relatively prime.

25. Let S be the set of all fractions $\frac{P}{Q} = \frac{xa+ye}{xb+yf}$ where x, y are relatively prime positive integers. Then every element of S lies between $\frac{a}{b}$ and $\frac{c}{f}$ and is in lowest terms. The first element of S to appear in a Farey series will have the smallest Q, i.e. $x = y = 1$. This fraction must be $\frac{c}{d}$ by hypothesis.

27. Since $\frac{a}{b} < \frac{a+c}{b+d} < \frac{c}{d}$, then $b + d > n$, or $\frac{a}{b}$ and $\frac{c}{d}$ would not be consecutive.

29. Since $\frac{a}{b} + \frac{c}{d} = \frac{ad+bc}{bd}$ is an integer, $bd \mid ad + bc$. Then, $bd \mid d(ad + bc) = ad^2 + cbd$. Now since $bd \mid cbd$, it must be that $bd \mid ad^2$. From this, $bdn = ad^2$ for some integer n, and it follows that $bn = ad$, or $b \mid ad$. Using Exercise 11, we see that $b \mid (a, b)(d, b) = (d, b) \mid d$. Similarly, we can find that $d \mid b$. Hence, as b and d are relatively prime, $b = d$.

31. Suppose that $ma + nb$ is not an integer multiple of (a, b); then by the division algorithm $ma + nb = j(a, b) + k$ where j, k are integers, with k less than (a, b). But we know that (a, b) can be written as the least linear combination, say $(a, b) = ra + sb$, where r, s are integers. But then $(m - jr)a + (n - js)b = ma + nb - j(ra + sb) = j(a, b) + k - j(a, b) = k$ is a linear combination of a and b which is less than (a, b), a contradiction. It follows that every linear combination a and b is a multiple of (a, b). To complete the proof, note that $j(a, b) = j(ra + sb) = (jr)a + (js)b$.

Section 2.2

1. **a.** 15. **b.** 6 **c.** 2 **d.** 5
3. **a.** $(-1)75 + (2)45$. **b.** $(6)222 + (-9)102$. **c.** $-138(666) + (65)1414$
   d. $-1707(20785) + 800(44350)$
5. **a.** 1. **b.** 7 **c.** 5
7. **a.** $(6, 10, 15) = 1 = (1)6 + (1)10 + (-1)15$. **b.** $70, 98, 105) = 7 = (0)70 + (-1)98 + (1)105$.
   c. $(280, 330, 405, 490) = 5 = (-13)280 + (0)330 + (9)405 + (0)490$
9. 2.
11. When $n = 1, a = 1$ and $b = 0$. Then $(a, b) = (1, 0) = 1$, so the algorithm takes $0 = 2 \cdot 1 - 2$ steps. Now assume that the algorithm uses $2n - 2$ steps. To find the g.c.d. of $\frac{2^{n+1}-(-1)^{n+1}}{3}$ and $\frac{2(2^n-(-1)^n)}{3}$, the first step reduces this to the g.c.d. of $\frac{2^{n+1}-(-1)^{n+1}}{3}$ and $\frac{2^n-(-1)^n}{3}$. The next step, as neither of these numbers is even, gives us $\frac{2^n-(-1)^n}{3}$ and $\frac{1}{3}(2^{n+1} - (-1)^{n+1} - 2^n + (-1)^n) = \frac{1}{3}(2^n + 2(-1)^n) = \frac{2}{3}(2^{n-1} - (-1)^{n-1})$. By the inductive hypothesis, the algorithm will take $2n - 2$ more steps, for a total of $2n = 2(n - 1) - 2$ steps.
13. Replace all 2's by 3's in the binary algorithm.
15. Show that if c and d are integers and $c = dq \pm r$ where q and r are integers, then $(c, d) = (d, r)$. then let $r_0 = a$ and $r_1 = b$ be positive integers with $a \geq b$. By successively applying the least-remainder division algorithm, we find that $r_0 = r_1 q_1 + e_2 r_2, \frac{-r_1}{2} < e_2 r_2 \leq \frac{r_1}{2}$, $\cdots, r_{n-2} = r_{n-1} q_{n-1} + e_n r_n, \frac{-r_{n-1}}{2} < e_n r_n \leq r_{n-1} 2, r_{n-1} = r_n q_n$. We assume that we eventually obtain a remainder of zero since the sequence of remainders $a = r_0 > r_1 > r_2 > \cdots \geq 0$ cannot contain more than a terms.

17. Let $v_2 = v_3 = 2$, and for $i \geq 4$, $v_i = 2v_{i-1} + v_{i-2}$. Thus the least remainder algorithm will proceed with $e_i = 1$ and $q_i = 2$ for all i. Assume that it takes n steps to find the g.c.d. of v_{n+1} and v_{n+2}. To find the g.c.d. of v_{n+2} and v_{n+3}, the first step will be: $v_{n+3} = 2v_{n+2} + v_{n+1}$ by the definition of our v_is. From this point, the algorithm will look identical to that for v_{n+1} and v_{n+2}. By our induction hypothesis, this will require n more steps. Hence, the total number of steps is $n + 1$.

19. Performing the Euclidean algorithm with $r_0 = m$ and $r_1 = n$, we find that $r_0 = r_1 q_1 + r_2, 0 \leq r_2 < r_1, r_1 = r_2 q_2 + r_3, 0 \leq r_3 < r_2, \ldots, r_{n-3} = r_{n-2}q_{n-2} + r_{n-1}, 0 \leq r_{n-1} < r_{n-2}$, and $r_{n-2} = r_{n-1}q_{n-1}$. We have $(m, n) = r_{n-1}$. We will use these steps to find the greatest common divisor $a^m - 1$ and $a^n - 1$. First, we show that if u and v are positive integers, then the least positive residue of $a^u - 1$ modulo $a^v - 1$ is $a^r - 1$ where r is the least positive residue of u modulo v. To see this, note that $u = vq + r$ where r is the least positive residue of u modulo v. It follows that $a^u - 1 = a^{vq+r} - 1 = (a^v - 1)$ $(a^{v(q-1)+r} + \cdots + a^{v+r} + a^r) + (a^r - 1)$. This shows that the remainder is $a^r - 1$ when $a^u - 1$ is divided by $a^v - 1$. Now let $R_0 = a^m - 1$ and $R_1 = a^n - 1$. When we perform the Euclidean algorithm starting with R_0 and R_1 we obtain $R_0 = R_1 Q_1 + R_2$, where $R_2 = a^{r_2} - 1, R_1 = R_2 Q_2 + R_3$ where $R_3 = a^{r_3} - 1, \ldots, R_{n-3} = R_{n-2} Q_{n-2} + R_{n-1}$ where $R_{n-1} = a^{r_{n-1}} - 1$. Hence the last nonzero remainder, $R_{n-1} = a^{r_{n-1}} - 1 = a^{(m,n)} - 1$ is the greatest common divisor of $a^m - 1$ and $a^n - 1$.

21. Note that $(x, y) = (x - ty, y)$, as any divisor of x and y is also a divisor of $x - ty$. So, every move in the game of Euclid preserves the g.c.d. of the two numbers. Since $(a, 0) = a$, if the game beginning with $\{a, b\}$ terminates, then it must do so at $\{(a, b), 0)\}$. Since the sum of the two numbers is always decreasing and positive, the game must terminate.

23. Note that a has $O(\log a)$ digits. Then apply Theorem 1.16.

Section 2.3

1. a. $36 = 2^2 \cdot 3^2$ **b.** $39 = 3 \cdot 13$ **c.** $100 = 2^2 \cdot 5^2$ **d.** $289 = 17^2$
   e. $222 = 2 \cdot 3 \cdot 37$ **f.** $256 = 2^8$ **g.** $515 = 5 \cdot 103$ **h.** $989 = 23 \cdot 43$
   i. $5040 = 2^4 \cdot 3^2 \cdot 5 \cdot 7$ **j.** $8000 = 2^6 \cdot 5^3$ **k.** $9555 = 3 \cdot 5 \cdot 7^2 \cdot 13$ **l.** $3^2 \cdot 11 \cdot 101$

3. $3 \cdot 5 \cdot 7 \cdot 11 \cdot 13 \cdot 17 \cdot 19$

5. p^2 where p is prime; pq or p^3 where p and q are distinct primes.

7. Let $n = p_1^{2a_1} p_2^{2a_2} \cdots p_k^{2a_k} q_1^{2b_1+3} q_2^{2b_2+3} \cdots q_l^{2b_l+3}$ be the factorization of a powerful number. Then $n = (p_1^{a_1} p_2^{a_2} \cdots p_k^{a_k} q_1^{b_1} q_2^{b_2} \cdots q_l^{b_l})^2 (q_1 q_2 \cdots q_l)^3$ is a product of a square and a cube.

9. Suppose that $p^a \parallel m$ and $p^b \parallel n$. Then $m = p^a Q$ and $n = p^b R$ where both Q and R are products of primes other than p. Hence $mn = (p^a Q)(p^b R) = p^{a+b} QR$. It follows that $p^{a+b} \parallel mn$ since p does not divide QR.

11. Suppose that $p^a \parallel m$ and $p^b \parallel n$ with $a \neq b$. Then $m = p^a Q$ and $n = p^b R$ where both Q and R are products of primes other than p. Suppose, without loss of generality, that $a = \min(a, b)$. Then $m + n = p^a Q + p^b R = p^{\min(a,b)}(Q + p^{b-\min(a,b)} R)$. Then $p \nmid (Q + p^{b-a} R)$ because $p \nmid Q$ but $p \mid p^{b-a} R$. It follows that $p^{\min(a,b)} \parallel (m + n)$.

13. $20! = 2^{18} \cdot 3^8 \cdot 5^4 \cdot 7^2 \cdot 11 \cdot 13 \cdot 17 \cdot 19$.

15. 300, 301, 302, 303, 304

17. The product of $4k + 1$ and $4l + 1$ is $(4k + 1)(4l + 1) = 16kl + 4k + 4l + 1 = 4(4kl + k + l) + 1 = 4m + 1$ where $m = 4kl + k + l$. Hence the product of two integers of the form $4k + 1$ is also of this form.

19. For the basis step note that 5 is a Hilbert prime. Let $n > 5$ and assume that all numbers in H less than or equal to n can be factored into Hilbert primes. The next greatest number in H is $n+4$. If $n+4$ is a Hilbert prime, then we are done. Otherwise, $n = hk$, where h and k are less than n and in H. By the inductive hypothesis, h and k can be factored into Hilbert primes.

21. 1,2,3,4,6,8,12, 24

23. a. 77 **b.** 36 **c.** 150 **d.** 33633 **e.** 605605 **f.** 277200

25. a. $2^2 3^5 5^3 7^2, 2^7 3^5 5^5 7^7$ **b.** $1, 2 \cdot 3 \cdot 5 \cdot 7 \cdot 11 \cdot 13 \cdot 17 \cdot 19 \cdot 23 \cdot 29$
c. $2 \cdot 5 \cdot 11, 2^3 5^7 11^{13} 3 \cdot 7 \cdot 13$ **d.** $101^{1000}, 41^{11} 47^{11} 79^{111} 83^{111} 101^{1001}$

27. The year 2121.

29. Let $a = p_1^{r_1} p_2^{r_2} \cdots p_k^{r_k}$ and $b = p_1^{s_1} p_2^{s_2} \cdots p_k^{s_k}$. $(a, b) = p_1^{\min(r_1, s_1)} \cdots p_k^{\min(r_k, s_k)}$ and $[a, b] = p_1^{\max(r_1, s_1)} \cdots p_k^{\max(r_k, s_k)}$. So $[a, b] = (a, b) p_1^{\max(r_1, s_1) - \min(r_1, s_1)} \cdots p_k^{\max(r_k, s_k) - \min(r_k, s_k)}$. Since $\max(r_i, s_i) - \min(r_i, s_i) \geq 0$, we now see that $(a, b) \mid [a, b]$.

31. If $[a, b] \mid c$, then since $a \mid [a, b], a \mid c$. Similarly, $b \mid c$. Now suppose that $a \mid c$ and $b \mid c$. Then $c = an = bm$, so c is a common multiple of a and b. Since all common multiples are a multiple of the least common multiple, $c = [a, b]k$ and so $[a, b] \mid k$.

33. If $n = 1$ then $p^1 \mid a$ implies $p \mid a$. Assume that $p \mid a^n = \pm \mid a \mid \cdot \mid a \mid \cdots \mid a \mid$. Then by Lemma 2.4, $p \mid |a|$ and so $p \mid a$.

35. a. Suppose that $(a, b) = 1$ and $p \mid (a^n, b^n)$ where p is a prime. It follows that $p \mid a^n$ and $p \mid b^n$. By Exercise 33, $p \mid a$ and $p \mid b$. But then $p \mid (a, b) = 1$, which is a contradiction.
b. Suppose that a does not divide b, but $a^n \mid b^n$. Then there is some prime power, say p^r that divides a but does not divide b. Thus, $a = p^r Q$, where Q is an integer. Now, $a^n = (p^r Q)^n = p^{rn} Q^n$, so that $p^{rn} \mid a^n$ and $a^n \mid b^n$. Then $b^n = m p^{rn}$, from which it follows that each of the n b's must by symmetry contain r p's. But this is a contradiction.

37. Suppose that $x = \sqrt{2} + \sqrt{3}$. Then $x^2 = 2 + 2\sqrt{2}\sqrt{3} + 3 = 5 + 2\sqrt{6}$. Hence $x^2 - 6 = 2\sqrt{6}$. It follows that $x^4 = 10x^2 + 25 = 24$. Consequently, $x^4 - 10x^2 + 1 = 0$. By Theorem 2.11 it follows that $\sqrt{2} + \sqrt{3}$ is irrational, since it is not an integer (we can see this since $3 < \sqrt{2} + \sqrt{3} < 4$).

39. Suppose that $\frac{m}{n} = \log_p b$. This implies that $p^{\frac{m}{n}} = b$, from which it follows that $p^m = b^n$. Since b is not a power of p, there must be another prime, say q, such that $q \mid b$. But then $q \mid b$ and $b \mid b^n = p^m = p \cdot p \cdots p$. By Lemma 2.4, $q \mid p$, which is impossible since p is a prime number.

41. Let p be a prime that divides a or b. Define s, t, u, v by $p^s \mid\mid a, p^t \mid\mid b, p^u \mid\mid (a + b)$, and $p^v \mid\mid a, b$. To show that $(a, b) = (a + b, a, b)$ we need to show that $\min(a, b) = \min(r, s)$ for any prime p. From the definition of v we have $v = \max(s, t) \geq \min(u, v)$. By Exercise 9 either $v = \min(s, t)$ or $v > \min(s, t)$ and $s = t$, so that if $v > \min(s, t)$ then $v = s = t = \min(s, t)$. In all cases $\min(s, t) = \min(u, v)$.

43. Let $a = p_1^{r_1} p_2^{r_2} \cdots p_k^{r_k}, b = p_1^{s_1} p_2^{s_2} \cdots p_k^{s_k}$, and $c = p_1^{t_1} p_2^{t_2} \cdots p_k^{t_k}$, with p_i prime and r_i, s_i, and t_i nonnegative. Observe that $\min(x, \max(y, z)) = \max(\min(x, y), \min(x, z))$. We also know that $[a, b] = p_1^{\max(r_1, s_1)} p_2^{\max(r_2, s_2)} \cdots p_k^{\max(r_k, s_k)}$, and so $([a, b], c) = p_1^{\min(t_1, \max(r_1, s_1))} p_2^{\min(t_2, \max(r_2, s_2))} \cdots p_k^{\min(t_k, \max(r_k, s_k))}$. We also know that $(a, c) = p_1^{\min(r_1, t_1)} p_2^{\min(r_2, t_2)} \cdots p_k^{\min(r_k, t_k)}$ and $(b, c) = p_1^{\min(s_1, t_1)} p_2^{\min(s_2, t_2)} \cdots p_k^{\min(s_k, t_k)}$. Then, $[(a, c), (b, c)] = p_1^{\max(\min(r_1, t_1), \min(s_1, t_1))} p_2^{\max(\min(r_2, t_2), \min(s_2, t_2))} \cdots p_k^{\max(\min(r_k, t_k), \min(s_k, t_k))}$. Therefore, $([a, b], c) = [(a, c), (b, c)]$. In a similar manner, noting that $\min(\max(x, z), \max(y, z)) = \max(\min(x, y), z)$, we find that $[(a, b), c] = ([a, c], [b, c])$.

45. Let $c = [a_1, \ldots, a_n]$, $d = [[a_1, \ldots, a_{n-1}], a_n]$, and $e = [a_1, \ldots, a_{n-1}]$. If $c \mid m$, then all a_i's divide m, hence $e \mid m$ and $a_n \mid m$, so $d \mid m$. Conversely, if $d \mid m$, then $e \mid m$ and $a_n \mid m$, so all a_i's divide m, thus $c \mid m$. Since c and d divide all the same numbers, they must be equal.

47. a. There are six cases, all handled the same way. So without loss of generality, suppose that $a \le b \le c$. Then $\max(a, b, c) = c$, $\min(a, b) = a$, $\min(a, c) = a$, $\min(b, c) = b$, and $\min(a, b, c) = a$. Hence $c = a + b + c - a - a - b + a = a + b + c - \min(a, b) - \min(a, c) - \min(b, c) + \min(a, b, c) = \max(a, b, c)$.

b. The power of a prime p occurs in the prime factorization of $[a, b, c]$ is $\max(a, b, c)$ where $a, b,$ and c are the powers of this prime in the factorizations of $a, b,$ and c, respectively. Also $a + b + c$ is the power of p in abc, $\min(a, b)$ is the power of p in (a, b), $\min(a, c)$ is the power of p in a, c), $\min(b, c)$ is the power of p in (b, c), and $\min(a, b, c)$ is the power of p in (a, b, c). It follows that $a + b + c - \min(a, b) - \min(a, c) - \min(b, c)$ is the power of p in $\frac{abc(a,b,c)}{(a,b)(a,c)(b,c)}$. Hence $[a, b, c] = \frac{abc(a,b,c)}{(a,b)(a,c)(b,c)}$.

49. Let $a = p_1^{r_1} p_2^{r_2} \cdots p_k^{r_k}$, $b = p_1^{s_1} p_2^{s_2} \cdots p_k^{s_k}$, and $c = p_1^{t_1} p_2^{t_2} \cdots p_k^{t_k}$, with p_i prime and $r_i, s_i,$ and t_i nonnegative. Then $p_i^{r_i + s_i + t_i} \| abc$, but $p_i^{\min\{r_i, s_i, t_i\}} \| (a, b, c)$ and $p_i^{r_i + s_i + t_i - \min\{r_i, s_i, t_i\}} \| [ab, ac, ab]$, and $p_i^{\min r_i, s_i, t_i} \cdot p_i^{r_i + s_i + t_i - \min\{r_i, s_i, t_i\}} = p_i^{r_i + s_i + t_i}$.

51. Let $a = p_1^{r_1} p_2^{r_2} \cdots p_k^{r_k}$, $b = p_1^{s_1} p_2^{s_2} \cdots p_k^{s_k}$, and $c = p_1^{t_1} p_2^{t_2} \cdots p_k^{t_k}$, with p_i prime and $r_i, s_i,$ and t_i nonnegative. Then, using that $(a, b, c) = p_1^{\min(r_1, s_1, t_1)} p_2^{\min(r_2, s_2, t_2)} \cdots p_k^{\min(r_k, s_k, t_k)}$, and $[a, b, c] = p_1^{\max(r_1, s_1, t_1)} p_2^{\max(r_2, s_2, t_2)} \cdots p_k^{\max(r_k, s_k, t_k)}$, we can write the prime factorization of $([a, b], [a, c], [b, c])$ and $[(a, b), (a, c), (b, c)]$. For instance, consider the case where $k = 1$. Then $([a, b], [a, c], [b, c]) = (p_1^{\max(r_1, s_1)}, p_1^{\max(r_1, t_1)}, p_1^{\max(s_1, t_1)}) = p_1^{\min(\max(r_1, s_1), \max(r_1, t_1), \max(s_1, t_1))}$. Similarly, $[(a, b), (a, c), (b, c)] = p_1^{\max(\min(r_1, s_1), \min(r_1, t_1), \min(s_1, t_1))}$. These two are equal.

53. $(nb + 2)! + 2, \ldots, (nb + 2)! + (nb + 2)$.

55. 103 cameras at 79 dollars each.

57. Let $a = \prod_{i=1}^s p_i^{\alpha_i}$ and $b = \prod_{i=1}^t p_i^{\beta_i}$. The condition $(a, b) = 1$ is equivalent to $\min(\alpha_i, \beta_i) = 0$ for all i and the condition $ab = c^n$ is equivalent to $n \mid (\alpha_i + \beta_i)$ for all i. Hence $n \mid \alpha_i$ and $\beta_i = 0$ or $n \mid \beta_i$ and $\alpha_i = 0$. Let d be the product of $p_i^{\frac{\alpha_i}{n}}$ over all i of the first kind, and let e be the product of $p_i^{\frac{\beta_i}{n}}$ over all i of the second kind. Then $d^n = a$ and $e^n = b$.

59. Suppose the contrary and that $a \le n$ is in the set. Then $2a$ cannot be in the set. Thus, if there are k elements in the set not exceeding n then, there are k integers between $n + 1$ and $2n$ which cannot be in the set. So there are at most $k + (n - k) = n$ elements in the set.

61. The fundamental theorem of arithmetic implies that m and n have the same prime divisors. So suppose that m and n have prime-power factorizations $m = p_1^{a_1} p_2^{a_2} \cdots p_k^{a_k}$ and $n = p_1^{b_1} p_2^{b_2} \cdots p_k^{b_k}$. From the equation $m^n = n^m$ it follows that $a_i n = b_i m$ for $i = 1, 2, \ldots, k$. We first assume that $n > m$. Then $a_i < b_i$ for $i = 1, 2, \ldots, k$. Hence n is divisible by m, so that $n = dm$ for some integer d. This implies that $m^{dm} = (dm)^m$. Taking the mth roots of both sides gives $m^d = dm$, which implies that $m^{d-1} = d$. Since $n > m$ we know that $d > 1$ so that $m > 1$. However $2^{2-1} = 2$ and when $d > 2$ it follows that $m^{d-1} > d$. When $d > 2$ and $m \ge 2$ we have $m^{d-1} \ge 2^{d-1} > d$ since $2^{3-1} > 3$ and when $d = 2$ and $m > 2$ we have $m^{d-1} = m > 2 = d$. Hence the only solution with $n > m$ has $m = 2$ and $n = 2d = 2 \cdot 2 = 4$. Consequently all solutions are given by $m = 2$ and $n = 4$, $m = 4$ and $n = 2$, or $m = n$.

63. First note that if $p \mid \binom{2n}{n}$, then $p \leq 2n$. This is true because every factor of the numerator of $\binom{2n}{n} = \frac{(2n)!}{(n!)^2}$ is less than or equal to $2n$. Let $\binom{2n}{n} = p_1^{r_1} p_2^{r_2} \cdots p_k^{r_k}$ be the factorization of $\binom{2n}{n}$ into distinct primes. By the definition of π, $k \leq \pi(2n)$. By Exercise 62, $p_i^{r_i} \leq 2n$. it now follows that $\binom{2n}{n} = p_1^{r_1} p_2^{r_2} \cdots p_k^{r_k} \leq (2n)(2n) \cdots (2n) \leq (2n)^{\pi(2n)}$.

65. Note that $\binom{2n}{n} \leq \sum_{a=0}^{2n} \binom{2n}{a} = (1+1)^{2n} = 2^{2n}$. Then from Exercise 64, $n^{\pi(2n)-\pi(n)} < \binom{2n}{n} \leq 2^{2n}$. Taking logarithms gives $(\pi(2n) - \pi(n)) \log n < \log(2^{2n} = n \log 4$. Now divide by $\log n$.

67. Note that $2^n = \prod_{a=1}^{n} 2 \leq \prod_{a=1}^{n} \frac{n+a}{a} = \binom{2n}{n}$. Then by Exercise 63, $2^n \leq (2n)^{\pi(2n)}$. Taking logs gives $\pi(2n) \geq \frac{n \log 2}{\log 2n}$. Hence, for a real number x, we have $\pi(x) \geq \left[\frac{x}{2}\right] \log 2 / \log \left[\frac{x}{2}\right] > \frac{c_1 x}{\log x}$. For the other half, Exercise 65 gives $\pi(x) - \pi(x/2) < a \frac{x}{\log x}$, where a is a constant. Then $\log \frac{x}{2^m} \pi(\frac{x}{2^m}) - \log \frac{x}{2^{m+1}} \pi(\frac{x}{2^{m+1}}) < a \frac{x}{2^m}$. Then, $\log x \pi(x) = \sum_{m=0}^{v} \left(\log \frac{x}{2^m} \pi(\frac{x}{2^m}) - \log \frac{x}{2^{m+1}} \pi(\frac{x}{2^{m+1}}) \right) < ax \sum_{m=0}^{v} \frac{1}{2^m} < c_2 x$. Where v is the largest integer such that $2^{v+1} \leq x$. Then $\pi(x) < c_2 \frac{x}{\log x}$.

Section 2.4

1. a. $3 \cdot 5^2 \cdot 7^3 \cdot 13 \cdot 101$ **b.** $11^3 \cdot 13 \cdot 19 \cdot 641$ **c.** $13 \cdot 17 \cdot 19 \cdot 47 \cdot 71 \cdot 97$.

3. a. $12^2 - 143 = 1$ is a square. So, $143 = 12^2 - 1 = (12+1)(12-1) = 13 \cdot 11$.
   b. $48^2 - 2279 = 25 = 5^2$. So, $2279 = 48^2 - 5^2 = (48+5)(48-5) = 53 \cdot 43$.
   c. Since $6 < \sqrt{43} < 7$, we begin by looking for a perfect square in the sequence $7^2 - 43 = 6, 8^2 - 43 = 21, 9^2 - 43 = 38, 10^2 - 43 = 57, 11^2 - 43 = 78, \ldots$. The smallest such perfect square is $22^2 - 43 = 21^2$. From this, it follows that $43 = (22+21)(22-21) = 43 \cdot 1$ which shows that 43 is prime.
   d. Since $106 < \sqrt{11413} < 107$, we begin by looking for a perfect square in the sequence $107^2 - 11413 = 36 = 6^2, \ldots$. Thus, $11413 = 107^2 - 6^2 = (107+6)(107-6) = 113 \cdot 101$.

5. The squares of the numbers from 0 to 25 end in the following: 0, 1, 4, 9, 16, 25, 36, 49, 64, 81, 0, 21, 44, 69, 96, 25, 56, 89, 24, 61, 0, 41, 84, 29, 76, 25.

7. Suppose that $x^2 - n$ is a perfect square with $x > \frac{n+p^2}{2p}$, say a^2. Now, $a^2 = x^2 - n > (\frac{n+p^2}{2p})^2 - n = (\frac{n-p^2}{2p})^2$. It follows that $a > \frac{n-p^2}{2p}$. From these inequalities for x and a, we see that $x + a > \frac{n}{p}$, or $n < p(x+a)$. Also, $a^2 = x^2 - n$ tells us that $(x-a)(x+a) = n$. Now, $(x-a)(x+a) = n < p(x+a)$. Cancelling, we find that $x - a < p$. But since $x - a$ is a divisor of n less than p, the smallest prime divisor of n, $x - a = 1$. In this case, $x = \frac{n+1}{2}$.

9. If $(2k+1) \mid n_k$, then $(2k+1) \mid r_k$ and it follows from $r_k < 2k+1$ that $r_k = 0$. Thus, $n_k = (2k+1)q_k$. Continuing, we see that $n = n + 2n_k - 2(2k+1)q_k = (2k+1)n + 2(n_k - kn) - 2(2k+1)q_k$. It follows from Exercise 8 that $n = (2k+1)n - 2(2k+1)\sum_{i=1}^{k-1} q_i - 2(2k+1)q_k = (2k+1)n - 2(2k+1)\sum_{i=1}^{k} q_i$. Using Exercise 8 again, we conclude that $n = (2k+1)(n - 2\sum_{i=1}^{k} q_i) = (2k+1)m_{k+1}$.

11. We know that u is even, since $a - c$ is the difference of odd numbers and $b - d$ is the difference of even numbers. That $(r, s) = 1$ follows from Theorem 2.1 (i). $a^2 + b^2 = c^2 + d^2$ implies that $(a + c)(a - c) = (b - d)(b + d)$. Dividing by u, we get $r(a + c) = s(d + b)$. Then. $s \mid r(a + c)$. But since $(r, s) = 1$, $s \mid a + c$.

13. To factor n, observe that $[(\frac{u}{2})^2 + (\frac{v}{2})^2](r^2 + s^2) = (\frac{ru+su}{2})^2 + (\frac{rv+sv}{2})^2$. Substituting $a - c, d - b, a + c$, and $d + b$ for ru, su, sv, and rv respectively, will allow everything to be simplified down to n. As u and v are both even, both of the factors are integers.

15. $2^{18} + 1 = 545 \cdot 481 = 5 \cdot 13 \cdot 37 \cdot 109$.

17. Note that $2^{2^2} = 16$ ends in a 6. Assume that the last decimal digit of 2^{2^n} is 6, that is $2^{2^n} \equiv 6 \pmod{10}$. It follows that $2^{2^{n+1}} = (2^{2^n})^{2^{n+1} - 2^n} \equiv 6^{2^{n+1} - 2^n} \equiv 6 \pmod{10}$.

19. Among the integers of the form $128k + 1$ less than $\sqrt{6700417}$, only 257, 641, 769, 1153, and 1409 are prime, and none of them divide 6700417. Hence 6700417 is prime and the prime factorization of F_5 is $641 \cdot 6700417$.

20. We have $2^{2^0} + 5 = 7$. This is the only prime of the form 2^{2^n} since $2^{2^n} + 5 \equiv (-1)^{2^n} + 5 \equiv 1 + 5 \equiv 0 \pmod 3$ when $n > 1$

21. $2^k \log_2 10$

23. See Exercise 25 of Section 1.9.

Section 2.5

1. **a.** $x = 33 - 5t, y = -11 + 2t$ **b.** $x = -300 + 13t, y = 400 - 17t$ **c.** $x = 21 - 14t, y = -21 + 21t$ **d.** no solution **e.** $x = 889 - 1969t, y = -633 + 1402t$

3. $39, 94$

5. $17, 23$

7. **a.** $x = 1, y = 16; x = 4, y = 14; x = 7, y = 12; x = 10, y = 10; x = 13, y = 8; x = 16, y = 6; x = 19, y = 4; x = 22, y = 2; x = 25, y = 0$.
   b. no solution
   c. $(0, 37), (6, 35), \ldots, (54, 1)$

9. **a.** $x = -5 - 3s - 2t, y = 5 - 2s, z = t$ **b.** no solution
   c. $x = -1 + 102s + t, y = 1 - 101s - 2t, z = t$

11. Let x be the number of pennies, y the number of dimes, and z the number of quarters. If $z = 0$, then $x = 9, y = 9; x = 19, y = 8; x = 29, y = 7; x = 39, y = 6; x = 49, y = 5; x = 59, y = 4; x = 69, y = 3; x = 79, y = 2; x = 89, y = 1; x = 99, y = 0$. If $z = 1$, then $x = 4, y = 7; x = 14, y = 6; x = 24, y = 5; x = 34, y = 4; x = 44, y = 3; x = 54, y = 2; x = 64, y = 1; x = 74, y = 1$. If $z = 2$, then $x = 9, y = 4; x = 19, y = 3; x = 29, y = 2; x = 39, y = 1; x = 49, y = 0$. If $z = 3$, then $x = 4, y = 2; x = 14, y = 1$; and $x = 24, y = 0$.

13. **a.** $x = 92 + 6t, y = 8 - 7t, x = t$ **b.** no solution **c.** $x = 50 - t, y = -100 + 3t, z = 150 - 3t, w = t$

15. $9, 19, 41$

17. The quadrilateral with vertices $(b, 0), (0, a), (b - 1, -1)$, and $(-1, a - 1)$, has area $a + b$. Pick's Theorem, from elementary geometry, states that the area of a simple polygon whose vertices are lattice points (points with integer coordinates) is given by $\frac{1}{2}x + y - 1$, where x is the number of lattice points on the boundary and y is the number of lattice points inside the polygon. Since $(a, b) = 1$, $x = 4$, and therefore, by Pick's Theorem, the quadrilateral contains $a + b - 1$ lattice points. Every point corresponds to a different value of n in the range $ab - a - b < n < ab$. Therefore every n in the range must get hit, so the equation is solvable.

19. The line $ax + by = ab - a - b$ bisects the rectangle with vertices $(-1, a - 1)$, $(-1, -1)$, $(b - 1, a - 1)$, and $(b - 1, -1)$ but contains no lattice points. Hence, half the interior points are below the line and half are above. The half below correspond to $n < ab - a - b$ and there are $(a - 1)(b - 1)/2$ of them.

21. $(25, 0, 25); (26, 4, 18); (27, 8, 11); (28, 12, 4)$

Section 3.1

1. **a.** $2 \mid (13 - 1) = 12$ **b.** $5 \mid (22 - 7) = 15$ **c.** $13 \mid (91 - 0) = 91$
   d. $7 \mid (69 - 62) = 7$ **e.** $3 \mid (-2 - 1) = -3$ **f.** $11 \mid (-3 - 30) = -33$
   g. $40 \mid (111 - (-9)) = 120$ **h.** $37 \mid (666 - 0) = 666$

3. **a.** $2, 11, 22$ **b.** $3, 9, 27, 37, 111, 333, 999$ **c.** $11, 121, 1331$

5. If $a = 2k + 1$, then $a^2 = (2k + 1)^2 = 4k^2 + 4k + 1 = 4k(k + 1) + 1$. If $k = 2l$, then $a^2 = 8l(2l + 1) + 1$. Hence $a^2 \equiv 1 \pmod 8$. If k is odd, then $k = 2l + 1$ when l is an integer. Then $a^2 = 4(2l + 1)(2l + 2) + 1 = 8(2l + 1)(l + 1)$. Hence $a^2 \equiv 1 \pmod 8$. It follows that $a^2 \equiv 1 \pmod 8$ whenever a is odd.

7. **a.** 1 **b.** 5 **c.** 9 **d.** 13

9. Since $a \equiv b \pmod m$, there exists an integer k such that $a = b + km$. Thus, $ac = (b + km)c = bc + k(mc)$. By Theorem 3.1, $ac \equiv bc \pmod{mc}$.

11. **a.** We proceed by induction on n. It is clearly true for $n = 1$. For the inductive step we assume that $\sum_{j=1}^{n} a_j \equiv \sum_{j=1}^{n} b_j \pmod m$ and that $a_{n+1} \equiv b_{n+1} \pmod m$. Now $\sum_{j=1}^{n+1} a_j = (\sum_{j=1}^{n} a_j) + a_{n+1} \equiv (\sum_{j=1}^{n} b_j) + b_{n+1} = \sum_{j=1}^{n+1} b_j \pmod m$ by Theorem 3.5(i). This completes the proof.

b. We use induction on n. For $n = 1$, the identity clearly holds. This completes the basis step. For the inductive step we assume that $\prod_{j=1}^{n} a_j \equiv \prod_{j=1}^{n} b_j \pmod m$ and $a_{n+1} \equiv b_{n+1} \pmod m$. Then $\prod_{j=1}^{n+1} a_j = a_{n+1}(\prod_{j=1}^{n} a_j) \equiv b_{n+1}(\prod_{j=1}^{n} b_j) = \prod_{j=1}^{n+1} b_j \pmod m$ by Theorem 3.5(iii). This completes the proof.

13.

·	a	0	1	2	3	4	5
b	$a - b$	$-b$	$1 - b$	$2 - b$	$3 - b$	$4 - b$	$5 - b$
0	a	0	1	2	3	4	5
1	$a - 1$	5	0	1	2	3	4
2	$a - 2$	4	5	0	1	2	3
3	$a - 3$	3	4	5	0	1	2
4	$a - 4$	2	3	4	5	0	1
5	$a - 5$	1	2	3	4	5	0

15. **a.** 4 **b.** 6 **c.** 4

17. $a \equiv \pm b \pmod p$

19. $1 + 2 + 3 + \cdots + (n + 1) = \frac{(n-1)n}{2}$. If n is odd, then $(n - 1)$ is even, so that $\frac{(n-1)}{2}$ is an integer. Hence $n \mid (1 + 2 + 3 + \cdots + (n - 1))$ if n is odd, and $1 + 2 + 3 + \cdots + (n - 1) \equiv 0 \pmod n$. Not true if n is even.

21. $\pm 1 \pmod 6$

23. Certainly $5^1 \equiv 1 + 4 \cdot 1 \pmod{16}$. Assume that $5^n \equiv 1 + 4n \pmod{16}$. Now $5^{n+1} \equiv 5^n 5 \equiv (1 + 4n)5 \pmod{16}$ by Theorem 3.3(iii). Further, $(1 + 4n)5 \equiv 5 + 20n \equiv 5 + 4n \pmod{16}$. Finally $5 + 4n = 1 + 4(n + 1)$. so, $5^{n+1} \equiv 1 + 4(n + 1) \pmod{16}$.

25. If $x \equiv 0 \pmod 4$ then $x^2 \equiv 0 \pmod 4$, if $x \equiv 1 \pmod 4$ then $x^2 \equiv 1 \pmod 4$, if $x \equiv 2 \pmod 4$ then $x^2 \equiv 4 \equiv 0 \pmod 4$, and if $x \equiv 3 \pmod 4$ then $x^2 \equiv 9 \equiv 1 \pmod 4$. Hence $x^2 \equiv 0$ or $1 \pmod 4$ whenever x is an integer. It follows that $x^2 + y^2 \equiv 0, 1$ or $2 \pmod 4$ whenever x and y are integers. We see that n is not the sum of two squares when $n \equiv 3 \pmod 4$.

27. By Theorem 3.1, for some integer a, $ap^k = x^2 - x = x(x-1)$. By the Fundamental Theorem of Arithmetic, p^k is a factor of $x(x-1)$. Since p cannot divide both x and $x - 1$, we know that $p^k \mid x$ or $p^k \mid x - 1$. Thus, $x \equiv 0$ or $x \equiv 1 \pmod{p^k}$.

29. There are m_i possibilities for a_i. Thus there are $m_1 m_2 \cdots m_k$ expressions of the form $M_1 a_1 + M_2 a_2 + \cdots + M_k a_k$ where $a_1, a_2, \ldots, a_k$ run through complete systems of residues modulo $m_1, m_2, \ldots, m_k$, respectively. Since this is exactly the size of a complete system of residues modulo M, the result will follow if we can show distinctness of each of these expressions modulo M. Suppose, by way of contradiction, that $M_1 a_1 + M_2 a_2 + \cdots + M_k a_k \equiv M_1 a_1' + M_2 a_2' + \cdots + M_k a_k' \pmod{M}$. Then $M_1 a_1 \equiv M_1 a_1' \pmod{m_1}$, and further $a_1 \equiv a_1' \pmod{m_1}$ since $(M_1, m_1) = 1$. Similarly $a_i \equiv a_i' \pmod{m_i}$. Thus all a_i' are in the same congruence class modulo m_i as a_i.

31. **a.** Let $\sqrt{n} = a + r$, where a is an integer, and $0 \le r < 1$. We now consider two cases, when $0 \le r < \frac{1}{2}$ and when $\frac{1}{2} \le r < 1$. For the first case, $T = [\sqrt{n} + \frac{1}{2}] = a$, and so $t = T^2 - n = -(2a + r^2)$. Thus $\mid t \mid = 2ar + r^2 < 2a(\frac{1}{2}) + (\frac{1}{2})^2 = a + \frac{1}{4}$. Since both T and n are integers, t is also an integer. It follows that $\mid t \mid t \le [a + \frac{1}{4}] = a = T$. For the second case, when $\frac{1}{2} \le r < 1$, we find that $T = [\sqrt{n} + \frac{1}{2}] = a + 1$ and $t = 2a(1 - r) + (1 - r^2)$. Since $\frac{1}{2} \le r < 1$, $0 < (1 - r) \le \frac{1}{2}$ and $0 < 1 - r^2 < 1$. It follows that $t \le 2a(\frac{1}{2}) + (1 - r^2)$. Because t is an integer, we can say that $t \le [a + (1 - r^2)] = a < T$.
    b. By the division algorithm, we see that if we divide x by T we get $x = aT + b$, where $0 \le b < T$. If a were negative, then $x = aT + b \le (-1)T + b < 0$; but we assumed x to be nonnegative. This shows that $0 \le a$. Suppose now that $a > T$. Then $x = aT + b \ge (T + 1)T = T^2 + T \ge (\sqrt{n} - \frac{1}{2})^2 + (\sqrt{n} - \frac{1}{2}) = n - \frac{1}{4}$ and, as x and n are integers, $x \ge n$. This is a contradiction, which shows that $a \le T$. Similarly, $0 \le c \le T$ and $0 \le d < T$.
    c. $xy = (aT + b)(cT + d) = acT^2 + (ad + bc)T + bd \equiv ac(T^2 - n) + zT + bd \equiv act + zt + bd \pmod{n}$.
    d. Use part (c) of this Exercise, substituting $eT + f$ for ac.
    e. The first half is identical to part (b) of this exercise; the second half follows by substituting $gT + h$ for $z + et$ and noting that $t^2 \equiv t \pmod{n}$.
    f. Certainly, ft and gt can be computed since all three numbers are less than T, which is less than $\sqrt{n} + 1$. So $(f + g)t$ is less than $2n < w$. Similarly, we can compute $j + bd$ without exceeding the word size. And, finally, using the same arguements, we can compute $hT + k$ without exceeding the word size.

33. **a.** 1 **b.** 1 **c.** 1 **d.** 1 **e.** This is Fermat's little theorem. See Section 5.1.

35. Since $f_{n-2} + f_{n-1} \equiv f_n \pmod{m}$, if two consequtive numbers recur in the same order, then the sequence must be repeating both as n increases and as it decreases. But there are only m residues, and so m^2 ordered sequence of two residues. As the sequence is infinite, some two elements of the sequence must recur by the pigeonhole principle. Thus the sequence of least positive residues of the Fibonacci numbers repeats. It follows that if mn divides some Fibonacci number, that is if $f_n \equiv 0 \pmod{m}$ then m divides infinitely many Fibonacci numbers. To see that this is so, note that the sequence must contain a 0, namely $f_0 \equiv 0 \pmod{m}$.

37. Let a and b be positive integers less than m. Then they have $O(\log m)$ digits (bits). Therefore by Theorem 1.13, we can multiply them using $O(\log^2 m)$ operations. Division by m takes $O(\log^2 m)$ operations by Theorem 1.16. Then, in all we have $O(\log^2 m)$ operations.

39. $(n - k)(n^{n+1} - 1)/(n - 1) + k$

Section 3.2

1. a. 6 **b.** 2, 5, 8 **c.** 10 **d.** 20 **e.** 111 **f.** $x \equiv 75 + 80k \pmod{1600}$ where $0 \le k \le 19$.

3. The 9163 solutions are given by $x \equiv 1074 + 3157k \pmod{28927591}$ where k is an integer such that $0 \le k \le 9162$.

5. 19 hours

7. $154x \equiv c \pmod{1001}$ has solutions if and only if $(1001, 154) = 77 \mid c$. When there are solutions, there are exactly 77 of them.

9. a. 13 **b.** 7 **c.** 5 **d.** 16

11. If $ax + by \equiv k \pmod{c}$, then there exists an integer k such that $ax + by - mk = c$. Since $d \mid ax + by - mk, d \mid c$. Thus there are no solutions when $d \nmid c$. Now, let $z = \frac{ax}{(a,b,m)} + \frac{by}{(a,b,m)}$.
Then $(a, b)z \equiv c \pmod{m}$. By Theorem 3.10, there are $d = ((a, b), m) = (a, b, m)$ solutions to this congruence. For each solution in z, though, there are m solutions in x and y. To see this, note that $z \equiv \frac{ax}{(a,b,m)} + \frac{by}{(a,b,m)} \pmod{m}$ has a solution for every x because $(\frac{b}{(a,b,m)}, m) = 1$. Thus $\frac{b}{(a,b,m)}$ has an inverse modulo m, and so $y \equiv z(\frac{b}{(a,b,m)})^{-1}) - \frac{a}{(a,b,m)}(\frac{b}{(a,b,m)})^{-1})x$. It follows there are dm solutions to $ax + by \equiv k \pmod{c}$ if $d \mid c$.

13. Suppose that $x^2 \equiv 1 \pmod{p^k}$ where p is an odd prime and k is a positive integer. Then $x^2 - 1 \equiv (x + 1)(x - 1) \equiv 0 \pmod{p^k}$. Hence $p^k \mid (x + 1)(x - 1)$. Since $(x + 1) - (x - 1) = 2$ and p is an odd prime, we know that p divides at most one of $(x - 1)$ and $(x + 1)$. It follows that either $p^k \mid (x + 1)$ or $p^k \mid (x - 1)$, so that $p \equiv \pm 1 \pmod{p^k}$.

15. To find the inverse of a modulo m, we must solve the Diophantine equation $ax + my = 1$, which can be done using the Euclidean Algorithm. Using Corollary 2.5.1, we can find the greatest common divisor in $O(\log^3 m)$ bit operations. The back substitution to find x and y will take no more than $O(\log m)$ multiplications, each taking $O(\log^2 m)$ operations. Therefore the total number of operations is $O(\log^3 m) + O(\log m)O(\log^2 m) = O(\log^3 m)$.

Section 3.3

1. $x \equiv 1 \pmod{6}$

3. 32

5. $x \equiv 1523 \pmod{2310}$.

7. 204

9. 1023

11. 2101

13. Consider the system of congruences $x \equiv 0 \pmod{p_1^2}$, $x \equiv -1 \pmod{p_2^2}$, $x \equiv -4 \pmod{p_3^2}, \ldots, x \equiv -k + 1 \pmod{p_k^2}$, where p_k is the kth prime. By the Chinese remainder theorem there is a solution to this simultaneous system of congruence since the moduli are relatively prime. It follows that there is a positive integer N that satisfies each of these congruences. Each of the k integers $n, N + 1, \ldots, N + k - 1$ is divisible by a square since p_j^2 divides $N + j - 1$ for $j = 1, 2, \ldots, k$.

15. Suppose that x is a solution. Then, following the hint, $x \equiv a_1 \pmod{m_1}$ implies that $x = a_1 + km_1$, where k is an integer. Then $x = a_1 + km_1 \equiv a_2 \pmod{m_2}$ and hence $a_1 + km_1 = a_2 + jm_2$, or $jm_2 - km_1 = a_1 - a_2$. Thus, $a_1 - a_2 = (m_1, m_2)(j\frac{m_2}{(m_1, m_2)} - k\frac{m_1}{(m_1, m_2)})$ because clearly (m_1, m_2) divides m_1 and m_2. It follows that there is a solution only if $(m_1, m_2) \mid (a_1 - a_2)$. Now suppose that $(m_1, m_2) \mid (a_1 - a_2)$. Then $a_1 - a_2$ can be written as a linear combination of m_1 and m_2, say as $jm_2 - km_1 = a_1 - a_2$, and so $a_1 + km_1 = a_2 + jm_2$. Let $x = a_1 + km_1$. Then clearly $x = a_1 + km_1$ is a solution to $x \equiv a_1 \pmod{m_1}$, and $x = a_1 + km_1 = a_2 + jm_2 \equiv a_2 \pmod{m_2}$. Note that if $x \equiv a_1 \pmod{m_1}$, then $x \equiv a_1 \frac{m_1}{(m_1, m_2)}$ as well. To see this observe that $x = a_1 + km_1 = a_1 + k(m_1, m_2)\frac{m_1}{(m_1, m_2)}$, and so $x \equiv a_1 \frac{m_1}{(m_1, m_2)}$. Hence, any solution of $x \equiv a_1 \pmod{m_1}$ and $x \equiv a_2 \pmod{m_2}$ is also a solution of $x \equiv a_1 \pmod{m_1}$ and $x \equiv a_2 \pmod{m_2}$. By the Chinese Remainder Theorem, there can be at most one solution modulo $\frac{m_1}{(m_1, m_2)}m_2 = [m_1, m_2]$.

17. **a.** $x = 430 + 2100j$ **b.** $x = 9102 + 10010j$

19. The basis step is given by Exercise 15. Suppose that the system of the first r congruences has a unique solution A modulo $M = [m_1, \ldots, m_r]$. Then by Exercise 15, the system $x \equiv A \pmod{M}, x \equiv a_{r+1} \pmod{m_{r+1}}$ has a unique solution modulo $[M, m_{r+1}] = [m_1, \ldots, m_{r+1}]$.

21. 2101

23. 0000, 0001, 0625, and 9376.

25. Every 85008 quarter-days starting with quarter-day 4257. (We must add or subtract one or two quarter-days to make the system solvable.)

27. Every 85008 quarter-days, starting at 0.

29. If the set of distinct congruences cover the integers modulo the least common multiple of the moduli, then that set will cover all integers. Examine the integers modulo 210, the l.c.m. of the moduli in this set of congruences. The first four congruences take care of all numbers containing a prime divisor of 2, 3, 5, or 7. The remaining numbers can be examined one at a time, and each can be seen to satisfy one (or more) of the congruences.

31. 26 feet and 6 inches long (or 52 feet and 9 inches long)

33. $x = 225a_1 + 1000a_2 + 576a_3 + 1800k$, where k is an integer and a_1 is 3 or 7, a_2 is 2 or 7, and a_3 is 14 or 18.

Section 3.4

1. **a.** $x \equiv 2, y \equiv 2 \pmod{5}$ **b.** no solution   **c.** $x \equiv 3y \equiv 0 \pmod{5}; x \equiv 4, y \equiv 1 \pmod{5}; x \equiv 0, y \equiv 2 \pmod{5}; x \equiv 1, y \equiv 3 \pmod{5}; x \equiv 2, y \equiv 4 \pmod{5}$.

3. 0, 1, p, or p^2

5. The basis step, where $k = 1$, is clear by assumption. For the inductive hypothesis assume that $A \equiv B \pmod{m}$ and $A^k \equiv B^k \pmod{m}$. Then, $A \cdot A^k \equiv A \cdot B^k \pmod{m}$ by Theorem 3.15. Further, $A^{k+1} = A \cdot A^k \equiv a \cdot B^k \equiv B \cdot B^k = B^{k+1} \pmod{m}$ by simple substitution. This completes the inductive proof.

7. Note that $1 = det(I) = det(A^2) = (det(A))^2 \pmod{m}$. So, $(det(A))^2 - 1 = (det(A) + 1)(det(A) - 1) \equiv 0 \pmod{m}$. It follows that $det(A) = \pm 1$.

9. **a.** $\begin{pmatrix} 4 & 4 & 3 \\ 4 & 3 & 4 \\ 3 & 4 & 4 \end{pmatrix}$

   b. $\begin{pmatrix} 2 & 0 & 6 \\ 2 & 1 & 4 \\ 3 & 4 & 0 \end{pmatrix}$

$$\text{c.} \begin{pmatrix} 5 & 5 & 5 & 4 \\ 5 & 5 & 4 & 5 \\ 5 & 4 & 5 & 5 \\ 4 & 5 & 5 & 5 \end{pmatrix}$$

11. **a.** Multiplying the first congruence by 2 gives $2x + 2y + 2z \equiv 2 \pmod 5$. Subtracting this from the second congruence gives $2y + z \equiv 4 \pmod 5$. There are five possible values for z modulo 5, and since $(2,5) = 1$, each of these leads to a unique value of y modulo 5, and substituting these values of y and z modulo 5 into the first congruence we obtain a unique value of x modulo 5. Hence there are exactly 5 incongruent solutions modulo 5. These are $x \equiv 4 \pmod 5, y \equiv 2 \pmod 5, z \equiv 0 \pmod 5; x \equiv 1 \pmod 5, y \equiv 4 \pmod 5, z \equiv 1 \pmod 5; x \equiv 3 \pmod 5, y \equiv 1 \pmod 5, z \equiv 2 \pmod 5; x \equiv 0 \pmod 5, y \equiv 3 \pmod 5, z \equiv 3 \pmod 5;$ and $x \equiv 2 \pmod 5, y \equiv 0 \pmod 5, z \equiv 4 \pmod 5$.

13. In Gaussian elimination, the chief operation is to subtract a multiple of one equation or row from another, in order to put a 0 in a desirable place. Given that an entry a must be changed to 0 by subtracting a multiple of b, we proceed as follows: Let $\bar b$ be the inverse for $b \pmod k$. Then $a - (a\bar b)b = 0$, and elimination proceeds as for real numbers. If $\bar b$ doesn't exist, and one cannot swap rows to get an invertible b, then the system is underdetermined.

15. Consider summing the ith row. Let $k = xn + y$, where $0 \le y < n$. Then x and y must satisfy the Diophantine equation $i \equiv a + cy + ex \pmod n$, if k is in the ith row. Then $x - ct$ and $y + et$ is also a solution for any integer t. By Exercise 14, there must be n positive solutions which yield n numbers k between 0 and n^2. Let $s, s+1, \ldots, s+n-1$ be the values for t that give these solutions. Then the sum of the ith row is $\sum_{r=0}^{n-1} (n(x - c(s + r)) + y + e(s + r)) = n(n + 1)$, which is independent of i.

Section 3.5

1. **a.** $133 = 7 \cdot 19$ **b.** $1189 = 29 \cdot 41$ **c.** $1927 = 41 \cdot 47$ **d.** $8131 = 47 \cdot 173$
   e. $36287 = 131 \cdot 277$ **f.** $48227 = 29 \cdot 1663$.
3. Numbers generated by linear functions where $a > 1$ will not be random in the sense that $x_{2s} - x_k = ax_{2s-1} + b - (ax_{s-1} + b) = a(x_{2s-1} - x_{s-1})$ is a multiple of a for all s. If $a = 1$, then $x_{2s} - x_s = x_0 + sb$. In this case, if $x_0 \ne 0$, then we will not notice if a factor of b that is not a factor of x_0 is a divisor of n.

Section 4.1

1. **a.** 2^8 **b.** 2^4 **c.** 2^{10} **d.** 2^1
3. **a.** 18381 is divisible by 3, but not by 9.   **b.** 65412351 is divisible by both 3 and 9.
   c. 987654321 is divisible by both 3 and 9.   **d.** 78918239735 is divisible by neither 3 nor 9.
5. **a.** $2^1 = 2$ **b.** $2^0 = 1$ **c.** $2^6 = 64$ **d.** $2^0 = 1$
7. **a.** doesn't divide   **b.** doesn't divide   **c.** divides   **d.** divides
9. **a.** neither   **b.** both   **c.** neither   **d.** only 5
11. This repunit is divisible by 3 if and only if n is divisible by 3 and is divisible by 9 if and only if n is divisible by 9.
13. The alternating sum of blocks of three digits of an n-digit repunit is 0 if $n \equiv 0 \pmod 6$, 1 if $n \equiv 1 \pmod 6$, 11 if $n \equiv 2 \pmod 6$, 111 if $n \equiv 3 \pmod 6$, 110 if $n \equiv 4 \pmod 6$, 100 if $n \equiv 5 \pmod 6$. Hence an n-digit repunit is divisible by 1001 if and only if $n \equiv 0 \pmod 6$. Since 7 divides this alternating sum if and only if $n \equiv 0 \pmod 6$, these are exactly the values of n for which this requnit is divisible by 7. Exactly the same reasoning and conclusion holds for divisibility by 13.

15. Let d be a divisor of $b - 1$. By Theorem 4.2, a number is divisible by d if and only if the sum of its digits is a multiple of d. Since the sum of the digits of a repunit is equal to the number of digits it has, a repunit is divisible by d if and only if it has a multiple of d digits.

17. A palindromic integer with $2k$ digits has the form $(a_k a_{k-1} \ldots a_1 a_1 a_2 \ldots a_k)_{10}$. Using the test for divisibility by 11 developed in this section, we find that $a_k - a_{k-1} + \cdots \pm a_1 \mp a_1 \pm a_2 \mp \cdots - a_k = 0 \equiv 0 \,(\mathrm{mod}\ 11)$ and so $(a_k a_{k-1} \ldots a_1 a_1 a_2 \ldots a_k)_{10}$ is divisible by 11.

19. Let $a_k a_{k-1} \ldots a_1 a_0$ be the decimal representation of an integer. Then $a_k a_{k-1} \ldots a_1 a_0 = a_0 a_1 a_2 + 10^3 a_3 a_4 a_5 + 10^3 (10^3 a_6 a_7 a_8) + \cdots$. So, $a_k a_{k-1} \ldots a_1 a_0 \equiv a_0 a_1 a_2 + a_3 a_4 a_5 + a_6 a_7 a_8 + \cdots \,(\mathrm{mod}\ 37)$. Thus $a_k a_{k-1} \ldots a_1 a_0$ is divisible by 37 if and only if $a_0 a_1 a_2 + a_3 a_4 a_5 + a_6 a_7 a_8 + \cdots$ is also. Hence, 443692 is divisible by 37 if and only if $443 + 692 = 1135$ is. And 1134 is divisible by 37 if and only if $1+135 = 136$ is. But 136 is not, and so 37 does not divide 443692. Further, 11092785 is divisible by 37 if and only if $11+092+785 = 888$ is. We know that $888 = 24 \cdot 37$, so 11092785 is a multiple of 37.

21. **a.** not divisible by 5   **b.** divisible by 5   **c.** divisible by neither   **d.** divisible by 101

23. 6

25. no, consider 19.

Section 4.2

3. once (August)

5. $W \equiv k + [2.6m - 0.2] - 2C + Y + [\frac{Y}{4}] + [\frac{C}{4}] - [\frac{N}{4000}] \quad (\mathrm{mod}\ 7)$

Section 4.3

1. **a.** 1-7, 2-6, 3-5, 4-bye. Round 2 : 2- 7, 3- 6, 4- 5, 1-bye. Round 3: 1-2, 3-7, 4-6, 5-bye. Round 4: 1-3, 4-7, 5-6, 2-bye. Round 5: 1-4, 2-3, 5-7, 6-bye. Round 6: 1-5, 2-4, 6-7, 3-bye. Round 7: 1-6, 2-5, 3-4, 7-bye.
   b. Teams i and j are paired in round k if and only if $i + j \equiv k \,(\mathrm{mod}\ 8)$. Team i draws a bye if $2i \equiv k \,(\mathrm{mod}\ 8)$.
   c. Teams i and j are paired in round k if and only if $i + j \equiv k \,(\mathrm{mod}\ 9)$. Team i draws a bye if $2i \equiv k \,(\mathrm{mod}\ 9)$.
   d. Teams i and j are paired in round k if and only if $i + j \equiv k \,(\mathrm{mod}\ 10)$. Team i draws a bye if $2i \equiv k \,(\mathrm{mod}\ 10)$.

3. For round 1, teams i and j are paired if $i + j \equiv 1 \,(\mathrm{mod}\ 5)$. Teams 1 and 5 are paired, and since $1 + 5 = 6$ is even, team 5 is the home team. Teams 2 and 4 are paired, and since $2 + 4 = 6$ is even, team 4 is the home team. Finally, in round 1 team 3 draws a bye.

Section 4.4

1. Let k be the six-digit number. Assign this car the space numbered $h(k) \equiv k \,(\mathrm{mod}\ 101)$. When a car is assigned the same space as another car assign it to the space $h(k) + g(k)$ where $g(k) \equiv k + 1 \,(\mathrm{mod}\ 99)$ and $0 < g(k) \le 98$. When this space is occupied next try $h(k) + 2g(k)$, then $h(k) + 3g(k)$, and so on.

3. **a.** It is clear that m memory locations will be probed as $j = 0, 1, 2, \ldots, m - 1$. To see that they are all distinct, and hence every memory location is probed, assume that $h_i(K) \equiv h_j(K) \,(\mathrm{mod}\ m)$. Then $h(K) + iq \equiv h(K) + jq \,(\mathrm{mod}\ m)$. It follows that $iq \equiv jq \,(\mathrm{mod}\ m)$, and as $(q, m) = 1, i \equiv j \,(\mathrm{mod}\ m)$ by Corollary 3.4.1. And so $i = j$ since i and j are both less than m.

b. Assume that $h_i(K) \equiv h_j(K) \pmod{m}$. Then $h(K) + iq \equiv h(K) + jq \pmod{m}$. From this it follows that $iq \equiv jq \pmod{m}$, and as $(q, m) = 1$, $i \equiv j \pmod{m}$ by Corollary 3.4.1. And so $i = j$ since i and j are both less than m.

5. $h(k_{11}) = 558, h(k_{12}) = 1002, h(k_{13}) = 2174, h(k_{14}) = 4$.

Section 4.5

1. **a.** 0 **b.** 0 **c.** 1 **d.** 1 **e.** 0 **f.** 1
3. **a.** 0 **b.** 1 **c.** 0
5. **a.** 7 **b.** 1 **c.** 1
7. Here, transposition means that adjacent digits are in the wrong order. Suppose, first, that the first two digits, x_1 and x_2, or equivalently, the fourth and fifth digits are exchanged, and the error is not detected. Then $x_7 \equiv 7x_1 + 3x_2 + x_3 + 7x_4 + 3x_5 + x_6 \equiv 7x_2 + 3x_1 + x_3 + 7x_4 + 3x_5 + x_6 \pmod{10}$. It follows that $7x_1 + 3x_2 \equiv 7x_2 + 3x_1 \pmod{10}$ or $4x_1 \equiv 4x_2 \pmod{10}$. By Corollary 3.4.1, we see that $x_1 \equiv x_2 \pmod 5$. This is equivalent to $| x_1 - x_2 | = 5$, as x_1 and x_2 are single digits. Similarly, if the second and third (or fifth and sixth) digits are transposed, $2x_2 \equiv 2x_3 \pmod{10}$, which reduces to $x_2 \equiv x_3 \pmod 5$ by Corollary 3.4.1. Similarly, if the third and fourth digits are transposed, $6x_3 \equiv 6x_4 \pmod{10}$ and $x_3 \equiv x_4 \pmod 5$.
9. **a.** 0 **b.** 3 **c.** 4 **d.** X
11. **a.** valid **b.** not valid **c.** valid **d.** valid **e.** not valid
13. **a.** yes. **b.** no.
15. **a.** 9,4
  b. If x_i is misentered as y_i, and the congruence defining x_{10} holds, then $ax_i \equiv ay_i \pmod{11}$ by setting the two definitions of x_{10} congruent. It follows by Corollary 3.4.1 that $x_i \equiv y_i \pmod{11}$ and so $x_i = y_i$. If the last digit, x_{11} is misentered as y_{11}, then the congruence defining x_{11} will hold if and only if $x_{11} = y_{11}$.
  c. Start by assuming $x_3 \ldots x_9$ are 0.

Section 5.1

1. -1
3. $(3^5)^2 \equiv 243^2 \equiv 1^2 \equiv 1 \pmod{11^2}$.
5. **a.** 9 **b.** 17
7. $-2 \equiv 2(p - 1)! \equiv 2(p - 3)!(p - 2)(p - 1) \equiv 2(p - 3)!(-1)(-2) \pmod{p}$. Since $(p, 2) = 1$, divide by 2.
9. $n^7 \equiv n \pmod 2$. Since $n^3 \equiv n \pmod 3$, $n^7 = (n^3)^2 \cdot n \equiv n^2 \cdot n \equiv n^3 \equiv n \pmod 3$. By Fermat's little theorem, $n^7 \equiv n \pmod 7$. Since $42 = 2 \cdot 3 \cdot 7$, it follows that $n^7 \equiv n \pmod{42}$.
11. 641 (at $k = 8$)
13. If $a^p \equiv b^p \pmod p$, then $a \equiv b \pmod p$, so $b = a + kp$ for some k. Hence, by the Binomial Theorem and Exercise 24 in Section 1.9, $b^p = a^p + \binom{p}{1} a^{p-1}kp + \cdots + (kp)^p \equiv a^p \pmod{p^2}$, since p^2 divides every other term.
15. Since $p - 1 \equiv -1, p - 2 \equiv -2, \ldots, \frac{p+1}{2} \equiv \frac{p-1}{2} \pmod p$, we have $\left(\frac{p-1}{2}\right)!^2 \equiv -(p - 1)! \equiv 1 \pmod p$, (since $p \equiv 3 \pmod 4$ the minus signs work out.) If $x^2 \equiv 1 \pmod p$, then $p \mid x^2 - 1 = (x - 1)(x + 1)$, so $x \equiv \pm 1 \pmod p$.

17. Suppose that $p \equiv 1 \pmod 4$. Let $y = \pm[\frac{(p-1)}{2}]!$. Then we have $y^2 \equiv [\frac{(p-1)}{2}]!^2 \equiv$
    $[\frac{(p-1)}{2}]!^2(-1)^{\frac{(p-1)}{2}} \equiv (1 \cdot 2 \cdot 3 \cdots \frac{(p-1)}{2})(-1 \cdot (-2) \cdot (-3) \cdots (-\frac{(p-1)}{2})) \equiv 1 \cdot 2 \cdot$
    $3 \cdots \frac{(p-1)}{2} \cdot \frac{(p+1)}{2} \cdots (p-3)(p-2)(p-1) = p! \equiv -1 \pmod p$, where we have used
    Wilson's theorem. Now suppose that $x^2 \equiv -1 \pmod p$. Then $x^2 \equiv y^2 \pmod p$ where
    $y = [\frac{(p-1)}{2}]!$. Hence $(x^2 - y^2) = (x - y)(x + y) \pmod p$. It follows that $p \mid (x - y)$ or
    $p \mid (x + y)$ so that $x \equiv \pm y \pmod p$.

19. Suppose that p is prime. Then by Fermat's little theorem for every integer a, $a^p \equiv a \pmod p$
    and by Wilson's theorem $(p - 1)! \equiv -1 \pmod p$ so that $a(p - 1)! \equiv -a \pmod p$. It
    follows that $a^p + (p - 1)!a \equiv a + (-a) \equiv 0 \pmod p$. Consequently $p \mid [a^p + (p - 1)!a]$

21. Suppose that n and $n + 2$ are twin primes. Then since n is prime by Wilson's theorem we
    know that $(n - 1)! \equiv -1 \pmod n$. Hence $4[(n - 1)! + 1] + n \equiv 4 \cdot 0 + n \equiv 0 \pmod n$.
    Also, since $n + 2$ is prime by Wilson's theorem it follows that $(n + 1)! \equiv -1 \pmod{n + 2}$,
    so that $(n + 1)n \cdot (n - 1)! \equiv (-1)(-2)(n - 1)! \equiv 2(n - 1)! \equiv -1 \pmod{n + 2}$. Hence
    $4[(n - 1)! + 1] + n \equiv 2(2 \cdot (n - 1)!) + 4 + n \equiv 2 \cdot (-1) + 4 + n = n + 2 \equiv 0 \pmod{n + 2}$. Since $(n, n + 2) = 1$ it follows that $4[(n - 1)! + 1] + n \equiv 0 \pmod{n(n + 2)}$.

23. $1 \cdot 2 \cdots (p - 1) \equiv (p + 1)(p + 2) \cdots (2p - 1) \pmod p$. Each factor is prime to p, so $1 \equiv \frac{(p+1)(p+2)\cdots(2p-1)}{1\cdot2\cdots(p-1)} \pmod p$. Thus $2 \equiv \frac{(p+1)(p+2)\cdots(2p-1)2p}{1\cdot2\cdots(p-1)p} \equiv \binom{2p}{p} \pmod p$.

25. We first note that $1^p \equiv 1 \pmod p$. Now suppose that $a^p \equiv a \pmod p$. then by Exercise 22
    we see that $(a + 1)^p \equiv a^p + 1 \pmod p$. But by the inductive hypothesis $a^p \equiv a \pmod p$
    we see that $a^p + 1 \equiv a + 1 \pmod p$. Hence $(a + 1)^p \equiv a + 1 \pmod p$. This completes
    the inductive step of the proof.

27. **a.** If $c < 26$ then c cards are put into the deck above the card, so it ends up in the 2cth
    position and $2c < 52$, so $b = 2c$. If $c \geq 26$ then $c - 1$ cards are put into the deck above the
    card, but 26 cards are taken away above it , so it ends up in the $b = (c - 26 + c - 1)$th place.
    Then $b = 2c - 25 \equiv 2c \pmod{53}$.
    b. 52

29. Assume without loss of generality that $a_p \equiv b_p \equiv 0 \pmod p$. Then, by Wilson's Theorem,
    $a_1 a_2 \cdots a_{p-1} \equiv b_1 b_2 \cdots b_{p-1} \equiv -1 \pmod p$. Then $a_1 b_1 \cdots a_{p-1} b_{p-1} \equiv (-1)^2 \equiv 1$
    $\pmod p$. If the set were a complete system, the last product would be $\equiv -1 \pmod p$.

31. The basis step is omitted. Assume $(p - 1)^{p^{k-1}} \equiv -1 \pmod{p^k}$. Then, $(p - 1)^{p^k} \equiv$
    $((p - 1)^{p^{k-1}})^p \equiv (-1 + mp^k)^p \equiv -1 + \binom{p}{1} mp^k + \cdots + (mp^k)^p \equiv -1$
    $\pmod{p^{k+1}}$, where we have used the fact that $p \mid \binom{p}{j}$ for $j \neq 0$ or p.

Section 5.2

1. We find that $3^{90} = (3^4)^{22} \cdot 3^2 = 81^4 \cdot 9 \equiv (-10) \cdot 9 = -90 \equiv 1 \pmod{91}$. Hence 91 is
   a pseudoprime modulo 3.

3. Note that $2^{262} \equiv 2 \pmod{161038}$. Then $2^{161038} \equiv 2^{262 \cdot 614 + 170} \equiv 2^{614 + 170} \equiv 2$
   $\pmod{161038}$.

5. From the Binomial Theorem, $(n - a)^n \equiv (-a)^n \equiv -(a^n) \equiv -a \equiv (n - a) \pmod n$,
   where we used $a^n \equiv a \pmod n$.

7. Raise the congruence $2^{2^n} \equiv -1 \pmod{F_n}$ to the $2^{2^n - n}$th power.

9. $b^n \equiv b \pmod n$ and $a_n \equiv a \pmod n$. It follows that $(ab)^n \equiv a^n b^n \equiv ab \pmod n$.

11. If $(ab)^{n-1} \equiv 1 \pmod n$, then $1 \equiv a^{n-1} b^{n-1} \equiv 1 \cdot b^n \pmod n$ which implies that n is
    a pseudoprime to the base b.

13. From $2^{18} \equiv 1 \pmod{1387}$ we get $2^{1387} \equiv 2 \pmod{1387}$ so 1387 is a pseudoprime. But $1387 - 1 = 2 \cdot 693$ and $2^{693} \equiv 512 \pmod{1387}$, which is all that must be checked, since $s = 1$. Thus 1387 fails Miller's test and hence is not a strong pseudoprime.

15. $25326001 = 2^4 1582875 = 2^s t$ and with this value of t, $2^t \equiv -1 \pmod{25326001}, 3^t \equiv -1 \pmod{25326001}$, and $5^t \equiv 1 \pmod{25326001}$.

17. For $7 \cdot 23 \cdot q$ to be a Carmichael number we must have $(7 - 1) = 6 \mid (7 \cdot 23 \cdot q - 1)$, $(23 - 1) = 22 \mid (7 \cdot 23 \cdot q - 1)$, and $(q - 1) \mid (7 \cdot 23 \cdot q - 1)$. We find that $-q - 1 \equiv 0 \pmod 6, 7q - 1 \equiv 0 \pmod{22}$, and $160 \equiv 0 \pmod{q - 1}$. Hence $q \equiv 5 \pmod 6, q \equiv 19 \pmod{22}$, and $(q - 1) \mid 160$. It follows from the two congruences that $q \equiv 41 \pmod{66}$. We find that only $q = 41$ satisfies all the conditions given.

19. We can assume that $b < n$. Then b has fewer than $\log_2 n$ bits. Also, $t < n$ so it has fewer than $\log_2 n$ bits. It takes at most $\log_2 n$ multiplications to calculate b^{2^s} so it takes $O(\log_2 n)$ multiplications to calculate $b^{2^{\log_2 t}} = b^t$. Each multiplication is of two $\log_2 n$ bit numbers, and so takes $O((\log_2 n)^2)$ operations. So all together we have $O((\log_2 n)^3)$ operations.

Section 5.3

1. a. 1,5   b. 1,2,4,5,7,8   c. 1,3,7,9   d. 1,3,5,9,11,13   e. 1,3,5,7,9,11,13,15
   f. 1,2,3,4,5,6,7,8,9,10,11,12,13,14,15,16

3. If $(a, m) = 1$, then $(-a, m) = 1$, so $-c_i$ must appear among the c_j. Also $c_i \not\equiv -c_i \pmod m$, else $2c_i \equiv 0 \pmod m$ and so $(c_i, m) \neq 1$.

5. 11

7. Since $a^2 \equiv 1 \pmod 8$ whenever a is odd, it follows that $a^{12} \equiv 1 \pmod 8$ whenever $(a, 32760) = 1$. Euler's theorem tells us that $a^{\phi(9)} = a^6 \equiv 1 \pmod 9$ whenever $(a, 9) = 1$, so that $a^{12} = (a^6)^2 \equiv 1 \pmod 9$ whenever $(a, 32760) = 1$. Furthermore, Fermat's little theorem tells us that $a^4 \equiv 1 \pmod 5$ whenever $(a, 5) = 1, a^6 \equiv 1 \pmod 7$ whenever $(a, 7) = 1$, and $a^{12} \equiv 1 \pmod{13}$ whenever $(a, 13) = 1$. It follows that $a^{12} \equiv (a^4)^3 \equiv 1 \pmod 5, a^{12} \equiv (a^6)^2 \equiv 1 \pmod 7$, and $a^{12} \equiv 1 \pmod{13}$ whenever $(a, 32760) = 1$. Since $32760 = 2^3 3^2 \cdot 5 \cdot 7 \cdot 13$ and the moduli $8, 9, 5, 7$, and 13 are pairwise relatively prime, we see that $a^{12} \equiv 1 \pmod{32760}$.

9. a. 9   b. 13   c. 7

11. 1

13. $\phi(10) = 4$, so $7^4 \equiv 1 \pmod{10}$ and $7^{1000} \equiv (7^4)^{250} \equiv 1^{250} \equiv 1 \pmod{10}$.

15. $\phi(13) = 12, \phi(14) = 6, \phi(15) = 8, \phi(16) = 8, \phi(17) = 16, \phi(18) = 6, \phi(19) = 18, \phi(20) = 8$

17. If $(a, b) = 1$ and $(a, b - 1) = 1$ then $a \mid (b^{k\phi(a)} - 1)/(b - 1)$ which is a base b repunit. If $(a, b - 1) = d > 1$, then d divides any repunit of length $k(b - 1)$, and $\frac{a}{d} \mid (b^{k\phi(a/d)} - 1)/(b - 1)$ and these sets intersect infinitely often.

Section 6.1

1. a. is   b. is not   c. is not   d. is not   e. is   f. is not   g. is not   h. is not   i. is

3. $\phi(5186) = \phi(5187) = \phi(5188) = 2592$.

5. If and only if n is a multiple of 3.

7. n is a power of 2.

9. If n is odd, then $(2, n) = 1$ and $\phi(2n) = \phi(2)\phi(n) = 1 \cdot \phi(n) = \phi(n)$. If n is even, say $n = 2^s t$ with t odd. Then we have $\phi(2n) = \phi(2^{s+1}t) = \phi(2^{s+1})\phi(t) = 2^s \phi(t) = 2(2^{s-1}\phi(t)) = 2(\phi(2^s)\phi(t)) = 2(\phi(2^s t)) = 2\phi(n)$.

11. $n = 2^k p_1 p_2 \cdots p_r$ where p_i is a Mersenne prime.

13. Let $p_1, \cdots, p_r$ be those primes dividing a but not b. Let $q_1, \cdots, q_s$ be those primes dividing b but not a. Let $r_1, \cdots r_t$ be those primes dividing a and b. Let $P = \prod(1 - \frac{1}{p_i}), Q = \prod(1 - \frac{1}{q_i})$ and $R = \prod(1 - \frac{1}{r_i})$. Then we have $\phi(ab) = abPQR = \frac{aPRbQR}{R} = \frac{\phi(a)\phi(b)}{R}$. But $\phi((a,b)) = (a,b)R$ so $R = \frac{\phi((a,b))}{(a,b)}$ and we have $\phi(ab) = \frac{\phi(a)\phi(b)}{R} = \frac{(a,b)\phi(a)\phi(b)}{\phi((a,b))}$ as desired.

15. If $n = p^r m$, then $\phi(p^r m) = (p^r - p^{r-1})\phi(m) \mid (p^r m - 1)$, hence $p \mid 1$ or $r = 1$. So n is square-free. If $n = pq$, then $\phi(pq) = (p-1)(q-1) \mid (pq - 1)$. Then $(p-1) \mid (pq-1) - (p-1)q = q - 1$. Similarly $(q-1) \mid (p-1)$, a contradiction.

17. Let $n = p_1^{a_1} p_2^{a_2} \cdots p_k^{a_k}$. Let P_i be the property that an integer is divisible by p_i. Let S be the set $\{1, 2, \cdots, n-1\}$. To compute $\phi(n)$ we need to count the elements of S with none of the properties $P_1, P_2, \cdots, P_k$. Let $n(P_{i_1}, P_{i_2}, \cdots, P_{i_m})$ be the number of elements of S with all of properties $P_{i_1}, P_{i_2}, \cdots, P_{i_m}$. Then $n(p_{i_1}, \cdots P_{i_m}) = \frac{n}{p_{i_1}} p_{i_2} \cdots p_{i_m}$. By Exercise 18 of Section 1.4, we have $\phi(n) = n - (\frac{n}{p_1} + \frac{n}{p_2} + \cdots + \frac{n}{p_k}) + (\frac{n}{p_1 p_2} + \cdots + \frac{n}{p_{k-1} p_k}) + \cdots + (-1)^k (\frac{n}{p_1 \cdots p_k}) = n(1 - \sum_{p_i \mid n} \frac{1}{p_i} + \sum_{p_{i_1} p_{i_2} \mid n} \frac{1}{p_{i_1} p_{i_2}} - \sum_{p_{i_1} p_{i_2} p_{i_3}} \frac{1}{p_{i_1} p_{i_2} p_{i_3}} + \cdots + (-1)^k \frac{n}{p_1 \cdots p_k})$. On the other hand, notice that each term in the expansion of $(1 - \frac{1}{p_1})(1 - \frac{1}{p_2}) \cdots (1 - \frac{1}{p_k})$ is obtained by choosing either 1 or $-\frac{1}{p_i}$ from each factor and multiply the choice together. This gives each term the form $\frac{(-1)^m}{p_{i_1} p_{i_2} \cdots p_{i_m}}$. Note that each term can occur in only one way. Thus $n(1 - \frac{1}{p_1})(1 - \frac{1}{p_2}) \cdots (1 - \frac{1}{p_k}) = n(1 - \sum_{p_i \mid n} \frac{1}{p_i} + \sum_{p_{i_1} p_{i_2}} \frac{1}{p_{i_1} p_{i_2}} - \cdots + (-1)^k \frac{n}{p_1 \cdots p_k}) = \phi(n)$.

19. Note that $1 \leq \phi(m) \leq m - 1$ for $m > 1$. Hence if $n \geq 2, n > n_1 > n_2 > \cdots \geq 1$ where $n_i = \phi(n)$ and $n_i = \phi(n_{i-1})$ for $i > 1$. Since $n_i, i = 1, 2, 3, \ldots$ is a decreasing sequence of positive integers, there must be a positive integer r such that $n_r = 1$.

21. Note that the definition of $f * g$ can also be expressed as $(f * g)(n) = \sum_{a \cdot b = n} f(a)g(b)$. Then the fact that $f * g = g * f$ is evident.

23. **a.** $(\iota)(nm) = 0 = \iota(n)\iota(m)$ if and only if one of n or m is greater than 1. Otherwise, all three parts of the equation are 1.

   b. $\iota(d)$ is nonzero only for $n = 1$, so $\iota * f(n) = \sum_{d \mid n} \iota(d)f(n/d) = \iota(1)f(n) = f(n)$.

25. Let $h = f * g$ and let $(m, n) = 1$. Then $h(mn) = \sum_{d \mid mn} f(d)g(\frac{mn}{d})$. Since $(m, n) = 1$, each divisor d of mn can be expressed in exactly one way as $d = ab$ where $a \mid m$ and $b \mid n$. Then $(a, b) = 1$ and $(\frac{m}{a}, \frac{n}{b}) = 1$. Then there is a one-to-one correspondence between the divisors d of mn and the pairs of products ab where $a \mid m$ and $b \mid n$. Then

$$h(mn) = \sum_{\substack{a \mid m \\ b \mid n}} f(ab)g(\frac{mn}{ab}) = \sum_{\substack{a \mid m \\ b \mid n}} f(a)f(b)g(\frac{m}{a})g(\frac{n}{b})$$

$$= \sum_{a \mid m} f(a)g(\frac{m}{a}) \sum_{b \mid n} f(b)g(\frac{n}{b}) = h(m)h(n)$$

as desired.

27. 0 in each case. See Exercise 29.

29. Suppose n is $n = p_1^{a_1} p_2^{a_2} \cdots p_m^{a_m}$. The only divisors d of n with $\mu(d) \neq 0$ are the divisors of d that are the product of zero or more distinct primes. the number of ways to choose k primes from the m primes in the factorization is $\binom{m}{k}$ and the value of μ at an integer d which is product of k primes is $(-1)^k$. It follows that $\sum_{d|n} \mu(d) = \sum_{j=0}^{m}(-1)^j \binom{m}{j}$ $= 0$, where we have used the result of Exercise 5 in Section 1.4.

31. Suppose that F is multiplicative where $F(n) = \sum_{d|n} f(d)$. Let $(m,n) = 1$. Any divisor d or mn can be written as $d = d_1 d_2$ where $d_1 \mid m$ and $d_2 \mid n$ and $(d_1, d_2) = 1$. Using the Möbius inversion formula proved in Exercise 26 we have $f(mn) = \sum_{d|mn} \mu(d)F(\frac{mn}{d}) = \sum_{d_1|m,d_2|n} \mu(d_1 d_2)F(\frac{mn}{d_1 d_2}) = \sum_{d_1|m,d_2|n} \mu(d_1)\mu(d_2)F(\frac{m}{d_1})F(\frac{n}{d_2}) = \sum_{d_1|m} \mu(d_1)F(\frac{m}{d_1}) \sum_{d_2|n} \mu(d_2)F(\frac{n}{d_2}) = f(m)f(n)$. Hence that f is multiplicative.

33. Let $f(n) = n^k$, where k is a real number. Then $f(mn) = (mn)^k = m^k n^k = f(m)f(n)$. It follows that $f(n)$ is completely multiplicative.

35. Let $f(n) = \sum_{d|n} \lambda(d)$. Then by Exercises 27 and 30, f is multiplicative. Now, $f(p^t) = \lambda(1) + \lambda(p) + \lambda(p^2) + \cdots + \lambda(p^t) = 1 - 1 + 1 - \cdots + (-1)^t = 0$ if t is odd and $= 1$ if t is even. Then $f(p_1^{a_1} p_2^{a_2} \cdots p_r^{a_r}) = \prod f(p_i^{a_i}) = 0$ if any a_i is odd and $= 1$ if all a_i are even and hence n is a square.

37. If f and g are completely multiplicative and m and n are positive integers we have $(fg)(mn) = f(mn)g(mn) = f(m)f(n)g(m)g(n) = f(m)g(m)f(n)g(n) = (fg)(m)(fg)(n)$, so fg is also completely multiplicative.

39. We have $f(mn) = \log mn = \log m + \log n = f(m) + f(n)$. Hence $f(n) = \log n$ is completely additive.

41. Let $(m, n) = 1$, then by the additivity of f we have $f(mn) = f(m) + f(n)$. Then $g(mn) = 2^{f(mn)} = 2^{f(m)+f(n)} = 2^{f(m)}2^{f(n)} = g(m)g(n)$, so g is multipicative.

Section 6.2

1. a. 48 **b.** 399 **c.** 2340 **d.** $2^{101} - 1$ **e.** 6912 **f.** 813404592 **g.** 5935776
  h. 26495791882560

3. n is a perfect square.

5. a. 6, 11 **b.** 10, 17 **c.** 14, 15, 23 **d.** 33, 35, 47 **e.** no solution **f.** 65, 44, 83

7. Note that $\tau(p^{k-1}) = k$ whenever p is prime and k is a positive integer $k > 1$.

9. Only p^2 where p is prime.

11. If n is not a perfect square $n^{\tau(n)/2}$. If n is a perfect square, $n^{\tau(n)/2}$.

13. $\sigma_k(p) = \sum_{d|p} d^k = 1^k = p^k = 1 + p^k$.

15. $\sum_{d|ab} d^k = \sum_{d_1|a,d_2|b} (d_1 d_2)^k = \sum_{d_1|a} d_1^k \sum_{d_2|a} d_2^k = \sigma_k(a)\sigma_k(b)$.

17. n is prime.

19. Let $n = p_1^{a_1} p_2^{a_2} \cdots p_r^{a_r}$ and let x and y be integers with $[x, y] = n$. Then $x \mid n$ and $y \mid n$ so we have $x = p_1^{b_1} p_2^{b_2} \cdots p_r^{a_r}$ and $y = p_1^{c_1} p_2^{c_2} \cdots p_r^{c_r}$, where b_i and $c_i = 0, 1, 2, \ldots$, a_i. Since $[x, y] = n$, we must have $\max\{b_i, c_i\} = a_i$ for each i. Then one of b_i and c_i must be equal to a_i and the other can range over $0, 1, \ldots, a_i$. Therefore we have $2a_i + 1$ ways to choose the pair (b_i, c_i) for each i. Then in total, we can choose the exponents $b_1, b_2, \ldots b_r$, $c_1, \ldots, c_r$ in $(2a_1 + 1)(2a_2 + 1) \cdots (2a_r + 1) = \tau(n^2)$ ways.

21. Let $n = ab$ where a and b are integers with $1 < a \le b < n$. Then either $a \ge \sqrt{n}$ or $b \ge \sqrt{n}$. Consequently $\sigma(n) \ge 1 + a + b + n > 1 + \sqrt{n} + n > n + \sqrt{n}$. Conversely, suppose that n is prime. Then $\sigma(n) = n + 1$ so that $\sigma(n) \le n + \sqrt{n}$. Hence $\sigma(n) > n + \sqrt{n}$ implies that n is composite.

23. The basis step is omitted. Suppose that $\sum_{j=1}^{n-1} \tau(j) = 2 \sum_{j=1}^{[\sqrt{n-1}]} \left[\frac{n-1}{j}\right] - [\sqrt{n-1}]^2$.

For the induction step, it suffices to show that $\tau(n) = 2 \sum_{j=1}^{[\sqrt{n-1}]} \left(\left[\frac{n}{j}\right] - \left[\frac{n-1}{j}\right]\right) =$

$2 \displaystyle\sum_{\substack{j \le [\sqrt{n-1}] \\ j|n}} 1$, which is true by the definition of $\tau(n)$, since there is one factor less than $\sqrt{n}$

for every factor greater than $\sqrt{n}$. Note that if n is a perfect square, we must add the term $(2\sqrt{n} - (2\sqrt{n} - 1) = 1$ to the last two sums. For $n = 100$, we have $\sum_{j=1}^{100} \tau(j) =$

$2 \sum_{j=1}^{10} \left[\frac{n}{j}\right] - 100 = 482$.

25. We use the identity $\sum_{j=0}^{\min\{a,b\}} (p^{a+b-j} + p^{a+b-j-1} + \cdots + p^j) = (p^a + p^{a-1} + \cdots +$

$1)(p^b + p^{b-1} + \cdots + 1)$. If $a = \prod p^{a_i}$ and $b = \prod p^{b_i}$, then $\sigma(a)\sigma(b) = \prod(p^{a_i} + p^{a_i-1} +$

$\cdots + 1)(p^{b_i} + p^{b_i-1} + \cdots + 1) = \prod \sum_{j=0}^{\min\{a_i,b_i\}} (p_i^{a_i+b_i-j} + p_i^{a_i+b_i-j-1} + \cdots + p_i^j)$

27. $\phi(1)\phi(2)\phi(3)\cdots\phi(n)$. See the Special Topics section in Niven and Zuckerman [28].

29. If $p, p + 2$ are prime, then $\phi(p + 2) = p + 1 = \sigma(p)$. If $2^p - 1$ is a Mersenne prime, then $\phi(2^{p+1}) = 2^p = \sigma(2^p - 1)$.

Section 6.3

1. 6, 28, 496, 8128, 33550336, 8589869056.

3. a. $2^5 - 1 = 31$,   **b.** $2^7 - 1 = 127$ **c.** $2^{11} - 1 = 2047$

5. 12, 18, 20, 24, 30, 36

7. If $n = p^k$ then $\sigma(p^k) = \frac{p^{k+1}-1}{p-1}$. Note that $2p^k - 1 < p^{k+1}$ since $p \ge 2$. It follows that

$p^{k+1} - 1 < 2(p^{k+1} - p^k) = 2p^k(p - 1)$, so that $\frac{(p^{k+1}-1)}{p-1} < 2p^k = 2n$. It follows that

$n = p^k$ is deficient.

9. Suppose that n is abundant or perfect. Then $\sigma(n) \ge 2n$. Suppose that $n \mid m$. Then $m = nk$ for some integer k. The divisors of m include the integers kd and $d \mid n$. Hence $\sigma(m) \ge \sum_{d|n} (k +$

$1)d = (k + 1)\sum_{d|n} d = (k + 1)\sigma(n) \ge (k + 1)2n > 2kn = 2m$. Hence m is abundant.

11. a. $\sigma(220) = 504 = \sigma(284)$.   **b.** $\sigma(1184) = 2394 = \sigma(1210)$.

   c. $\sigma(79750) = 168480 = \sigma(88730)$

13. $\sigma(120) = 360 = 3 \cdot 120$, so 120 is 3-perfect.

15. $\sigma(2^7 3^4 5 \cdot 7 \cdot 11^2 \cdot 17 \cdot 19) = 5 \cdot 14182439040$.

17. Suppose that n is 3-perfect and 3 does not divide n. Then $\sigma(3n) = \sigma(3)\sigma(n) = 4 \cdot 3n = 12n = 4 \cdot 3n$. Hence $3n$ is 4-perfect.

19. 908107200

21. $\sigma(\sigma(16)) = \sigma(31) = 32 = 2 \cdot 16$.

23. If r and s are integers, then $\sigma(rs) \ge rs + r + s + 1$. Suppose $n = 2^q t$ is superperfect with t odd and $t > 1$. Then $2n = 2^{q+1}t = \sigma(\sigma(2^q t)) = \sigma\left(\left(2^{q+1} - 1\right)\sigma(t)\right) \ge (2^{q+1} - 1)$

$\sigma(t) + (2^{q+1} - 1) + \sigma(t) + 1 > 2^{q+1}\sigma(t) \ge 2^{q+t}(t + 1)$. Then $t > t + 1$, a contradiction.

25. a. prime   **b.** prime   **c.** prime   **d.** not prime

27. a. $M_n(M_n + 2) = (2^n - 1)(2^n + 1) = 2^{2n} - 1$. If $2n + 1$ is prime then $\phi(2n + 1) = 2n$ and $2^{2n} \equiv 1 \pmod{2n + 1}$. Then $(2n + 1) \mid 2^{2n} - 1 = M_n(M_n + 2)$. Therefore $(2n + 1) \mid M_n$ or $(2n + 1) \mid (M_n + 2)$.

   b. $23 \nmid 2049$ so $23 \mid 2047 = M_{11}$; $47 \nmid 8388609$ so $47 \mid 8388607 = M_{23}$.

29. Since m is odd, $m^2 \equiv 1 \pmod 8$, so $n = p^a m^2 \equiv p^a \pmod 8$. By Exercise 26(a), $a \equiv 1 \pmod 4$, so $p^a \equiv p^{4k} p \equiv p \pmod 8$, since p^{4k} is an odd square. Thus $n = p \pmod 8$.

31. First suppose that $n = p^a$ where p is prime and a is a positive integer. Then $\sigma(n) = \frac{p^{a+1} - 1}{p - 1} < \frac{p^{a+1}}{p - 1} = \frac{np}{p - 1} = \frac{n}{1 - \frac{1}{p}} \leq \frac{n}{2/3} < \frac{3n}{2}$ so that $\sigma(n) \neq 2n$ and n is not perfect.

Next suppose that $n = p^a q^b$ where a and b are primes and a and b are positive integers. Then $\sigma(n) = \frac{p^{a+1} - 1}{p - 1} \cdot \frac{q^{b+1} - 1}{q - 1} < \frac{p^{a+1} q^{b+1}}{(p-1)(q-1)} = \frac{npq}{(p-1)(q-1)} = \frac{n}{(1 - \frac{1}{p})(1 - \frac{1}{q})} \leq \frac{n}{(\frac{2}{3})(\frac{4}{5})} = \frac{15n}{8} < 2n$. Hence $\sigma(n) \neq 2n$ and n is not perfect.

33. By Exercise 11 of Section 6.2, the product of all positive divisors of an integer n is $n^{\frac{\tau(n)}{2}}$. If the product of all divisors of n other than n is n^2 then $n^{\frac{\tau(n)}{2} - 1} = n^2$ so that $\frac{\tau(n)}{2} = 3$. This implies that $\tau(n) = 6$. The integers with $\tau(n) = 6$ are those of the form p^5 and $p^2 q$ where p and q are primes.

35. See Exercise 25 of Section 1.9

Section 7.1

1. DWWDF NDWGD ZQ.

3. IEXXK FZKXC UUKZC STKJW.

5. READM YLIPS

7. $k = 12$.

9. ANIDE AISLI KEACH ILDNO NEISB ETTER THANY OUROW NFROM CHINE SEFOR TUNEC OOKIE

11. $a \equiv 9, b \equiv 12 \pmod{26}$.

13. THISM ESSAG EWASE NCIPH EREDU SINGA NAFFI NETRA NSFOR MATIO N

15. $C \equiv 7P + 16 \pmod{26}$.

17. VSPFXH HIPLKB KIPMIE GTG.

19. First look for a set of letters repeated several times and note how far apart their occurances are. The greatest common divisor n of these distances will likely be the length (period) of the cipher. Then do a frequency count on the sets of letters which are n apart and do the usual frequency-count analysis on them.

Section 7.2

1. RL OQ NZ OF XM CQ KG QI VD AZ.

3. TO SL EE PP UR CH AN CE TO DR EA MX

5. $\begin{pmatrix} a & b \\ c & d \end{pmatrix} \equiv \begin{pmatrix} 3 & 24 \\ 24 & 25 \end{pmatrix} \pmod{26}$.

7. We have $C \equiv AP \pmod{26}$. Multiplying both sides on the left by A gives $AC \equiv a^2 P \equiv IP \equiv P \pmod{26}$. The congruence $A^2 \equiv I \pmod{26}$ follows since A is involutory. It follows that A is also a deciphering matrix.

9. The product cipher is given by $C \equiv AP \pmod{26}$ where $A = \begin{pmatrix} 11 & 6 \\ 2 & 13 \end{pmatrix}$.

11. If the plaintext is grouped into blocks of size m, we may take $\frac{[m,n]}{m}$ of these blocks to form a super-block of size $[m, n]$. If $\mathbf{A}$ is the $m \times m$ enciphering matrix, form the $[m, n] \times [m, n]$ matrix $\mathbf{B}$ with $\frac{[m,n]}{m}$ copies of $\mathbf{A}$ on the diagonal and zeros elsewhere: $\mathbf{B} =$

$$\begin{pmatrix} A & 0 & \cdots & 0 \\ 0 & A & \cdots & 0 \\ \vdots & & \ddots & \vdots \\ 0 & & \cdot & A \end{pmatrix}.$$ Then $\mathbf{B}$ will encipher $\frac{[m,n]}{m}$ blocks of size m at once. Similarly, if $\mathbf{C}$ is

the $n \times n$ enciphering matrix, form the corresponding $[m, n] \times [m, n]$ matrix $\mathbf{D}$. Then by Exercise 8, $\mathbf{BD}$ is an $[m, n] \times [m, n]$ enciphering matrix which does everything at once.

13. Multiplication of $(0 \cdots 010 \cdots 0)$ $\begin{pmatrix} P_1 \\ P_2 \\ \vdots \\ P_n \end{pmatrix}$ with the 1 in the ith place yields the 1×1 matrix

(P_i). So if the jth row of a matrix $\mathbf{A}$ is $(0 \cdots 010 \cdots 0)$ then $\mathbf{A} \begin{pmatrix} P_1 \\ \vdots \\ P_n \end{pmatrix} = \begin{pmatrix} C_1 \\ \vdots \\ C_n \end{pmatrix}$ gives

$C_j = P_i$. So if every row of $\mathbf{A}$ has its 1 in a different column, then each C_j is equal to a different P_i. Hence $\mathbf{A}$ is a "permutation" matrix.

15. $\mathbf{P} \equiv \begin{pmatrix} 17 & 4 \\ 1 & 7 \end{pmatrix} \mathbf{C} + \begin{pmatrix} 22 \\ 15 \end{pmatrix}$ (mod 26)

17. TOXIC WASTE

19. Make a frequency count of the trigraphs and use a published English language count of frequencies of trigraphs.

21. Let $\mathbf{A}$ be an $m \times m$ matrix, $\mathbf{B}$ be an $m \times 1$ matrix, $\mathbf{D}$ be an $n \times n$ matrix, and $\mathbf{E}$ be an $n \times 1$ matrix. Form $mn \times mn$ matrices $\mathbf{X}$ and $\mathbf{Y}$ by placing n copies of $\mathbf{A}$ along the diagonal of $\mathbf{X}$ and m copies of $\mathbf{D}$ along the diagonal of $\mathbf{Y}$. Form $mn \times 1$ matrices $\mathbf{Z}$ and $\mathbf{W}$ by stringing n copies of $\mathbf{B}$ together and m copies of $\mathbf{E}$, respectively. Then the product tranformation is given by $\mathbf{C} = \mathbf{YXP} + \mathbf{YZ} + \mathbf{W}$ which is an affine transformation based on a block size of mn.

Section 7.3

1. 14 17 17 27 11 17 65 76 07 76 14.
3. BEAM ME UP
5. 11 is the enciphering and deciphering exponent, since $11 = \overline{11}$ modulo 30 = $\phi(31)$.
7. $K \equiv 97 \, (\text{mod } 101)$.
9. Each individual chooses k_i relatively prime to $p - 1$. The common key $K = a^{k_1 k_2 \cdots k_n}$. One way for each person to know K is for each person to compute $y_i \equiv a^{k_i} \, (\text{mod } p)$, $0 < y_i < p$, and make it known to the others. Then each person can compute $K = y_1 y_2 \cdots y_n$.

Section 7.4

1. $p = 151$ and $q = 97$.
3. Since a block of ciphertext p is less than n, we must have $(p, n) = p$ or q.
5. 1215 1224 1471 0023 0116.
7. GR EE TI NG SX
9. **a.** 0371 0354 0858 0858 0087 1369 0354 0000 0087 1543 1797 0535.
   b. 0833 0475 0074 0323 0621 0105 0621 0865 0421 0000 0746 0803 0105 0621 0421.

11. **a.** If $n_i < n_j$, the block sizes are chosen small enough so that each block is unique modulo n_i. Since $n_i < n_j$, each block will be unique modulo n_j after applying the transformation D_{k_j}. Therefore we can apply E_{k_j} to $D_{k_i}(P)$ and retain uniqueness of blocks. If $n_i > n_j$ the arguement is similar.

    b. If $n_i < n_j$, individual j receives $E_{k_j}(D_{k_i}(P))$ and knows an inverse for e_j modulo $\phi(n_i)$. So he can apply $D_{k_j}(E_{k_j}(D_{k_i}(P))) = D_{k_i}(P)$. Since he also knows e_i, he can apply $E_{k_i}(D_{k_i}(P)) = P$ and discover the plaintext P. If $n_i > n_j$, individual j receives $D_{k_i}(E_{k_j}(P))$. Since he knows e_i he can apply $E_{k_i}(D_{k_i}(E_{k_j}(P))) = E_{k_j}(P)$. Since he also knows $\overline{e_j}$ he can apply $D_{k_j}(E_{k_j}(P)) = P$ and discover the plaintext P.

    c. Since only individual i knows $\overline{e_i}$, only he can apply the transfomation D_{k_i} and thereby make $E_{k_i}(D_{k_i}(P))$ intelligible.

    d. 1609 1802 0790 2508 1949 0267.

13. 242, 60, 0, 672, 60, 570, 350, 726, 312

Section 7.5

1. **a.** super-increasing. **b.** not. **c.** super-increasing. **d.** not.
3. If $\sum_{j=1}^{n-1} a_j < a_n$. Then $\sum_{j=1}^{n} a_j = \sum_{j=1}^{n-1} a_j + a_n < a_n + a_n = 2a_n < a_{n+1}$.
5. (17,51,85,8,16,49,64)
7. NUTS
9. If the multipliers and moduli are $(w_1, m_1), (w_2 < m_2), \ldots, (w_r, m_r)$, the inverses $\overline{w_1}, \overline{w_2}, \ldots, \overline{w_r}$ can be computed with respect to their corresponding moduli. Then we multiply and reduce succesively by $(\overline{w_r}, m_r), (\overline{w_{r-1}}, m_{r-1}), \ldots, (\overline{w_1}, m_1)$. The result will be the plaintext sequence of easy knapsack problems.
11. $95 \cdot 8 \cdot 21$
13. For $i = 1, 2, \ldots, n$, we have $b^{\alpha_i} \equiv a_i \pmod{m}$. Then $b^S \equiv P \equiv (b^{\alpha_1})^{x_1} (b^{\alpha_2})^{x_2} \cdots (b^{\alpha_n})^{x_n} \equiv b^{\alpha_1 x_1 + \cdots + \alpha_n x_n}$. Then $S \equiv \alpha_1 x_1 + \cdots + \alpha_n x_n \pmod{\phi(m)}$. Since $S + k\phi(m)$ is also a logarithm of P to the base b we may take the congruence to be an equation. Since the $x_i = 0$ or 1, this becomes an additive knapsack problem on the sequence $(\alpha_1, \alpha_2, \ldots, \alpha_n)$.

Section 7.6

1. **a.** $e_1 = 3696, e_2 = 264, e_3 = 5600, e_4 = 385.$ **b.** $C = 5389.$
3. The shadows are $k_1 \equiv 4 \pmod{8}, k_2 \equiv 5 \pmod{9}$, and $k_3 \equiv 2 \pmod{11}$.

Section 8.1

1. **a.** 4 **b.** 4 **c.** 6 **d.** 3
3. **a.** $\phi(6) = 2$, and $5^2 \equiv 1 \pmod 6$.
    b. $\phi(11) = 10, 2^2 \equiv 4, 2^5 \equiv -1, 2^{10} \equiv 1 \pmod{11}$.
5. Only 1, 5, 7, 11 are prime to 12. Each one squared is congruent to 1, but $\phi(12) = 4$.
7. There are 2: 3 and 5.
9. Suppose that $a^t \equiv 1 \pmod n$. Then $\overline{a}^t \equiv (\overline{a})^t (a^t)(a^t) \equiv (a\overline{a})^t a^t \equiv 1^t \cdot 1 \equiv 1 \pmod n$.
11. $\frac{[r,s]}{(r,s)} \le \text{ord}_n ab \le [r, s]$
13. Let $r = \text{ord}_m a^t$, then $a^{tr} \equiv 1 \pmod m$, hence $tr \ge ts$ and $r \ge s$. Since $1 \equiv a^{st} \equiv (a^t)^s \pmod n$, we have $s \ge r$.

15. Suppose that r is a primitive root modulo the odd prime p. Then $r^{\frac{(p-1)}{q}} \not\equiv 1 \pmod{p}$ for all prime divisors q of $p-1$ since no smaller power than the $(p-1)$st of r is congruent to 1 modulo p. Conversely, suppose that $r^{\frac{(p-1)}{q}} \not\equiv 1 \pmod{p}$ for all prime divisors of $p-1$. Suppose that r is not a primitive root of p. Then there is an integer t such that $r^t \equiv 1 \pmod{p}$ with $t < p-1$. Since t must divide $p-1$, we have $p-1 = st$ for some positive integer s greater than 1. Then $\frac{(p-1)}{s} = t$. Let q be a prime divisor of s. Then $\frac{(p-1)}{q} = t(\frac{s}{q})$, so that $r^{\frac{(p-1)}{q}} = r^{t(\frac{s}{q})} = (r^t)^{\frac{s}{q}} \equiv 1 \pmod{p}$. This contradicts the original assumption, so r is a primitive root modulo p.

17. Since $2^{2^2} + 1 \equiv 0 \pmod{F_n}$, then $2^{2^2} \equiv -1 \pmod{F_n}$. Squaring gives $(2^{2^2})^2 \equiv 1 \pmod{F_n}$. Thus, $\mathrm{ord}_{F_n} 2 \le 2^n 2 = 2^{n+1}$.

19. Note that $a^t < m = a^n - 1$ whenever $1 \le t < n$. Hence a^t cannot be congruent to 1 modulo m when t is a positive integer less than n. However, $a^n \equiv 1 \pmod{m}$ since $m = (a^n - 1) \mid (a^n - 1)$. It follows that $\mathrm{ord}_m a = n$. Since $\mathrm{ord}_m a \mid m$, we see that $n \mid \phi(m)$.

21. If $2^{pq} \equiv 2 \pmod{pq}$, then $2^{pq} \equiv 2 \pmod{p}$. Since $2^p \equiv 1 \pmod{M_p}$, we have $2^{M_p M_q} \equiv 2^{(M_p - 1)M_q + M_q} \equiv 2^{M_q} \equiv 2 \pmod{M_p}$. Similarly, $2^{M_p M_q} \equiv 2 \pmod{M_q}$. Now if r is prime and $M_p \equiv M_q \equiv 0 \pmod{r}$, then $2^p \equiv 2^q \equiv 1 \pmod{r}$, but then $2^{(p,q)} \equiv 2^1 \equiv 1 \pmod{r}$, a contradiction. Therefore $(M_p, M_q) = 1$ and we have $2^{M_p M_q} \equiv 2 \pmod{M_p M_q}$. Conversely, $2^{M_p M_q} \equiv 2 \pmod{M_p M_q}$ implies $2^{M_p M_q} \equiv 2^{(M_p - 1)M_q + (M_q - 1)+1} \equiv 2^{(M_q - 1)+1} \equiv 2 \pmod{M_p}$, so $M_q \equiv 1 \pmod{p}$ since 2 has order p modulo M_p. Hence, $2^{pq} \equiv 2^q \equiv 2 \pmod{p}$ and similarly, $2^{pq} \equiv 2 \pmod{q}$. Therefore $2^{pq} \equiv 2 \pmod{pq}$.

23. Let $j = \mathrm{ord}_{\phi(n)} e$. Then $e^j \equiv 1 \pmod{\phi(n)}$. Since $\mathrm{ord}_n P \mid \phi(n)$, we have $e^j \equiv 1 \pmod{\mathrm{ord}_n P}$. Then by Theorem 8.2, $P^{e^j} \equiv P \pmod{n}$, so $C^{e^{j-1}} \equiv (P^e)^{e^{j-1}} \equiv P^{e^j} \equiv P \pmod{n}$ and $C^{e^j} \equiv P^e \equiv C \pmod{n}$.

Section 8.2

1. a. 2   b. 2   c. 0   d. 3
3. a. 2   b. 4   c. 8   d. 6   e. 12   f. 22
5. 2, 6, 7, 11
7. 2, 3, 10, 13, 14, 15
9. By Lagrange's Theorem there are at most two solutions to $x^2 \equiv 1 \pmod{p}$, and we know $x \equiv \pm 1$ are the two solutions. Since $p \equiv 1 \pmod{4}, 4 \mid (p-1) = \phi(p)$ so there is an element x of order 4 modulo p. Then $x^4 = (x^2)^2 \equiv 1 \pmod{p}$, so $x^2 \equiv \pm 1 \pmod{p}$. If $x^2 \equiv 1 \pmod{p}$ then x does not have order 4. Therefore $x^2 \equiv -1 \pmod{p}$.
11. a. This statement is the contrapositive of Lagrange's Theorem.
    b. For any integer a not divisible by p, $f(a) = (a-1)\cdots(a-a)-(a-p+1)- a^{p-1}+1 \equiv 0 - a^{p-1}+1 \equiv -1+1 \equiv 0 \pmod{p}$ using Euler's Theorem and $f(p) \equiv (-1)(-2)\cdots(-p+1)+1 \equiv 0 \pmod{p}$ by Wilson's Theorem. Hence $f(x)$ is a $(p-1)$th degree polynomial with p incongruent roots modulo p, namely $1,2,3,\ldots,p-1,p$.
    c. From part (b) $f(0) \equiv 0 \pmod{p}$, but $f(0) - (-1)(-2)\cdots(-(p-1))+1 \equiv (-1)^{p-1}(p-1)!+1 \equiv (p-1)!+1 \equiv 0 \pmod{p}$ is just Wilson's Theorem.
13. a. Since $q_i^{t_i} \mid \phi(p) = p-1$, by Theorem 8.8 there exists $\phi(q_i^{t_i})$ elements of order $q_i^{t_i}$ for each $i = 1,2,\ldots,r$. Let a_i be a fixed element of this order.

b. Using induction and Exercise 6 of Section 8.1, we have $\text{ord}_p(a) =$
$\text{ord}_p(a_1 a_2 \cdots a_r) = \text{ord}_p(a_1 \cdots a_{r-1})\text{ord}_p(a_r) = \cdots = \text{ord}_p(a_1) \cdots \text{ord}_p(a_r)$ since
$\{\text{ord}_p(a_1), \text{ord}_p(a_2), \ldots, \text{ord}_p(a_r)\} = \{q_1^{t_1}, \ldots, q_r^{t_r}\}$ are pairwise relatively prime.

c. $\phi(29) = 28 = 2^2 7$, and $12^4 \equiv 1 \pmod{29}$, so $\text{ord}_{29}(12) = 4$. Also, $16^7 \equiv 1$
$\pmod{29}$ so $\text{ord}_{29}(16) = 7$. Then by part (b), $\text{ord}_{29}(12 \cdot 16) = 4 \cdot 7 = 28$. Therefore $12 \cdot$
$16 = 192 \equiv 18 \pmod{29}$ is a primitive root modulo 29.

15. If n is odd, composite and not a power of 3, then the product in Exercise 8 is $\prod_{j=1}^{r}(n - 1, p_j -$
$1) \geq (n-1, 3-1)(n-1, 5-1) \geq 2 \cdot 2 = 4$. So there must be two bases other than -1 and $+1$.

17. a. Suppose that $f(x)$ is a polynomial with integer coefficients of degree $n - 1$. Suppose that
$x_1, x_2, \cdots, x_n$ are incongruent modulo p where p is prime. Consider the polynomial $g(x) =$
$f(x) - \sum_{j=1}^{n}\left(f(x_j)\prod_{i \neq j}(x - x_i)\overline{(x_j - x_i)}\right)$. Note that for each $j = 1, 2, \cdots, n$, x_j,
is a root of this polynomial modulo p since its value at x_j is $f(x_j) - [0 + 0 + \cdots +$
$f(x_j)\prod_{i \neq j}(x_j - x_i)(x_j - x_i) + \cdots + 0] \equiv f(x_j) - f(x_j) \cdot 1 \equiv 0 \pmod{p}$. Since
$g(x)$ has n incongruent roots modulo p and since it is of degree $n - 1$ or less, we can use
Lagrange's theorem (Theorem 8.6) to see that $g(x) \equiv 0 \pmod{p}$ for every integer x.
b. 6

19. By Exercise 18 of Section 8.1, $j \mid \text{ord}_{\phi(n)}e$. $\phi(pq) = 4p'q'$, so $j \mid \phi(4p'q') = 2(p' - 1)$
$(q' - 1)$. Choose $e = p' - 1, d = 2(q' - 1)$.

Section 8.3

1. 4, 10, 22
3. a. 2 **b.** 2 **c.** 2 **d.** 3
5. a. 2 **b.** 2 **c.** 2 **d.** 3
7. a. 3 **b.** 3 **c.** 3 **d.** 3
9. 7, 13, 17, 19.
11. 3, 13, 15, 21, 29, 33
13. Suppose that r is a primitive root of m and that $x^2 \equiv 1 \pmod{m}$. Let $x \equiv r^t \pmod{m}$
where $0 \leq t \leq p - 1$. Then $r^{2t} \equiv 1 \pmod{m}$. Since r is a primitive root, it follows that
$\phi(m) \mid 2t$ so that $2t = k\phi(m)$ and $t = k\phi(m)/2$ for some integer k. We have $x \equiv r^t =$
$r^{k\phi(m)/2} = r^{(\phi(m)/2)k} \equiv (-1)^k \equiv \pm 1 \pmod{p}$, since $r^{\phi(m)/2} \equiv -1 \pmod{p}$.

15. By Theorem 8.12 we know that $\text{ord}_{2^k} 5 = \phi(2^k)/2$. Hence the 2^{k-2} integers 5^j, $j =$
$0, 1, \cdots, 2^{k-2} - 1$, are incongruent modulo $2k$. Similarly the 2^{k-2} integers -5^j, $j =$
$0, 1, \cdots, 2^{k-2} - 1$, are incongruent modulo 2^k. Note that 5^j cannot be congruent to -5^i
modulo 2^k where i and j are integers, since $5^j \equiv 1 \pmod 4$ but $-5^i \equiv 3 \pmod 4$. It
follows that the integers $1, 5, \cdots, 5^{2^{k-2}-1}, -1, -5, \cdots, -5^{2^{k-2}-1}$ are 2^{k-1} incongruent
integers modulo 2^k. Since $\phi(2^k) = 2^{k-1}$ and every integer of the form $(-1)^\alpha 5^\beta$ is
relatively prime to 2^k, it follows that every odd integer is congruent to an integer of this form
with $\alpha = 0$ or 1 and $0 \leq \beta = 2^{k-2} - 1$.

Section 8.4

1. $\text{ind}_5 1 = 22, \text{ind}_5 2 = 2, \text{ind}_5 3 = 16, \text{ind}_5 4 = 4, \text{ind}_5 5 = 1, \text{ind}_5 6 = 18, \text{ind}_5 7 = 19,$
$\text{ind}_5 8 = 6, \text{ind}_5 9 = 10, \text{ind}_5 10 = 3, \text{ind}_5 11 = 9, \text{ind}_5 12 = 20, \text{ind}_5 13 = 14, \text{ind}_5 14 =$
$21, \text{ind}_5 15 = 17, \text{ind}_5 16 = 8, \text{ind}_5 17 = 7, \text{ind}_5 18 = 12, \text{ind}_5 19 = 15, \text{ind}_5 20 = 5,$
$\text{ind}_5 21 = 13, \text{and ind}_5 22 = 11.$
3. a. 7 or 18 $\pmod{22}$ **b.** no solution

5. $8, 9, 20, 21 \pmod{29}$

7. $1, 12, 45, 47, 78, 91, 93, 100, 137, 139, 144, 183, 185, 188, 210, 229, 231, 232, 252, 254, 275,$
   $277, 321, 323, 367, 369, 386, 413, 415, 430, 459, 461, 496 \pmod{506}$

9. Taking indices base r gives $4 \operatorname{ind}_r x \equiv \operatorname{ind}_r(-1) \equiv \frac{p-1}{2} \pmod{\phi(p)}$. Thus $4 \mid \frac{p-1}{2}$.

11. $(1,2), (0,2)$

13. $x \equiv 29 \pmod{32}, x \equiv 4 \pmod 8$

15. The index system of 17 modulo 120 is $(0, 0, 1, 1)$. The index system of 41 modulo 120 is
    $(0,0,1,4)$.

17. $x \equiv 17 \pmod{60}$

19. If r is a primitive root of 2^e, we take indices base r of $x^k \equiv a \pmod{2^e}$. Then $k \operatorname{ind}_r x \equiv$
    $\operatorname{ind}_r a \pmod{2^{e-1}}$. Solve for $\operatorname{ind}_r x$ and then get x.

21. First we show that $\operatorname{ord}_{2^e} 5 = 2^{e-2}$. Indeed, $\phi(2^e) = 2^{e-1}$, so it suffices to show that the
    highest power of 2 dividing $5^{2^{e-2}} - 1$ is 2^e. We proceed by induction. The basis step is the
    case $e = 2$, which is true. Note that $5^{2^{e-2}} - 1 = (5^{2^{e-3}} - 1)(5^{2^{e-3}} + 1)$. The first factor is
    exactly divisible by 2^{e-1} by the induction hypothesis. The second factor differs from the first
    by 2, so it is exactly divisible by 2, therefore $5^{2^{e-2}} - 1$ is exactly divisible by 2^e, as desired.
    Hence, if k is odd, the numbers $\pm 5^k, \pm 5^{2k}, \ldots, \pm 5^{2^{e-2}k}$ are 2^{e-1} incongruent kth power
    residues, which is the number given by the formula. If 2^m exactly divides k, then $5^k \equiv -5^k$
    $\pmod{2^e}$, so the formula must be divided by 2, hence the factor $(k, 2)$ in the denominator.
    Further, 5^{2^m} has order $2^{e-2}/2^m$ if $m \le e - 2$ and order 1 if $m > e - 2$, so the list must
    repeat modulo 2^e every $\operatorname{ord}_{2^e} 5^{2^m}$ terms, whence the other factor in the denominator.

23. a. From the first inequality in the proof, if n is not square-free, the probability is strictly less
    than $2n/9$, which is subsantially smaller than $(n - 1)/4$ for large n. If n is square-free, the
    arguement following (8.3) shows that if n has 4 or more factors, then the probability is less
    than $n/8$. The next inequality shows that the worst case for $n = p_1 p_2$ is when $s_1 = s_2$ and
    s_1 is as small as possible, which is the case stated in the exercise.
    b. $0.24999249\ldots$

Section 8.5

1. We have $2^2 \equiv 4 \pmod{101}, 2^5 \equiv 32 \pmod{101}, 2^{10} \equiv (2^5)^2 \equiv 32^2 \equiv 14 \pmod{101},$
   $2^{20} \equiv (2^{10})^2 \equiv 14^2 \equiv 95 \pmod{101}, 2^{25} \equiv (2^5)^5 \equiv 32^5 \equiv (32^2)^2 32 \equiv 1024^2 \cdot 32 \equiv$
   $14^2 \cdot 32 \equiv 196 \cdot 32 \equiv -6 \cdot 32 \equiv -192 \equiv 10 \pmod{101}, 2^{50} \equiv (2^{25})^2 \equiv 10^2 = 100 \equiv -1$
   $\pmod{101}, 2^{100} \equiv (2^{50})^2 \equiv (-1)^2 \equiv 1 \pmod{101}$. Since $2^{(101-1)/q} \not\equiv 1 \pmod{101}$
   for every proper divisor q of 100, and $2^{(101-1)/q} \equiv 1 \pmod{101}$ it follows that 101 is prime.

3. $233 - 1 = 2^3 29, 3^{116} \equiv -1 \pmod{233}, 3^8 \equiv 27 \not\equiv 1 \pmod{233}$

5. The first condition implies $x^{F_n - 1} \equiv 1 \pmod{F_n}$. The only prime dividing $F_n - 1 = 2^{2^n}$
   is 2, and $\frac{F_n - 1}{2} = 2^{2^n - 1}$, so the second condition implies $2^{\frac{F_n - 1}{2}} \not\equiv 1 \pmod{F_n}$. Then by
   Theorem 8.18, F_n is prime.

7. See Lenstra [117].

Section 8.6

1. a. 20   b. 12   c. 36   d. 48   e. 180   f. 388080   g. 8640   h. 125411328000

3. 65520

5. Follows directly from Theorem 8.21

7. If $p \mid m$, but $p \nmid n$, then $pn > n$, but $\lambda(pn) = \lambda(n)$, a contradiction.

9. Suppose that $ax \equiv b \pmod{m}$. Multiplying both sides of this congruence by $a^{\lambda(m)-1}$ gives $a^{\lambda(m)}x \equiv a^{\lambda(m)-1}b \pmod{m}$. Since $a^{\lambda(m)} \equiv 1 \pmod{m}$, it follows that $x \equiv a^{\lambda(m)-1}b \pmod{m}$.

11. **a.** First suppose that $m = p^a$. Then we have $x(x^{c-1} - 1) \equiv 0 \pmod{p^a}$. Let s be a primitive root for p^a, then the solutions to $x^{c-1} \equiv 1$ are exactly the powers s^k with $(c - 1)k \equiv 1 \pmod{\phi(p^a)}$, and there are $(c - 1, \phi(p^a))$ of these. Also, 0 is a solution, so we have $1 + (c - 1, \phi(p^a))$ solutions all together. Now if $m = p_1^{a_1} \cdots p_r^{a_r}$, we can count the number of solutions modulo $p_i^{a_i}$ for each i. There is a one-to-one correspondence between solutions modulo m and the set of r-tuples of solutions to the system of congruences modulo each of the prime powers. The correspondence is given by the Chinese Remainder Theorem.

 b. Suppose $(c - 1, \phi(m)) = 2$, then $c - 1$ is even. Since $\phi(p^a)$ is even for all prime powers, except 2, we have $(c - 1, \phi(p_i^{a_i})) = 2$ for each i. Then by part (a), we have the number of solutions $= 3^r$. If 2^1 is a prime factor, then $\phi(m) = \phi(m/2)$, and since x^c and x have the same parity, x is a solution modulo m if and only if it is a solution modulo $m/2$, so the proposition still holds.

13. Use the solution to Exercise 15 with $p = 3$. The last inequality leaves only finitely many cases.

15. By Theorem 8.22 pqr is a Carmichael number if and only if $p - 1 \mid qr - 1$, $q - 1 \mid pr - 1$, and $r - 1 \mid pq - 1$. Hence, $q - 1 = \frac{pr-1}{n}$, and $r - 1 = \frac{pq-1}{m}$ for some integers $n, m > 1$. Solving for q and r, we find $q = \frac{p(m-1)+m(n-1)}{mn-p^2}$, and $r = \frac{p(n-1)+n(m-1)}{mn-p^2}$. Hence, $mn > p^2$, so $n \geq \frac{p^2+1}{n}$. So $q = \frac{pr-1+n}{n}$, and $r \leq \frac{(pq-1)n}{p^2+1} + 1$. Then $q \leq \frac{(pq-1)p}{p^2+1} + \frac{p}{n} + \frac{n-1}{n}$, hence $q \leq \frac{n+p-1}{n}(p^2 + 1) - p \leq \frac{p^3+p^2-p+1}{2}$. By symmetry, the same holds for r, therefore, there are only finitely many choices for q and r.

17. $q_n(ab) \equiv ((ab)^{\lambda(n)} - 1)/n = (a^{\lambda(n)}b^{\lambda(n)} - a^{\lambda(n)} - b^{\lambda(n)} + 1 + a^{\lambda(n)} + b^{\lambda(n)} - 2)/n = (a^{\lambda(n)} - 1)(b^{\lambda(n)} - 1)/n + ((a^{\lambda(n)} - 1) + (b^{\lambda(n)} - 1))/n \equiv q_n(a) + q_n(b) \pmod{n}$. At the last step, we use the fact that n^2 must divide $(a^{\lambda(n)} - 1)(b^{\lambda(n)} - 1)$, since $\lambda(n)$ is the universal exponent.

Section 8.7

1. 69, 76, 77, 92, 46, 11, 12, 19, 36, 29, 84, 05, 02, 00, ...
3. 10
5. **a.** $a \equiv 1 \pmod{20}$ **b.** $a \equiv 1 \pmod{30030}$ **c.** $a \equiv 1 \pmod{111111}$ **d.** $a \equiv 1 \pmod{2^{25} - 1}$
7. **a.** 31 **b.** 715827882 **c.** 31 **d.** 195225786 **e.** 1073741823 **f.** 1073741823
9. 1, 24, 25, 18, 12, 30, 11, 10, 21.
11. 3 has maximal order modulo $2^{25} - 1$. To make a large enough multiplier, raise 3 to a power relatively prime to $\phi(2^{25} - 1) = 32400000$, for example, to the 11th power.
13. 665
15. **a.** 8, 4, 2, 1, 8, 4, 2, 1, ...
 b. 9, 12, 16, 6, 8, 3, 4, 13, 2, 18, 1, 9, 12, ...

Section 8.8

1. **a.** 8 **b.** 5 **c.** 2 **d.** 6 **e.** 30 **f.** 20
3. **a.** $s = 3$ **b.** $s = 21$ **c.** $s = 2$

Section 9.1

1. **a.** 1   **b.** 1, 4   **c.** 1,3,4,9,10,12   **d.** 1,4,5,6,7,9,11,16,17
3. $1, -1, -1, 1$, respectively.
5. **a.** $\left(\frac{7}{11}\right) \equiv 7^{\frac{11-1}{2}} \equiv -1 \, (\text{mod } 11)$.
   b. $(7, 14, 21, 28, 35) \equiv (7, 3, 10, 6, 2) \, (\text{mod } 11)$ and three of these are greater than $\frac{11}{2}$, so
   $\left(\frac{7}{11}\right) \equiv \equiv (-1)^3 = -1$.
7. $\left(\frac{-2}{p}\right) = \left(\frac{-1}{p}\right) \left(\frac{2}{p}\right)$ by Theorem 9.3. Using Theorems 9.4 and 9.5 we have: If $p \equiv 1$
   $(\text{mod } 8)$ then $\left(\frac{-2}{p}\right) = 1 \cdot 1 = 1$. If $p \equiv 3 \, (\text{mod } 8)$ then $\left(\frac{-2}{p}\right) = (-1)(-1) = 1$. If $p \equiv -1$
   $(\text{mod } 8)$ then $\left(\frac{-2}{p}\right) = -1 \cdot 1 = -1$. If $p \equiv -3 \, (\text{mod } 8)$ then $\left(\frac{-2}{p}\right) = 1 \cdot (-1) = -1$.
9. Since $p - 1 \equiv -1, p - 2 \equiv -2, \ldots, \frac{p+1}{2} \equiv \frac{p-1}{2} \, (\text{mod } p)$, we have $\left(\frac{p-1}{2}\right)!^2 \equiv$
   $-(p-1)! \equiv 1 \, (\text{mod } p)$, (since $p \equiv 3 \, (\text{mod } 4)$ the minus signs cancel.) By Euler's
   Criterion, $\left(\frac{p-1}{2}\right)!^{\frac{p-1}{2}} \equiv \left(\frac{1}{p}\right) \left(\frac{2}{p}\right) \cdots \left(\frac{\frac{p-1}{2}}{p}\right) \equiv (-1)^t \, (\text{mod } p)$, by definition of the
   Legendre symbol. Since $\frac{p-1}{2}! \equiv \pm 1 \, (\text{mod } p)$, and $\frac{p-1}{2}$ is odd, we have the result.
11. If $p \equiv 1 \, (\text{mod } 4)$, $\left(\frac{-a}{p}\right) = \left(\frac{-1}{p}\right) \left(\frac{a}{p}\right) = 1 \cdot 1 = 1$. If $p \equiv 3 \, (\text{mod } 4)$, $\left(\frac{-a}{p}\right) =$
   $\left(\frac{-1}{p}\right) \left(\frac{a}{p}\right) = (-1) \cdot 1 = -1$.
13. **a.** $2, 4 \, (\text{mod } 7)$ **b.** $1 \, (\text{mod } 7)$ **c.** no solution
15. If neither 2 nor 3 is a quadratic residue of p then $2 \cdot 3 = 6$ is a quadratic residue of p. So the
    pair is one of $(2,4), (4,6), (1,3)$.
17. **a.** Since $p = 4n + 3, 2n + 2 = \frac{p+1}{2}$. Then $x^2 \equiv (\pm a^{n+1})^2 \equiv a^{2n+2} \equiv a^{p+1/2} \equiv$
    $a^{(p-1)/2}a \equiv 1 \cdot a \equiv a \, (\text{mod } p)$ using the fact that $a^{(p-1)/2} \equiv 1 \, (\text{mod } p)$ since a is a
    quadratic residue of p.
    b. By Lemma 9.1, there are exactly two solutions to $y^2 \equiv 1 \, (\text{mod } p)$, namely $y \equiv \pm 1$
    $(\text{mod } p)$. Since $p \equiv 5 \, (\text{mod } 8)$, -1 is a quadratic residue of p and 2 is a quadratic
    nonresidue of p. Since $p = 8n + 5$, we have $4n + 2 = \frac{(p-1)}{2}$ and $2n + 2 = \frac{(p+3)}{4}$. Then
    $(\pm a^{n+1})^2 \equiv a^{(p+3)/4} \, (\text{mod } p)$ and $(\pm 2^{2n+1}a^{n+1})^2 \equiv 2^{(p-1)/2}a^{(p+3)/4} \equiv -a^{(p+3)/4}$
    $(\text{mod } p)$ by Euler's Criterion. We must show that one of $a^{(p+3)/4}$ or $-a^{(p+3)/4} \equiv a$
    $(\text{mod } p)$. Now a is a quadratic residue of p, so $a^{(p-1)/2} \equiv 1 \, (\text{mod } p)$ and therefore
    $a^{(p-1)/4}$ solves $x^2 \equiv 1 \, (\text{mod } p)$. But then $a^{(p-1)/4} \equiv \pm 1 \, (\text{mod } p)$, that is $a^{(p+3)/4} \equiv$
    $\pm a \, (\text{mod } p)$ or $\pm a^{(p+3)/4} \equiv a \, (\text{mod } p)$ as desired.
19. $1, 4, 11, 14 \, (\text{mod } 15)$.
21. $47, 96, 135, 278, 723, 866, 905, 954$
23. If $x_0^2 \equiv a \, (\text{mod } p^{e+1})$ then $x_0^2 \equiv a \, (\text{mod } p^e)$. Conversely, if $x_0^2 \equiv a \, (\text{mod } p^e)$ then
    $x_0^2 = a + bp^e$ for some integer b. We can solve the linear congruence $2x_0 y \equiv -b \, (\text{mod } p)$,
    say $y = y_0$. Let $x_1 = x_0 + y_0 p^e$. Then $x_1^2 = x_0^2 + 2x_0 y_0 p^e \equiv a \, (\text{mod } p^{e+1})$ since $p \mid$
    $2x_0 y_0 + b$. This is the induction step in showing that $x^2 \equiv a \, (\text{mod } p^e)$ has solutions if and
    only if $\left(\frac{a}{p}\right) = 1$.
25. **a.** 4   **b.** 8   **c.** 8   **d.** 16
27. Suppose $p_1, p_2, \ldots, p_n$ are the only primes of the form $4k + 1$. Let $N =$
    $4(p_1, p_2, \ldots p_n)^2 + 1$. Let q be an odd prime factor of N. Then $q \neq p_i, i = 1, 2, \ldots, n$.
    $N \equiv 0 \, (\text{mod } q)$, so $4(p_1, p_2, \ldots, p_n)^2 \equiv -1 \, (\text{mod } q)$ and therefore $\left(\frac{-1}{q}\right) = 1$, so $q \equiv 1$
    $(\text{mod } 4)$ by Theorem 9.4.

29. If $x_0^2 \equiv a \pmod{p^n}$ then $x_0^2 \equiv a \pmod{p}$, and so $\left(\frac{a}{p}\right) = 1$. Conversely, if $\left(\frac{a}{p}\right) = 1$, then $x^2 \equiv a \pmod{p}$ has a solution. Suppose $x^2 \equiv a \pmod{p^n}$ has a solution x_0. Then $x_0^2 = a + bp^k$ for some integer b. Let $y = y_0$ be a solution to $2x_0 y \equiv -b \pmod{p}$. Check that $x_1 = x_0 + y_0 p^n$ is a solution to $x^2 \equiv a \pmod{p^{n+1}}$. By induction, we are done.

31. Let r be a primitive root of the odd prime p. Suppose that r were a quadratic residue of p. Then there would be an integer x such that $x^2 \equiv r \pmod{p}$. It then follows that $r^{(p-1)/2} \equiv (x^2)^{(p-1)/2} = x^{p-1} \equiv 1 \pmod{p}$. This contradicts the fact that $\text{ord}_p r = p - 1$. Hence r is a quadratic nonresidue of p.

33. If r is a primitive root of q, then the set of all primitive roots is given by $\{r^k : (k, \phi(q)) = (k, 2p) = 1\}$. So the $p - 1$ numbers $\{r^k : k$ is odd and $k \neq p, 1 \leq k < 2p\}$ are all the primitive roots of q. On the other hand, q has $(q - 1)/2 = p$ quadratic residues, which are given by $\{r^2, r^4, \ldots, r^{2p}\}$. This set has no intersection with the set of primitive roots listed.

35. If $p = 2^{2^n} + 1$, then $\phi(p) = 2^{2^n}$. Then a is a nonresidue if and only if $a = r^k$ with k odd. But then $(k, \phi(p)) = 1$, so a is also a primitive root. Conversely, whenever k is odd, r^k is a nonresidue and a primitive root. Then $(k, \phi(p)) = 1$ for $k = 1, 3, 5, \ldots, p - 2$. Hence $\phi(p)$ must be a power of 2, and so $p = 2^k + 1$. For $2^k + 1$ to be a prime, $k = 2^n$ for some n.

37. **a.** $q = 2^p + 1 = 2(4k + 3) = 8k + 7$, so $\left(\frac{2}{q}\right) = 1$ by Theorem 9.5. Then by Euler's Criterion, $2^{(q-1)/2} \equiv 2^p \equiv 1 \pmod{q}$. Therefore $q \mid 2^p - 1$.
   b. $11 = 4 \cdot 2 + 3$ and $23 = 2 \cdot 11 + 1$, so $23 \mid 2^{11} - 1 = M_{11}$, by part (a). $23 = 4 \cdot 5 + 3$ and $47 = 2 \cdot 23 + 1$, so $47 \mid M_{23}$. $251 = 4 \cdot 62 + 3$ and $503 = 2 \cdot 251 + 1$, so that $503 \mid M_{251}$.

39. Let $q = 2k + 1$. Since q does not divide $2^p + 1$, we must have, by Exercise 38, that $k \equiv 0$ or 3 $\pmod 4$. That is, $k \equiv 0, 3, 4$ or 7 $\pmod 8$. Then $q \equiv 2(0, 3, 4$ or $7) + 1 \equiv \pm 1 \pmod 8$.

41. We multiply $\left(\frac{j(j+1)}{p}\right)$ by $\left(\frac{\overline{j}}{p}\right) = 1$. We find that $\left(\frac{j(j+1)}{p}\right) = \left(\frac{\overline{j}}{p}\right)^2 \left(\frac{j(j+1)}{p}\right) = \left(\frac{\overline{j}j}{p}\right)\left(\frac{\overline{j}(j+1)}{p}\right) = \left(\frac{1}{p}\right) = 1 \cdot \left(\frac{1+\overline{j}}{p}\right) = \left(\frac{\overline{j}+1}{p}\right)$.

43. Let r be a primitive root of p. Then $x^2 \equiv a \pmod{p}$ has a solution if and only if $2\,\text{ind}_r x \equiv \text{ind}_r a \pmod{p-1}$ has a solution for $\text{ind}_r x$. Since $p - 1$ is even, the last congruence is solvable if and only if $\text{ind}_r a$ is even, which happens when $a = r^2, r^4, \ldots, r^{p-1}$, i.e. $\frac{p-1}{2}$ times.

45. $q = 2(4k + 1) + 1 = 8k + 3$, so 2 is a quadratic nonresidue of q. By Exercise 29, 2 is a primitive root.

47. Check that $q \equiv 3 \pmod 4$, so -1 is a quadratic nonresidue of q. Since $4 = 2^2$, we have $\left(\frac{-4}{q}\right) = \left(\frac{-1}{q}\right)\left(\frac{2^2}{q}\right) = (-1)(1) = -1$. Therefore -4 is a nonresidue of q. By Exercise 29, -4 is a primitive root.

49. **a.** By adding $(\overline{2}b)^2$ to both sides, we complete the square.
   b. There are 4 solutions to $x^2 \equiv C + a \pmod{pq}$. From each, subtract $\overline{2}b$.
   c. DETOUR

51. **a.** By noting this, the second player can tell which cards dealt are quadratic residues, since the ciphertext will also be quadratic residues modulo p.
   b. All ciphers will be quadratic residues modulo p.

Section 9.2

1. **a.** -1 **b.** -1 **c.** -1 **d.** -1 **e.** 1 **f.** 1

3. If $p \equiv 1 \pmod 6$ there are 2 cases: If $p \equiv 1 \pmod 4$ then $\left(\frac{-1}{p}\right) = 1$ and $\left(\frac{3}{p}\right) = \left(\frac{p}{3}\right) =$
$\left(\frac{1}{3}\right) = 1$. So $\left(\frac{-3}{p}\right) = 1$. If $p \equiv 3 \pmod 4$ then $\left(\frac{-1}{p}\right) = -1$ and $\left(\frac{3}{p}\right) = -\left(\frac{p}{3}\right)$, so
$\left(\frac{-3}{p}\right) = (-1)(-1) = 1$. If $p \equiv -1 \pmod 6$ and $p \equiv 1 \pmod 4$, then $\left(\frac{-3}{p}\right) =$
$\left(\frac{-1}{p}\right)\left(\frac{3}{p}\right) = 1 \cdot \left(\frac{p}{3}\right) = \left(\frac{-1}{3}\right) = -1$. If $p \equiv 3 \pmod 4$, then $\left(\frac{-3}{p}\right) = \left(\frac{-1}{p}\right)\left(\frac{3}{p}\right) =$
$(-1)\left(-\left(\frac{p}{3}\right)\right) = \left(\frac{p}{3}\right) = \left(\frac{-1}{3}\right) = -1$.

5. $p \equiv 1, 3, 9, 19, 25$ or $27 \pmod{28}$.

7. a. $3^{(F_1-1)/2} \equiv -1 \pmod 5$.

 b. $3^{(F_3-1)/2} \equiv -1 \pmod{257}$.

 c. $3^{255} \equiv 94 \pmod{F_4}$. $3^{32768} \equiv 3^{255 \cdot 128} 3^{128} \equiv 94^{128} 3^{128} \equiv -1 \pmod{F_4}$.

9. a. The lattice points in the rectangle are the points (i, j) where $0 < i < \frac{p}{2}$ and $0 < j < \frac{q}{2}$.
These are the lattice points (i, j) with $i = 1, 2, \ldots, (p-1)/2$ and $j = 1, 2, \ldots, (q-1)/2$.
Consequently, there are $\frac{p-1}{2} \cdot \frac{q-1}{2}$ such lattice points.

 b. The points on the diagonal connecting O and C are the points (x, y) where $y = \left(\frac{q}{p}\right)x$.
Suppose that x and y are integers with $y = \left(\frac{q}{p}\right)x$. Then $py = qx$. Since $(p, q) = 1$ it follows
that $p \mid x$ which is impossible if $0 < x < \frac{p}{2}$. Hence there are no lattice points on this diagonal.

 c. The number of lattice points in the triangle with vertices $O, A,$ and C is the number of
lattice points (i, j) with $i = 1, 2, \ldots, \frac{(p-1)}{2}$ and $1 \le j \le \frac{iq}{p}$. For a fixed value of i in the
indicated range, there are $[\frac{iq}{p}]$ lattice points (i, j) in the triangle. Hence the total number of
lattice points in the triangle is $\sum_{i=1}^{\frac{(p-1)}{2}} [\frac{iq}{p}]$.

 d. The number of lattice points in the triangle with vertices $O, B,$ and C is the number of
lattice points (i, j) with $j = 1, 2, \ldots, \frac{(q-1)}{2}$ and $1 \le i < \frac{jp}{q}$. For a fixed value of j in the
indicated range, there are $[\frac{jp}{q}]$ lattice points (i, j) in the triangle. Hence the total number of
lattice points in the triangle is $\sum_{j=1}^{(q-1)/2} [\frac{jp}{q}]$.

 e. Since there are $\frac{p-1}{2} \frac{q-1}{2}$ lattice points in the rectangle, and no points on the diagonal OC,
the sum of the numbers of lattice points in the triangles OBC and OAC is $\frac{p-1}{2} \frac{q-1}{2}$. By
parts (b) and (c), it follows that $\sum_{j=1}^{\frac{(p-1)}{2}} [\frac{jq}{p}] + \sum_{j=1}^{\frac{(q-1)}{2}} [\frac{jp}{q}] = \frac{p-1}{2} \cdot \frac{q-1}{2}$. By Lemma 9.3
it follows that $\left(\frac{p}{q}\right) = (-1)^{T(p,q)}$ and $\left(\frac{q}{p}\right) = (-1)^{T(q,p)}$ where $T(p, q) = \sum_{j=1}^{\frac{(p-1)}{2}} [\frac{jp}{q}]$
and $T(q, p) = \sum_{j=1}^{\frac{(q-1)}{2}} [\frac{jq}{p}]$. We conclude that $\left(\frac{p}{q}\right)\left(\frac{q}{p}\right) = (-1)^{\frac{p-1}{2} \cdot \frac{q-1}{2}}$.

Section 9.3

1. a. 1   b. −1   c. 1   d. 1   e. −1   f. 1

3. $1, 7, 13, 17, 19, 29, 37, 49, 61, 67, 71, 83, 91, 101, 103, 107, 109, 113, 119 \pmod{120}$

5. Let b be a quadratic nonresidue of p_1. Let a be a solution to the system of linear congruences:
$x \equiv b \pmod{p_1}, x \equiv 1 \pmod{p_2}, \ldots, x \equiv 1 \pmod{p_r}$.

7. Note that $(a, b) = (b, r_1) = (r_1, r_2) = \cdots = (r_{n-1}, r_n) = 1$ and since the q_i are even,

the r_i are odd. Since $r_0 = b$ and $a \equiv \epsilon_1 r_1 \pmod{b}$, by Theorem 9.9, we have $\left(\frac{a}{b}\right) =$

$\left(\frac{\epsilon_1 r_1}{r_0}\right) = \left(\frac{\epsilon_1}{r_0}\right)\left(\frac{r_1}{r_0}\right) = \left(\frac{\epsilon_1}{r_0}\right) \cdot \left(\frac{r_0}{r_1}\right) \cdot (-1)^{\frac{r_0-1}{2} \cdot \frac{r_1-1}{2}}$. If $\epsilon_1 = 1$, then $\left(\frac{a}{b}\right) =$

$(-1)^{\frac{r_0-1}{2} \cdot \frac{\epsilon_1 r_1-1}{2}}\left(\frac{r_0}{r_1}\right)$ If $\epsilon_1 = -1$, then $\left(\frac{\epsilon_1}{r_0}\right) = (-1)^{\frac{r_0-1}{2}}$ and we have $\left(\frac{a}{b}\right) =$

$(-1)^{\frac{r_0-1}{2} \cdot \frac{r_1+1}{2}}\left(\frac{r_0}{r_1}\right) = (-1)^{\frac{r_0-1}{2} \cdot \frac{-r_1-1}{2}}\left(\frac{r_0}{r_1}\right) = (-1)^{\frac{r_0-1}{2} \cdot \frac{\epsilon_1 r_1-1}{2}}\left(\frac{r_0}{r_1}\right)$ since

$\frac{r_1+1}{2}$ and $\frac{-r_1-1}{2}$ have the same parity. Similarly, $\left(\frac{r_0}{r_1}\right) = (-1)^{\frac{r_1-1}{2} \cdot \frac{\epsilon_2 r_2-1}{2}}\left(\frac{r_1}{r_2}\right)$, so

$\left(\frac{a}{b}\right) = (-1)^{\frac{r_0-1}{2} \cdot \frac{\epsilon_1 r_1-1}{2} + \frac{r_1-1}{2} \cdot \frac{\epsilon_2 r_2-1}{2}}\left(\frac{r_1}{r_2}\right)$. Proceed inductively until the last step,

when $\left(\frac{r_n}{r_{n-1}}\right) = \left(\frac{1}{r_{n-1}}\right) = 1$.

b. If either $r_{i-1} \equiv 1 \pmod{4}$ or $\epsilon_i r_i \equiv 1 \pmod{4}$, then $\frac{r_{i-1}-1}{2} \cdot \frac{\epsilon_i r_i-1}{2}$ is even.

Otherwise, that is, if $r_{i-1} \equiv \epsilon_i r_i \equiv 3 \pmod{4}$, then $\frac{r_{i-1}-1}{2} \cdot \frac{\epsilon_i r_i-1}{2}$ is odd. Then

$\frac{r_{n-1}-1}{2} \cdot \frac{\epsilon_n r_n-1}{2}$ the exponent in part (a) is even or odd as T is even or odd.

9. **a.** -1 **b.** -1 **c.** -1

11. Let $n_1 = p_1^{a_1} p_2^{a_2} \cdots p_r^{a_r}$ and $n_2 = q_1^{b_1} q_2^{b_2} \cdots q_s^{b_s}$. Then $\left(\frac{a}{n_1 n_2}\right) =$

$\left(\frac{a}{p_1}\right)^{a_1} \cdots \left(\frac{a}{p_r}\right)^{a_r}\left(\frac{a}{q_1}\right)^{b_1} \cdots \left(\frac{a}{q_s}\right)^{b_s} = \left(\frac{a}{n_1}\right)\left(\frac{a}{n_2}\right)$.

13. If a is odd, then by Exercise 12, we have $\left(\frac{a}{n_1}\right) = \left(\frac{n_1}{|a|}\right)$. By Theorem 9.8$(i)$, we have $\left(\frac{n_1}{|a|}\right) =$

$\left(\frac{n_2}{|a|}\right) = \left(\frac{a}{n_2}\right)$, using Exercise 12 again. If a is even, say $a = 2^s t$ with t odd, Exercise 12

gives $\left(\frac{a}{n_1}\right) = \left(\frac{2}{n_1}\right)^s (-1)^{\frac{t-1}{2} \cdot \frac{n_1-1}{2}}\left(\frac{n_1}{|t|}\right)$ and $\left(\frac{a}{n_2}\right) = \left(\frac{2}{n_2}\right)^s (-1)^{\frac{t-1}{2} \cdot \frac{n_2-1}{2}}\left(\frac{n_2}{|t|}\right)$.

Since $n_1 \equiv n_2 \pmod{|t|}$, we have $\left(\frac{n_1}{|t|}\right) = \left(\frac{n_2}{|t|}\right)$, and since $4 \mid a, n_1 \equiv n_2 \pmod{4}$

and so $(-1)^{\frac{t-1}{2} \cdot \frac{n_1-1}{2}} = (-1)^{\frac{t-1}{2} \cdot \frac{n_2-1}{2}}$. Now $a \equiv 0 \pmod{4}$, so $s \geq 2$. If s is 2, then

certainly $\left(\frac{2}{n_1}\right)^2 = \left(\frac{2}{n_2}\right)^2$. If $s > 2$, then $8 \mid a$ and $n_1 \equiv n_2 \pmod{8}$, so $\left(\frac{2}{n_1}\right) =$

$(-1)^{(n_1^2-1)/8} = (-1)^{(n_2^2-1)/8} = \left(\frac{2}{n_2}\right)$. Therefore $\left(\frac{a}{n_1}\right) = \left(\frac{a}{n_2}\right)$.

15. If a is odd, by Exercise 12 we have $\left(\frac{a}{|a|-1}\right) = \left(\frac{|a|-1}{|a|}\right) = \left(\frac{-1}{a}\right) = (-1)^{\frac{|a|-1}{2}} = 1$ if

$a > 0$ and $= -1$ if $a < 0$. If a is even, $a = 2^s t$ with t odd, then by Exercise 12 $\left(\frac{a}{|a|-1}\right) =$

$\left(\frac{2}{|a|-1}\right)^s (-1)^{\frac{t-1}{2}}\left(\frac{|a|-1}{|t|}\right)$. Since $b \geq 2$, check that $\left(\frac{2}{|a|-1}\right)^b = 1, (|a|-1 \equiv 7$

$\pmod{8}$ if $b > 2)$. Also $(-1)^{\frac{t-1}{2}}\left(\frac{|a|-1}{|t|}\right) = (-1)^{\frac{t-1}{2}}\left(\frac{-1}{|t|}\right) = (-1)^{\frac{n-1}{2} + \frac{|a|-1}{2}} = 1$

if $a > 0$ and $= -1$ if $a < 0$.

Section 9.4

1. $2^{(561-1)/2} = 2^{280} = (2^{10})^{28} \equiv (-98)^{28} \equiv (-98^2)^{14} \equiv 67^{14} \equiv (67^2)^7 \equiv 1^7 = 1$

$\pmod{561}$. Furthermore, $\left(\frac{2}{561}\right) = 1$ since $561 \equiv 1 \pmod{8}$.

3. Suppose that n is an Euler pseudoprime to both the bases a and b. Then $a^{(n-1)/2} \equiv \left(\frac{a}{n}\right)$

and $b^{(n-1)/2} \equiv \left(\frac{b}{n}\right)$. It follows that $(ab)^{(n-1)/2} \equiv \left(\frac{a}{n}\right)\left(\frac{b}{n}\right) = \left(\frac{ab}{n}\right)$. Hence n is an

Euler pseudoprime to the base ab.

5. Suppose that $n \equiv 5 \pmod 8$ and n is an Euler pseudoprime to the base 2. Since $n \equiv 5$ $\pmod 8$ we have $\left(\frac{2}{n}\right) = -1$. Since n is an Euler pseudoprime to the base 2, we have $2^{(n-1)/2} \equiv \left(\frac{2}{n}\right) = -1 \pmod n$. Write $n - 1 = 2^2 t$ where t is odd. Since $2^{(n-1)/2} \equiv 2^{2t}$ $\equiv -1 \pmod n$, n is a strong pseudoprime to the base 2.

7. $n \equiv 5 \pmod{20}$

9. There are 80 different values for b.

Section 9.5

1. 1229

3. Since $p, q \equiv 3 \pmod 4$, -1 is not a quadratic residue modulo p or q. If the four square roots are found using the method in Example 9.19, then only one of each possibility for choosing $+$ or $-$ can yield a quadratic residue in each congrunce, so there is only system which results in a square.

5. Paula sends $x = 1226, y = 625$. After receiving a 1, she sends $u\overline{r} = 689$.

7. The prover sends $x = 1403^2 = 1968409 \equiv 519 \pmod{2491}$. Verifier sends $\{1, 5\}$. Prover sends $y = 1425$. Verifier computes $y^2 z = 1425^2 \cdot 197 \cdot 494 \equiv 519 \equiv x \pmod{2491}$

9. a. 959, 1730, 2895, 441, 2900, 2684
   b. 1074
   c. $1074^2 \cdot 959 \cdot 1730 \cdot 441 \cdot 2684 \equiv 336 \equiv 403^2 \pmod{3953}$

Section 10.1

1. a. $.4$, b. $.41\overline{6}$, c. $.\overline{923076}$, d. $.5\overline{3}$, e. $.\overline{009}$, f. $.\overline{000999}$

3. a. $3/25$, b. $11/90$ c. $4/33$

5. $b = 2^r 3^s 5^t 7^u$

7. a. pre-period 1, period length 1 b. pre-period 3, period 0 c. pre-period 1, period 4
   d. pre-period 2, period 0 e. pre-period 1, period 1 f. pre-period 2, period 4

9. a. 3 b. 11 c. 37 d. 101 e. 41, 271 f. 7, 13

11. By induction: $c_{k+1} = [b\gamma_k] = \left[\frac{kb^2 - bk + b}{(b-1)^2}\right] = \left[\frac{k(b-1)^2 + (k+1) - k}{(b-1)^2}\right] = k$, and $\gamma_{k+1} = (k+1)b - k$, if $k \neq b - 2$. If $k = b - 2$, we have $c_{b-2} = b$, so we have determined $b - 1$ consecutive digits.

13. The base b expansion $(.100100001\ldots)$ is non-repeating.

15. This is Exercise 14 with $c_n = n$.

17. In the proof of Theorem 10.2, the numbers $p\gamma_n$ are the remainders of b^n upon division by p. The process recurs as soon as some γ_i repeats a value. Since $1/p = (.\overline{c_1 c_2 \ldots c_{p-1}})$ has period length $p - 1$, by Theorem10.4, we have that $\text{ord}_p b = p - 1$, so there is an integer k such that $b^k \equiv m \pmod p$. So the remainders of mb^n upon division by p are the same as the remainders of $b^k b^n$ upon division by p. Hence the nth digit of the expansion of m/p is determined by the remainder of b^{k+n} upon division by p, therefore, it will be the same as the $(k + n)$th digit of $1/p$.

19. n must be prime with 10 a primitive root.

21. Suppose $e = h/k$. Then $k!(e - 1 - 1/1! - 1/2! - \cdots - 1/k!)$ is an integer. But this is equal to $k!(1/(k + 1)! + 1/(k + 2)! + \cdots) = 1/(k + 1) + 1/(k + 1)(k + 2) + \cdots < 1/(k + 1) + 1/(k + 1)^2 + \cdots = 1/k < 1$. But $k!(e - 1 - 1/1! - 1/2! - \cdots - 1/k!)$ is positive, and therefore cannot be an integer, a contradiction.

Section 10.2

1. **a.** $\frac{15}{7}$ **b.** $\frac{10}{7}$ **c.** $\frac{6}{31}$ **d.** $\frac{355}{113}$ **e.** 2 **f.** $\frac{3}{2}$ **g.** $\frac{5}{3}$ **h.** $\frac{8}{5}$

3. **a.** $[1; 2, 1, 1, 2]$ **b.** $[1; 1, 7, 2]$ **c.** $[2; 9]$ **d.** $[3; 7, 1, 1, 1, 1, 2]$
   e. $[-1; 13, 1, 1, 2, 1, 1, 2]$ **f.** $[0, 9, 1, 3, 6, 2, 4, 1, 2]$

5. **a.** $1, \frac{3}{2}, \frac{4}{3}, \frac{7}{5}, \frac{18}{13}$ **b.** $1, 2, \frac{15}{8}, \frac{32}{17}$ **c.** $2, \frac{19}{9}$ **d.** $3, \frac{22}{7}, \frac{25}{8}, \frac{47}{15}, \frac{72}{23}, \frac{119}{38}, \frac{310}{99}$
   e. $-1, -\frac{12}{13}, -\frac{13}{14}, -\frac{25}{27}, -\frac{63}{68}, -\frac{88}{95}, -\frac{151}{163}, -\frac{390}{421}, -\frac{931}{1005}$
   f. $0, \frac{1}{9}, \frac{1}{10}, \frac{4}{39}, \frac{25}{244}, \frac{54}{527}, \frac{241}{2352}, \frac{295}{2879}, \frac{831}{8110}$

7. For Exercise 5: **a.** $\frac{3}{2} > \frac{7}{5}$ and $1 < \frac{4}{3} < \frac{18}{13}$ **b.** $2 > \frac{32}{17}$ and $1 < \frac{15}{8}$ **c.** Trivially true.
   d. $\frac{22}{7} > \frac{47}{15} > \frac{119}{38}$ and $3 < \frac{25}{8} < \frac{72}{23}, < \frac{310}{99}$ **e.** $-\frac{12}{13} > -\frac{25}{27} > -\frac{88}{95} > -\frac{390}{421}$ and
   $-1 < -\frac{13}{14} < -\frac{63}{68}, < -\frac{151}{163} < -\frac{931}{1005}$ **f.** $\frac{1}{9} > \frac{4}{39} > \frac{54}{527} > \frac{295}{2879}$ and $0 < \frac{1}{10} <$
   $\frac{25}{244} < \frac{241}{2352} < \frac{831}{8110}$

9. Let $\alpha = \frac{r}{s}$. The Euclidean Algorithm for $\frac{1}{\alpha} = \frac{s}{r} < 1$ gives $s = 0 \cdot r + s; r = a_0 \cdot s + a_1$,
   and continues just like for $\frac{r}{s}$.

11. Note that $f_1 = 1 \le q_1$, which must be a positive integer. Assume $q_j \ge f_j$ for $j < k$. Then
    $q_k = a_k q_{k-1} + q_{k-2} \ge a_k f_{k-1} + f_{k-2} \ge f_{k-1} + f_{k-2} = f_k$.

13. By Exercise 6, we have $\frac{p_n}{p_{n-1}} = [a_n; a_{n-1}, \ldots, a_0] = [a_0; a_1, \ldots, a_n] = \frac{p_n}{q_n} = \frac{r}{s}$ if the
    continued fraction is symmetric. Then, $q_n = p_{n-1} = s$ and $p_n = r$, so by Theorem 10.8 we
    have $p_n q_{n-1} - q_n p_{n-1} = rq_{n-1} - s^2 = (-1)^{n-1}$. Then $rq_{n-1} = s^2 + (-1)^{n-1}$ and
    so $r | s^2 + (-1)^n$. Conversely, if $r | s^2 + (-1)^{n-1}$, then $(-1)^{n-1} = p_n q_{n-1} - q_n p_{n-1}$
    $= rq_{n-1} - p_{n-1}s$. So $r | p_{n-1}s + (-1)^{n-1}$ and hence $r | (s^2 + (-1)^{n-1}) -$
    $(p_{n-1}s + (-1)^{n-1}) = s(s - p_{n-1})$. Since $s, p_{n-1} < r$ and $(r, s) = 1$, we have $s =$
    p_{n-1}. Then $[a_n; a_{n-1}, \ldots, a_0] = \frac{p_n}{p_{n-1}} = \frac{r}{s} = [a_0; a_1, \ldots, a_n]$.

15. Assume the statement is true for k odd and prove it for $k + 2$. Define $a_k' = [a_k; a_{k+1}, a_{k+2}]$
    and check that $a_k' < [a_k; a_{k+1}, a_{k+2} + x] = a_k' + x'$. Then $[a_0; a_1, \ldots, a_{k+2}] =$
    $[a_0; a_1, \ldots, a_k'] > [a_0; a_1, \ldots, a_k' + x'] = [a_0; a_1, \ldots, a_{k+2} + x]$. Proceed similarly for
    k even.

Section 10.3

1. **a.** $[1; 2, 2, \ldots]$ **b.** $[1; 1, 2, 1, 2, \ldots]$ **c.** $[2; 4, 4, \ldots]$ **d.** $[1; 1, 1, \ldots]$.
3. $\frac{312689}{99532}$.
5. If $a_1 > 1$, let $A = [a_2; a_3, \ldots]$. Then $[a_0; a_1, \ldots] - [-a_0 - 1; 1, a_1 - 1, a_2, a_3, \ldots] =$
   $a_0 + \frac{1}{a_1 + \frac{1}{A}} + \left(-a_0 - 1 + \frac{1}{1 + \frac{1}{a_1 - 1 + \frac{1}{A}}}\right) = 0$. Similarly if $a_1 = 1$.
7. If $\alpha = [a_0; a_1, a_2, \ldots]$, then $\frac{1}{\alpha} = \frac{1}{[a_0; a_1, a_2, \ldots]} = 0 + \frac{1}{a_0 + \frac{1}{a_1 + \cdots}} = [0; a_0, a_1, a_2, \ldots]$.
   Then the kth convergent of $\frac{1}{\alpha}$ is $[0; a_0, a_1, a_2, \ldots, a_{k-1}] = \frac{1}{[a_0; a_1, a_2, \ldots, a_{k-1}]}$, which is
   the reciprocal of the $(k - 1)$th convergent of α.
9. By Theorem 10.16, such a $\frac{p}{q}$ is a convergent of α. $\frac{\sqrt{5}+1}{2} = [1; 1, 1, \ldots]$, so $q_n =$
   f_n (Fibonacci) and $p_n = q_{n+1}$. Then $\lim_{n \to \infty} \frac{q_{n-1}}{q_n} = \lim_{n \to \infty} \frac{q_{n-1}}{p_{n-1}} =$

$\frac{2}{\sqrt{5}+1} = \frac{\sqrt{5}-1}{2}$. So $\lim_{n\to\infty}\left(\frac{\sqrt{5}+1}{2} + \frac{q_{n-1}}{q_n}\right) = \frac{\sqrt{5}+1}{2} + \frac{\sqrt{5}-1}{2} = \sqrt{5}$. So $\frac{\sqrt{5}+1}{2} +$

$\frac{q_{n-1}}{q_n} > c$ only finitely often. Whence, $\frac{1}{\left(\frac{\sqrt{5}+1}{2} + \frac{q_{n-1}}{q_n}\right)q_n^2} < \frac{1}{cq_n^2}$. The following identity

finishes the proof. Note that $\alpha_n = \alpha$ for all n. Then $\left|\alpha - \frac{p_n}{q_n}\right| =$

$\left|\frac{\alpha_{n+1}p_n + p_{n-1}}{\alpha_{n+1}q_n + q_{n-1}} - \frac{p_n}{q_n}\right| = \left|\frac{-(p_n q_{n-1} - p_{n-1}q_n)}{q_n(\alpha q_n + q_{n-1})}\right| = \frac{1}{q_n^2\left(\alpha + \frac{q_{n-1}}{q_n}\right)}$.

11. If β is equivalent to α, then $\beta = \frac{a\alpha+b}{c\alpha+d}$. Solving for α gives $\alpha = \frac{-d\beta+b}{c\beta-a}$, so α is equivalent to β.

13. If $a \neq 0$, then $\frac{r}{s} = \frac{(rb)a+0}{(sa)b+0}$, so $\frac{r}{s}$ and $\frac{a}{b}$ are equivalent. If $a = 0$ then $\frac{r}{s} = \frac{1\cdot a+r}{0\cdot b+s}$.

15. Note that $p_{k,t}q_{k-1} - q_{k,t}p_{k-1} = t(p_{k-1}q_{k-1} - q_{k-1}p_{k-1}) + (p_{k-2}q_{k-1} - p_{k-1}q_{k-2}) = \pm1$. Thus $p_{k,t}$ and $q_{k,t}$ are relatively prime.

17. See, for example, the classic work by O. Perron, *Die Lehre von den Kettenbrüchen*, Leipzig, Teubner (1929).

19. $179/57$

Section 10.4

1. a. $[2; \overline{1,1,1,4}]$　**b.** $[3; \overline{3,6}]$　**c.** $[4; \overline{1,3,1,8}]$　**d.** $[6; \overline{1,5,1,12}]$
　e. $[7; \overline{1,2,7,2,1,14}]$　**f.** $[9; \overline{1,2,3,1,1,5,1,8,1,5,1,1,3,2,1,18}]$

3. a. $[2; \overline{2}]$　**b.** $[1; \overline{12,1,2,2,2,1}]$　**c.** $[0; 1,1,\overline{2,3,10,3}]$

5. a. $\frac{23+\sqrt{29}}{10}$　**b.** $\frac{-1+3\sqrt{5}}{2}$　**c.** $\frac{8+\sqrt{82}}{6}$

7. a. $\sqrt{10}$　**b.** $\sqrt{17}$　**c.** $\sqrt{26}$　**d.** $\sqrt{37}$

9. a. $\alpha_0 = \sqrt{d^2-1}, a_0 = d-1, P_0 = 0, Q_0 = 1, P_1 = (d-1)\cdot1 - 0 = d-1, Q_1 = \frac{(d^2-1)-(d-1)^2}{1} = 2d-2, \alpha_1 = \frac{d-1+\sqrt{d^2-1}}{2(d-1)} = \frac{1}{2} + \frac{1}{2}\sqrt{\frac{d+1}{d-1}}, a_1 = 1, P_2 = 1(2d-2) - (d-1) = d-1, Q_2 = \frac{d^2-1-(d-1)^2}{2d-2} = 1, \alpha_2 = \frac{d-1+\sqrt{d^2-1}}{1}, a_2 = 2d-2, P_3 = 2(d-1)\cdot1 - (d-1) = d-1 = P_1, Q_3 = \frac{(d^2-1)-(d-1)^2}{1} = 2d-2 = Q_1$, so $\alpha = [d-1; \overline{1,2(d-1)}]$.

b. $\alpha_0 = \sqrt{d^2-d}, a_0 = [\sqrt{d^2-d}] = d-1$, since $(d-1)^2 < d^2-d < d^2$. Then $P_0 = 0, Q_0 = 1, P_1 = d-1, Q_1 = d-1, \alpha_1 = \frac{(d-1)+\sqrt{d^2-d}}{d-1} = 1 + \sqrt{\frac{d}{d-1}}, a_1 = 2, P_2 = d-1, Q_2 = 1, \alpha_2 = \frac{(d-1)+\sqrt{d^2-d}}{1}, a_2 = 2(d-1), P_3 = P_1, Q_3 = Q_1$. Therefore, $\sqrt{d^2-d} = [d-1; \overline{2,2(d-1)}]$.

c. $[9; \overline{1,18}], [10; \overline{2,20}], [16; \overline{2,32}], [24; \overline{2,48}]$

11. a. Note that $d < \sqrt{d^2+4} < d+1$. We compute $\alpha_0 = \sqrt{d^2+4}, a_0 = d, P_0 = 0, Q_0 = 1, P_1 = d, Q_1 = 4, \alpha_1 = \frac{d+\sqrt{d^2+4}}{4}, a_1 = \left[\frac{2d}{4}\right] = \frac{d-1}{2}$, since d is odd. $P_2 = d-2, Q_2 = d, \alpha_2 = \frac{d-2+\sqrt{d^2+4}}{d}, \frac{d-2+d}{d} < \alpha_2 < \frac{d-2+d+1}{d}$, so $a_2 = 1, P_3 = 2, Q_3 = d, \alpha_3 = \frac{2+\sqrt{d^2+4}}{d}, a_3 = 1, P_4 = d-2, Q_4 = 4, \alpha_4 = \frac{d-2+\sqrt{d^2+4}}{4}, \frac{d-2+d}{4} = \frac{d-1}{2} < \alpha_4 < \frac{d-2+d+1}{4}$, so $a_4 = \frac{d-1}{2}, P_5 = d, Q_5 = 1, \alpha_5 = \frac{d+\sqrt{d^2+4}}{1}, a_5 = 2d, P_6 = d = P_1, Q_6 = 4 = Q_1$, so $\alpha = [d; \overline{\frac{d-1}{2},1,1,\frac{d-1}{2},2d}]$.

b. Note that $d - 1 < \sqrt{d^2 - 4} < d$. We compute $\alpha_0 = \sqrt{d^2 - 4}, a_0 = d - 1, P_0 = 0, Q_0 = 1, P_1 = d - 1, Q_1 = 2d - 5, \alpha_1 = \frac{d-1+\sqrt{d^2-4}}{2d-5}, \frac{d-1+d-1}{2d-5} < \alpha_0 < \frac{d-1+d}{2d-5}$ and $d > 3$ so $a_1 = 1, P_2 = d - 4, Q_2 = 4, a_2 = \frac{d-4+\sqrt{d^2-4}}{4}, a_2 = \frac{d-3}{2}, P_3 = d - 2, Q_3 = d - 2, \alpha_3 = \frac{d-2+\sqrt{d^2-4}}{d-2}, a_3 = 2, P_4 = d - 2, Q_4 = 4, \alpha_4 = \frac{d-2+\sqrt{d^2-4}}{4}, a_4 = \frac{d-3}{2}, P_5 = d - 4, Q_5 = 2d - 5, \alpha_5 = \frac{d-4+\sqrt{d^2-4}}{2d-5}, a_5 = 1, P_6 = d - 1, Q_6 = 1, \alpha_6 = \frac{d-1+\sqrt{d^2-4}}{1}, a_6 = 2d - 2, P_7 = d - 1 = P_1, Q_7 = 2d - 5 = Q_1$, so $\alpha = [d - 1; \overline{1, \frac{d-3}{2}, 2, \frac{d-3}{2}, 1, 2d - 2}]$.

13. Suppose $\sqrt{d}$ has period length two. Then $\sqrt{d} = [a; \overline{c, 2a}]$ from the discussion preceeding Example 10.15. Then $\sqrt{d} = [a; y]$ with $y = [\overline{c; 2a}] = [c; 2a, y] = c + \frac{1}{2a+\frac{1}{y}} = \frac{2acy+c+y}{2ay+1}$. Then $2ay^2 - 2acy - c = 0$, and since y is positive, we have $y = \frac{2ac+\sqrt{(2ac)^2+4(2a)c}}{4a} = \frac{ac+\sqrt{(ac)^2+2ac}}{2a}$. Then $\sqrt{d} = [a; y] = a + \frac{1}{y} = a + \frac{2a}{ac+\sqrt{(ac)^2+2ac}} = \sqrt{a^2 + \frac{2a}{c}}$, so $d = a^2 + \frac{2a}{c}$, and $b = \frac{2a}{c}$ is an integral divisor of $2a$. Conversely, let $\alpha = \sqrt{a^2 + b}$ and $b|2a$, say $kb = 2a$, then $a_0 = [\sqrt{a^2 + b}] = a$, since $a^2 < a^2 + b < (a + 1)^2$. Then $P_0 = 0, Q_0 = 1, P_1 = a, Q_1 = b, \alpha_1 = \frac{a+\sqrt{a^2+b}}{b}, a_1 = 4k, P_2 = a, Q_2 = 1, \alpha_2 = \frac{a+\sqrt{a^2+b}}{1}, a_2 = 2a, P_3 = a = P_1, Q_3 = b = Q_1$, so $\alpha = [a; \overline{4k, 2a}]$, which has period length two.

15. **a.** not **b.** purely periodic **c.** purely periodic **d.** not **e.** purely periodic **f.** not

17. Let $\alpha = \frac{a+\sqrt{b}}{c}$. Then $-\frac{1}{\alpha'} = -\frac{c}{a-\sqrt{b}} = \frac{ca+\sqrt{bc^2}}{b-a^2} = \frac{A+\sqrt{B}}{C}$, say. By Exercise 12, $0 < a < \sqrt{b}$ and $\sqrt{b} - a < c < \sqrt{b} + a < 2\sqrt{b}$. Multiplying by c gives $0 < ca < \sqrt{bc^2}$ and $\sqrt{bc^2} - ca < c^2 < \sqrt{bc^2} + ca < 2\sqrt{bc^2}$. That is, $0 < A < \sqrt{B}$ and $\sqrt{B} - A < c^2 < \sqrt{B} + A < 2\sqrt{B}$. Multiply $\sqrt{b} - a < c$ by $\sqrt{b} + a$ to get $C = b - a^2 < \sqrt{bc^2} + ca = A + \sqrt{B}$. Multiply $c < \sqrt{b} + a$ by $\sqrt{b} - a$ to get $\sqrt{B} - A = \sqrt{bc^2} - ac < b - a^2 = C$. So, $-\frac{1}{\alpha'}$ satisfies all the inequalities in Exercise 12, and therefore is reduced.

19. Start with $\alpha_0 = \sqrt{D_k} + 3^k + 1$ (this will have the same period since it differs from $\sqrt{D_k}$ by an integer) and use induction. Apply the continued fraction algorithm to show $\alpha_{3i} = \sqrt{D_k} + 3^k - 2 \cdot 3^{k-i} + 2/2 \cdot 3^{k-i}, i = 1, 2, \ldots, k, \alpha_{3k+3i} = \sqrt{D_k} + 3^k - 2/2 \cdot 3^i, i = 1, 2, \ldots, k - 1$, and $\alpha_{6k} = \sqrt{D_k} + 3^k + 1 = \alpha_0$ Since $\alpha_i \neq \alpha_0$ for $i < 6k$ we see that the period is $6k$.

Section 10.5

1. 7, 17
3. $3119 \cdot 4261$
5. We have $17^2 = 289 \equiv 3 \pmod{143}$ and $19^2 = 361 \equiv 3 \cdot 5^2 \pmod{143}$. Combining these, we have $(17 \cdot 19)^2 \equiv 3^2 5^2 \pmod{143}$. Hence, $323^2 \equiv 15^2 \pmod{143}$. It follows that $(323^2 - 15^2) = (323 - 15)(323 + 15) \equiv 0 \pmod{143}$. This produces the two factors $(323 - 15, 143) = (308, 143) = 11$ and $(323 + 15, 143) = (338, 143) = 13$ of 143.
7. $3001 \cdot 40001$

Section 11.1

1. **a.** (3,4,5), (5,12,13), (15,8,17), (7,24,25), (21,20,29), and (35,12,37).

   b. In addition to part (a), (6,8,10), (9,12,15), (12,16,20), (15,20,25), (18,24,30), (21,28,35), (24,32,40), (10,24,26), (15,36,39), (30,16,34)

3. By Lemma 11.1, 5 divides at most one of x, y, and z. If $5 \nmid x$ or y, then $x^2 \equiv \pm 1 \pmod 5$ and $y^2 \equiv \pm 1 \pmod 5$. Then, $z^2 \equiv 0, 2,$ or $-2 \pmod 5$. But ± 2 is not a quadratic residue modulo 5, so $z^2 \equiv 0 \pmod 5$, whence $5 \nmid z$.

5. Let k be an integer ≥ 3. If $k = 2n + 1$, let $m = n + 1$. Then m and n have opposite parity, $m > n$ and $m^2 - n^2 = 2n + 1 = k$, so m and n define the desired triple. If k has an odd divisor $d > 1$, then use the construction given for d and multiply the result by $\frac{k}{d}$. If k has no odd divisors, then $k = 2^j$ for some integer j. Let $m = 2^{j-1}$ and $n = 1$. Then $k = 2mn$, $m > n$, and m and n have opposite parity, so m and n define the desired triple.

7. The values for x and z determine the triple. Solve the equations in Exercise 6 for x_n and z_n to get $x_n = 3x_{n+1} - 2z_{n+1} + 1$, $z_n = -4x_{n+1} + 3z_{n+1} - 2$. Check that if x_{n+1} and z_{n+1} are a solution, then so are x_n and z_n as determined by the equations. By the triangle inequality, $z_{n+1} < 2x_{n+1} + 1$, thus, $z_n = -4x_{n+1} + 3z_{n+1} - 2 = 3z_{n+1} - 2(2x_{n+1} + 1) < z_{n+1}$. Also, $z_{n+1} = \sqrt{2x_{n+1}^2 + 2x_{n+1} + 1} > \sqrt{2}x_{n+1} > \frac{7}{5}x_{n+1}$, so $z_n \geq 3z_{n+1} - 4(\frac{7}{5}z_{n+1}) = \frac{1}{21}z_{n+1} > 0$. Therefore, $0 < z_n < z_{n+1}$. Similarly, $-2 < x_n < x_{n+1}$. Therefore, if x, $y = x+1$, z is a solution, we may apply our equations recursively to get a smaller solution. Our inequalities imply that in finite steps we will have $z = 1$ and $x = -1$. Applying the equations in Exercise 6 to these values twice, gives the triple (3,4,5).

9. See Exercise 15 with $p = 3$.

11. (9,12,15), (35,12,37), (5,12,13), (12,16,20).

13. For m positive, $x = 4m^2$, $y = m^2 - 1$, $z = m^2 + 1$.

15. Check that if $m > \sqrt{p} \cdot n$ then $x = \frac{m^2 - pn^2}{2}$, $y = mn$, $z = \frac{m^2 + pn^2}{2}$ is a solution. Conversely, if x, y, z is a primitive solution, then $y^2 = \frac{z^2 - x^2}{p}$, so $p | (z \pm x)$. Take $m^2 = z \mp x$ and $n^2 = \frac{z \pm x}{p}$.

Section 11.2

1. $x^k + y^k = x^2 x^{k-2} + y^2 y^{k-2} < (x^2 + y^2)y^{k-2} = z^2 y^{k-2} < z^2 z^{k-2} = z^k$.

3. **a.** If p divides x, y, or z, then certainly $p | xyz$. If not, then by Fermat's Little Theorem, $x^{p-1} \equiv y^{p-1} \equiv z^{p-1} \equiv 1 \pmod p$. Hence, $1 + 1 \equiv 1 \pmod p$, which is impossible.

   b. $a^p \equiv a \pmod p$ for every integer a. Then $x^p + y^p \equiv z^p \pmod p$ implies $x + y \equiv z \pmod p$, so $p | x + y - z$.

5. Let x and y be the lengths of the legs and z be the length of the hypotenuese. Then $x^2 + y^2 = z^2$. If the area is a perfect square, we have $A = \frac{1}{2}xy = r^2$. Then, if $x = m^2 - n^2$, and $y = 2mn$, we have $r^2 = mn(m^2 - n^2)$. All of these factors are relatively prime, so $m = a^2$, $n = b^2$, and $m^2 - n^2 = c^2$, where a, b, and c are positive integers. Then, $a^4 - b^4 = c^2$, which contradicts Exercise 4.

7. Write $z^2 + 2(2y^2)^2 = (x^2)^2$ and use Exercise 8 of Section 11.1. $z^2 = \frac{m^2 - n^2}{2}$, $2y^2 = mn$, and $x^2 = \frac{m^2 + 2n^2}{2}$. From $2y^2 = mn$, $m = \frac{r^2}{2}$ and $n = s^2$, so $x^2 = 4(\frac{r}{2})^4 + s^4$, which contradicts Exercise 6.

9. Follows by Exercise 4 and Theorem 11.2

11. Assume $n \nmid xyz$, and $(x, y, z) = 1$. Now $(-x)^n = y^n + z^n = (y + z)(y^{n-1} + \cdots + z^{n-1})$, and these factors are relatively prime, so they are nth powers, say $y + z = a^n$, and $(y^{n-1} + \cdots + z^{n-1}) = \alpha^n$, whence $x = a\alpha$. Similarly, $z + x = b^n$, and $(z^{n-1} + \cdots + x^{n-1}) = \beta^n$, $y = b\beta$, $x + y = c^n$, and $(x^{n-1} + \cdots + y^{n-1}) = \gamma^n$, and $z = c\gamma$. Since $x^n + y^n + z^n \equiv 0 \pmod{p}$, we have $p \mid xyz$, say $p \mid x$. Then $2x \equiv b^n + c^n + (-a)^n \equiv 0 \pmod{p}$, so by the condition on p, we have $p \mid abc$. If $p \mid b$ then $y = -b\beta \equiv 0 \pmod{p}$, but then $p \mid x$ and y, a contradiction. Similarly, p cannot divide c. Therefore, $p \mid a$, so $y \equiv -z \pmod{p}$, and so $\alpha^n \equiv (y^{n-1} + \cdots + z^{n-1}) \equiv ny^{n-1} \equiv n\gamma^n \pmod{p}$. Let g be the inverse of $\gamma \pmod{p}$, then $(ag)^n \equiv n \pmod{p}$, which is a contradiction.

13. 3, 4, 5, 6

Section 11.3

1. a. $19^2 + 4^2$ b. $23^2 + 11^2$ c. $37^2 + 9^2$ d. $137^2 + 9^2$

3. a. $5^2 + 3^2$ b. $9^2 + 3^2$ c. $10^2 + 0^2$ d. $21^2 + 7^2$ e. $133^2 + 63^2$
   f. $448^2 + 352^2$

5. a. $1^2 + 1^2 + 1^2$ b. not possible   c. $3^2 + 1^2 + 1^2$ d. $3^2 + 3^2 + 1^2$ e. not possible   f. not possible

7. Let $n = x^2 + y^2 + z^2 = 4^m(8k + 7)$. If $m = 0$, see Exercise 6. If $m \geq 1$, then n is even so 0 or 2 of x, y, z are odd. If 2 are odd, $x^2 + y^2 + z^2 \equiv 2$ or $6 \pmod 8$, but then $4 \nmid (n - 7)$, a contradiction, so all of x, y, z are even. Then $4^{m-1}(8k + 1) = (\frac{x}{2})^2 + (\frac{y}{2})^2 + (\frac{z}{2})^2$ is the sum of 3 squares. Repeat until $m = 0$ and use Exercise 6 to get a contradiction.

9. a. $10^2 + 1^2 + 0^2 + 2^2$ b. $22^2 + 4^2 + 1^2 + 3^2$ c. $14^2 + 4^2 + 1^2 + 5^2$
   d. $56^2 + 12^2 + 17^2 + 1^2$

11. Let $m = n - 169$. Then m is the sum of 4 squares: $m = x^2 + y^2 + z^2 + w^2$. If, say, x, y, z are zero, then $n = w^2 + 169 = w^2 + 10^2 + 8^2 + 2^2 + 1^2$. If, say, x, y are zero, then $n = z^2 + w^2 + 169 = z^2 + w^2 + 12^2 + 4^2 + 3^2$. If, say, x is zero, then $n = y^2 + z^2 + w^2 + 169 = y^2 + z^2 + w^2 + 12^2 + 5^2$. If none are zero, then $n = x^2 + y^2 + z^2 + w^2 + 13^2$.

13. If k is odd, then 2^k is not the sum of 4 positive squares.

15. If $p = 2$ the theorem is obvious. Else, $p = 4k + 1$, whence -1 is a quadratic residue modulo p, say $a^2 \equiv -1 \pmod p$. Let x and y be as in Thue's Lemma. Then $x^2 < p$ and $y^2 < p$ and $-x^2 \equiv (ax)^2 \equiv y^2 \pmod p$. Thus $p \mid x^2 + y^2 < 2p$, therefore $p = x^2 + y^2$ as desired.

17. The left sum runs over every pair of integers $i < j$, for $1 \leq i < j \leq 4$, so there are 6 terms. Each integer 1,2,3, and 4 appears in exactly 3 pairs, so $\sum_{1 \leq i < j \leq 4}[(x_i + x_j)^4 + (x_i - x_j)^4] = \sum_{1 \leq i < j \leq 4}(2x_i^4 + 12x_i^2x_j^2 + 2x_j^4), = \sum_{k=1}^{4} 6x_k^4 + \sum_{1 \leq i < j \leq 4} 12x_i^2x_j^2 = 6\left(\sum_{k=1}^{4} x_k^2\right)^2$.

19. If m is positive, then $m = \sum_{k=1}^{4} x_k^2$, for some x_k's. Then $6m = 6\sum_{k=1}^{4} x_k^2 = \sum_{k=1}^{4} 6x_k^2$. By Exercise 18, each term of the last sum is the sum of 12 4th powers. Therefore $6m$ is the sum of 48 4th powers.

21. For $n = 1, 2, \ldots, 50$, $n = \sum_{1}^{n} 1^4$. For $n = 51, 52, \ldots, 81$, $n - 48 = n - 3(2^4) = \sum_{1}^{n-48} 1^4$, so $n = 2^4 + 2^4 + 2^4 + \sum_{1}^{n-48} 1^4$ is the sum of $n - 45$ squares.

Section 11.4

1. **a.** $(\pm 2, 0), (\pm 1, \pm 1)$ **b.** no solution   **c.** $(\pm 1, \pm 2)$.
3. **a.** solutions   **b.** no solution   **c.** solutions   **d.** solutions   **e.** solutions   **f.** no solution
5. 73, 12; 10657,1752; 128766, 21169
7. 6239765965720528801, 798920165762623330040
9. Reduce modulo p to get $x^2 \equiv -1 \pmod{p}$, but -1 is a quadratic nonresidue modulo p if $p = 4k + 3$, so there is no solution.
11. Following the hint, solve $a^2 - 2b^2 = \pm 1$. Solving $s - t = a, t = b$ for s and t and substituting in the expressions for x and y gives: $x = (a + b)^2 - b^2 = a^2 + 2ab$ and $y = 2ab + 2b^2$.
13. x and y are odd. Add $2x^2$ to each side. This will reduce to $4 \equiv 0 \pmod{8}$, a contradiction.

Bibliography

BOOKS

Number Theory

1. W. W. Adams and L. J. Goldstein, *Introduction to Number Theory*, Prentice-Hall, Englewood Cliffs, New Jersey, 1976.

2. G. E. Andrews, *Number Theory*, W. B. Saunders, Philadelphia, 1971.

3. T. A. Apostol, *Introduction to Analytic Number Theory*, Springer-Verlag, New York, 1976.

4. R. G. Archibald, *An Introduction to the Theory of Numbers*, Merrill, Columbus, Ohio, 1970.

5. I. A. Barnett, *Elements of Number Theory*, Prindle, Weber, and Schmidt, Boston, 1969.

6. A. H. Beiler, *Recreations in the Theory of Numbers*, 2nd ed., Dover, New York, 1966.

7. E. D. Bolker, *Elementary Number Theory*, Benjamin, New York, 1970.

8. Z. I. Borevich and I. R. Shafarevich, *Number Theory*, Academic Press, New York, 1966.

9. D. M. Burton, *Elementary Number Theory*, Allyn and Bacon, Boston, 1976.

10. R. D. Carmichael, *The Theory of Numbers and Diophantine Analysis*, Dover, New York, 1959 (reprint of the original 1914 and 1915 editions).

11. H. Davenport, *The Higher Arithmetic*, 5th ed., Cambridge University Press, Cambridge, England, 1982.

12. L. E. Dickson, *Introduction to the Theory of Numbers*, Dover, New York 1957 (reprint of the original 1929 edition).

13. U. Dudley, *Elementary Number Theory*, 2nd ed., Freeman, New York, 1969.

14. H. M. Edwards, *Fermat's Last Theorem*, Springer-Verlag, New York, 1977.

15. D. Flath, *Introduction to Number Theory*, Wiley, New York, 1989.

16. C. F. Gauss, *Disquisitiones Arithmeticae,* Yale, New Haven, Connecticut, 1966.

17. A. A. Gioia, *The Theory of Numbers*, Markham, Chicago 1970.

18. E. Grosswald, *Topics from the Theory of Numbers*, 2nd ed., Birkhäuser, Boston, 1982.

19. H. Gupta, *Selected Topics in Number Theory*, Abacus Press, Kent, England, 1980.

20. R. K. Guy, *Unsolved Problems in Number Theory*, Springer-Verlag, New York, 1981.

21. G. H. Hardy and E. M. Wright, *An Introduction to the Theory of Numbers*, 5th ed., Oxford University Press, Oxford, 1979.

22. L. Hua, *Introduction to Number Theory*, Springer-Verlag, New York 1982.

23. K. Ireland and M. I. Rosen, *A Classical Introduction to Modern Number Theory*, Springer-Verlag, New York, 1982.

24. E. Landau, *Elementary Number Theory*, Chelsea, New York, 1958.

25. W. J. LeVeque, *Fundamentals of Number Theory*, Addison-Wesley, Reading, Massachusetts, 1977.

26. W. J. LeVeque, editor, *Reviews in Number Theory* [1940-1972], and R. K. Guy, editor, *Reviews in Number Theory* [1973-1983], six volumes each, American Mathematical Society, Washington, D.C., 1974 and 1984, respectively.

27. C. T. Long, *Elementary Introduction to Number Theory*, 2nd ed., Heath, Lexington, Massachusetts, 1972.

28. G.B. Matthews, *Theory of Numbers*, Chelsea, New York (no date).

29. I. Niven, H.S. Zuckerman, and H.L. Montgomery, *An Introduction to the Theory of Numbers*, 5th ed., Wiley, New York, 1991.

30. O. Ore, *An Invitation to Number Theory*, Random House, New York, 1967.

31. O. Ore, *Number Theory and its History*, McGraw-Hill, New York, 1948.

32. A. J. Pettofrezzo and D. R. Byrkit, *Elements of Number Theory*, Prentice-Hall, Englewood Cliffs, New Jersey, 1970.

33. H. Rademacher, *Lectures on Elementary Number Theory*, Blaisdell, New York 1964, reprint Krieger, 1977.

34. P. Ribenboim, *The Book of Prime Number Records*, 2nd ed., Springer-Verlag, New York, 1989.

35. P. Ribenboim, *13 Lectures on Fermat's Last Theorem,* Springer-Verlag, New York, 1979.

36. J. Roberts, *Elementary Number Theory*, MIT Press, Cambridge, Massachusetts, 1977.

37. M. R. Schroeder, *Number Theory in Science and Communication*, 2nd ed., Springer-Verlag, Berlin, 1986.

38. D. Shanks, *Solved and Unsolved Problems in Number Theory*, 3rd ed., Chelsea, New York, 1985.

39. H. S. Shapiro, *Introduction to the Theory of Numbers*, Wiley, New York, 1983.

40. J. E. Shockley, *Introduction to Number Theory*, Holt, Rinehart, and Winston, 1967.

41. W. Sierpiński, *Elementary Theory of Numbers*, Polski Akademic Nauk, Warsaw, 1964.

42. W. Sierpiński, *A Selection of Problems in the Theory of Numbers*, Pergammon Press, New York, 1964.

43. W. Sierpiński, *250 Problems in Elementary Number Theory*, Polish Scientific Publishers, Warsaw, 1970.

44. H. M. Stark, *An Introduction to Number Theory*, Markham, Chicago, 1970; reprint MIT Press, Cambridge, Massachusetts, 1978.

45. B. M. Stewart, *The Theory of Numbers*, 2nd ed., Macmillan, New York, 1964.

46. J. V. Uspensky and M. A. Heaslet, *Elementary Number Theory*, McGraw-Hill, New York, 1939.

47. S. Vajda, *Fibonacci & Lucas Numbers and the Golden Section: Theory and Applications*, Ellis Horwood, Chichester, England, 1989.

48. C. Vanden Eyden, *Number Theory*, International Textbook, Scranton, Pennsylvania, 1970.

49. I. M. Vinogradov, *Elements of Number Theory*, Dover, New York, 1954.

50. H. N. Wright, *First Course in Theory of Numbers*, Wiley, New York, 1939.

Number Theory with Computer Science

51. A. M. Kirch, *Elementary Number Theory: A Computer Approach*, Intext, New York, 1974.

52. D. G. Malm, *A Computer Laboratory Manual for Number Theory*, COMPress, Wentworth, New Hampshire, 1979.

53. D. D. Spencer, *Computers in Number Theory*, Computer Science Press, Rockville, Maryland, 1982.

Factorization and Primality Testing

54. D.M. Bressoud, *Factorization and Primality Testing*, Springer-Verlag, New York, 1989.

55. J. Brillhart, D.H. Lehmer, J.L. Selfridge, B. Tuckerman, S.S. Wagstaff, Jr., *Factorizations of $b^n \pm 1$, $b = 2,3,5,6,7,10,11,12$ up to high powers*, revised ed., American Mathematical Society, Providence, Rhode Island, 1988.

56. E. Kranakis, *Primality and Cryptography*, Wiley-Teubner, Stuttgart, West Germany, 1986.

57. C. Pomerance, *Lecture Notes on Primality Testing and Factoring*, Mathematical Association of America, Washington, D.C., 1984.

58. H. Riesel, *Prime Numbers and Computer Methods for Factorization*, Birkhäuser, Boston, 1985.

History

59. E. T. Bell, *Men of Mathematics*, Simon & Schuster, New York, 1965.

60. C. B. Boyer, *A History of Mathematics*, Wiley, New York, 1968.

61. D. M. Burton, *The History of Mathematics*, Allyn and Bacon, Boston, 1985.

62. L. E. Dickson, *History of the Theory of Numbers*, three volumes, Chelsea, New York, 1952 (reprint of the 1919 original).

63. *Dictionary of Scientific Biography*, Scribners, New York, 1970.

64. H. Eves, *An Introduction to the History of Mathematics* , 5th ed., Saunder, Philadelphia, 1983.

65. M. Kline, *Mathematical Thought from Ancient to Modern Times*, Oxford University, New York, 1972.

66. U. Libbrecht, *Chinese Mathematics in the Thirteenth Century, The Shu-shu chiu-chang of Ch'in Chiu-shao:'* MIT, 1973.

67. W. Scharlau and H. Opolka, *From Fermat to Minkowski, Lectures on the Theory of Numbers and its Historical Development*, Springer-Verlag, New York, 1985.

68. A. Weil, *Number Theory: An approach through history from Hummurapi to Legendre*, Birkhäuser, Boston,1984.

Cryptology

69. H. Beker and F. Piper, *Cipher Systems*, Wiley, New York, 1982.

70. B. Bosworth, *Codes, Ciphers, and Computers*, Hayden, Rochelle Park, New Jersey, 1982.

71. D. Chaum, R. L. Rivest, A. T. Sherman, eds., *Advances in Cryptology - Proceedings of Crypto 82*, Plenum, New York, 1983.

72. D. Chaum, ed., *Advances in Cryptology - Proceedings of Crypto 83*, Plenum, New York, 1984.

73. D.E.R. Denning, *Cryptography and Data Security*, Addison-Wesley, Reading, Massachusetts, 1982.

74. W.F. Friedman, *Elements of Cryptanalysis*, Aegean Park Press, Laguna Hills, California, 1978.

75. A. Gersho, ed., *Advances in Cryptography*, Dept. of Electrical and Computer Engineering, Univ. Calif. Santa Barbara, 1982.

76. D. Kahn, *The Codebreakers, the Story of Secret Writing*, Macmillan, New York, 1967.

77. N. Koblitz, *A Course in Number Theory and Cryptography*, Springer-Verlag, New York, 1987.

78. A.G. Konheim, *Cryptography: A Primer*, Wiley, New York, 1981.

79. S. Kullback, *Statistical Methods in Cryptanalysis*, Aegean Park Press, Laguna Hills, California, 1976.

80. J.H. Loxton, editor, *Number Theory and Cryptography*, Cambridge University Press, Cambridge, England, 1990.

81. R.C. Merkle, *Secrecy, Authentication, and Public Key Systems*, UMI Research Press, Ann Arbor, Michigan, 1982.

82. C.H. Meyer and S. M. Matyas, *Cryptography: A New Dimension in Computer Data Security*, Wiley, New York, 1982.

83. C.P. Pfleeger, *Security in Computing*, Prentice Hall, Englewood Cliffs, New Jersey, 1989.

84. C. Pomerance, ed., *Cryptology and Computational Number Theory*, American Mathematical Society, Providence, Rhode Island, 1990.

85. A. Salomaa, *Public-Key Cryptography*, Springer-Verlag, New York, 1990.

86. J. Seberry and J. Pieprzyk, *Cryptography: An Introduction to Computer Security*, Prentice Hall, New York, 1989.

87. G. J. Simmons, ed., *Secure Communications and Asymmetric Cryptosystems*, AAAS Selected Symposium Series 69, Westview Press, Boulder, Colorado, 1982.

88. A. Sinkov, *Elementary Cryptanalysis*, Mathematical Association of America, Washington, D.C., 1966.

89. H. C. Williams, ed., *Advances in Cryptology - CRYPTO '85*, Springer-Verlag, Berlin, 1986.

Computer Science

90. K. Hwang, *Computer Arithmetic: Principles, Architecture and Design*, Wiley, New York, 1979.

91. D.E. Knuth, *Art of Computer Programming: Semi-Numerical Algorithms* Volume 2, 2nd ed., Addison Wesley, Reading Massachusetts, 1981.

92. D.E. Knuth, *Art of Computer Programming: Sorting and Searching*, Volume 3, Addison-Wesley, Reading, Massachusetts, 1973.

93. L.Kronsjö, *Algorithms: Their Complexity and Efficiency*, Wiley, New York, 1979.

94. J.H. McClellan and C. M. Rader, *Number Theory in Digital Signal Processing*, Prentice-Hall, Englewood Cliffs, New Jersey, 1979.

95. M.A. Soderstrand, et. al., editors, *Residue Number System Arithmetic: Modern Applications in Digital Signal Processing*, IEEE Press, New York, 1986.

96. N. S. Szabó and R. J. Tanaka, *Residue Arithmetic and its Applications to Computer Technology*, McGraw-Hill, 1967.

General

97. H. Anton, *Elementary Linear Algebra*, 3rd ed., Wiley, New York, 1981.

98. E. Landau, *Foundations of Analysis*, 2nd ed., Chelsea, New York, 1960.

99. K. H. Rosen, *Discrete Mathematics and its Applications*, 2nd ed., McGraw-Hill, New York, 1991.

100. W. Rudin, *Principles of Mathematical Analysis*, 2nd ed., McGraw-Hill, New York 1964.

ARTICLES

Number Theory

101. L. M. Adleman, C. Pomerance, and R. S. Rumely, "On distinguishing prime numbers from composite numbers," *Annals of Mathematics*, Volume 117 (1983), 173-206.

102. R. P. Brent, "Improved Techniques for Lower Bounds for Odd Perfect Numbers," *Mathematics of Computation*, Volume 57 (1991), 857-868.

103. J. Brillhart, "Fermat's factoring method and its variants," *Congressus Numerantium*, Volume 32 (1981), 29-48.

104. B. Cipra, "Big Number Breakdown,", *Science*, Volume 248 (1990), 1608.

105. B. Cipra, "PCs Factor a 'Most Wanted' Number," *Science*, Volume 242 (1988), 1634-1635.

106. W. N. Colquitt and L. Welsh, Jr., "A New Mersenne Prime," *Mathematics of Computation*, Volume 56 (1991), 867-870.

107. J. D. Dixon, "Factorization and primality tests," *American Mathematical Monthly*, Volume 91 (1984), 333-353.

108. J. Ewing, " $2^{86243} - 1$ is prime," *The Mathematical Intelligencer*, Volume 5 (1983), 60.

109. J. E. Freund, "Round robin mathematics," *American Mathematical Monthly*, Volume 63 (1956), 112-114.

110. R. K. Guy, "How to factor a number" *Proceedings of the Fifth Manitoba Conference on Numerical Mathematics*, Utilitas, Winnepeg, Manitoba, 1975, 49-89.

111. A. K. Head, "Multiplication modulo n," *BIT*, Volume 20 (1980), 115-116.

112. P. Hagis, Jr., "Sketch of a proof that an odd perfect number relatively prime to 3 has at least eleven prime factors," *Mathematics of Computations*, Volume 46 (1983), 399-404.

113. G. Kolata, "Factoring Gets Easier," *Science*, Volume 222 (1983), 999-1001.

114. J. C. Lagarias and A. M. Odlyzko, "New algorithms for computing $\pi(x)$," Bell Laboratories Technical Memorandum TM-82-11218-57.

115. H. P. Lawther, Jr., "An application of number theory to the splicing of telephone cables," *American Mathematical Monthly*, Volume 42 (1935), 81-91.

116. D. H. Lehmer and R. E. Powers, "On factoring large numbers," *Bulletin of the American Mathematical Society*, Volume 37 (1931), 770-776.

117. H. W. Lenstra, Jr., "Primality testing," *Studieweek Getaltheorie en Computers*, 1-5 September 1980, Stichting Mathematisch Centrum, Amsterdam, Holland.

118. G. L. Miller, "Riemann's hypothesis and tests for primality," *Proceedings of the Seventh Annual ACM Symposium on the Theory of Computing*, 234-239.

119. L. Monier, "Evaluation and comparison of two efficient probablistic primality testing algorithms, *Theoretical Computer Science*, Volume 11 (1980), 97-108.

120. J. M. Pollard, "Theorems on Factorization and Primality Testing," *Proceedings of the Cambridge Philosophical Society*, Volume 76 (1974), 521-528.

121. J. M. Pollard, "A Monte Carlo Method for Factorization," *Nordisk Tidskrift for Informationsbehandling (BIT)*, Volume 15 (1975), 331-334.

122. C. Pomerance, "Recent developments in primality testing," *The Mathematical Intelligencer*, Volume 3 (1981), 97-105.

123. C. Pomerance, "The search for prime numbers," *Scientific American*, Volume 247 (1982), 136-147.

124. M. O. Rabin, "Probabilistic algorithms for testing primality," *Journal of Number Theory*, Volume 12 (1980), 128-138.

125. H. Riesel, "Modern factorization methods," *BIT*(1985), 205-222.

126. R. Rumely, "Recent advances in primality testing," *Notices of the American Mathematical Society*, Volume 30 (1983), 475-477.

127. D. Slowinski, "Searching for the 27th Mersenne prime," *Journal of Recreational Mathematics*, Volume 11 (1978/9), 258-261.

128. R. Solovay and V. Strassen, "A fast Monte Carlo test for primality," *SIAM Journal for Computing*, Volume 6 (1977), 84-85 and erratum, Volume 7 (1978), 118.

129. S. Wagon, "Primality testing," *The Mathematical Intelligencer*, Volume 8, Number 3 (1986), 58-61.

130. S. S. Wagstaff, "Some uses of microcomputers in number theory research," *Computers and Mathematics with Applications*, Volume 19 (1990), 53-58.

131. S. S. Wagstaff, "Using computers to teach number theory," *SIAM News*, Volume 19 (1986), 14 and 18.

132. S. S. Wagstaff and J. W. Smith, "Methods of factoring large integers," in *Number Theory, New York, 1984-1985*, Lecture Notes in Mathematics, Volume 1240, Springer-Verlag, Berlin, 1987, 281-303.

133. H. C. Williams, "An overview of factoring," in *Advances in Cryptology, Proceedings of Crypto 83*, Plenum, New York, 1984, 87-102.

134. H. C. Williams, "The influence of computers in the development of number theory," *Computers and Mathematics with Applications*, Volume 8 (1982), 75-93.

135. H. C. Williams, "Primality testing on a computer", *Ars Combinatorica*, Volume 5 (1978), 127-185.

136. M. C. Wunderlich, "Implementing the continued fraction algorithm on parallel machines," *Mathematics of Computation*, Volume 44 (1985), 251-260.

137. M. C. Wunderlich and J.M. Kubina, "Extending Waring's conjecture to 471,600,000," *Mathematics of Computation*, Volume 55 (1990), 815-820.

Cryptography

138. L. M. Adleman, "A subexponential algorithm for the discrete logarithm problem with applications to cryptography," *Proceedings of the 20th Annual Symposium on the Foundations of Computer Science*, 1979, 55-60.

139. M. Blum, "Coin-flipping by telephone - a protocol for solving impossible problems," *IEEE Proceedings, Spring Compcon.*, 133-137.

140. W. Diffie and M. Hellman, "New directions in cryptography," *IEEE Transactions on Information Theory*, Volume 22 (1976), 644-655.

141. D. R. Floyd, "Annotated bibliographical in conventional and public key cryptography," *Cryptologia*, Volume 7 (1983), 12-24.

142. J. Gordon, "Use of intractable problems in cryptography," *Information Privacy*, Volume 2 (1980), 178-184.

143. M. E. Hellman, "The mathematics of public-key cryptography," *Scientific American*, Volume 241 (1979) 146-157.

144. L. S. Hill, "Concerning certain linear transformation apparatus of cryptography," *American Mathematical Monthly*, Volume 38 (1931), 135-154.

145. A. Lempel, "Cryptology in transition," *Computing Surveys*, Volume 11 (1979), 285-303.

146. R. J. Lipton, "How to cheat at mental poker," and "An improved power encryption method," unpublished reports, Department of Computer Science, University of California, Berkeley, 1979.

147. R. C. Merkle and M. E. Hellman, "Hiding information and signatures in trapdoor knapsacks," *IEEE Transactions in Information Theory*, Volume 24 (1978), 525-530.

148. S. Pohlig and M. Hellman, "An improved algorithm for computing logarithms over GF(p) and its cryptographic significance," *IEEE Transactions on Information Theory*, Volume 24 (1978), 106-110.

149. M. O. Rabin, "Digitalized signatures and public-key functions as intractable as factorization," MIT Laboratory for Computer Science Technical Report LCS/TR-212, Cambridge, Massachusetts, 1979.

150. R. L. Rivest, A. Shamir, and L. M. Adleman, "A method for obtaining digital signatures and public-key cryptosystems," *Communications of the ACM*, Volume 21 (1978), 120-126.

151. A. Shamir, "A polynomial time algorithm for breaking the basic Merkle-Hellman cryptosystem," *Proceedings of the 23rd Annual Symposium of the Foundations of Computer Science*, 145-152.

152. A. Shamir, "How to share a secret," *Communications of the ACM*, Volume 22 (1979), 612-613.

153. A. Shamir, R. L. Rivest, and L. M. Adleman, "Mental Poker," *The Mathematical Gardner*, ed. D. A. Klarner, Wadsworth International, Belmont, California, 1981, 37-43.

Index

List of Symbols